Brazilian Medicinal Plants

Natural Products Chemistry of Global Plants
Editor: Raymond Cooper

This unique book series focuses on the natural products chemistry of botanical medicines from different countries such as Sri Lanka, Cambodia, Brazil, China, Africa, Borneo, Thailand and Silk Road countries. These fascinating volumes are written by experts from their respective countries. The series will focus on the pharmacognosy, covering recognized areas rich in folklore as well as botanical medicinal uses as a platform to present the natural products and organic chemistry. Where possible, the authors will link these molecules to pharmacological modes of action. The series intends to trace a route through history from ancient civilizations to the modern day showing the importance to man of natural products in medicines, foods and a variety of other ways.

Recent Titles in this Series:

Traditional Herbal Remedies of Sri Lanka
Viduranga Y. Waisundara

Medicinal Plants of Bangladesh and West Bengal
Botany, Natural Products, and Ethnopharmacology
Christophe Wiart

Brazilian Medicinal Plants
Luzia Modolo and Mary Ann Foglio

https://www.crcpress.com/Natural-Products-Chemistry-of-Global-Plants/book-series/CRCNPCGP

Brazilian Medicinal Plants

Edited by

Luzia Valentina Modolo
Mary Ann Foglio

CRC Press
Taylor & Francis Group
Boca Raton London New York

CRC Press is an imprint of the
Taylor & Francis Group, an **informa** business

CRC Press
Taylor & Francis Group
6000 Broken Sound Parkway NW, Suite 300
Boca Raton, FL 33487-2742

First issued in paperback 2021

© 2020 by Taylor & Francis Group, LLC
CRC Press is an imprint of Taylor & Francis Group, an Informa business

No claim to original U.S. Government works

ISBN 13: 978-1-03-208503-6 (pbk)
ISBN 13: 978-1-138-09375-1 (hbk)

Library of Congress Cataloging-in-Publication Data

Names: Modolo, Luzia V., author.
Title: Brazilian medicinal plants / Luzia V. Modolo, Mary Ann Foglio.
Description: Boca Raton, Florida : CRC Press, 2019. | Series: Natural products chemistry of global plants | Includes bibliographical references and index. | Summary: "The vast and exciting Brazilian flora biodiversity is still underexplored. Several research groups are devoted to the study of the chemical structure richness found in the different Biomes. This volume presents a comprehensive account of the research collated on natural products produced from Brazilian medicinal plants and focuses on various aspects of the field. The authors describe the key natural products and their extracts with emphasis upon sources, an appreciation of these complex molecules and applications in science. Many of the extracts are today associated with important drugs, nutrition products, beverages, perfumes, cosmetics and pigments, and these are highlighted"-- Provided by publisher.
Identifiers: LCCN 2019031670 | ISBN 9781138093751 (hardback) | ISBN 9781315106427 (ebook)
Subjects: LCSH: Medicinal plants--Brazil. | Medicinal plants--Brazil--Identification.
Classification: LCC QK99.B6 M63 2019 | DDC 581.6/340981--dc23
LC record available at https://lccn.loc.gov/2019031670

Visit the Taylor & Francis Web site at
http://www.taylorandfrancis.com

and the CRC Press Web site at
http://www.crcpress.com

Publisher's Note
The publisher has gone to great lengths to ensure the quality of this reprint but points out that some imperfections in the original copies may be apparent.

Contents

Introduction to Book Series – Natural Products Chemistry of Global Plants

CRC Press is publishing a new book series on the Natural Products Chemistry of Global Plants. This new series will focus on pharmacognosy, covering recognized areas rich in folklore and botanical medicinal uses as a platform to present the natural products and organic chemistry and, where possible, link these molecules to pharmacological modes of action. This book series on the botanical medicines from different countries include, but are not limited to, Brazil, Bangladesh, Borneo, Cambodia, Cameroon, Iran, Madagascar, The Silk Road, Sri Lanka, Thailand, Turkey, Uganda, Vietnam, Western Cape (South Africa) and Yunnan Province (China), written by experts from each country. The intention is to provide a platform to bring forward information from under-represented regions.

Medicinal plants are an important part of human history, culture and tradition. Plants have been used for medicinal purposes for thousands of years. Anecdotal and traditional wisdom concerning the use of botanical compounds is documented in the rich histories of traditional medicines. Many medicinal plants, spices and perfumes changed the world through their impact on civilization, trade and conquest. Folk medicine is commonly characterized by the application of simple indigenous remedies. People who use traditional remedies may not understand in our terms the scientific rationale for why they work but know from personal experience that some plants can be highly effective.

This series provides rich sources of information from each region. An intention of the series of books is to trace a route through history from ancient civilizations to the modern day showing the important value to humankind of natural products in medicines, in foods and in many other ways. Many of the extracts are today associated with important drugs, nutrition products, beverages, perfumes, cosmetics and pigments, which will be highlighted.

The series is written both for chemistry students who are at university level and for scholars wishing to broaden their knowledge in pharmacognosy. Through examples of the chosen botanicals, herbs and plants, the series will describe the key natural products and their extracts with emphasis upon sources, an appreciation of these complex molecules and applications in science.

In this series the chemistry and structure of many substances from each region are presented and explored. Often books describing folklore medicine do not describe the rich chemistry or the complexity of the natural products and their respective biosynthetic building blocks. By drawing on the chemistry of these functional groups to show how they influence the chemical behavior of the building blocks, which make up large and complex natural products, the story becomes more fascinating. Where possible it will be advantageous to describe the pharmacological nature of these natural products.

In this book on medicinal plants in Brazil, the great diversity of the Brazilian flora is documented and presented for the unique biological and chemical diversity. The book emphasizes the biodiversity and the critical importance of the fragile ecosystem and can contribute to a better understanding of Brazilian diversity, as well as the effect of the anthropological and environmental conditions in this ecosystem.

R. Cooper, Ph.D., Editor-in-chief
Dept. Applied Biology & Chemical Technology
The Hong Kong Polytechnic University
Hong Kong
rcooperphd@aol.com

Preface

According to the Brazilian Institute for Geography and Statistics (IBGE) the country is divided into six biomes: Amazon, Pantanal, Cerrado, Caatinga, Atlantic Forest and Pampa. The different biomes reveal an impressive variety of plants, microorganisms and animals. Indeed, Brazil has an outstanding vast biodiversity. The country has approximately 20% of the world's known plant species. With such an immense territory and complex ecosystems, many other species remain yet to be discovered. Overall, Brazil is in a privileged situation in terms of natural products, with biodiversity being one of the major bases for the progressive discovery of new compounds with great potential for both biological and non-biological applications, such as biofuels, textiles and others.

When did this all begin? Some will say with Adam and Eve, others with the native Indian populations, who indeed deserve credit for all the knowledge that they have passed on from one generation to the next, giving rise to extremely important data. The country's rich cultural and biological diversity is a priceless treasure to be wisely used.

Many scientists have given their share of time to explore and understand the wonders that the ecosystems have to offer. Among the many important contributions throughout time are those of Otto Gottlieb, who was an enthusiastic pioneer on the study of natural products, together with Ernest Wenkert, Jayr de Paiva Campello and Giuseppe Cilento, who paved the way for many others. The awareness of the importance of studying these natural resources permitted the expansion of the Brazilian scientific community triggering the work by Walter B. Mors, Mauro Taveira Magalhães, Carl Djerassi, Frederick Sanger, Margaret Joan Sanger, Raimundo Braz-Filho, Alaíde Braga de Oliveira, Anita Marsaioli, Francisco José de Abreu Matos, Maria Iracema Lacerda Machado Madruga, David W. Cochran, Hugo E. Gottlieb, Edward W. Hagaman, Arnaldo Felisberto Imbiriba da Rocha, Angelo da Cunha Pinto, Nidia Franca Roque, Sebastião Ferreira Fonseca, José Rego de Sousa, Afrânio Aragão Craveiro, Tanus Jorge Nagem and so many others. Now the new science generations have the obligation to address aspects that will guarantee equilibrium of wildlife with mankind for all to benefit from these ecosystems.[1]

Brazilian Medicinal Plants gives a snapshot of the marvelous research work being performed by Brazilian scientists. Among the chapters represented herein, we cover regulation issues, different biome treasures and microorganisms of medicinal plants found in the Brazilian biomes. Some concepts behind forward and reverse genetics and potential use of genome editing technologies on Brazilian medicinal plants and an integrated view of plant-environment interaction strategies to improve target metabolite yields are presented. Also, aspects of physicochemical methods for the quality control of medicinal plants, plant derivatives and herbal medicines in Brazil and some approaches used for the synthesis of natural products of Brazilian medicinal plants origin are discussed.

Brazilian flora is undoubtedly one of the most plentiful sources of inspiration for the development of new drugs. However, various natural products that have therapeutic properties are not always available in sufficient amounts for sustainable use and/or for the development of new derivatives by modifying such substances. Moreover, obtaining a renewable supply of active compounds from biological sources may be problematic, especially with respect to perennial plant species. The complexity of many natural products can also limit the scope of chemical modifications necessary to optimize therapeutic use. Despite these barriers, the total synthesis of various bioactive natural products and analogs has proven that organic synthesis is a powerful tool for increasing the availability of valuable natural products of limited supply or with very complex structures.

Plant natural products evolved as key elements in adaptive responses to stress, both biotic and abiotic, in close connection with the sessile habit. During this process, an intricate relationship

---

[1] Angelo C. Pinto; Fernando de C. da Silva; Vitor F. Ferreira; Otto R. Gottlieb e as conexões com o Brasil de Ernest Wenkert; Quím. Nova vol.35 no.11 São Paulo 2012

between dedicated metabolic pathways and structural features of plants was established, affording high efficiency in metabolic competence and generating great metabolite diversity. This metabolic array has proven a major reservoir of bioactive compounds for treating and preventing human diseases. Considering the defense-related role of natural products in plants and the signaling pathways that trigger their biosynthesis upon stress exposure, it may be advantageous to use environmental signals or their transduction elements for enriching biomass with pharmacologically interesting metabolites. Among the environmental factors that promote natural product accumulation when applied at moderate intensity are the following: heat, cold, drought, herbivory, pathogens, UV radiation, osmotic stress and heavy metals.

Some specific regions are described. The Caatinga biome is a semi-arid ecosystem found exclusively in Brazil, referred to as "a mosaic of scrubs and patches of seasonally dry forest." This unique biome occupies a large geographic area of the country, spread from the state of Ceará (CE) to the north of the state of Minas Gerais (MG). Despite being one of the largest Brazilian biomes, the scientific knowledge of this biodiversity is poorly understood.

Also, a review of the natural products from the Goiás Cerrado provides the importance of ethnobotany and Brazilian traditional medicine, used as a strategy to focus on isolation, structural elucidation and biological evaluation and use of modern mass spectrometry (MS) techniques to analyze plants from this region.

The biogeographic region of the Atlantic Forest covers a part of the Brazilian coast and parts of Paraguay, Argentina and Uruguay. This biome is considered an important hotspot and a priority for biodiversity conservation. The huge plant diversity accounts for 5% of the world's flora. Although only small fragments of the Atlantic Forest remain due to intense deforestation across the past five centuries, this area is biologically and chemically very rich.

Among challenges to consider for the next 20 years are food security and nutrition issues worldwide. Insights into the sustainable use of biodiversity are crucial factors to provide new food products from unused fruits and vegetables with high nutritional gain. Four Amazon fruits from trees, *Spondias mombin*, *Myrciaria dubia*, *Genipa americana* and the well-known Brazilian nut (*Bertholletia excelsa*) are presented. Also palm tree fruits from *Astrocaryum vulgare*, *Mauritia flexuosa*, *Bactris gasipaes* and the well-known açaí (*Euterpe oleracea*) are described. These fruits and nuts are the most abundant sources of bioactive compounds with antioxidant activity as a result of phenolic compounds, carotenoids, tocopherols, vitamin C, unsaturated fatty acids (UFA), terpenoids and steroids. Characteristic compounds, present in a higher amount, are a highlight for some fruits, such as vitamin C in camu-camu fruit, carotenoids in the peach palm and tucuma fruits, iridoids in genipap, and selenium and UFA in the Brazil nut. The synergistic effect of these compounds leads to many health benefits.

Many research groups are working on different aspects of natural products in Brazil. This volume presents an important account of ongoing research on natural products produced from Brazilian medicinal plants. The vast and exciting Brazilian flora biodiversity is still underexplored, yet several research groups are devoted to the study of the chemical structural richness found in the different biomes. The authors described the key natural products and their extracts with emphasis upon sources, an appreciation of these complex molecules and applications in science. Many of the extracts are today associated with important drugs, nutrition products, beverages, perfumes, cosmetics and pigments, and these are highlighted. Specifically, Brazilian biodiversity, its flora, its people and its research are described. With an emphasis on the increasing global interests in botanical drugs, this volume may help the international natural product communities to better understand the herbal resources in Brazil and the regulations and legislation to work with native plants. Recent achievements on plant research of regionally different groups are presented to give the reader the tip of the iceberg, recognizing that much more research and funding is required.

Mary Ann Foglio
Brazil

Editors

Luzia Valentina Modolo received her PhD degree in Functional and Molecular Biology in 2004 from the State University of Campinas (SP, Brazil). She is a faculty member of the Department of Botany at the Federal University of Minas Gerais (MG, Brazil) and currently holds Associate Professor Position. Dr. Modolo is the coordinator of the Network for the Development of Novel Urease Inhibitors (www.redniu.org) and her research interests include plant nutrition and secondary metabolism and signalling processes in plant tissues triggered by environmental stress.

Mary Ann Foglio is Senior Researcher at the Faculty of Pharmaceutical Science (FPS) at the University of Campinas (UNICAMP). Her early training was in chemistry with a Bachelor's degree in Chemistry (1982), master's (1987) and PhD (1996) at UNICAMP in Organic Chemistry. From 1987–2015 she was a researcher at the Multidisciplinary Chemistry, Biology and Agricultural Research Center. Thereafter she took a position as full professor at FPS School of Pharmacy. Her work focuses on translational research of natural products. Over the years her work has involved standardization of plant inputs and the development of products that meet safety, efficacy and reproducibility criteria, resulting in fifteen patent deposits. She supervises undergraduate and graduate students, and post-docs on bioactive natural product research. Her lab is sponsored by CNPq, CAPES, FINEP, and FAPESP governmental agencies.

Contributors

Wellyda Rocha Aguiar
Medicinal Herb Garden "Francisco José
  de Abreu Matos"
Federal University of Ceará, Campus do Pici
Fortaleza, Ceará, Brazil

Rosemeire Brondi Alves
Grupo de Estudos em Química Orgânica
  e Biológica (GEQOB)
Departamento de Química, Instituto
  de Ciências Exatas
Universidade Federal de Minas Gerais
  (UFMG)
Belo Horizonte, Minas Gerais, Brazil

Jéssica Cristina Amaral
Chemistry Department
São Carlos Federal University
São Carlos, São Paulo, Brazil

Ângela Regina Araújo
Universidade Estadual Paulista
Núcleo de Bioensaios, Biossíntese
  e Ecofisiologia de Produtos
  Naturais – NuBBE
Araraquara, São Paulo, Brazil

Mary Anne Medeiros Bandeira
Medicinal Herb Garden "Francisco José
  de Abreu Matos"
Federal University of Ceará
Campus do Pici
Fortaleza, Ceará, Brazil

Francisco Geraldo Barbosa
Department of Organic and Inorganic
  Chemistry
Federal University of Ceará
Campus do Pici
Fortaleza, Ceará, Brazil

Carolina Rabal Biasetto
Universidade Estadual Paulista
Núcleo de Bioensaios, Biossíntese
  e Ecofisiologia de Produtos
  Naturais – NuBBE
Araraquara, São Paulo, Brazil

Vanderlan da Silva Bolzani
Universidade Estadual Paulista
Núcleo de Bioensaios, Biossíntese e
  Ecofisiologia de Produtos
  Naturais – NuBBE
Araraquara, São Paulo, Brazil

Paula Carolina Pires Bueno
Instituto de Química
Universidade Estadual Paulista (UNESP)
Araraquara, São Paulo, Brazil

Ana Cecília Bezerra Carvalho
General Office of Drugs and Biological
  Products
Brazilian Health Regulatory
  Agency – ANVISA
Brasília, DF, Brazil

Ian Castro-Gamboa
Universidade Estadual Paulista
Núcleo de Bioensaios, Biossíntese
  e Ecofisiologia de Produtos Naturais – NuBBE
Araraquara, São Paulo, Brazil

Alberto José Cavalheiro
Instituto de Química
Universidade Estadual Paulista (UNESP)
Araraquara, São Paulo, Brazil

Samuel Chaves-Silva
Departamento de Botânica
Instituto de Ciências Biológicas
Universidade Federal de Minas Gerais
Belo Horizonte, Minas Gerais, Brazil

Lilian Cherubin Correia
Universidade Estadual Paulista
Núcleo de Bioensaios, Biossíntese
  e Ecofisiologia de Produtos
  Naturais – NuBBE
Araraquara, São Paulo, Brazil

Maria Fátima das Graças Fernandes da Silva
Chemistry Department
São Carlos Federal University
São Carlos, São Paulo, Brazil

Michelli Massaroli da Silva
Chemistry Department, São Carlos Federal
    University
São Carlos, São Paulo, Brazil

Thamara Ferreira da Silva
Departamento de Botânica
Instituto de Ciências Biológicas
Universidade Federal de Minas Gerais
Belo Horizonte, Minas Gerais, Brazil

Sylvain Darnet
Centre for Valorization of Amazonian
    Bioactive Compounds & Federal University
    of Pará
Belém, Pará, Brazil

Cristiane Jovelina da-Silva
Departamento de Botânica
Instituto de Ciências Biológicas
Universidade Federal de Minas Gerais
Belo Horizonte, Minas Gerais, Brazil

Wellington Alves de Barros
Grupo de Estudos em Química Orgânica
    e Biológica (GEQOB)
Departamento de Química, Instituto
    de Ciências Exatas
Universidade Federal de Minas Gerais (UFMG)
Belo Horizonte, Minas Gerais, Brazil

Fernanda de Costa
Plant Physiology Laboratory
Center for Biotechnology and Department
    of Botany
Federal University of Rio Grande do Sul (UFRGS)
Campus do Vale
Porto Alegre, Rio Grande do Sul, Brazil

Ângelo de Fátima
Grupo de Estudos em Química Orgânica
    e Biológica (GEQOB)
Departamento de Química, Instituto
    de Ciências Exatas
Universidade Federal de Minas Gerais (UFMG)
Belo Horizonte, Minas Gerais, Brazil

Cecilia Maria Alves de Oliveira
Instituto de Química
Universidade Federal de Goias
Goiânia, Goiás, Brazil

Lidiane Gaspareto Felippe
Universidade Estadual Paulista
Núcleo de Bioensaios, Biossíntese e
    Ecofisiologia de Produtos Naturais – NuBBE
Araraquara, São Paulo, Brazil

João Batista Fernandes
Chemistry Department
São Carlos Federal University
São Carlos, São Paulo, Brazil

Arthur Germano Fett-Neto
Plant Physiology Laboratory
Center for Biotechnology and Department of
    Botany
Federal University of Rio Grande do Sul
    (UFRGS)
Campus do Vale
Porto Alegre, Rio Grande do Sul, Brazil

Moacir Rossi Forim
Chemistry Department
São Carlos Federal University
São Carlos, São Paulo, Brazil

Maysa Furlan
Universidade Estadual Paulista
Núcleo de Bioensaios, Biossíntese e
    Ecofisiologia de Produtos Naturais – NuBBE
Araraquara, São Paulo, Brazil

Chirlei Glienke
Federal University of Paraná
Department of Genetics
Curitiba, Paraná, Brazil

**Núbia Alves Mariano Teixeira Pires
Gomides**
Unidade Acadêmica Especial
    de Biotecnologia
Universidade Federal de Goias
Catalão, Goiás, Brazil

Camila Fernanda de Oliveira Junkes
Plant Physiology Laboratory
Center for Biotechnology and Department
    of Botany
Federal University of Rio Grande do Sul
    (UFRGS)
Campus do Vale
Porto Alegre, Rio Grande do Sul, Brazil

Lucilia Kato
Instituto de Química
Universidade Federal de Goias
Goiânia, Goiás, Brazil

Melina Cossote Kumoto
General Office of Drugs and Biological
    Products
Brazilian Health Regulatory
    Agency – ANVISA
Brasília, DF, Brazil

Lucas Vieira Lima
Universidade Federal de Minas Gerais
Laboratório de Sistemática Vegetal
Departamento de Botânica
Instituto de Ciências Biológicas
Belo Horizonte, Minas Gerais, Brazil

Mary Anne Sousa Lima
Department of Organic and Inorganic
    Chemistry
Federal University of Ceará
Campus do Pici
Fortaleza, Ceará, Brazil

Rita de Cássia Lemos Lima
Department of Drug Design and
    Pharmacology
University of Copenhagen
Copenhagen, Denmark

Adaíses Simone Maciel-Silva
Universidade Federal de Minas Gerais
Laboratório de Sistemática Vegetal
Departamento de Botânica
Instituto de Ciências Biológicas
Belo Horizonte, Minas Gerais, Brazil

Jair Mafezoli
Department of Organic and Inorganic
    Chemistry
Federal University of Ceará
Campus do Pici
Fortaleza, Ceará, Brazil

Rebeca Previate Medina
Universidade Estadual Paulista
Núcleo de Bioensaios, Biossíntese
    e Ecofisiologia de Produtos Naturais – NuBBE
Araraquara, São Paulo, Brazil

Luzia Valentina Modolo
Departamento de Botânica
Instituto de Ciências Biológicas
Universidade Federal de Minas Gerais
Belo Horizonte, Minas Gerais, Brazil

Afif Felix Monteiro
Universidade Estadual Paulista
Núcleo de Bioensaios, Biossíntese
    e Ecofisiologia de Produtos
    Naturais – NuBBE
Araraquara, São Paulo, Brazil

Aline Pereira Moraes
Instituto de Química
Universidade Federal de Goias
Goiânia, Goiás, Brazil

Franciele Antonia Neis
Plant Physiology Laboratory
Center for Biotechnology and Department of
    Botany
Federal University of Rio Grande do Sul
    (UFRGS)
Campus do Vale
Porto Alegre, Rio Grande do Sul,
    Brazil

Leonardo da Silva Neto
Grupo de Estudos em Química Orgânica
    e Biológica (GEQOB)
Departamento de Química, Instituto
    de Ciências Exatas
Universidade Federal de Minas Gerais
    (UFMG)
Belo Horizonte, Minas Gerais,
    Brazil

Instituto Federal Farroupilha
Alegrete, Rio Grande do Sul,
    Brazil

Breno Germano de Freitas Oliveira
Grupo de Estudos em Química Orgânica
    e Biológica (GEQOB)
Departamento de Química, Instituto
    de Ciências Exatas
Universidade Federal de Minas Gerais
    (UFMG)
Belo Horizonte, Minas Gerais,
    Brazil

Maria da Conceição Ferreira de Oliveira
Department of Organic and Inorganic
   Chemistry
Federal University of Ceará
Campus do Pici
Fortaleza, Ceará, Brazil

João Paulo Silvério Perfeito
General Office of Drugs and Biological
   Products
Brazilian Health Regulatory
   Agency – ANVISA
Brasília, DF, Brazil

Elaine Pessoa
Centre for Valorization of Amazonian
   Bioactive Compounds & Federal University
   of Pará
Belém, Pará, Brazil

Hervé Rogez
Centre for Valorization of Amazonian
   Bioactive Compounds & Federal University
   of Pará
Belém, Pará, Brazil

Adão Aparecido Sabino
Grupo de Estudos em Química Orgânica
   e Biológica (GEQOB)
Departamento de Química, Instituto
   de Ciências Exatas
Universidade Federal de Minas Gerais
   (UFMG)
Belo Horizonte, Minas Gerais,
   Brazil

Daiani Cristina Savi
Federal University of Paraná
Department of Genetics
Curitiba, Paraná, Brazil

Josilene Lima Serra
Centre for Valorization of Amazonian
   Bioactive Compounds & Federal University
   of Pará
Belém, Pará, Brazil

Vanessa Gisele Pasqualotto Severino
Instituto de Química
Universidade Federal de Goias
Goiânia, Goiás, Brazil

Dulce Helena Siqueira Silva
Univ. Estadual Paulista
Núcleo de Bioensaios, Biossíntese
   e Ecofisiologia de Produtos Naturais – NuBBE
Araraquara, São Paulo, Brazil

Aristônio Magalhães Teles
Instituto de Ciências Biológicas,
Universidade Federal de Goias
Goiânia, Goiás, Brazil

Marília Valli
Universidade Estadual Paulista
Núcleo de Bioensaios, Biossíntese e
   Ecofisiologia de Produtos Naturais – NuBBE
Araraquara, São Paulo, Brazil

Vinicius Galvão Wakui
Instituto de Química
Universidade Federal de Goias
Goiânia, Goiás, Brazil

Anna Carolina Alves Yendo
Plant Physiology Laboratory
Center for Biotechnology and Department
   of Botany
Federal University of Rio Grande do Sul (UFRGS)
Campus do Vale
Porto Alegre, Rio Grande do Sul, Brazil

1 Unraveling the Complexities of Brazilian Regulations for Medicinal Plants and Herbal Medicinal Products

Ana Cecília Bezerra Carvalho,
Melina Cossote Kumoto, and João Paulo Silvério Perfeito
General Office of Drugs and Biological Products.
Brazilian Health Regulatory Agency – ANVISA.
Brasília, DF, Brazil

CONTENTS

1.1 INTRODUCTION

The regulation of medicinal plants and herbal medicinal products (HMPs) is quite complex since it involves several factors and rules. There are several steps to be considered, such as cultivation, harvesting, manufacturing or manipulation in compounding pharmacies, commercialization and prescription, being essential that adequate control be performed so that safe and effective products are available to the population.

The laws involved in this regulation include those published at the federal government level, such as the National Congress, the Presidency of the Republic, ANVISA (Brazilian Health Regulatory Agency), the National Health Council, the Ministry of Health and the Ministry of the Environment, among other governmental bodies, as well as state and municipal legislatures. This legislative framework is often complex to follow because it is based on the legal definitions of medicines, which encompass rigid rules of control, production and evidence of safety and efficacy. Conversely, there is a lot of informal trade in HMPs in Brazil, where there are minimal controls at best to guarantee plant identification and the safe level of contaminants.

Unfortunately, a large part of the Brazilian population does not understand that these products require a license before being prescribed or sold, and this fact generates doubts and uncertainties for producers and consumers. Thus, discussion of all possible regulatory paths of product release, from the simplest one to the most complex – the latter being industrial production, which encompasses more steps and is more regulated is important to discuss.

Before starting the technical-regulatory discussion, the roles of normative acts in the Brazilian sanitary regulation discussed in this chapter need to be clarified:

- Law: published at the federal level, by the National Congress, or in the state or municipal sphere, aims to introduce a new subject to be regulated;
- Decree: published by the head of the Executive Power – in the federal, state or municipal sphere. Usually regulates and complements a law;
- Ordinance: published by public administration bodies, such as ministries, to guide compliance with legal provisions. In the case of HMP, ordinances have legal force and are usually issued by the Ministry of Health;
- Resolution- Resolução da Diretoria Colegiada (RDC) (Collegiate Board Resolution): This is the highest act published by ANVISA, after consideration from its board of directors and public discussion of the matter. The RDC can be accompanied by Normative Instructions (IN), which complement and detail it. ANVISA may also publish Specific Resolutions (RE) usually for concessions of manufacturing and marketing authorization.

ANVISA was created in 1999, through Law 9,782 (Brazil, 1999). Nowadays, the Agency is the main public body responsible for regulating medicines, including HMP. Thus, since 1999, RDC, RE and IN have been observed in these products' regulations. Before that, the norms were published by the Ministry of Health, mainly through ordinances.

1.2   VARIOUS DEFINITIONS AND APPROACHES

1.2.1   DEFINITIONS

Products obtained from plant species can have a different classification, depending on the processing level: medicinal plant, herbal drug, herbal preparation or HMP.

Medicinal plant is defined as the fresh or dry plant species, usually collected at the time of use, used for therapeutic purposes.

Herbal drug consists of a plant that has gone through collection/harvesting, stabilization, when applicable, and drying, that can be found whole, cut or powdered form. The term "drug", as defined by law, gives the product a medicinal or sanitary purpose (Ministério da Saúde, 2018a).

Herbal preparation is the product obtained from the extraction of fresh medicinal plant or herbal drug, which contains the substances responsible for the therapeutic action, and may occur in the form of extract, fixed and volatile oil, wax, exudate and others (Ministério da Saúde, 2018a).

The herbal raw material that can be used in the production of HMPs comprises the medicinal plant, the herbal drug and the herbal preparation (Ministério da Saúde, 2018a).

Finally, HMPs are medicines, technically elaborated, which use herbal raw materials as the active ingredient, for prophylactic, curative, palliative or diagnostic purposes. These are divided into two categories: herbal medicine (HM) and traditional herbal product (THP) (Ministério da Saúde, 2018a).

HMPs can be single or in combination, when elaborated with one or more plant species, respectively. The isolated active substances of any origin (plant or synthetic) and their association with herbal preparations are not considered HMPs; instead, these types of assets are registered in Brazil in a different class of medicines in accordance with RDC 24/2011 (Ministério da Saúde, 2011a).

The commercial establishments legally involved in the production chain of medicinal plants and HMPs are *ervanaria*, pharmacy, drugstore, pharmaceutical ingredient suppliers and pharmaceutical industry.

Ervanaria is an herbal shop, where the dispensation of medicinal plants is carried out. This establishment does not need to have a pharmacist authorized by the Regional Pharmacy Council

(CRF) responsible for the dispensary, which is not the case with the others listed. However, it cannot dispense medicines or any product other than medicinal plants. Also, the dispensed medicinal plants cannot have therapeutic claims, and the correct storage and botanical classification must be respected, as determined by Law 5,991/1973 (Brazil, 1973).

Drugstore is the establishment authorized to sell and dispense medicines, special foods and health products directly to the consumer (Brazil, 1973).

Pharmacy is where compounded medications are made, according to prescription formula (issued by a professional allowed to prescribe drugs) or officinal formula (included on national forms or international forms recognized by ANVISA). It is authorized to trade drugs, medicines and pharmaceutical ingredients, as well as carrying dispensation of medicines in hospital units or other equivalent healthcare facility.

Both pharmacy and drugstore need an authorization for regular operation. (Autorização de Funcionamento de Empresas [AFE]).

In the context of pharmacies, in 2010, the *Farmácia Viva* project was established by the Brazilian Ministry of Health, under SUS (the acronym for Brazil's public health system 'Único de Saúde'), for the exclusive compounding of medicinal plants and HMPs, in order to provide pharmaceutical social assistance (Brazil, 2010). With the intention of regulating this institution, ANVISA published RDC 18/2013, which addresses the good practices of processing and storage of medicinal plants, preparation and dispensing of magistral and officinal HMPs (Ministério da Saúde, 2013). This approach brings all the requirements for the public compounding of HMPs, according to the claims brought by the Brazilian Ministry of Health and set out in the National Policy on Medicinal Plants and Herbal Medicines (PNPMF) (Ministério da Saúde, 2006a).

Although several categories of health professionals regulate the prescription of medicinal plants and HMPs, only the council of each specific category determine regulations that determine their professionals' attributions. Even an HMP considered exempt from medical prescription can only be prescribed by a professional qualified by the respective professional council.

The pharmaceutical ingredient supplier is important in the supply chain of the pharmaceutical ingredients, both active and excipient, used in the production of medicines by the industry and compounding pharmacies. They need to receive an AFE before producing the raw materials (Ministério da Saúde, 2018a).

Pharmaceutical industries, in turn, are establishments that produce large amounts of drugs, in batches, according to the good manufacturing practices (GMP). They must have authorization for manufacturing medicines, an AFE and certificate of GMP (Ministério da Saúde, 2018a). There is a specific authorization to produce THP, which is issued according to RDC 13/2013 (Ministério da Saúde, 2018a).

In 2006, two important public policies were published to expand the access of the Brazilian population to medicinal plants and HMPs and to increase their use: the National Policy on Integrative and Complementary Practices (PNPIC) and the National Policy on Medicinal Plants and Herbal Medicines (PNPMF).

The PNPIC was published by Ordinance 971/2006, with the goal to introduce, in the public health system – SUS – services and products related to traditional knowledge, such as Phytotherapy, Acupuncture, Homeopathy and Social thermotherapy, guaranteeing integral coverage in health care, through practices that previously were only accessible in private care. PNPIC brought actions to be implemented in SUS and in other government healthcare bodies, such as the Ministry of Health, municipal and state health secretariats, ANVISA and Fiocruz (Ministério da Saúde, 2006b). In 2018, the number of practices available to the population was expanded, reaching 29 (Ministério da Saúde, 2018b).

The PNPMF was published by Decree 5.813/2006 establishing the guidelines of priority actions to promote the safe and effective use of medicinal plants and HMPs, aiming to consolidate relevant initiatives in the country and national and international recommendations on the subject. The PNPMF promotes transversal actions involving different areas such as health, the environment and economic and social development. This broadens the therapeutic options available for the prevention and treatment of diseases; values the knowledge of traditional communities; encourages the sustainable use of Brazilian biodiversity; and stimulates the expansion and strengthening of the productive chain and the national industry (Ministério da Saúde, 2006c).

As a regulatory agency, ANVISA plays an important role in PNPMF, executing activities like monitoring and overseeing the commercialization, compounding and distribution of herbal raw materials and HMPs. The Agency also grants marketing authorization to new HMPs, through registration or notification (Ministério da Saúde, 2006c).

The publication of these policies urged ANVISA to renew its legislative framework, in order to adapt to the new national and international scenarios that were emerging. In consideration, several guidelines were published in subsequent years, as discussed below.

There are several ways in which to produce and regulate HMPs in Brazil. According to Brazilian legislation, medicines can be compounded or industrialized. They also can be aimed at human or veterinary use, thus, are regulated by ANVISA or by the Ministry of Agriculture, Livestock and Supply (MAPA), respectively.

The following different types of production and commercialization of HMPs are recognized for human use:

1.2.2 Medicinal Plants Trade, as Established by Law 5,991/1973

The trade of medicinal plants is regulated by Law 5,991/1973, and article 7 establishes that their dispensation is exclusive to pharmacies and *ervanarias* and that the plants must be properly identified, in botanical terms, and in adequate packing. So, the rules settled by this law do not address medicines, but only medicinal plants. The latter cannot have therapeutic indications on their packaging or advertising material (Ministério da Saúde, 2018a). Although this article was published more than 40 years ago, it has not been regulated yet, leaving open the requirements of quality, safety and efficacy for medicinal plants.

1.2.3 Compounding of HMPs

In Table 1.1 the main rules related to the compounding of HMPs in Brazil are presented.

Nowadays there are two types of compounding pharmacies authorized to compound HMP in Brazil: the compounding pharmacy and the *Farmácia Viva*. Compounding pharmacies are regulated by RDC 67/2007, which was updated by RDC 87/2008. This type of establishment is authorized to handle a wide range of medicinal products, depending on the authorization granted, such as low therapeutic substances, hormones, antibiotics, cytostatics, substances under special control, sterile products and homeopathic medicines. Parenteral nutrition, enteral and polyelectrolyte concentrate for hemodialysis (Concentrado polieletrolítico para hemodiálise [CPHD]) solutions are

TABLE 1.1

Main Rules Related to the Compounding of Herbal Medicinal Products (HMPs) in Brazil

Resolution	Addresses the Following Issue
RDC 67/2007	Good compounding practices of magistral and officinal preparations for human use in pharmacies
RDC 18/2013	Good practices of processing and storage of medicinal plants, preparation and dispensing of magistral and officinal products of medicinal plants and HMPs in *Farmácia Viva* under SUS
RDC 87/2008	Good practices of magistral and officinal preparations for human use in pharmacies, updating RDC 67/2007

excluded from its scope. The *Farmácia Viva*, in turn, is a public establishment, set up under the aegis of SUS, and has the scope of compounding only medicinal plants and HMPs.

Compounded HMPs for human use can be obtained in two different ways: magistral or officinal preparation. The magistral preparation is the one prepared in the pharmacy, from a prescription of a qualified professional, aimed at an individual patient, and that establishes in detail its composition, pharmaceutical form and posology. The official preparation must follow the formula registered in the Herbal Medicines National Formulary of the Brazilian Pharmacopoeia (FFFB) – or other codes recognized by ANVISA.

The FFFB was initially published in 2011 and updated in 2018, through the publication of its first supplement. The first publication of the FFFB contained 47 monographs on herbal drugs for infusions and decoctions, 17 tinctures, 1 syrup, 5 gels, 5 balms, 1 soap and 2 creams. With the publication of the first supplement, the chapter on tinctures was updated with the inclusion of new monographs, totaling 40, and a chapter on capsules with herbal preparations (containing 28 new monographs) was added. All formulations presented in the FFFB are considered officinal and may be handled in the compounding pharmacy or the *Farmácia Viva*, without the need for any individualized prescription (Ministério da Saúde, 2011b).

Of note, a pharmacy can maintain a minimum stock of preparations as listed in the FFFB, duly identified, according to the technical and management needs of the establishment, since the quality and stability of the herbal drugs and their preparations can be assured. Quality control when producing a minimum stock is similar to that stipulated for the pharmaceutical industry, because it is a batch production (Ministério da Saúde, 2011b, 2018a).

Compounded medicines have their quality controlled during their production in the pharmacy facilities and are exempted from registration in ANVISA.

The *Farmácia Viva* has somewhat different regulations from compounding pharmacies, since it must be regulated by both ANVISA, within the scope of compounding medicines, and the Ministry of Agriculture, which has the competence to regulate cultivation and harvesting of herbs. However, at the time of writing, the regulations of the Ministry of Agriculture have not yet been published.

Lastly, the *Farmácia Viva* is a compounding pharmacy which is authorized to produce its own inputs; that is, it carries out the cultivation, harvesting, processing and storage of native or acclimatized medicinal plants (Ministério da Saúde, 2018c). Moreover, it is authorized to dispense HMPs in other health facilities of the SUS network, such as outpatient clinics, hospitals and healthcare units, while other compounding pharmacies can only dispense in their own facilities.

1.2.4 INDUSTRIALIZED HMPs

In Table 1.2 the main rules applied for industrialized HMPs in Brazil (Ministério da Saúde, 2018a) are presented.

Together with the rising use of industrialized herbal products, a concern has emerged about updating this subject in the Brazilian regulatory framework. The first Brazilian legislation for the registration of HMs was established in 1967 (Perfeito, 2012); this normative has already been revised four times through RDC 17/2001, RDC 48/2004, RDC 14/2010 and RDC 26/2014. It is noticeable that the Brazilian regulation on the topic has been improving and evolving, mainly in the last decade, in a process of international convergence that resulted in the publication of a new regulatory milestone in 2014 – RDC 26.

Nowadays, the normative ruling registration and post-registration of HMPs are RDC 26/2014, RDC 38/2014, IN 2/2014 and IN 4/2014 – their subjects are detailed in Table 1.2. The ruling GMPs are RDC 17/2010 (regarding medicines), RDC 13/2013 (regarding THP) and RDC 69/2014 (regarding active pharmaceutical ingredient) (Brazil, 2018a).

Besides these normatives, there are other regulations common to all medicines licensed in Brazil, such as the rules for leaflets, packaging, labeling, clinical research, validation of analytical methodologies, stability studies and others (Brazil, 2018a). They are mentioned also in Table 1.2.

TABLE 1.2

Main Rules Applied for Industrialized Herbal Medicinal Products (HMPs) in Brazil (Ministério da Saúde, 2018a)

Resolution	Addresses the Following Issue
RDC 26/2014	Registration of HMs and registration and notification of THPs
RDC 38/2014	Post-approval changes in HMs and THPs
IN 2/2014	"List of HMs for simplified licensing" and "List of THPs for simplified licensing"
IN 4/2014	Guide for registration of HMs and registration and notification of THPs
RDC 13/2013	Good manufacturing practices for THPs
RDC 69/2014	Good manufacturing practices for active pharmaceutical ingredient
Law 6,360/1976	Measures on sanitary surveillance of medicines, drugs, pharmaceutical ingredient and related materials, cosmetics, sanitizers and other products
Decree 8,077/2013	Regulates the conditions for the operation of companies subject to sanitary licensing, and the registration, control and monitoring, within the sanitary surveillance, of the products referred to in Law No. 6,360/1976
Law 5,991/1973	Sanitary control of trade in drug, medicine and pharmaceutical ingredients
RDC 17/2010	Good manufacturing practice of medicines
RDC 47/2009	Elaboration, harmonization, updating, publication and availability of medicine leaflet for patients and health professionals
RDC 71/2009	Rules for labeling of medicines
RE 1/2005	Guide for conducting stability studies
RDC 166/2017	Criteria for the validation of analytical methods
RDC 37/2009	Admissibility of foreign pharmacopoeias
RDC 59/2014	Medicine name, its complements, and medicine family
RDC 234/2018	Outsourcing of production steps, analysis of quality control, transportation and storage of medicines and biological products
RDC 4/2009	Pharmacovigilance of medicinal products for human use
RDC 9/2015	Rules for conducting clinical trials in Brazil
RDC 98/2016	Provides for the criteria and procedures for the framework of medicinal products classified as over the counter (OTC)

RDC 26/2014 divided HMP into two subclasses: HM, which is the product that should demonstrate its safety and efficacy through non-clinical and clinical studies; and THP, when the proof of safety and effectiveness occurs through the traditional use. Thus, when the term HMP is used, both HMs and THPs are covered.

HMPs produced on an industrial scale must have authorization provided by ANVISA before their commercialization, and this can be granted by registration or notification. In the case of HMs, registration is necessary, which is the regulatory process that depends on evaluation and favorable manifestation of ANVISA, prior to releasing the product. THPs, in turn, can be registered or notified. Notification is a simplified process as explained below. Thus, ANVISA's licensing authorization is essential to confirm the quality of an HMP, since it is preceded by a technical evaluation, in which the safety and efficacy of the product must have been demonstrated.

Notification is a simplified manufacturing and marketing authorization process, implemented by ANVISA, to reduce bureaucracy in the process of licensing products. Since they are produced within predefined technical criteria, by authorized companies that comply with GMP, this offers a low health risk to patients. In this process, the product release to the market can occur immediately after the communication to ANVISA, and this is accomplished simply through a specific electronic system, with the control provided later through regulatory inspections in the companies' sites.

RDC 26/2014 determines that the notification procedure is only applicable to THP in the following manner: (a) they are obtained from an herbal active pharmaceutical ingredient (IFAV) listed in the latest edition of the FFFB; and (b) they have a specific monograph of quality control published in a pharmacopoeia recognized by ANVISA (Ministério da Saúde, 2018a).

As required by RDC 26/2014, all THPs should be nonprescription products (over the counter [OTC]). Therefore, for the notification of THPs, it is necessary that the therapeutic indication cited in the FFFB is considered a nonprescription one, in accordance with current regulations, namely RDC 98/2016 and IN 11/2016.

The registration of HMs and THPs, on the other hand, constitutes a more detailed process, in which the requesting company presents all the technical-scientific evidence related to the efficacy/effectiveness, safety and quality of the HM for evaluation, aiming at obtaining authorization for its production and marketing.

The registration dossier of an HMP consists of a documentary part, a technical report, a stability report, a production and quality control report, and a safety and efficacy (for HM)/effectiveness (for THP) report.

Regarding the safety and efficacy of HMPs, the traditional use was already stipulated in RDC 17/2000, but this was not a usual path adopted by companies. Since the advent of RDC 26/2014, this alternative path has been expanded, and now it is much more detailed, basing much of its requirements on international legislation, mainly the European, Canadian and Australian.

To be considered a THP, the herbal product must prove a continued safe use for a period longer than 30 years; the administration route cannot be injectable or ophthalmic; the claims must be coherent with the traditional usage and must be appropriate for use without the supervision of a physician for diagnosing, prescribing and monitoring. It is important to note that a THP cannot comprise ingredients which have a known toxic hazard or toxic chemical substances above safe limits.

HMs, on the other hand, should demonstrate their safety and efficacy through the presentation of non-clinical and clinical studies. Since there are no specific rules for conducting these studies in HMPs, these should be conducted according to the general regulations of ANVISA and the Brazilian National Health Council.

According to RDC 26/2014, when there are no non-clinical tests proving the safety of the HMP, they should be carried out in accordance with the latest version of the ANVISA Guide for conducting non-clinical studies of toxicology and pharmacological safety necessary to the development of medicinal products, where applicable to HMs. Also, when there are no clinical trials demonstrating the HMP's safety and efficacy, these should be performed following good clinical practice (GCP); RDC 9/2015 – the current standard for conducting clinical trials; the Guide "Operational guidance: information needed to support clinical trials of herbal products" published in 2008, by WHO/Ministry of Health (Ministério da Saúde [MS]); and the determinations of the National Health Council (CNS), established through Resolution 446/2011, and Resolution 251/1997 (Ministério da Saúde, 2018a).

RDC 26/2014 also determines that when there are non-clinical and clinical trials available in scientific and technical literature, these can be presented to ANVISA for evaluation of their quality and representativeness. If valid, it is not necessary to carry out new studies by the company which intends to register the HM. The company should send to ANVISA copies of all technical and scientific documentation corresponding to them. The studies presented must have been carried out with the same herbal drug (when this is the finished product) or herbal preparation, in the same dose and therapeutic indications presented by the registrant of the HMP (Ministério da Saúde, 2018a).

Besides the notification and ordinary registration, there is the procedure of simplified registration. In this modality, it is not necessary to present a safety and efficacy report for the HMP, since these items have been previously evaluated by ANVISA or by the European Community and European Medicines Agency (EMA), for a particular HMP. Hence, there are two possibilities for simplified registration: presence of the HMP in the lists of HMs or THPs, published by IN 2/2014; or among those published by the European Community (Community herbal monographs based on well-established or traditional use) and elaborated by the Committee on Herbal Medicinal Products – HMPC of the EMA.

RDC 26/2014 has improved the technical requirements for quality assurance, which are better suited to the control of raw materials and complex products such as IFAV and HMPs, bringing the Brazilian legislation requirements closer to the international framework. It is noteworthy that regarding quality control, regulatory requirements are the same for HMs and THPs.

A primary requirement to ensure the quality of HMs and THPs is the compliance with GMP of IFAV, whose requirements are specified in RDC 69/2014.

Manufacturers of HMPs can also produce the IFAV for themselves. But for this, they must follow the norm applied to this activity, as well as possessing an AFE for the manufacturing of active pharmaceutical ingredients (APIs).

RDC 26/2014 also defines that the production of HMs and THPs must follow the GMP, regulated by RDC 17/2010 or by RDC 13/2013, the latter being applied to THPs. Therefore, HM manufacturers must be GMP-certified by RDC 17/2010, and THP manufacturers must be certified by RDC 17/2010 or by RDC 13/2013.

HMP quality control comprises assessments of the herbal raw material (the herbal drug, the herbal preparation) and the finished product (HM and THP), as well as the stability of the medicine during the proposed shelf life time.

For the herbal drug, it is necessary to confirm its botanical identity; its integrity; organoleptic characteristics; humidity; ash content and presence of foreign material, such as micro- and macroscopic contaminants, including fungi, bacteria, mycotoxins and heavy metals. The harvesting site and methods for eliminating contaminants, if used, must be stated, together with the investigation of possible residues. Finally, the qualitative and quantitative analysis of markers must be presented, and quantitative control can be replaced by biological control of the therapeutic activity.

For the herbal preparation, the solvents, excipients and vehicles used in its extraction should be reported; the extraction methods employed; the part(s) of the plant used; approximate drug: extract ratio and presence of residual solvents. The results of the physical-chemical tests of the herbal preparation are also requested, as described in RDC 26/2014 and IN 4/2014.

Regarding the finished product (HM and THP), the control requirements depend on the dosage form and are focused in the evaluation of the HMP's integrity and stability, which requires the chromatographic profile, marker assays and control of microbiological contamination, among others.

The methods used to control HMP quality should be present in the Brazilian Pharmacopoeia, current edition, or one of the pharmacopoeias recognized by ANVISA, according to RDC 37/2009 (German, American, Argentinian, British, European, French, International, Japanese, Mexican and Portuguese). Another option is to validate the methodology according to the provisions stated by RDC 166/2017. The source of the method (internal development or compendial) must be properly indicated.

Considering the absence of a methodology in an official pharmacopoeia, validation should be carried out in order to demonstrate that the method is appropriate for the intended purpose, and this can be a qualitative, quantitative or semi-quantitative determination. To do so, the methodology must be challenged according to the validation parameters explained in RDC 166/2017.

Regarding the validation of HMP methods, specific caveats are made in the evaluation of the parameter accuracy. In order to perform the tests related to this parameter, the chemical reference substance (CRS) must be added to a diluted solution of the finished product, allowing the complex matrix effect to be considered. This matrix effect should also be evaluated by comparing the angular coefficients of the calibration curves constructed with the CRS of the analyte in the solvent and with the sample fortified with the CRS of the analyte. The parallel approach of these lines indicates an absence of interference from the matrix constituents.

In addition to the control of raw materials and finished product, stability studies are required in order to verify if the physical, chemical, biological and microbiological characteristics of the HMP remain within the specifications, during the period of shelf life proposed. The results are used to establish or confirm this period and recommend storage conditions. The tests to be carried out during this study are determined by RE 1/2005 and comprise, among others, the product's physical characteristics, chromatographic profile, qualitative and quantitative analysis of markers and micro-biological control.

Stability studies and validation of analytical procedures for HMs and THPs follow the general regulations established for medicines by ANVISA. However, due to the complexity of HMP composition, specific guidelines have been adopted and are detailed in IN 4/2014.

In addition to the possibility of a company producing and controlling the HMP quality in its own facilities, it is also possible to outsource these activities, as recently established by RDC 234/2018 (Ministério da Saúde, 2018d).

All registered or notified HMPs must renew their marketing authorization every five years, in order to demonstrate that the product remains safe and effective, according to market data (Brazil, 2015a). Also, after a registration has been granted, the requesting company may need to make changes in its product, and this can be done using RDC 38/2014. This norm guides the submission of post-approval changes in HMs and THPs. For notified THPs, post-approval changes do not apply, so if any change in the product is required, the notification must be canceled and resubmitted, including the proposed change in the new application.

The regulations already discussed and presented in Table 1.2 are constantly improved and revised. Since the establishment of PNPIC and PNPMF, virtually all the HMP regulatory framework was republished, aiming to promote and develop the HMP national market.

In 2017, there were 359 HMPs licensed in Brazil, of which 332 were single HMPs and 27 in combination. At that time there were no THPs notified. Between 2006 and 2012, there was a 31% decrease in the number of HMPs registered in Brazil (a reduction of 159 products). The number of HMPs that left the market during this period is equivalent to about half the number of products registered today. This decrease took place specifically between 2008 and 2011. These products were withdrawn from the market due to lack of technical sanitary requirements (Perfeito, 2012) or lack of commercial interest.

There are 101 plant species with herbal preparations registered as HMP actives in Brazil: 35 are native, naturalized or exotic plants cultivated in Brazil; that is, only about 35% of the total registered species are obtained on Brazilian soil. Thus, most HMPs produced in Brazil have as active ingredients plant species that come from abroad. One possible reason for this is that there is more published scientific information about these exotic species (Santos et al., 2011), both in terms of their safety and efficacy as well as quality control. Clinical studies on native Brazilian herbal species are rare, and even if there is a long traditional use of some species, this tradition in human use is not well documented (Carvalho et al., 2018).

Another aspect that could have contributed to the reduction of HMPs on the Brazilian market is related to sales restrictions. Prescription medicines, unlike nonprescription (or OTC), cannot be advertised to people other than prescribers and cannot be displayed over the counter. Considering the 359 HMPs registered in Brazil, 214 (59.6%) are OTCs and 145 (40.4%) are prescription medicines. The number of HMPs classified as under prescription is considerably higher in Brazil than in most other countries (WHO, 2005, 2011). Many products designated as for sale under prescription in Brazil are sold in other countries as OTCs or supplements. It is worth noting that there are no specific rules on sales restrictions of HMPs, so the general determination for medicines, established by RDC 98/2016, must be followed. This regulation states that a list will be published specifying the therapeutic indications and sales restrictions for each API. This list is still under development, and the best classification for each HMP will be discussed individually (Ministério da Saúde, 2016a).

In 2017, 77 companies had market authorization in Brazil. These companies are distributed in 11 Brazilian states, with most of them concentrated in the southeastern region. There is no company producing HMPs in the northern region, despite the biodiversity of this area, showing how this market niche is so little exploited in Brazil (Carvalho et al., 2018).

It is essential to emphasize that together with the current reduction in HMPs on the market there is also a decrease in the production of herbal raw material. National companies of API are not complying with the GMP to produce herbal drugs and herbal preparations; hence a high number of IFAVs used in the country are imported (Branco, 2015).

Many companies that hold licenses for HMP products in Brazil do not follow the changes in legislation and do not adapt to them. Also, they rarely take advantage of the financing offered by governmental programs and barely invest in research and technological development. Often, registration requests are rejected because of technical problems that have been repeated for years and should already have been overcome.

Unfortunately, there is little interaction between companies and national research centers; thus, many studies carried out in Brazilian universities do not reach the possibility of pharmaceutical development and registration as a regulated product. There is a lack of investigation with native plants, and a dearth of professionals specialized in the production of drugs from complex sources, such as HMPs, which requires equally complex control techniques (Araújo et al., 2013; Perfeito, 2012).

Although the number of HMP companies has decreased, there was no reduction in the turnover values in the sector. This is because there is no price regulation of HMPs in Brazil, so it is up to the companies to define the prices, unlike the practice in synthetic medicines. Nevertheless, Brazilian values, in terms of product numbers and market value, are very low when compared to other markets: 80% of Germans, 70% of Canadians and 49% of French people use HMs (Carvalho et al., 2018; WHO, 2011). Despite this, according to the Ministry of Health, between 2013 and 2015, there was a 161% increase in demand for herbal and medicinal treatments in SUS (Ministério da Saúde, 2018e).

There are few registered products containing native species in Brazil, as mentioned above, but taking into account informal trade the number of native products is quite large. It is necessary to raise awareness in the Brazilian population about the need for HMPs to be evaluated for their quality, safety and efficacy, prior to commercialization. It is rare for users to verify whether the herbal product is authorized, and people assume that if it is on the market it must be legal, or if it is "natural" it does not pose any danger to health. From January 2015 to June 2018, ANVISA published 440 resolution – RE related to drug irregularities, of which 14% were related to HMP, 80% of which were unauthorized products (Ministério da Saúde, 2018f). Additionally, there are few cases of adverse events related to the use of HMPs, and the number of adverse events reported with regard to medicinal plants is very low (Balbino and Dias, 2010).

It is important to note that, in a similar manner to the international legislation, according to Brazilian law, the same herbal species can be licensed for human use not only in medicines, but also in food, cosmetic and health products, often with effects in a similar way to those approved for drugs and often coexisting on the market with the same dosages (Minghetti et al., 2016). Examples of products obtained from medicinal plants that are regulated in foods are *Allium sativum* and isoflavones from soybean (*Glycine max*). Several products of topical use found in Brazilian pharmacies are regulated in Brazil in the area of cosmetics, such as gels and shampoos of *Calendula officinalis* or *Matricaria recutita*. There are also products registered in the category of health products, such as vegetable oils, including sunflower (*Helianthus annuus*) and orange (*Citrus aurantium*) (Ministério da Saúde, 2016b). Regulation for these other classes of products is more lenient than that required for medicines, causing many products that have lost their registration as medicines to migrate to these other categories.

1.3   CONCLUSION

Brazilian sanitary regulation has changed considerably over recent years in order to deal with the peculiarities and complexity of herbal materials, such as the herbal active ingredient and HMPs. The new regulations have brought internationally harmonized concepts of quality control, safety and efficacy to the national framework. Since 2010, a procedure of medicinal notification has been put in place, which allows the rapid release of products into the market. Also, with the expansion of the scope of official compendia, like the FFFB, the notification should be expanded. So, the new directions established by ANVISA, together with public policies, converge to foster the development of the national HMP production chain, as well as encouraging financial support for research on the topic.

The requirements for using Brazil's rich socio-biodiversity are now better delineated, both in health legislation and in the new biodiversity law. Now, it is up to the commercial sector to dedicate efforts on the development of new products containing new medicinal plants, taking into consideration the traditional knowledge of Brazilian communities and the wide acceptance of natural products in the treatment or prevention of diseases, by the population. Therefore, it is expected that all these factors will contribute to increase the number of safe and effective HMPs traded in Brazil and will also promote a viable and successful alternative to the national industry.

REFERENCES

Araújo, R. F. M.; Rolim-Neto, P. R.; Soares-Sobrinho, J. L.; Amaral, F. M. M.; Nunes, L. C. C. 2013. Phytomedicines: legislation and market in Brazil. Revista Brasileira de Farmacognosia, 94, 331–341.

Balbino, E. E.; Dias, M. F. 2010. Farmacovigilância: um passo em direção ao uso racional de plantas medicinais e fitoterápicos. Revista Brasileira de Farmacognosia, 20, 992–1000.

Branco, P. F. 2015. Boas práticas de fabricação de insumos de origem vegetal: evolução das normas que norteiam a produção e o panorama do parque fabril brasileiro. Master dissertation, Universidade Federal do Ceará.

Brazil. 1973. Lei n° 5.991, de 17 de dezembro de 1973. Dispõe sobre o controle sanitário do comércio de drogas, medicamentos, insumos farmacêuticos e correlatos, e dá outras providências. http://www.planalto.gov.br/ccivil_03/LEIS/L5991.htm.

Brazil. 1999. Lei n° 9.782, de 26 de janeiro de 1999. Define o sistema nacional de vigilância sanitária, cria a Agência Nacional de Vigilância Sanitária, e dá outras providências. http://www.planalto.gov.br/ccivil_03/LEIS/L9782.htm.

Brazil. 2010. Portaria n° 886, de 20 de abril de 2010. Institui a Farmácia Viva no âmbito do Sistema Único de Saúde (SUS). http://bvsms.saude.gov.br/bvs/saudelegis/gm/2010/prt0886_20_04_2010.

Carvalho, A. C. B.; Lana, T. N.; Perfeito, J. P. S.; Silveira, D. 2018. The Brazilian market of herbal medicinal products and the impacts of the new legislation on traditional medicines. Journal of Ethnopharmacology, 212, 29–35.

Minghetti, P.; Franzè, S.; Raso, F.; Morazzoni, P. 2016. Innovation in phytotherapy: Is a new regulation the feasible perspective in Europe? Planta Medica, 82, 591–595.

Ministério da Saúde. 2006a. Política nacional de plantas medicinais e fitoterápicos. http://bvsms.saude.gov.br/bvs/publicacoes/politica_nacional_fitoterapicos.pdf.

Ministério da Saúde. 2006b. Decreto n° 971, de 3 de maio de 2006. Aprova a política nacional de práticas integrativas e complementares (pnpic) no sistema único de saúde. http://bvsms.saude.gov.br/bvs/saudelegis/gm/2006/prt0971_03_05_2006.html.

Ministério da Saúde. 2006c. Decreto n° 5.813, de 22 de junho de 2006. Aprova a política nacional de plantas medicinais e fitoterápicos e dá outras providências. http://www.planalto.gov.br/ccivil_03/_Ato2004-2006/2006/Decreto/D5813.htm.

Ministério da Saúde. 2018c. Glossário temático: práticas integrativas e complementares em saúde. http://portalarquivos2.saude.gov.br/images/pdf/2018/marco/12/glossario-tematico.pdf.

Ministério da Saúde. 2018e. Medicina alternativa: Uso de plantas medicinais e fitoterápicos sobe 161%. http://www.brasil.gov.br/editoria/saude/2016/06/uso-de-plantas-medicinais-e-fitoterapicos-sobe-161.

Ministério da Saúde. 2018f. Imprensa Nacional. http://www.in.gov.br (accessed on September 4, 2018).

Ministério da Saúde. Agência Nacional de Vigilância Sanitária (ANVISA) – Brasil. 2007. Resolução da Diretoria Colegiada – RDC. https://www20.anvisa.gov.br/segurancadopaciente/index.php/legislacao/item/rdc-67-de-8-de-outubro-de-2007.

Ministério da Saúde. Agência Nacional de Vigilância Sanitária (ANVISA) – Brasil. 2008. Resolução da Diretoria Colegiada – RDC http://bvsms.saude.gov.br/bvs/saudelegis/anvisa/2008/res0087_21_11_2008.

Ministério da Saúde. Agência Nacional de Vigilância Sanitária (ANVISA) – Brasil. 2011a. Resolução da Diretoria Colegiada – RDC n° 24 http://portal.anvisa.gov.br/documents/33836/2957213/RDC+2411+-+atualizada.pdf/592f6198-85c5-4c95-b0af-0e6a05a36122.

Ministério da Saúde. Agência Nacional de Vigilância Sanitária (ANVISA) – Brasil. 2011b. Formulário de fito-terápicos da Farmacopeia Brasileira. http://portal.anvisa.gov.br/documents/33832/259456/Formulario_de_Fitoterapicos_da_Farmacopeia_Brasileira.pdf/c76283eb-29f6-4b15-8755-2073e5b4c5bf.

Ministério da Saúde. Agência Nacional de Vigilância Sanitária (ANVISA) – Brasil. 2013. Resolução da Diretoria Colegiada – RDC n° 18. http://bvsms.saude.gov.br/bvs/saudelegis/anvisa/2013/rdc0018_03_04_2013.pdf (accessed August 22, 2018).

Ministério da Saúde. Agência Nacional de Vigilância Sanitária (ANVISA) – Brasil. 2016a. Resolução da Diretoria Colegiada – RDC n° 98. http://portal.anvisa.gov.br/documents/10181/2921766/RDC_98_2016.pdf/32ea4e54-c0ab-459d-903d-8f8a88192412.

Ministério da Saúde. Agência Nacional de Vigilância Sanitária (ANVISA) – Brasil. 2018a. Consolidado de nor-mas de fitoterápicos e dinamizados. http://portal.anvisa.gov.br/registros-e-autorizacoes/medicamentos/produtos/medicamentos-fitoterapicos/orientacoes.

Ministério da Saúde. Agência Nacional de Vigilância Sanitária (ANVISA) – Brasil. 2018b. Resolução da Diretoria Colegiada – RDC n° 234. http://imprensanacional.gov.br/materia/-/asset_publisher/Kujrw0TZC2Mb/content/id/27128992/do1-2018-06-25-resolucao-rdc-n-234-de-20-de-junho-de-2018-27128955.

Ministério da Saúde. Brasil. 2016b. Consulta de produtos. http://portal.anvisa.gov.br/consulta-produtos-registrados.

Ministério da Saúde. Brasil. 2018. Práticas integrativas. http://dab.saude.gov.br/portaldab/ape_pic.php?conteudo=praticas_integrativas.

Perfeito, J. P. S. 2012. O registro sanitário de medicamentos fitoterápicos no Brasil: uma avaliação da situação atual e das razões de indeferimento. Master dissertation, Universidade de Brasília. http://repositorio.unb.br/bitstream/10482/10429/1/2012_JoaoPauloSilverioPerfeito.pdf.

Santos, R.L.; Guimaraes, G.P.; Nobre, M.S.C.; Portela, A.S. 2011. Análise sobre a fitoterapia como prática integrativa no Sistema Único de Saúde. Revista Brasileira de Plantas Medicinais, 13, 486–491. http://www.scielo.br/scielo.php?script=sci_arttext&pid=S1516-05722011000400014&lng=en&nrm=iso>. ISSN 1516-0572.

World Health Organization – WHO. 2005. National policy on traditional medicine and regulation of herbal medicines - report of a WHO global survey, Genebra. http://apps.who.int/medicinedocs/pdf/s7916e/s7916e.pdf.

World Health Organization – WHO. 2011. The world medicines situation 2011 – Traditional medicines: global situation, issues and challenges. http://digicollection.org/hss/documents/s18063en/s18063en.pdf.

2 Physico-Chemical Methods for the Quality Control of Medicinal Plants, Plant Derivatives and Phytomedicines in Brazil

Paula Carolina Pires Bueno and Alberto José Cavalheiro
Instituto de Química, Universidade Estadual Paulista
(UNESP), Araraquara, São Paulo, Brazil

CONTENTS

2.1   INTRODUCTION

Worldwide, the population uses a plethora of plant species as homemade medicines for the treat-ment and prevention of several diseases. Therefore, they represent an important niche of the global drug market and health care system, which emphasizes the need to ensure their quality, safety and efficacy. Many factors impact directly or indirectly on the quality of plant raw materials and their derived products. These factors include the selection of matrices for cultivation, domesti-cation, establishment of cultivation conditions, harvesting, storage and the operation procedures for the production of extracts and other derivatives. Regardless of being widely used in different disciplines, such as medicinal, alimentary or industrial, many publications in international high impact journals have brought attention to the need of performing a critical analyses, considering the variation of weather and soils in each cultivation area, the seasonality, the time of harvesting, the conditions of the drying and stabilization process, and the accurate botanical identification of the plant species.

For this reason, in the last decades, there has been a growing concern with the standardization for this type of product in order to guarantee a safe and reliable use. For this purpose, the World Health Organization (WHO), several international regulatory agencies and the Brazilian National Health Surveillance Agency (ANVISA) have been emphasizing the importance of guaranteeing the quality of the plant material, using modern and adequate techniques of analysis and suitable standards.

In order to efficiently characterize these factors while conducting the quality control of herbal materials, an important aspect is for the analysts to take a multidisciplinary approach and pursue areas of pharmacognosy, natural products chemistry, botany, plant morphology, plant physiology, organic and analytical chemistry. Thereafter, during the development and definition of the final pharmaceutical preparation, aspects of pharmaceutical technology and industrial physics also are essential.

Therefore, the objective of this chapter is to (a) introduce the regulatory framework in Brazil, (b) present the understanding of the main physico-chemical tests for assessing the quality of plant raw materials, intermediate products (for example, essential oils and extracts) and phytomedicines in a comprehensive and integrated way and (c) their potential applications to the important com-merce of natural products use in foods and medicines in Brazil.

2.2   IMPORTANT PUBLICATIONS

The World Health Organization (WHO) provides free access to the *Quality control methods for herbal materials* (World Health Organization, 2011), a wide-ranging publication concerning the applicable analytical procedures to products of herbal origin (available at http://www.who.int/iris/handle/10665/44479). Another free-access WHO publication, the *WHO monographs on selected medicinal plants,* provides detailed information on selected medicinal plants, with to date, 116 monographs, distributed in 4 volumes (available at http://apps.who.int/medicinedocs/en/d/Js2200e/).

These monographs present chemical and pharmacological data, parameters and specifications of quality control for each plant species described there. Regarding the international context, other monographs and procedures available at the American and European Pharmacopoeias are also very useful and important to the understanding and harmonization of analytical methods described in this chapter.

In Brazil, the National Health Surveillance Agency (ANVISA), through the Resolução da Diretoria Colegiada (RDC) Resolution, n. 26 from May 13, 2014 establishes the registration of herbal medicines, and the registration and notification of traditional herbal medicines. The resolution defines the required tests that must compose the analysis report of the plant raw material, the corresponding derivatives and final products. Along with the results, the report must also indicate the methods used and the technical specification defined for each batch.

Other ANVISA publications include requirements for quality control, registration and good manufacturing practices (GMP) applicable to phytomedicines. Among them, there is the Normative Instruction IN n. 04, from June 18, 2014 that determines the publication of the Orientation guide to the registration of phytomedicines and registration and notification of a traditional phytotherapeutic product; the RDC n. 13, from March 14, 2013, which displays the GMP for traditional phytotherapeutic products; and the RDC n. 69, from December 08, 2014 that relates to the GMP for active pharmaceutical ingredients. Accordingly, the Brazilian Pharmacopoeia, another ANVISA publication, provides the methods, general analysis procedures, official reference substances and monographs applicable to plant materials and phytomedicines, some of which are native from Brazil.

Looking at the definitions available at RDC n. 26 from March 13, 2014, the understanding of three of them is crucial for the correct interpretation of the nomenclature used in the process of registration and division of the physico-chemical tests presented in this chapter: (i) plant crude drug (or plant raw material), (ii) plant derivatives, which are plant active raw materials (or plant active pharmaceutical ingredients) and (iii) herbal medicine product (or phytomedicine).

Plant crude drug or plant raw material is the medicinal plant or its parts, which contain compounds responsible for the therapeutic action, after collection, harvesting, stabilization and drying processes. The plant can be whole, fragmented, crushed or pulverized. **Plant derivative** is the product obtained from the extraction of the fresh medicinal plant, or plant raw material that contains the substances responsible for the therapeutic action. The product can be in the form of extract, fixed and volatile oil, wax, exudate and others. **Herbal medicine product or phytomedicine** is the product obtained from an active plant raw material, except isolated substances, with prophylactic, curative or palliative finalities, including herbal medicines and traditional herbal products. The product can be simple, when the active ingredient originates from a single medicinal plant species, or combined, when the active ingredient originates from more than one plant species.

The physico-chemical tests established for the quality control testing routine of the plant raw material, the corresponding derivatives and phytomedicines can be basically divided into four parts: (I) identification tests, (II) quantitative analysis, (III) purity and integrity tests and (IV) characterization tests, which are described in Table 2.1. Such tests, when interpreted together, supply a comprehensive physico-chemical profile and the overall quality of the product under evaluation.

Except for the purity and integrity tests, and the characterization tests, which are applicable to any plant species, some determinations are very specific, for example, the qualitative and quantitative analysis of a certain chemical marker or active principle of a plant species. If such a species does not yet have an official monograph, the development and validation of the method for chemical analysis is necessary. Because of the high chemical complexity of such a matrix, the analyst or researcher must have good multidisciplinary knowledge, and an accurate critical sense to select the best sample preparation, as well as to interpret the final results.

Apart from being necessary for the reliability of the results in the context of the quality system, the validation of analytical methods is justified for legal, technical and commercial reasons. The international bodies and agencies, such as IUPAC (International Union of Pure and Applied Chemistry), FDA (Food and Drug Administration) and ICH (International Conference

TABLE 2.1

List of the Main Physico-Chemical Tests Applicable to the Quality of Plant Raw Materials and Plant Derivatives

Type	Tests	Plant Raw Material	Liquid Extracts	Dry Extracts	Essential Oils	Fixed Oils
I	Organoleptic analysis	X	X	X	X	X
	Macroscopic botanical identification	X				
	Microscopic botanical identification	X				
	Chemical identification	X				
	Chromatographic fingerprinting	X	X	X	X	X
II	Quantitative analysis	X	X	X	X	X
III	Water and volatile material	X		X		
	Total ash	X				
	Acid-insoluble ash	X				
	Foreign material	X				
	Extractable material	X				
	Content of essential oils	X				
	Pesticide residues	X	X	X	X	X
	Heavy metals	X	X	X	X	X
	Radioactivity contamination	X				
	Aflatoxins	X	X	X	X	X
	Solvent residues		X	X	X	X
	Dry residue		X			
IV	pH	X				
	Relative density	X			X	X
	Apparent density			X		
	Alcohol content	X				
	Refraction index				X	
	Optical rotatory power				X	
	Solubility			X		
	Viscosity		X			X
	Saponification value					X
	Acidic value					X
	Ester value					X
	Iodine value					X

(I) identification tests; (II) quantitative analysis; (III) purity and integrity tests; (IV) characterization tests.

on Harmonization), and the Brazilian organizations such as INMETRO (National Institute of Metrology, Quality and Technology) and ANVISA, provide guidelines and recommendations for the execution of validation procedures, a fundamental requirement for demonstration of process technical competence.

In Brazil, ANVISA displays the resolution RDC n. 166 from July 24, 2017, which establishes criteria for the validation of analytical methods applicable to pharmaceutical ingredients, medicines and biologic products in every production step. Among the parameters to be evaluated, are selectivity, linearity, accuracy, precision (repeatability and intermediate precision), detection limit, quantification limit, linear response range and robustness. The experimental procedures, limits and criteria of acceptation must be defined according to the method to be validated, and according to the values recommended in the guidelines mentioned above.

2.3 RECEIVING AND SAMPLING PLANT RAW MATERIALS AND PLANT DERIVATIVES

Usually, plant raw materials are dehydrated, which allows for better handling and storage, together with improving the conservation and durability of the crude drug. Liquid extracts can be commercialized concentrated or diluted. In both cases, in the act of receiving, the analyst must pay attention to the packaging conditions, verifying the conformity regarding the type of material used (plastic, paper, cardboard), hygiene (presence of dirt and stains in the external part) and sealing. When opening a crucial aspect is to pay attention to the level of division and compaction of the plant (in the case of plant raw materials), the presence of physical (shards of glass, pieces of paper, sand and rocks) and biologic contaminants (mold, insects, etc.). Only then, the following initial organoleptic analysis, verifying the aspect, texture, color and smell of the material should take place.

The next step is the sampling, which should be executed carefully, so that the results express representative values of the total quality of the batch. Accordingly, establishing a sampling plan according to the norms of the defined quality system, considering the procedures of the pharmacopoeia or those described in publications such as ISO 2859 is recommended– Sampling procedures for inspection by attributes, the WHO Guidelines for Sampling of Pharmaceutical Products and Related Materials, or the NBR 5426 from 1985 – Sampling plans and procedures in the inspection of attributes, among others.

For plant raw materials, the Brazilian Pharmacopoeia 5th edition suggests a sampling plan, which considers the amount of sample available, the numbers of containers and the level of division of the drug (Table 2.2). Overall, for drugs finely fragmented, pulverized or with dimensions smaller than 1.0 cm, the sample must be at least 250.0 g, considering a total amount batch up to 100 kg, and/or at least an amount, which is sufficient for the execution of every physico-chemical test with further amount for sample. For drugs with dimensions greater than 1.0 cm, the sample must be at least 500.0 g. If the total amount of the received batch is smaller than 10 kg, the final sample must be at least 125.0 g. The samples must be taken in similar amounts from the superior part, middle and inferior part of each container to be sampled. In the end, every portion must be joined, homogenized and divided in four equal parts, separating the sample over a square area. Then, the two diagonal parts must be put together followed by the remaining parts. If necessary, the process must be repeated until there is no accentuated difference in the dimensions of the fragments in question. Finally, a part of this sampled and homogenized material must be retained as a reference sample, which might be used in any retesting procedures if necessary.

The physico-chemical analysis of the plant sample properly homogenized, standardized and labeled must be initiated by an organoleptic analysis, verifying the aspect, color and smell of the sample. Also, before initiating the assays, the plant material must be pulverized and the size of the particle must be standardized with a standard sieve, aiming to guarantee reproducible and reliable results.

TABLE 2.2
Sampling Plan for Plant Raw Materials Preconized by the Brazilian Pharmacopoeia, 5th Edition

Number of Packages (units)	Number of Packages to Be Sampled
1 – 3	All
4 – 10	3
11 – 20	5
21 – 50	6
51 – 80	8
81 – 100	10
More than 100	10% of the total of packages

2.4 PHYSICO-CHEMICAL TESTS FOR THE QUALITY CONTROL OF PLANT RAW MATERIALS

The physico-chemical tests for the quality control of plant raw materials can be divided into identification tests, quantitative analysis purity and integrity tests (Table 2.1), so that the compilation of the results allows a very careful evaluation on the authenticity and quality of the material under investigation.

The identification tests aim to determine the authenticity of the plant material. Among the most important tests are the macroscopic and microscopic botanical examination, the qualitative analysis by phytochemical prospection (chemical identification) and qualitative analysis of the chemical profile by chromatographic fingerprinting.

The purity and integrity tests may be interpreted as general tests, since the methodologies and principles are applicable to any sample, regardless of the species concerned. Such tests provide information on the drying process, cultivation and harvesting conditions, pesticides contamination, storage, possibility of adulteration and others. Among the physico-chemical tests recommended, there are the determination of the water and volatile matter, total ash content, acid-insoluble ash content, presence of foreign matter, essential oils content (if applicable), determination of extractable matter, radioactive contamination, determination of pesticide residues, heavy metals and aflatoxins.

Finally, the quantitative analysis of metabolites groups or chemical markers depends on the expression of the biosynthetic apparatus characteristic of each plant species. For this purpose, spectrophotometric methods for total determinations (such as determination of total flavonoids or total phenols) and various chromatographic methods for quantitative determinations are very useful and valuable for the complete characterization of the plant raw material.

2.4.1 IDENTIFICATION TESTS

2.4.1.1 Macroscopic and Microscopic Botanical Examination

The botanical identification of the raw plant material is based on the comparison of the macroscopic and microscopic botanical features of the sample with an authentic material, or with information presented in monographs or reference literature. The macroscopic identification of the plant raw material, if whole depends on the plant organ (flowers, leaves, skin, stem, roots and seeds) and is based on the analysis of the flavor and smell (if possible), color, size (length, width, thickness), surface, texture, fracture and appearance of the fracture or other characteristics described in monographs. Because of the subjective and probably misleading because of the presence of similar or contaminating materials, the association of the analysis with the microscopic analysis is crucial, being indispensable if the material is broken or powdered.

The procedure for the microscopic examination of plant raw materials consists in the exam of histologic cuts of the plant material, whether intact or fragmented, or directly in the sample if the plant is pulverized. For this purpose, microscopes equipped with 4×, 10× and 40× magnifying lenses are typically used. Briefly, the preparation of the slides consists in (a) softening or re-hydrating the dehydrated material (using water for fine tissues such as flowers and leaves, or mixtures of water, ethanol and glycerol for rigid and dense tissues such as barks and stems); (b) execution of the histological cuts (freehand or with the help of a microtome); (c) clarification with chloral hydrate test solution (TS) or hypochlorite TS; (d) staining (simple or combined using reagents such as safranin 1% (w/v) solution for observation of cutin, lignin and suberin; iodinated zinc chloride TS for observation of cellulosic cell wall; phloroglucinol TS, for lignified cell wall staining, among many others available); (e) slide assembly. Finally, the botanical features observed in the microscope must be compared with slides prepared with authentic material, or with information available in monographs of pharmacopoeias, specialized literature or in virtual databases of images and illustrations.

2.4.1.2 Identification by Phytochemical Prospection or Chemical Identification

The phytochemical prospection is a fast, simple and low-cost test, which allows the identification of chemical compounds or functional groups characteristic of the plant species. Briefly, a small portion of the plant material is pulverized and extracted, aiming to obtain an extract rich in substances of interest, such as alkaloids, anthraquinones, phenolic compounds, steroids, etc. Thereafter, specific reagents solutions are added to the extract, in which color changes or precipitation will evidence the class of the existing substances. Some of the most characteristic reactions and observed results follow, which are described in Table 2.3.

TABLE 2.3

Common Tests for Phytochemical Prospection of Plant Raw Materials

Alkaloids

Acid/base extraction of the dewaxed material and partition with chloroform

Dragendorff (potassium iodide and bismuth subnitrate): orange or orange/red precipitate indicates the presence of alkaloids

Mayer (potassium iodide and mercuric chloride): white precipitate indicates the presence of alkaloids

Wagner/Bouchardat (iodine in potassium iodide): brown precipitate indicates the presence of alkaloids

Bertrand (silico-tungstic acid): white precipitate indicates the presence of alkaloids

Hager (saturated solution of picric acid): yellow precipitate indicates the presence of alkaloids

Vitali-Morin reaction (acetone and potassium hydroxide in methanolic solution): violet color develops in the presence of tropane alkaloids

Wasicky reaction (*p*-dimethylaminobenzaldehyde in concentrated sulfuric acid): red color develops in the presence of tropane alkaloids

Mandelin reaction (ammonium vanadate in concentrated sulfuric acid): violet to red color develops in the presence of indole alkaloids

Otto reaction (potassium dichromate in concentrated sulfuric acid): violet color indicates the presence of indole alkaloids

Murexide reaction (potassium chlorate and concentrated hydrochloric acid, followed by evaporation and exposition to ammonia vapor): pink/violet color indicates the presence of purine alkaloids (methylxanthines)

Anthraquinones

Hot hydroalcoholic extraction, followed by hydrolysis and partitioning with chloroform or *n*-hexanes

Bornträger reaction (diluted ammonium hydroxide): red color indicates the presence of 1,8-hydroxy-anthraquinones; violet/blue color indicates the presence of 1,2-hydroxyl-anthraquinones

Cardiac Glycosides

Hydroalcoholic extraction followed by partition with chloroform

Keller-Killiani reaction (glacial acetic acid, ferric chloride and concentrated sulfuric acid): brown-reddish color in the contact zone of the two liquids and bluish-green (gradual) in the layer containing acetic acid indicates the presence of deoxy sugars from cardiac glycosides

Pesez reaction (glacial acetic acid, xanthydrol and heating): red color indicates the presence of deoxy sugars

Kedde reaction (3,5-dinitrobenzoic acid in ethanolic solution and alkaline environment): dark red/violet color indicates the presence of the lactonic ring

Baljet reaction (picric acid solution in alkaline environment): orange color indicates the presence of the lactonic ring

Legal reaction (sodium nitroprusside in alkaline environment): dark red color indicates the presence of the lactonic ring of cardenolides

Raymond-Marthoud reaction (*m*-dinitrobenzene in alkaline environment): orange or violet color indicates the presence of the lactonic ring of cardenolides

Coumarins

Alkaline extraction with re-acidification and recovery in organic solvents

Detection by UV light (360 nm): a shiny blue or green fluorescence develops in the presence of coumarins

Reaction with alkaline solution (potassium hydroxide in methanol): yellow color develops due to the rupture of the lactonic ring

(Continued)

TABLE 2.3 *(Continued)*
Common Tests for Phytochemical Prospection of Plant Raw Materials

Flavonoids

Hydroalcoholic or alcoholic extraction

Shinoda reaction (concentrated hydrochloric acid and magnesium): orange/red color develops in the presence of flavonoids

Pew reaction (concentrated hydrochloric acid and zinc in methanol): red color develops in the presence of flavonoids

Boric acid complexation (boric acid and oxalic acid in acetone followed by evaporation, resuspension in ethyl ether and observation under UV light): a greenish-yellow fluorescence indicates the presence of flavonoids

Reaction with aluminum chloride (alcoholic solution of aluminum chloride and observation under UV light): intensification of the fluorescence/formation of a greenish-yellow color indicate the presence of flavonoids

Saponins

Hot aqueous extraction for general detection; acid hydrolysis of the aqueous extract and extraction with chloroform for detection of sapogenins

Frothing Test: formation of stable persistent foam after the agitation indicates the presence of saponins

Haemagglutination test (red blood cell suspension): precipitation indicates the presence of saponins

Liebermann-Burchard reaction (acetic anhydride and concentrated sulfuric acid in chloroform): blue/green color develops in the presence of steroid nucleus and dark pink/red color develops in the presence of triterpene nucleus

Salkowski reaction (concentrated sulfuric acid): red or reddish-brown ring indicates positive result for triterpene nucleus, while a pink /violet ring indicates the presence of steroid nucleus

Steroids and Triterpenes

Alcoholic extraction followed by partition with chloroform or ether

Liebermann-Burchard reaction (acetic anhydride and concentrated sulfuric acid in chloroform): blue/green color develops in the presence of steroid nucleus and dark pink/red color develops in the presence of triterpene nucleus

Salkowski reaction (concentrated sulfuric acid): red or reddish-brown ring indicates positive result for triterpene nucleus, while a pink/violet ring indicates the presence of steroid nucleus

Tannins

Hot aqueous extraction

Haemagglutination test (red blood cell suspension): precipitation indicates the presence of tannins

Reaction with gelatin (gelatin solution or skin powder): precipitation indicates the presence of tannins

Reaction with alkaloids (quinine sulfate in acid environment): white precipitate indicates the presence of tannins

Reaction with copper acetate (copper acetate solution): precipitation or turbidity indicates the presence of tannins

Lead acetate reaction (solution of lead acetate in acid environment): precipitation or turbidity indicates the presence of hydrolysable tannins

Reaction with bromine water: yellow or red precipitate indicates the presence of condensed tannins

Reaction with ferric chloride (aqueous solution of ferric chloride): blue color indicates the presence of hydrolysable tannins and green color indicates the presence of condensed tannins

Stiasny reaction (concentrated hydrochloric acid and formaldehyde): red precipitate indicates the presence of condensed tannins; hydrolysable tannins in solution can be detected with the addition of sodium acetate and methanolic solution of ferric chloride (formation of blue color)

2.4.1.3   Identification by Chromatographic Fingerprinting

The chromatographic fingerprinting is a very versatile test, which supports the chemical identification of the plant raw material. The analysis must be performed not only for plant raw materials, but fundamentally also for the characterization of plant derivatives (extracts and essential oils) and finished products (phytomedicines). Because of the inherent characteristics of each type of matrix, the first step is to pay attention to the sample preparation, which should include an efficient extraction procedure, using the adequate solvents and, if necessary, followed by a pre-concentration step. Apart from helping in the authenticity and characterization of the plant material, the chromatographic fingerprinting allows the detection of eventually contamination or adulteration, which can occur by the presence of other plant species or even by the addition of isolated substances. Among the most used techniques, high performance liquid chromatography (mainly HPLC-UV, HPLC-UV-DAD or HPLC-MS), gas chromatography (GC-FID or GC-MS) and thin-layer chromatography (TLC or HPTLC) are important due to the methods low complexity, easy accessibility, dissemination in several pharmacopoeias and low-cost. Besides, in the case of TLC the use of adequate staining solutions (which can be general or specific) is important as well as making sure the bands have good separation and resolution for the calculation of the Rf values. Regardless of the chosen chromatographic technique, the comparison of the results with reference standards or materials is important, which will allow the correct authentication of the plant material concerned.

2.4.2 DETERMINATION OF WATER AND VOLATILE MATERIAL

Usually medicinal plants are dehydrated, a practice that prevents enzymatic reactions responsible for the modification of the original chemical constitution, deterioration, and microbial and mycotoxin contamination. In quality control, the determination of the water content allows to verify if the batch was dehydrated properly and to calculate the content of active principles on a dry weight basis.

Methods of analysis such as the azeotropic (distillation with toluene) and the volumetric methods (also known as Karl Fischer method, which is based on the reaction between water and an anhydrous solution of iodine and sulfur dioxide in methanol) can be used for the determination of water content. However, the most used method is the determination of water by loss on drying. For that, 2.0–5.0 g of the herbal material accurately weighed is placed in a previously dried and tared weighing flask, and then dried in an oven at 100–105°C. Alternatively, a protocol using balances attached to an infrared irradiation drying system is feasible. However, aside the water percentage, the gravimetric method also considers the loss of volatile material, as essential oils. For plants that contain resins or volatile substances that are altered in high temperatures, the measurementof loss by desiccation in a desiccator containing phosphorus pentoxide, under atmospheric pressure, or reduced pressure at room temperature is recommended.

2.4.3 DETERMINATION OF TOTAL ASH CONTENT

The determination of total ash content allows the measurement of inorganic compounds content, which can be originated from the plant tissue itself, such as carbonates, phosphates, chlorides, or originated by exogenous adulterants such as sand, rocks and soil attached to the surface of the plant. Furthermore, this test provides information about the hygiene conditions during the harvesting and drying process. In general, the method of analysis consists of transferring 2.0–4.0 g of the pulverized plant material accurately weighed to a previously ignited and tared crucible. Then the material is ignited in an oven at 500–600°C, until obtaining a whitish residue, which indicates the absence of organic matter. The percentage of remaining ashes in the sample is calculated by gravimetry.

2.4.4 DETERMINATION OF ACID-INSOLUBLE ASH CONTENT

The determination of acid-insoluble ash provides the silica content (especially sand), present in the sample. The procedure consists in boiling the residue obtained in the total ash determination with diluted hydrochloric acid, filtration and ignition in an oven at 500–600°C. In the same way, the determination of the acid-insoluble ash percentage is calculated by gravimetry.

2.4.5 DETERMINATION OF FOREIGN MATTER

The determination of foreign matter may be divided in three main levels: (i) if there are parts of the herbal material other than those described in the monograph of a given plant or if they are present above the limits specified by the monograph; (ii) if there is the presence of any organisms, parts or sub-products other than those specified in the plant monograph; (iii) if there are impurities of mineral or inorganic nature, such as rocks, sand, soil, etc. The determination can be made by weighing the sample before and after the visual inspection and manual separation (with the help of tweezers and magnifying lenses, if necessary) of any foreign material present. The results must be registered in terms of percentage. For example, the Brazilian native species Espinheira Santa (*Maytenus ilicifolia* Mart. ex Reissek – Celastraceae, Brazilian Pharmacopoeia, 5th edition) must be constituted by dry leaves and might contain the maximum of 2.0% of foreign material, whether they are branches or pieces of stems, insects, parts of other plants, rocks, sand, etc.

2.4.6   DETERMINATION OF ESSENTIAL OILS CONTENT

For those plants rich in essential oils the determination of the essential oils content is recommended, since frequently they are responsible for the pharmacological effect, for example *Cordia verbenacea* DC (Boraginaceae), whose essential oil is the active principle of a phytomedicine in Brazil recommended as a topical anti-inflammatory agent. If the amount of essential oil is lower than specified in the monographthat can be an indication that the plant material was dehydrated at high temperatures, impairing the quality of the raw material and the corresponding derivatives. Besides the negative impact in the essential oil yield, the qualitative and quantitative chemical profile are considerably modified, since lower boiling point mono- and sesquiterpenes volatilize more quickly than the higher boiling point ones. Also, if a plant rich in essential oils is dehydrated under high temperatures, this can lead to the formation of undesirable products through oxidation or degradation. The most used method for the determination of the essential oil's content in plants is steam distillation, which allows the separation and determination of the percentage of oil in relation to the dehydrated crude drug.

2.4.7   EXTRACTABLE MATERIAL

This method determines the quantity of active compounds that can be extracted with a certain solvent from a given amount of the plant raw material on a laboratory scale. Although considered a characterization test, the results provide information regarding the quality from the performance point of view and, consequently, about the expected yield on the industrial scale.

The extraction can be made using portions of 2.0–4.0 g of the plant material and 100 mL of solvent applying at least three methods: (i) hot extraction, using water; (ii) by cold maceration using ethanol or other solvent specified in the drug monograph; (iii) Sohxlet extraction using ethanol. Also, for research and industrial scaling up purposes, the processes of maceration and percolation using all the different proportions of hydroalcoholic mixtures (or other solvents) can be used. The result is obtained through the determination of the dry residue of the obtained extract by gravimetry, using an oven at 100–105°C. The final result is calculated in mg/g or in % w/w.

2.4.8   QUANTITATIVE ANALYSIS OF THE ACTIVE PRINCIPLES AND/OR CHEMICAL MARKERS

The quantification of the active principles or chemical markers (which may or may not be responsible for the pharmacologic effect) is an obligatory quality requirement, being mandatory to be performed for plant raw materials, plant derivatives (extracts, essential oils or others) and for the evaluation of the final phytomedicine.

Such analysis is considered very complex due to the chemical variability of these phytochemicals, the inherent biological variability of most of the species, and the dynamic limitations of the different methodologies that can be applied. The great diversity of existing metabolites requires, most times, the use of separation techniques coupled with detectors that provide some type of spectral information. Also, the sample preparation must allow the extraction of the substances of interest with good recovery. The quantification itself is achieved using analytical curves prepared with reference standards in known concentrations.

For that reason, the best analytical technique should be considered for the development of an analytical methodology for the quantification of phytochemicals, which must be sensitive and selective for the compound or compounds of interest. Several chromatographic arrays, colorimetric, spectrophotometric, titrimetric and densitometric techniques are often used. Techniques such as GC-FID or GC-MS and liquid chromatography (HPLC-UV, HPLC-UV-DAD, HPLC-MS) are widespread and allow a very advantageous detection, in terms of both specificity and selectivity, lowering the interference of other compounds. However, the use of other techniques like infrared spectroscopy (IR), nuclear magnetic resonance (NMR), sequential mass spectrometry (MS/MS), high performance liquid chromatography hyphenated to NMR or fluorescence detectors, capillary electrophoresis (CE) and others may be very useful, especially if very little is known about the chemical profile of the plant species concerned.

Total quantification is another approach for the quantitative analysis, which does not require a chromatographic separation step. However, the sum of all substances belonging to a given chemical class, such as flavonoids, alkaloids, tannins, anthraquinones, etc. is considered. Total flavonoids expressed in quercetin equivalents (or expressed in rutin) and total polyphenols expressed in gallic acid equivalents are good examples of widespread methodologies. A classic example of this type of quantification is described in the pharmacopeia monograph of Barbatimão (*Stryphnodendron adstringens* Mart. Coville – Fabaceae; Brazilian Pharmacopoeia, 5th edition). The plant raw material, composed by the dry stem barks, must contain, at least, 8.0% of total tannins expressed in pyrogallol equivalents, determined by spectrophotometry in the visible region of the spectrum (760 nm).

Another similar example is described in Espinheira Santa (*Maytenus ilicifolia* Mart. ex Reissek – Celastraceae, Brazilian Pharmacopoeia, 5th edition) monograph. The plant raw material, composed by the dry leaves, must contain, at least, 2.0% of total tannins expressed in pyrogallol equivalents, of which at least 2.8 mg/g equals epicatechin. Except for differences in the sample preparation, the methodology for the determination of total tannins is the same as that used in the determination of total tannins of Barbatimão. However, the determination of epicatechin is achieved by HPLC-UV, with the aid of an analytical curve of epicatechin.

2.4.9 Determination of Heavy Metals

Naturally, plants accumulate minerals as part of the nutritional or as a detoxification mechanism. Elements such as aluminum, arsenic, cadmium, cobalt, lead and mercury are very common. However, the over-accumulation is caused mainly due to the presence of metals in contaminated soils, sediments or air, by cross contamination and sequestering of the pesticides. In general, heavy metals are toxic not only for the plant organism, but also for human beings. Therefore, regulatory agencies worldwide, including in Brazil, recommend the verification and control of the presence of heavy metals in medicinal plants.

The determination of heavy metals can be made in terms of limit tests for specific metals or total metals, using techniques such as colorimetry, voltammetry or atomic absorption. However, the heavy metal determination using atomic absorption spectrometry provides more accurate and reliable results.

2.4.10 Determination of Pesticides Residues

Pesticide residues include insecticides, herbicides, fungicides, fumigants for pest control during storage and viral disease control agents. Because of the high toxicity for human beings and environment, the use of these agents has been restrained worldwide. In Brazil, the use of pesticides in medicinal plants plantations is prohibited. However, control is required due to the possibility of cross- or accidental-contamination originating from other close plantations or from contaminated water, air and soil. Therefore considering the history of pesticides' persistent use in the area of cultivation and to promote alternative actions for the pest control in agriculture is needed.

Often, the determination of pesticide residues is achieved by liquid or gas chromatography with mass spectrometry hyphenation techniques preferably. The sample preparation procedure requires a standard protocol in which impurities are removed by partition or adsorption (clean up) while residues of a wide range of pesticides are concentrated.

2.4.11 RADIOACTIVITY CONTAMINATION

The radioactivity determination in plant raw materials must be performed when the plant species is cultivated in a place of probable radioactive contamination or in the proximity. Also the radioisotope activity concentration and the types of radioactive contamination that might be present must be considered. For this purpose, the measurements must be carried out by official laboratories in accordance with the international organizations recommendations, such as the Codex Alimentarius, the International Atomic Energy Agency (IAEA), the Food and Agriculture Organization of the United Nations (FAO) and the World Health Organization (WHO). In Brazil, the Institute of Radiation Protection and Dosimetry (IRD) of the Brazilian Nuclear Energy Commission (under the authority of the Ministry of Science, Technology, Innovation and Communications) is the official government reference body and the center of expertise in radiation protection, dosimetry and metrology of ionizing radiation in the country.

2.4.12 DETERMINATION OF AFLATOXINS

Aflatoxins are secondary metabolites belonging to difurano-coumarins class, mainly produced by *Aspergillus flavus* and *A. parasiticus*, fungi that contaminate foods and herbal medicines under high temperature and humidity conditions. Since they are highly carcinogenic, their control is required by the regulatory agencies worldwide. Among the aflatoxins found in plant matrices, aflatoxins, B1, B2, G1 and G2 are the most important and must be controlled. Aflatoxins from group B (B1 and B2) contain a cyclopentane ring and show a blue fluorescence under ultraviolet light. Whereas aflatoxins from group G (G1 and G2) contain a lactone ring and exhibit a green and yellow green fluorescence, respectively. Such fluorescence capacity allows the identification and differentiation between aflatoxins from G and B groups.

According to the American Pharmacopoeia, herbal materials with history of contamination by mycotoxins must be tested regarding the presence of aflatoxins. The maximum limit for the presence of aflatoxins according to this pharmacopoeia is 5 µg of aflatoxin B1 per kg of plant material or 20 µg/kg for the sum of the aflatoxins B1, B2, G1 and G2. Also, the American Pharmacopoeia highlights three possible methods for the quantitative analysis of aflatoxins, which are described in Chapter 561 (Articles of Botanical Origin). The method of choice should be Method I, which applies thin layer chromatography and the use of standard solutions for the aflatoxins. Alternatively, Methods II (HPTLC) or III (HPLC) can also be used (The United States Pharmacopoeia, 2014). The European Pharmacopoeia recommends the use of high performance liquid chromatography and the maximum limit of aflatoxin B1 is 2 µg/kg, or 4 µg/kg for the sum of the aflatoxins B1, B2, G1 and G2, unless there is a specific value in the monograph (European Directorate for the Quality of Medicines and Health Care, 2007). The Brazilian legislation follows the same limits and procedures recommended by the European Pharmacopoeia.

2.5 PHYSICO-CHEMICAL TESTS FOR THE QUALITY CONTROL OF PLANT DERIVATIVES: LIQUID EXTRACTS, SOFT EXTRACTS AND DRY EXTRACTS

The physico-chemical tests for the quality control of liquid extracts (fluid extracts and tinctures), soft extracts and dry extracts can be divided in identification tests, purity and integrity tests, quantitative analysis and, additionally, characterization tests.

Fluid extracts and tinctures are preparations of liquid consistency that are usually obtained from dried plant parts by maceration or percolation. They concentrate the active ingredients extracted from the plant matrix and the main difference between them is the extraction ratio: fluid extracts contain a 1:1 ratio of dried plant raw material/extract and tinctures contain a 1:5 or 1:10 ratio dried plant raw material/extract. Because of the liquid nature, the organoleptic analysis must evaluate the following: aspect (presence of particles in the liquid, turbidity level, transparency), color (usually amber, yellowish or greenish, light or dark), smell and flavor (characteristic of the plant, being bitter or sweet, spicy or astringent, etc.).

Soft extracts are semi-solid preparations obtained by partial solvent evaporation of liquid extracts. The color is usually dark amber, and the smell and flavor are characteristic of the plant species. On the other hand, dry extracts are solid preparations obtained by total evaporation of the solvent used for their production, usually with the aid of a drying agent or excipient. The analysis of the aspect refers to homogeneity of the sample, presence of particles or other foreign matters, the color, smell and flavor if applicable.

For extracts, the identification tests are the same as those used for raw plant materials, except for the macroscopic and microscopic botanical examination. One of the most used methods is chromatographic fingerprinting by TLC, HPLC or GC, as previously described in the identification tests of the raw plant material. However, in some cases, this test does not require a previous and exhaustive extraction step since the active principles or chemical markers are already extracted from the plant matrix, facilitating the procedure.

Likewise, the quantitative analysis of the active principles/chemical markers is considered one of the main tests for assessing the quality of such plant derivatives and should be performed using the same methodologies developed for the plant raw material, except for any inadequacies regarding sample preparation. Thereby, the spectrophotometric and titrimetric methods are often used for total determinations (as for example, determination of total flavonoid or total phenolic content), as well as chromatographic methods, such as GC-FID, GC-MS, HPLC-UV-DAD or HPLC-MS, mainly.

The purity and integrity tests determine the dry residue (also denominated soluble solids) for liquid and soft extracts. For dry extracts, the water and volatile material can be efficiently determined by loss on drying. The determination of solvents residues other than hydroalcoholic solutions must be performed if hazardous extraction solvents are used in the extraction process. Also, depending on the need, the determination of the presence of pesticide residues, heavy metals and aflatoxins, as described for raw plant materials is required.

Finally, the characterization tests of liquid extracts contemplate the determination of the relative density, pH, viscosity and alcohol content. For dry extracts the characterization tests comprise the granulometry, apparent density and solubility.

2.5.1 Determination of Dry Residue (or Soluble Solids)

The classic extraction methods for the production of fluid extracts and tinctures apply a given amount of the plant raw material, which can be fresh or dried, for a certain final product volume, usually using hydroalcoholic solutions. Aiming to guarantee a minimum content of active principles per milliliter of product, the determination of soluble solids content is an important tool for the extraction process monitoring and to assess the quality of the final product. Moreover, the determination of the dry residue is very useful when the content of active principles or chemical markers present in the final extract is not possible to quantify.

The simplest method of analysis is the evaporation of the extract solvent in an oven at 100–105°C and calculation of the percentage of dry residue by gravimetry. Results are usually expressed in mg/mL or in % w/v.

2.5.2 Determination of Solvent Residues

Liquid extracts are usually produced with the use of hydroalcoholic solutions. However, obtaining some standardized extracts requires the use of less friendly solvents, such as methanol, ethyl acetate, *n*-hexanes or acetone, among others. The use of such solvents can be necessary, depending on the characteristics of polarity and solubility of certain classes of plant metabolites, or to increase the efficacy of the manufacturing process.

However, even if the final product has the solvent removed, as in the case of dry extracts, there is the difficulty of trace residues. Therefore, regulatory agencies worldwide have established limits of tolerance of solvent residues in liquid or dry plant extracts. For the qualitative and quantitative determination of these residues, the most applied method is gas chromatography attached to flame ionization detector (GC-FID) or mass detector (GC-MS): the latter being the more advantageous regarding specificity and selectivity.

2.5.3 RELATIVE DENSITY

The determination of the relative density of an extract can be an indicator for the type of extraction solvent used in the process of maceration and percolation. When hydroalcoholic combinations are used, the relative density values are lower than 1.0 (which is the relative density value of water at 20°C) and tend to be lower, as the ethanol percentage increases. Whereas, when propylene glycol or glycerin is used as extraction solvents, the values of relative density are higher than 1.0 and decrease if water or ethanol is added.

The most common methods used to determine the relative density of an extract include the use of hydrostatic balance, densimeter or pycnometer in which the latter is simpler and less expensive.

2.5.4 APPARENT DENSITY (OR TAPPED DENSITY)

The apparent density is an important parameter for the characterization of dry extracts. By definition, this corresponds to the volume occupied by a determined mass of solid, powder or grainy material, considering the total volume of the sample, including any empty space between the particles. This differs from the real density, which is the real volume occupied by a determined solid, not taking into consideration the porosity. The apparent density is influenced by the powder's particle shape, size, and size distribution, among others. The analytical procedure for the determination is a quick, cost-effective and standardized method. The test consists in measuring the volume occupied by a given amount of the sample introduced in a test tube after tapping the tube against a rigid bulkhead until little further volume change is observed. The measured value is generally expressed in g/mL.

2.5.5 GRANULOMETRY

The granulometry is also a characterization test of samples in solid state (powder or grains) that is determined with the help of sieves operated by mechanical devices. Depending on the characteristics of the sample, a set of sieves is assembled so that the ones with the greatest opening overlap the sieve with the smaller opening. Approximately 25.0 g of the sample (depending on the nature of the material, density and diameter of the sieves to be used) is transferred to the superior sieve and the set is subjected to mechanical vibration. After this process ends, the sample resting in each sieve is removed, weighed and the retaining percentage is calculated.

2.5.6 SOLUBILITY

The determination of the solubility of a dry extract in any given solvent and temperature constitutes another characterization test. This information is important both to assess the quality of the received batch and as a parameter in formulation studies. The analytical procedure is simple and consists adding increasing portions of the sample to constant volumes of solvent in which the extract must be analyzed. The total amount of solute is determined in the supernatant liquid. Results are calculated by gravimetry and usually expressed in mg/g.

2.5.7 pH

The determination of the pH is important, not only as a characterization parameter of liquid or solid extracts, but also as a crucial factor for their use in formulations, as this affects both solubility and stability. The pH measures the acidity of a substance, which depends on the concentration [H$^+$] or [H$_3$O$^+$] ions dissociated in the solution. Depending on the extract, the pH can be measured directly with the insertion of the electrode of the pH meter in the sample. Alternatively, a previous water dilution step may be required, especially for highly concentrated extracts and dry extracts.

2.5.8 Viscosity

The viscosity test expresses the resistance of a liquid to the flow: in a given temperature, the less viscous the liquid, more fluidly . If the temperature increases, the kinetic energy of the molecules also increases, decreasing the time in which the molecules remain cohesive. Consequently, viscosity is reduced. Therefore, since this is a property related to all liquids, this determination is a very useful parameter for the quality control of liquid extracts and tinctures. The results provide indications regarding the type of extraction solvent used in the percolation (for example, extractions performed with 70% ethanol provide extracts less viscous than extractions made with propylene glycol). Also, depending on the plant species used, the final extract may be more or less viscous. For example, plants rich in mucilage provide more viscous extracts than those deficient in this chemical group. Finally, rapid information concerning the concentration of soluble solids in the extract can also be obtained: an extract at 2.5% w/v is less viscous than an extract at 20.0% w/v.

The simplest and most cost-effective method measures the time required for the test liquid to flow through capillary tubes (using for example the viscometers of Ostwald, Ubbelohde, Baumé and Engler) or Ford cup orifice viscometer. The result is expressed in seconds (s). Similarly, the sphere viscometer measures the velocity of falling of a sphere inside a tube containing the sample. The greater the viscosity of the sample, the lower the velocity of falling. Finally, other more sophisticated methods determine the viscosity measuring the resistance of the rotation movement in metallic axes when immerse in the liquid (for example using the Brookfield viscometer), with results expressed in centipoise (cP) or centistokes (cSt).

2.5.9 Alcohol Content

Most of the liquid extracts produced for medicinal purposes are obtained using hydroalcoholic combinations. The choice of the alcohol concentration of the extraction solvent is made according to the class of the compounds that is going to be extracted, and the intended use. On that account, concentration standardization is important in order to guarantee the quality and efficacy of the final product.

The determination of the alcohol content, along with other tests, provide important information regarding the physico-chemical profile of the extract. The greater the ethanol proportion in the extraction solvent combination, the greater the alcohol content, the lower the relative density and consequently the lower viscosity. Furthermore, deviations in the alcohol content in the final product may indicate problems in the extraction process, such as temperature out of control, moisture content above the specified limits for the plant raw material, errors during the preparation of the extraction solvent, among others.

The simplest and most inexpensive method consists in determining the alcohol content by direct measurement using an alcoholmeter. This is a simple and direct method, very useful for process control. However, depending on the alcohol content of the extract, at levels lower than 30%, the distillation method is more precise. This method consists in distilling a portion of the sample along with an equal quantity of water and determining the density of the distilled liquid at 20°C. The results are calculated with the aid of an alcoholometric table.

2.6   PHYSICO-CHEMICAL TESTS FOR THE QUALITY CONTROL OF ESSENTIAL OILS

Essential or volatile oils are products obtained from parts of the plant raw material (leaves, fruits, flowers, inflorescences or stem barks) and constitute complex combinations of volatile, lipophilic, odoriferous and liquid substances. They are obtained especially through steam distillation, by pressing citric fruit pericarps, or even by lipophilic extraction using organic solvents (with the disadvantage of also extracting fixed oils and other lipophilic substances). They can also be obtained through procedures of extraction such as *Enfleurage* (very common for the extraction of essential oils of flowers) or still by supercritical CO_2 extraction. Chemically, most of the essential oils are composed by derivatives of phenylpropanoids (especially phenols and its ethers), constituting a group of metabolites which have very peculiar physical and physico-chemical characteristics.

The common tests for the quality control of essential oils include the identification tests (organoleptic analysis and chromatographic fingerprinting), purity and integrity tests (solvents residues, heavy metals, radioactivity, pesticides residues) and quantitative analysis. The main characterization tests include the determination of the refraction index, rotatory power and relative density, which also contribute to the identification and determination of purity and integrity of essential oils.

The identification of the essential oils is enhanced by the organoleptic characteristics. They have an acrid and spicy flavor and a smell that is very characteristic of each plant species from which the essential oil was extracted. Frequently, an expert analyst is able to differentiate essential oils obtained from different plant species from the same genera. For example: the essential oil of *Eucalyptus globulus* Labill (*Myrtaceae*) has a strong smell of camphor, totally different from the essential oil of another species from the *Eucalyptus* genera, *E. citriodora* Hook (*Myrtaceae*), which is characteristic of citronellal. Regarding coloration, normally they are colorless or slightly yellowish, apart from some exceptions, as is the case of the essential oil of *Matricaria recutita* L. Rauschert (Asteraceae), which is blue due to its azulene content.

A noteworthy aspect of essential oils is the volatility. Further feature, when evaporated, they should not leave residues. This factor is important when one suspects that the sample was adulterated with fixed oils, which leaves oil residues when subjected to evaporation. In practical terms, a drop of the sample is applied on a dry paper filter and left to stand at room temperature or in an oven at controlled temperature. If a translucent stain remains, this is an indication of the adulteration by fixed oils.

The identification using chromatographic techniques as TLC allows a quick, simple and affordable analysis, also allowing the detection of contaminants and adulterants. For this purpose, the identification of some compounds can be made through staining and Rf calculation of the bands. The results are compared with reference standards, photographs. Semi-quantitative analysis is also possible to be performed by TLC, using standard solutions prepared in known concentrations, followed by densitometry.

The use of gas chromatography, although of higher cost, is extremely useful and advantageous for the analysis of essential oils. The method is performed using capillary columns, which allows great separation of compounds and consequently excellent efficiency. The most used detection systems are the flame ionization detection (GC-FID) and the mass spectrometry detection (GC-MS), the last providing precise information regarding the identity and purity of the peaks. Approaches like the comparison with the retention times of reference standards, the calculation of retention index, or the comparison with databases and libraries containing information regarding the molecular mass and fragmentation profile, allow a very fast and reliable qualitative analysis of essential oils.

For the quantitative analysis, as well as for other quantitative methods, the need to establish analytical curves using reference standards prepared in known concentrations is necessary. The calculations are made as a function of the areas, which are directly proportional to the quantity of the corresponding compounds in the sample.

2.6.1 DETERMINATION OF THE REFRACTION INDEX

The refraction index of a pure substance under given temperature and pressure conditions is an indicative parameter for identification. For the essential oils, the refraction index varies from 1.450 to 1.590 when compared to water at 20°C, which is 1.333. If the sample is adulterated with fixed oils, ethanol, water or other solvent, the refraction value index is modified, indicating the presence of adulterants or contaminants.

2.6.2 DETERMINATION OF THE OPTICAL ROTATORY POWER

Essential oils show optical activity due to their chemical constitution: most of the compounds have stereogenic centers. They have the capability to rotate the plane of polarized light in a way that the transmitted light is deviated in a certain angle in relation to the incident one. For a certain substance, if one of the enantiomers deflects the plane of polarized light to the right (+), this is called dextrorotatory, while the other, deflecting the plane of polarized light to the left (−) is called levorotatory. The angle of this reflection is the same in magnitude for the enantiomers, however with opposite signs.

The determination of the rotational power in essential oils is performed by polarimetry, and the resulting value is equal to the sum of all substances present in the oil. The analysis of the experimental values when compared with the reference values allows to obtain information about the authenticity of the essential oil concerned, possible adulterations, or even if the sample is originated from a natural or synthetic source.

2.6.3 DETERMINATION OF THE RELATIVE DENSITY

The determination of the relative density is a simple procedure, as described for the evaluation of liquid extracts, and can be an indicative of the quality of the sample. For essential oils, the relative density varies from 0.690 to 1.118, so that this is characteristic to each type of oil. The most used method is the determination of the relative density by pycnometry.

2.7 PHYSICO-CHEMICAL TESTS FOR THE QUALITY CONTROL OF THE FIXED OILS

Fixed oils consist mainly of triacylglycerols, which are esters of long-chain fatty acids (identical or different) and glycerol, and their corresponding derivatives. Such compounds can be biosynthesized by both animals and plants, and, along with fats and waxes, they compose the group of lipids, whose main biological function is the storage of energy. In general, oils can be extracted by cold pressing or extraction with organic solvents from many parts of the plant; however, the seeds contain greater quantities.

The importance of studying and controlling the quality of such compounds is explained by the great economic value for the food, energy, pharmaceutical and cosmetic industry. Classic examples of the high-value commercial products can be highlighted: olive oil (*Olea europaea L.;* Oleaceae), soy oil (*Glycine max L. Merril,* Fabaceae), castor oil (*Ricinus communis,* Euphorbiaceae), argan oil (*Argania spinosa L.;* Sapotaceae), almond oil (*Prunus amygdalus Batsch,* Rosaceae), among others.

Physico-chemically these oils are liquids at room temperature, non-volatiles (fixed), insoluble in water and soluble in organic solvents. Exceptions may occur, such as the coconut oil (*Cocos nucifera L,* Arecaceae), which tends to be solid at room temperature. According to the chemical composition of the fatty acids present in each oil, they may be classified as saturated fixed oils, which are mainly composed by glyceryl esters of saturated fatty acids, such as lauric, myristic, palmitic and stearic acids; monounsaturated fatty acids, mainly composed by glyceryl esters of monounsaturated fatty acids, such as oleic and euric acids; and polyunsaturated fatty acids, mainly consisting of glyceryl

TABLE 2.4

Main Physico-Chemical Tests for the Characterization of Fixed Oils

Test	Definition	Finality
Saponification value	Amount of potassium hydroxide (mg) required to neutralize the free acids and saponify the esters present in 1.0 g of sample	The saponification value provides indication about adulterations of the fats with unsaponifiable substances (mineral oil, for example)
Acid value	Amount of potassium hydroxide (mg) required to neutralize the free fatty acids in 1.0 g of sample	High acid value suggests hydrolysis of the esters which constitute the fatty matter. It may be caused by chemical industrial treatments during extraction and purification, bacterial activity, catalytic action (heat, light), inappropriate storage and presence of impurities such as humidity, among others
Ester value	Amount of potassium hydroxide (mg) required to saponify the esters present in 1.0 g of sample. It can be calculated by the difference between the saponification value and the acid value	Ester value equal or very close to the saponification value indicates a good quality oil
Iodine value	Amount of iodine (g) absorbed by 100 g of substance	Provides the level of unsaturation of the esterified and free fatty acids. Suggests the level of purity and highlights the presence of adulterants

esters of unsaturated fatty acids, such as linoleic and linolenic acids. Therefore, the degree of unsaturation will directly reflect the distinct physico-chemical characteristics of each oil.

The physico-chemical tests required for the quality control of such products are the same as recommended for the analysis of essential oils and include the identification tests such as TLC and gas chromatography (the last requiring a previously sample derivatization step), quantitative analysis, and purity and integrity tests. However, the characterization of fixed oils includes the determination of the saponification value, acid value, ester value, and iodine value, as summarized in Table 2.4. Furthermore, the pharmacopoeia characterization tests include the determination of the melting temperature, solidification temperature, refraction index, unsaponifiable matter, fatty acid composition, peroxide value and hydroxyl value, which should also be performed according to the monograph of each oil and destination of the final product.

2.8 PHYSICO-CHEMICAL TESTS FOR THE QUALITY CONTROL OF PHYTOMEDICINES

Pharmaceutical formulations can be classified in solid formulations (powders, granules, tablets, capsules, suppositories), semi-solids (ointments, gels, creams, pastes) and liquids (syrup, suspension, solutions). The physical and physico-chemical tests applicable are defined according to the pharmaceutical formulation in which the extract of natural origin was incorporated. Therefore, the quality control of phytomedicines is based on the same general methods for assessing the conformity of formulations of any other medicines containing synthetic drugs. Such tests are required not only for the routine quality control, but also for the registration and stability studies. Table 2.5 summarizes the main physical and physico-chemical tests for the quality control of very common phytomedicines formulations while Table 2.6 lists all plant species exemplified in this chapter.

Although such tests are exhaustively described in the pharmacopoeias, special attention must be given to the qualitative and quantitative analysis of the active principles or chemical markers, since they are specific to the plant material responsible for the pharmacological action. Very often, for pharmacopoeial formulations, these methods are described in the respective monographs. However,

TABLE 2.5

List of the Main Physico-Chemical Tests for the Quality Control of Phytomedicines

Tests	Tablets	Capsules	Suppositories	Ointments, Creams, Liniments and Gels	Syrups, Solutions and Suspensions
Aspect, colour, smell, dimensions, texture	X	X	X	X	X
Qualitative analysis (chromatographic profile)	X	X	X	X	X
Content of active ingredients or chemical markers	X	X	X	X	X
Average mass	X	X	X	X	
Determination of volume					X
Uniformity of the unitary doses	X	X	X	X	X
Desintegration	X	X	X		
Dissolution	X	X	X		
Hardness	X				
Friability	X				
Content of water	X	X			
Softening temperature			X		
pH			X	X	X
Relative density					X
Viscosity				X	X

in the absence of official monographs, the method for the qualitative and quantitative analysis must be developed and validated according to the same validation standards suggested for medicines containing synthetic drugs.

Finally, the importance of regulations procedures and monographs, both in Brazil and worldwide are highlighted, which are available and must be practiced extensively to ensure quality of phytomedicines. For a country such as Brazil which has such a rich biodiversity and depends on the commerce of phyto-medicinal products, the quality control of medicinal plants, plant derivatives and phytomedicines is very important.

TABLE 2.6

Summary of the Names of All Plant Species, Their Respective Family and Common Names Discussed in This Chapter

Scientific Name	Family	Common Names
Argania spinosa (sin. *Sideroxylon argan*)	Sapotaceae	Argan tree, tree of iron
Cocos nucifera	Arecaceae	Coconut palm
Cordia verbenaceae (sin. *Cordia curassavica*)	Boraginaceae	Erva baleeira and maria-milagrosa
Eucalyptus citriodora (sin. *Corymbia citriodora*)	Myrtaceae	Lemon-scented gum
Eucalyptus globulus	Myrtaceae	Southern blue gum
Glycine max	Fabaceae	Soybean
Matricaria recutita (sin. *Matricaria chamomilla*)	Asteraceae	German chamomile
Maytenus ilicifolia	Celastraceae	Espinheira santa
Olea europaea (sin. *Phillyrea lorentii*)	Oleaceae	Wild olive, brown olive, Indian olive, olienhout
Prunus amygdalus	Rosaceae	Almond
Ricinus communis	Euphorbiaceae	Castor bean
Stryphnodendron adstringens (sin. *Stryphnodendron barbatimam*)	Fabaceae	Barbatimão

REFERENCES

Allen, L. V.; Ansel, H. C. 2014. Ansel's Pharmaceutical Dosage Forms and Drug Delivery Systems, 10th ed. Philadelphia: Wolters Kluwer.

Associação Brasileira de Normas Técnicas – ABNT. 1985. Planos de amostragem e procedimentos na inspeção por atributos. NBR 5426. http://www.saude.rj.gov.br/comum/code/MostrarArquivo.php?C= Njg1Nw%2C%2C.

Batalha, M. O.; Ming, L. C. 2003. Plantas medicinais e aromáticas: um estudo de competitividade no Estado de São Paulo. São Paulo: Sebrae.

Calixto, J. B. 2000. Efficacy, safety, quality control, marketing and regulatory guidelines for herbal medicines (phytotherapeutic agents). Brazilian Journal of Medical and Biological Research, 33, 179–189.

Di Stasi, L. C. 1996. Plantas medicinais: arte e ciência. Um guia de estudo interdisciplinar. São Paulo: Editora da Universidade Estadual Paulista.

European Directorate for the Quality of Medicines and Health Care. 2007. European Pharmacopeia, 6th ed. Strasbourg: European Pharmacopoeia Commission.

Evans, W. C. 1996. Trease and Evans' Pharmacognosy, 14th ed. London: WB Saunders Company.

Instituto Nacional de Metrologia – INMETRO. 2016. Orientação sobre validação de métodos de ensaios químicos, DOQ-CGCRE-008. http://www.inmetro.gov.br/Sidoq/Arquivos/CGCRE/DOQ/DOQ-CGCRE-8_05.pdf.

International Organization for Standardization. 1999. ISO 2859-1 – Sampling procedures for inspection by attributes. Sampling schemes indexed by acceptance quality limit (AQL) for lot-by-lot inspection. Geneva: British Standard BS 6001-1.

Ministério da Saúde. Agência Nacional de Vigilância Sanitária (ANVISA) – Brasil. 2010. Resolução da Diretoria Colegiada – RDC n° 49. http://portal.anvisa.gov.br/documents/33832/259143/RDC+n% C2%BA+49-2010.pdf/06bfd0e9-c2d2-4fa5-8c9e-bce42fe92592.

Ministério da Saúde. Agência Nacional de Vigilância Sanitária (ANVISA) – Brasil. 2013. Resolução da Diretoria Colegiada – RDC n° 13. http://bvsms.saude.gov.br/bvs/saudelegis/anvisa/2013/rdc0013_14_03_2013.html.

Ministério da Saúde. Agência Nacional de Vigilância Sanitária (ANVISA) – Brasil. 2014a. Resolução da Diretoria Colegiada – RDC N° 26. http://bvsms.saude.gov.br/bvs/saudelegis/anvisa/2014/rdc0026_13_05_2014.pdf.

Ministério da Saúde. Agência Nacional de Vigilância Sanitária (ANVISA) – Brasil. 2014b. Resolução da Diretoria Colegiada – RDC n° 69. https://www20.anvisa.gov.br/coifa/pdf/rdc69.pdf.

Ministério da Saúde. Agência Nacional de Vigilância Sanitária (ANVISA) – Brasil. 2017. Resolução da Diretoria Colegiada – RDC n° 166. https://www20.anvisa.gov.br/coifa/pdf/rdc166.pdf.

Ribani, M; Bottoli, C. B. G.; Collins, C. H.; Jardim, I. C. S. F.; Melo, L. F. C. 2004. Validação em métodos cromatográficos e eletroforéticos. Química Nova, 27, 771–780.

Robbers, J. E.; Speedie, M. K.; Tyler, V. E. 1996. Pharmacognosy and Pharmacobiotechnology. Baltimore: Williams & Wilkins.

Simões C. M. O.; Schenkel, E. P.; Gosmann, G.; Mello, J. C. P.; Mentz, L. A.; Petrovick, P. R. 2007. Farmacognosia: da planta ao medicamento, 6th ed. Porto Alegre: Editora da Universidade Federal do Rio Grande do Sul/Editora da Universidade Federal de Santa Catarina.

Thompson, M.; Ellison, S. L. R.; Wood, R. 2002. Harmonized guidelines for single-laboratory validation of methods of analysis (IUPAC technical report). Pure and Applied Chemistry, 74, 835–855.

United States Pharmacopeial Convention. 2013. The United States Pharmacopoeia – USP37, NF32. Rockville: United States Pharmacopeia.

Wagner, H.; Bladt, S. 1996. Plant Drug Analysis – a Thin Layer Chromatography Atlas, 2nd ed. Berlin: Springer.

World Health Organization – WHO. 2005. WHO Guidelines for sampling of pharmaceutical products and related materials. WHO Technical Report Series, No. 929, Annex 4. Spain: WHO Press. https://www.who.int/ medicines/areas/quality_safety/quality_assurance/GuidelinesSamplingPharmProductsTRS929Annex4. pdf?ua=1.

World Health Organization – WHO. 2007. WHO Guidelines for assessing quality of herbal medicines with reference to contaminants and residues. Spain: WHO Press. http://apps.who.int/medicinedocs/documents/ s14878e/s14878e.pdf.

World Health Organization – WHO. 2011. WHO Quality control methods for herbal materials. Spain: WHO Press. http://apps.who.int/medicinedocs/documents/h1791e/h1791e.pdf.

3 The Widening Panorama of Natural Products Chemistry in Brazil

Maria Fátima das Graças Fernandes da Silva,
João Batista Fernandes, Moacir Rossi Forim,
Michelli Massaroli da Silva, and Jéssica Cristina Amaral
Chemistry Department, São Carlos Federal
University, São Carlos, Brazil

CONTENTS

3.1 INTRODUCTION AND AN OVERVIEW OF NATURAL PRODUCTS CHEMISTRY AND CHEMOSYSTEMATICS IN BRAZIL

Secondary metabolites (natural products) from different sources such as plants and microorganisms have been and will continue to be one of the most important and promising sources in the discovery of raw materials for the development of pharmaceuticals, agrochemicals and other bioproducts. Over thousands of years, mankind has used natural products for the treatment and prevention of various diseases. Earliest reports of the use of natural products date back to 2600 BC, in Mesopotamia. Among the various plant-derived substances described *Cupressus sempervirens* (cypress), *Commiphora species* (myrrh) and *Papaver somniferum* (poppy juice) were mentioned, which are indeed still used nowadays in the treatment of parasitic infections, colds and inflammations (Cragg and Newman, 2005; Cragg and Newman, 2013; Dias et al., 2012). To date, nature in its vast biodiversity has proven to be a continuing and rich source of substances with great potential not only for the treatment of countless diseases but also for non-biological purposes, highly relevant to the world economy. However, it is estimated that only 15% of the existing higher plants (about 300,000 to 500,000) have already been studied phytochemically and even more dramatically, only 6% of them have undergone pharmacological investigation (Cragg and Newman, 2013). The value of natural products to humanity is invaluable. From the knowledge acquired from generation to generation and the exchange of experiences between different peoples, mankind uses bio-derived and bio-inspired products in the most diverse areas ranging from food conservation and tinctures

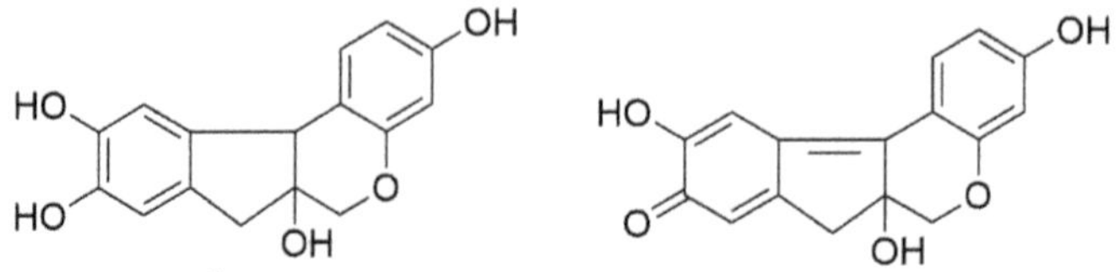

FIGURE 3.1   Structures of dyes isolated from *C. echinata* (Pau-Brasil).

to prevention and treatments of diseases since the sixteenth century. This fact, together with other existing organisms on the planet, clearly demonstrates the potential that nature has in providing new substances with high added value.

Brazil stands out especially for the vast biodiversity. Different biomes, such as Caatinga, Pantanal, Atlantic Forest, Cerrado, Pampa and Amazon, reveal an impressive variety of plants, microorganisms and animals (Silva and Rodrigues, 2014). An estimation of 20% of the world-known species are in Brazil, and due to the immense territory and complex ecosystems, many other species remain yet to be discovered (Silva et al., 2010; Silva and Rodrigues, 2014). Overall, Brazil is in a privileged situation in terms of natural products, with a biodiversity that is one of the major basis for the progressive discovery of new compounds with great potential for both biological and non-biological applications, such as biofuels, textiles, and others.

Since the sixteenth century, Brazil has drawn attention to the immense and extraordinary biodiversity. Among examples is the red dye Braziline, **1** and the oxidation product, Brazileine **2**, (Figure 3.1) extracted from *Caesalpinia echinata* (also known as Fabaceae; Pau-Brasil) that was the main export product of colonial Brazilian time (Pinto, 1995; Pinto et al., 2002; Viegas Jr and Bolzani, 2006).

During the scientific expeditions, the first Portuguese physicians soon recognized the importance of local plants used by indigenous tribes in healing numerous diseases (Pinto et al., 2002). The zoologist Johann Baptist Spix and botanist and physician Carl Friederich von Martius carried out systematic studies of the Brazilian fauna and flora (Pinto et al., 2002). Von Martius was crucial in the first steps of Brazilian phytochemistry, and through him that the pharmacist Theodoro Peckolt came to Brazil in 1847 to study the Brazilian flora. Due to his exceptional contributions, many consider Peckolt the father of Brazilian phytochemistry (Pinto et al., 2002).

Among several important investigations, the isolation of the first iridoid from *Plumeria lancifolia* Mart, agoniadin, stands out today known as plumierid (**3**, Figure 3.2). The chemical structure

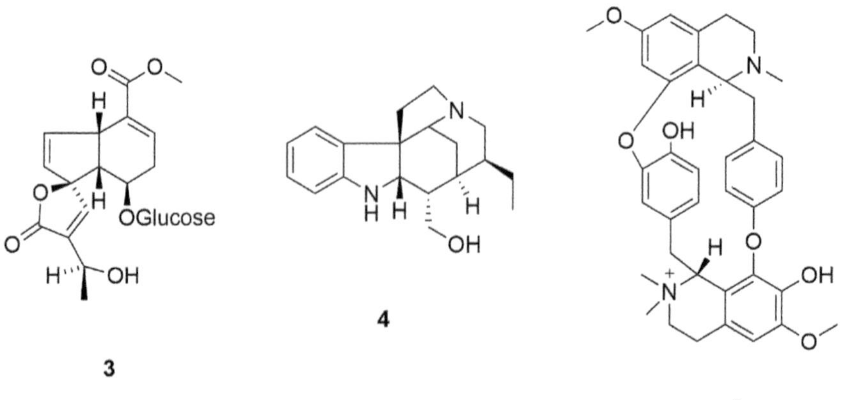

FIGURE 3.2   Structures of some natural products from Brazilian plants.

was elucidated 88 years after the compound was isolated, by Halpern and Schmid (Halpern and Schmid, 1958; Pinto et al., 2002; Silva et al., 2010; Santos et al., 1998). In 1838, Ezequiel Correia dos Santos isolated the first alkaloid from Brazilian plants, namely pereirine from *Geissospermum laeve* (Vell.) Baill ("Pau Pereira"). The structure was elucidated many years later and was called geissoschizoline (**4**, Figure 3.2) (Santos, 2007). When discovered, this plant had already long been used by indigenous people for the treatment of various diseases such as inflammation, malaria and others (Almeida, 2017; Bolzani et al., 2012; Silva and Rodrigues, 2014). Another example of a substance, isolated from Brazilian plants, is tubocurarine (**5**, Figure 3.2), found in *Chondodendron tomentosum* (Menispermaceae), widely used as a muscle relaxant during the preoperative period. Furthermore, South American Indians have used it in the past for subsistence purposes such as hunting and fishing. Over the years, several side effects were observed with the use of tubocurarine, leading to the replacement with new synthesized anesthetics inspired in that compound (Bolzani et al., 2012; Silva et al., 2010).

In Brazil, modern phytochemistry was introduced in the 1950s mainly by Walter Baptist Mors and Otto Richard Gottlieb (Pinto et al., 2002; Valli et al., 2018). The former was responsible for the creation of the current Research Center for Natural Products in Rio de Janeiro, one of the most prestigious in Brazil. From the extensive work list done by Mors, it is worth highlighting the isolation and identification of the active substance, 14,15-epoxygeranylgeraniol, from the oil obtained from the fruits of *Ptedoron pubescens Benth*, which protects the mammalian skin from the penetration of the cercariae of *Schistosoma mansoni*. In addition, he contributed intensively to the training of numerous researchers in the field of natural products. Otto R. Gottlieb was considered the greatest twentieth century phytochemist in Latin America due to his numerous contributions in terms of articles, books and patents published, as well as in the training of numerous researchers. His trajectory can be translated in more than 700 published articles, some patents on neolignans and lignans of Lauraceae and Myristicaceae and several books (Caparica, 2012; Gottlieb, 1996; Silva and Bolzani, 2012; Valli et al., 2018). He created a chemosystematics system that classifies plants by their chemical characteristics, bringing rational methods for the search of bioactive compounds in plants (Gottlieb, 1992; Silva and Bolzani, 2012). His group studied lignans and neolignans under the chemosystematics, phytochemical and biological aspects (Gottlieb and Mors, 1980; Gottlieb and Yoshida, 1989), leading to numerous publications, which correspond to approximately 75% of all literature on these substances of the Brazilian flora. Lignan and neolignan derivatives have a broad spectrum of biological activities such as antileishmanial, antitumor, antichagasic, and antimalarial (Gottlieb and Mors 1980; Gottlieb, 1988; Silva Filho et al., 2008; Souza, 2012).

Gottlieb's chemosystematic studies also show that there is a tendency for morphological diversification to occur in parallel with chemical diversification. Thus, the morphological classification of a genus is questioned, and characterizes it as a strong candidate for the search of substances different from those of the family chemical profile and with greater probability of being unpublished. An example is the Brazilian species of *Hortia*, which is a neotropical genus of the Rutaceae, traditionally included in the subfamily Toddalioideae, subtribe Toddaliinae (Engler, 1931; ibid. Severino et al., 2012). Historically, De Candolle (1824, ibid. Severino et al., 2012) described *H. brasiliana* Vand., which was collected in the southeastern part of the state of Minas Gerais and in the state of Rio de Janeiro, and Saint-Hilaire (1824, ibid. Severino et al., 2012) described a shrub collected in Goiás and in western Minas Gerais, and also named it *H. brasiliana*. Engler (1874, ibid. Severino et al., 2012) also attributed the name *H. brasiliana* to the shrubby species of central Brazil. In the same study, he described a new arborescent species, *H. arborea* Engl. Recently, Groppo et al. (2005, ibid. Severino et al., 2012) showed that *H. brasiliana* and *H. arborea* represent the same arborescent species, and that *H. brasiliana* Vand. ex DC. is the correct name. Albeit well known, the shrubby species found in central Brazil remained unnamed, so Groppo et al. (2005, ibid. Severino et al., 2012) called it *H. oreadica*. The taxonomic interest in the Rutaceae motivated an investigation of the taproots and stems of *H. oreadica*, and the result of this study and literature data showed that *Hortia* produce highly specialized limonoids (**6–16**) that are similar to those from the *Flindersia*

(Flindersioideae). The taxonomy of *Hortia* has been debatable, with most authors placing this sub-family in the Toddalioideae. Considering the complexity of the isolated limonoids, *Hortia* does not show any close affinity to the genera of Toddalioideae (Severino et al., 2012; Severino et al., 2014). Indeed, many chemists were perhaps interested in alkaloids from the Rutaceae and rarely identify all potentially systematic important classes of compounds present in the plant. The low concentration of limonoids is also likely to be responsible for the complex limonoids having remained undiscovered for many years in the genera of the Toddalioideae. Thus, complex limonoids appear to be of little value in resolving the taxonomic situation of *Hortia* (Figure 3.3).

A well-known example of alkaloids in Rutaceae is pilocarpine, which is isolated from *Pilocarpus* sp. (**17**, Figure 3.4), that is used in the relief of xerostomia (dry mouth), one of the side effects of radiotherapy for the treatment of neck and head cancers (Almeida, 2010; Valli et al., 2018).

Among other taxonomic markers, we can mention the indole monoterpene alkaloids from Apocynaceae and Rubiaceae, which may reveal evolutionary clues and help in the delimitation of subfamilies (da Silva et al., 2010; Bolzani et al., 2012). These are some examples of several other studies that demonstrate the importance of Brazil's natural products in chemosystematics.

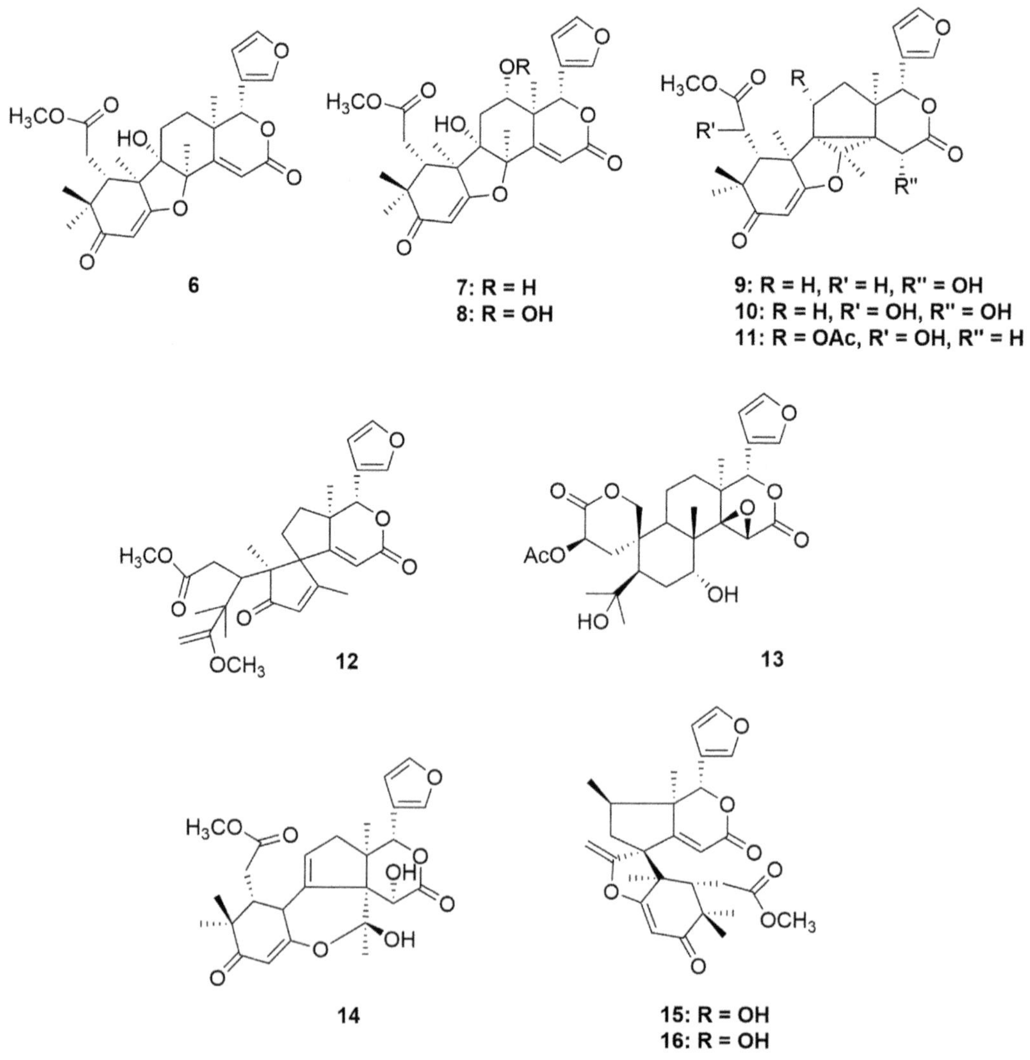

6

7: R = H
8: R = OH

9: R = H, R' = H, R" = OH
10: R = H, R' = OH, R" = OH
11: R = OAc, R' = OH, R" = H

12

13

14

15: R = OH
16: R = OH

FIGURE 3.3 Typical limonoids of *Hortia* species (Rutaceae).

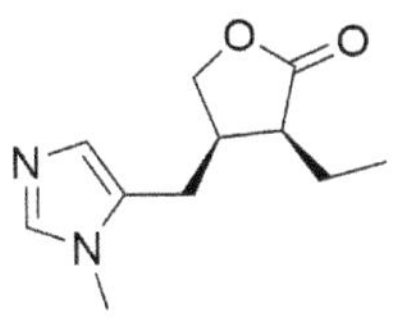

17

FIGURE 3.4   Structure of pilocarpine (**17**).

Currently, research groups in Brazil have been tackling many different scientific puzzles in several fields. Among them are the following: chemical ecology, chemosystematics, biological activity, biosynthesis, analytical methodology, biotechnology, green chemistry, natural products with added value, natural products of marine origin, chemical products of natural micro-organisms. One of the factors that led to the diversification of research lines was the Brazilian researchers' awareness of the value of Brazilian biodiversity, which holds tremendous potentiality for the discovery of new substances with added value and of biological interest. One of the extremely relevant initiatives to advance the chemistry of natural products in Brazil was the implementation of the BIOTA-FAPESP program supported by the Foundation for Research Support of the State of São Paulo (FAPESP). Created in 1999, with the aims to report the diversity and distribution of the fauna and flora of the State of São Paulo, as well as a tool to evaluate the possibilities of sustainable exploitation with economic potential (Joly et al., 2008, 2010). Another very important initiative was the development of a free access database of natural products coming from the Brazilian biodiversity created in 2013, through a collaborative project between the Nuclei of Bioassays, Biosynthesis and Ecophysiology of Natural Products (NuBBE, São Paulo State University UNESP – Araraquara) and Computational and Medicinal Chemistry of São Paulo University (São Carlos, SP) groups (NuBBe_{DB}). Since its inception, more than 2,000 compounds isolated from microorganisms, plants, animals, marine organisms, biotransformation and semi-synthetic products have been deposited in the database. In addition to the compounds' structures, biological and pharmacological information, spectroscopic and physico-chemical properties are also compiled (Pilon et al., 2017; Valli et al., 2013). The information contained in NuBBe_{DB} help researchers and those interested in bioprospecting secondary metabolites, chemosystematics, metabolomics, among others.

In recent decades, there has been a significant advance in the study of natural products derived from different biological sources. A search conducted in the NuBBe database shows that of the 2,218 compounds of Brazilian biodiversity, 78.6% were isolated from plants, 4.9% from microorganisms, 0.36% from marine organisms and 0.18% from animals (Figure 3.5). Other 15.6% are semi-synthetic products and 1.53% are biotransformation products.

Another project worth highlighting is the development of the global natural products social molecular networking (GNPS) platform. This project was created at the University of San Diego by several researchers, among them some Brazilians such as Norberto Peporine Lopes of the Nucleus of Research in Natural and Synthetic Products (Faculty of Pharmaceutical Sciences of Ribeirão Preto, USP) (Wang et al., 2016). In this open access platform, any researcher in the world can enter their data enlarging the library. This type of information will be extremely useful for those who work in the field of natural products.

These data demonstrate that the main biological source of research in natural products is still from plants. However, in recent years there has been a considerable increase in the diversification of these sources. Research on microorganisms is the second most explored source and the trend is that this number will increase even more. Interest in other biological sources is due, in part, to the structural diversity found in other sources, such as microorganisms and marine organisms.

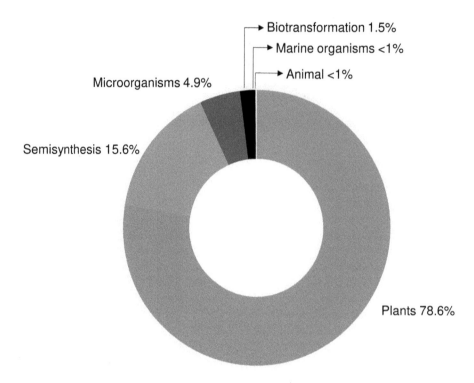

FIGURE 3.5 Sources of substances isolated from Brazilian biodiversity and derivatives. Data available in NuBBE database.

An example from Brazilian researchers was the isolation and/or UPLC-HRMS detection of bromotyrosine-derived alkaloids, 11-hydroxyaerothionin (**18**), fistularin-3 (**19**), verongidoic acid (**20**), aplysinamisine II (**21**), aerothionin (**22**), purealidin L (**23**), homopurpuroceratic acid B (**24**) (Figure 3.6) from the culture of *Pseudovibrio denitrificans* Ab134, bacterium isolated from the tissue of the marine sponge *Arenosclera brasiliensis*. These substances are known only in sponges of the order Verongida, making this publication pioneer in the literature (Nicacio et al., 2017).

Silva-Junior et al. (2018), aiming to understand the relationship between the bacterium *Serratia marcescens* 3B2 and cutter ants of the genus *Atta sexdens rubropilosa*, carried out the chemical study of the bacteria and isolated several pyrazines (**25–30**, Figure 3.7). Some of these compounds were identified as alarm and trail pheromones of leafcutter ants. This work will help in the elaboration of future studies that aim at the control of these ants.

Fruit waste provides an opportunity to obtain useful and valuable chemicals. For example, 15 million tons of citrus waste accumulates annually from the food and drink processing industry and other minor contributors. Often simply disposed of or incorporated into animal feed, this huge and naturally occurring resource is not being fully taken advantage of, although the means to do so have been successfully demonstrated. It has long been recognized that orange peel represents a promising source of hesperidin, a flavonoid. One million metric tons of peel residues are generated as a result of fruit processing, and thus, an extract of this residue could be considered for the isolation of hesperidin, other flavonoids, and many other compounds. Bellete et al. (2018) showed that several flavonoids present in citrus waste can be isolated using a faster and greener methodology. They observed the presence of flavonoids hesperitin (**31**), narigenin (**32**), nobiletin (**33**), chrysoerythol (**34**), sinensetin (**35**), isosakuranetin (**36**), 3,5,6,7,3′,4′- hexamethoxylflavone (**37**), 3,5,6,7,8,3′,4′-heptamethoxyflavone

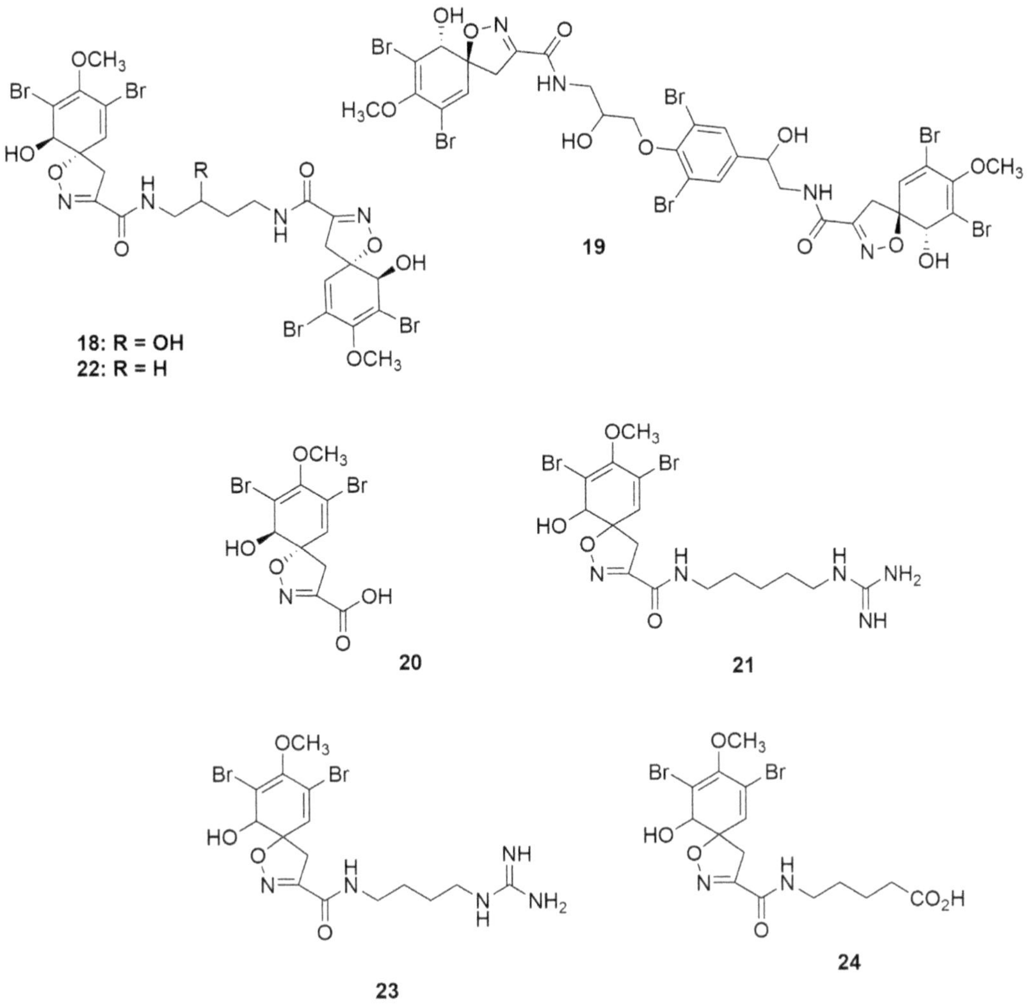

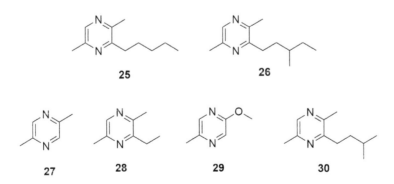

OCH3
Br
Br
HO
R
H
N
O
N
H
18: R = OH
22: R = H
19
20
21
23
24
NH
NH2
CO2H
OH

FIGURE 3.6 Bromotyrosine derivatives isolated from *P. denitrificans*.

N
25
26
27
28
29
30

FIGURE 3.7 Substances belonging to the pyrazine family produced by *S. marcescens* 3B2.

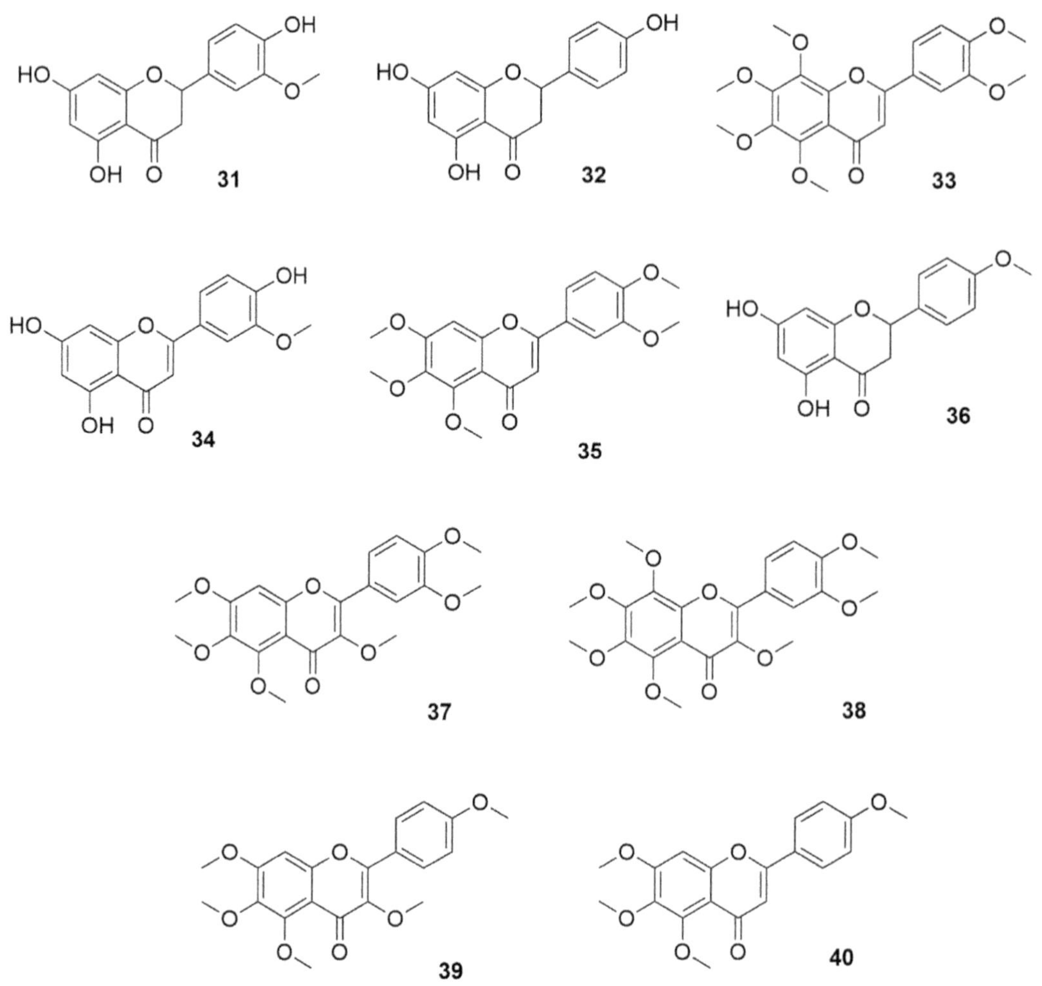

FIGURE 3.8   Flavonoids present in Brazilian citrus fruit waste.

(38), 3,5,6,7,4′-pentamethoxyflavone (39) and 5-methoxysalvigenin (40) (Figure 3.8). This class of compounds presents a broad spectrum of biological activities as antioxidant, antimicrobial, antiviral, anti-inflammatory, etc. (Cao et al., 1997; Cushnie and Lamb, 2005). Lani et al. (2016) showed the potential of certain flavonoids against the *Chikungunya* virus, making this study a good example of the potential of agro-industrial waste to produce bioactive substances.

Recently, the Quézia Cass group has shown the importance of determining the chirality of isolated natural products, which is often very difficult. An example of the structural complexity is palytoxin (41, Figure 3.9) isolated from *Palythoa* sp. corals, which contains 64 stereogenic centers (Viegas Jr and Bolzani, 2006). A significant amount of racemates or enantiomerically enriched mixtures has been reported from natural sources. This number is estimated to be even larger according to the Cass group, since the enantiomeric purity of secondary metabolites is rarely checked. This latter fact may have significant effects on the evaluation of the biological activity of chiral natural products. A second bottleneck is the determination of their absolute configurations. Despite the widespread use of optical rotation and electronic circular dichroism, most of the stereochemical assignments are based on empirical correlations with similar compounds reported in the literature. As an alternative, the Cass group suggest the combination of vibrational circular dichroism and quantum chemical calculations,

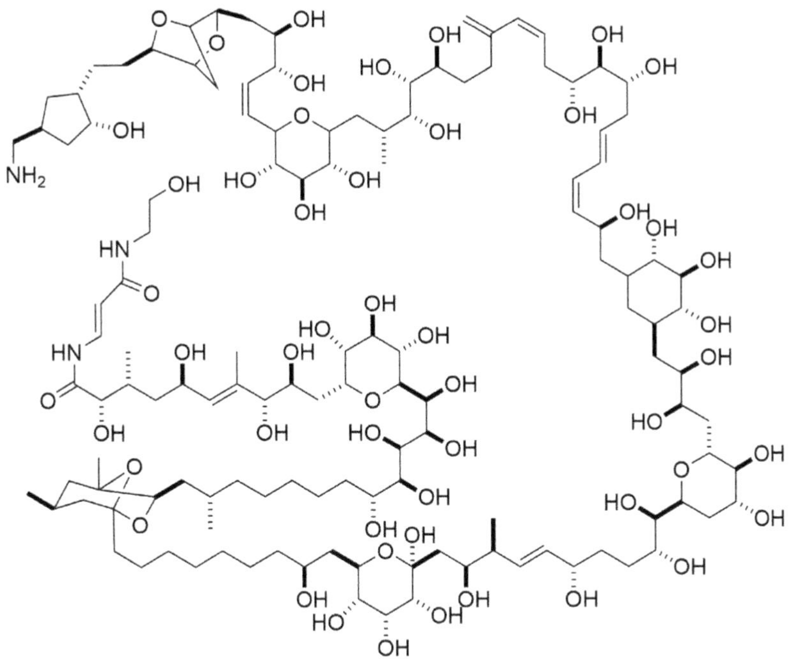

41

FIGURE 3.9   Palytoxin (**41**) structure isolated from *Palythoa* sp.

which has emerged as a powerful and reliable tool for both conformational and configurational analysis of natural products, even for those lacking UV-Vis chromophores (Batista et al., 2018).

It is well-known that nature produces countless secondary metabolites at different concentrations, for instance a single plant can produce 5,000 to 25,000 metabolites (Leme et al., 2014). To identify and quantify the set of secondary metabolites produced by an organism as well as to understand the role of these substances in the relationships and interactions of these organisms with the environment is practically impossible using traditional techniques. However, with more sensitive equipment of nuclear magnetic resonance (NMR), mass spectrometry (MS), and chemometrics applied to Liquid Chromatography (LC)-MS and/or LC-NMR data we can evaluate the similarities and differences between the chromatographic profiles and perform a metabolomics analysis. Oliviera. used mass spectrometry and molecular networking to identify 63 flavonoids from 19 extracts of *Adenocalymma imperatoris-maximilianni* (Oliveira et al., 2017). Angolini et al. (2016) after investigation of *Streptomyces wadayamensis* A23 genome, actinobacteria isolated from *Citrus reticulata*, showed that this lineage has biosynthetic machinery capable of generating various antibiotics. The metabolic profile from this lineage analyzed by mass spectrometry showed the production of several bioactive substances already predicted by the genome mining. The pioneer results presented by Angolini et al. (2016) open up exciting opportunities for different research fields from genomics to the production of bioactive substances.

3.2   STRUCTURAL DIVERSITY AND DISTRIBUTION OF SECONDARY METABOLITES IN BRAZILIAN PLANTS

Analyzing all the chemical data recorded in the NuBBe database for the species of a plant family, we verified that the Rutaceae were the most studied in Brazil with 12.5% of the total metabolites reported, followed by Meliaceae (9.1%), and a large part of the work was developed by da Silva's group. There is a general agreement between taxonomists that both these families should be placed in the same order, Sapindales (da Silva et al., 2010). The Rutaceae family includes about 150 genera with more than 1,500 species, which are distributed throughout the tropical and temperate regions of the world, being most abundant in tropical America, South Africa, Asia and Australia. Rutaceae constitute the largest group of Sapindales and are characterized by a great diversity of secondary metabolites not common in other families of the order. The most representative are alkaloids derived from anthranilic acid, coumarins, flavonoids and limonoids. There are 32 genera with about 200 species currently recognized in Brazil, and species of 24 genera have already been studied by da Silva's group. Meliaceae consists of about 550 species distributed between approximately 51 genera. There are six genera currently recognized in Brazil, and species of all genera have already been studied by the da Silva group. Limonoids, biosynthetically related compounds, are found in Meliaceae and Rutaceae. There are consistent differences between the limonoids of the Rutaceae and those of the Meliaceae.

In agreement with the NuBBE database the other families most studied in Brazil are in order of their decreasing total metabolites reported: Lauraceae (8.9%), Rubiaceae (7.5%), Fabaceae (6.8%), Euphorbiaceae (4.2%), Piperaceae (4.1%), Myrtaceae (2.2%), Verbenaceae (1.5%), Anarcadiaceae and Celastraceae (1.2%), Annonaceae; Asteraceae; Moraceae and Sapindacea (<1%) and others corresponding to 38.5%. These families and others discussed in this chapter, their respective species and common names can be seen in Table 3.1.

Regarding the class of compounds reported in the database of chemical and biological information of Brazilian biodiversity, the most representative are terpenes (34%). Another 15.3% are

TABLE 3.1
Summary of the Names of All Plant Species, Their Respective Family and Common Names Presented in This Chapter

Species	Family	Common Names
Cupressus sempervirens	Cupressaceae	Cipreste-dos-cemitérios, cipreste, cipreste-comum, cipreste-de-Itália, falso-cedro
Commiphora species (sin. *Commiphora voensis*)	Burseraceae	–
Papaver somniferum	Papaveraceae	Papoula, papoula do ópio, dormideira
Caesalpinia echinata (sin. *Guilandina echinata*)	Fabaceae	Pau-brasil, ibirapitanga, orabutã, brasileto, ibirapiranga, ibirapita, ibirapitã, muirapiranga, pau-rosado, pau-de-pernambuco
Plumeria lancifolia	Apocynaceae	Agonia, agonium, arapou, arapuê, arapuo, colônia, guina-mole, jasminmanga, quina-branca, quina-mole, sacuíba, sucuba, sucuriba, sucuúba, tapioca, tapouca, tapuoca
Geissospermum laeve (sin. *Geissospermum martianum*)	Apocynaceae	Pau-pereiro, pinguaciba, pau-de-pente e pau-para-toda-obra
Chondrodendron tomentosum (sin. *Chondrodendron hypoleucum*)	Menispermaceae	Curare, pareira-brava, pareira, uva-da-serra, uva-do-mato
Pterodon emarginatus (sin. *Pterodon pubescens*)	Fabaceae	Sucupira branca
Hortia oreadica	Rutaceae	Quina-do-campo, para-tudo, quina
Citrus reticulata	Rutaceae	Tangerina, mexerica, laranja-mimosa, mandarina, fuxiqueira, poncã, manjerica, laranja-cravo, mimosa, bergamota, clementina
Swinglea glutinosa (sin. *Chaetospermum glutinosum*)	Rutaceae	Limão swinglea, limão ornamental
Maytenus ilicifolia (sin. *Maytenus aquifolium*)	Celastraceae	Espinheira-santa,

(Continued)

TABLE 3.1 *(Continued)*
Summary of the Names of All Plant Species, Their Respective Family and Common Names Presented in This Chapter

Species	Family	Common Names
Salacia campestris	Hippocrateaceae	Bacupari-do-campo, japicuru, capirucu
Citrus sinensis (sin. *Citrus aurantium*)	Rutaceae	Laranja, laranja-doce, laranja-de-umbigo, laranja-bahianinha, laranja-lima, laranja-natal, laranja-pera-do-Rio, laranja-rubi, laranja-valência, laranja-hamlin, laranja-bahia, laranja-sangüinea, laranja-pêra, laranjeira
Citrus limonia	Rutaceae	Limão-rosa, limão-cavalo, limão-egua, limão-francês, limão-china, limão-vinagre, limão tambaqui
Alternanthera littoralis (sin. *Achyranthes maritima*)	Amaranthacea	Periquito-da-praia
Erythrina verna (sin. *Erythrina mulungu*)	Fabaceae	Suinã, suiná, sapatinho-de-judeu, canivete, amansa-senhor, bico-de-papagaio, comedoi, molongo, murungu, corticeira, sananduva, pau-imortal
Annona hypoglauca (sin. *Annona tessmannii*)	Annonaceae	Biribá
Annona crassiflora (sin. *Annona rodriguesii*)	Annonaceae	Acanga, araticum, araticum do mata, tapanahuacanga
Duguetia furfuracea (sin. *Duguetia hemmendorffii*)	Annonaceae	Ata-brava, ata-de-lobo, pinha do campo, pinha brava
Croton grandivelum (sin. *Croton echinocarpus*)	Euphorbiaceae	Sangra d'água, sangue-de-drago
Chimarrhis turbinata (sin. *Pseudochimarrhis diformis*)	Rubiaceae	Pau de remo
Tabernaemontana catharinensis (sin. *Tabernaemontana australis*)	Apocynaceae	Leiteiro, leiteiro-de-folha-fina, leiteiro de vaca
Psychotria laciniata (sin. *Psychotria kleinii*)	Rubiaceae	Gandiúva-dánta, café-dánta, pimenteira-do-mato, buta, cravo-de-negro, erva-de-anta, pasto-de-anta
Piper tuberculatum (sin. *Piper arboreum* subsp. *tuberculatum*)	Piperaceae	Pimenta longa, pimenta-d'arda
Senna spectabilis (sin. *Cassia spectabilis*)	Fabaceae	Cássia, Cássia-do-nordeste, Cássia-macranta, Cássia-macrantera, Fedegoso, Fedegoso do Rio, Macrantera
Senna multijuga (sin. *Cassia multijuga*)	Fabaceae	Pau-cigarra, aleluia-amarela, árvore-da-cigarra, canafístula, canudeiro.
Hippeastrum psittacinum (sin. *Hippeastrum illustre*)	Amaryllidaceae	Amarílis, açucena, flor-da-imperatriz
Crotalaria retusa (sin. *Crotalaria retusifolia*)	Fabaceae	Amendoim bravo, chique-chique, guizo de cascavel, chocalho de cascavel
Calycophyllum spruceanum	Rubiaceae	Pau-mulato, mulateiro, mulateiro-da-várzea, escorrega-macaco, pau-mulato-da-várzea, capirona, pau-marfim
Laurencia dendroidea (sin. *Laurencia obtusa* var. *dendroidea*)	Rhodomelaceae	–
Vellozia graminifolia	Velloziaceae	–
Stemodia foliosa (sin. *Stemodiacra foliosa*)	Scrophulariaceae	Meladinha
Croton cajucara (*sin. Oxydectes cajucara*)	Euphorbiaceae	Sacaca, cajuçara, casca-sacaca, muirasacaca
Casearia sylvestris (sin. *Casearia subsessiliflora*)	Flacourtiaceae	Guaçatonga, baga-de-pomba, bugre-branco, café-bravo, cafezeiro-bravo, cafezeiro-do-mato, carvalhinho, pau-de-bugre, erva-de-pontada, erva-de-bugre, erva-de-lagarto, erva-de-teiú, gaibim
Casearia rupestris	Flacourtiaceae	Pururuca, pururuba

(Continued)

TABLE 3.1 *(Continued)*

Summary of the Names of All Plant Species, Their Respective Family and Common Names Presented in This Chapter

Species	Family	Common Names
Guarea macrophylla (sin. *Guarea demerarana*)	Meliaceae	Pau-de-arco, camboatá, catiguá-morcego
Baccharis retusa (sin. *Baccharis affinis*)	Asteraceae	Carqueja
Cedrela fissilis (sin. *Cedrela elliptica*)	Meliaceae	Cedro, cedro-rosa, cedro-cetim, acaiacá, acaiacatinga, acajá-catinga, acajatinga, acaju, acaju-caatinga, capiúva, cedrinho
Cabralea canjerana (sin. *Cabralea lacaziana*)	Meliaceae	Canjarana, canjerana, canjerana de prego, cajarana, canharana, cedro canjerana, pau de santo, caieira, canjarana do litoral, caja espúrio
Typha domingensis (sin. *Typha salgirica*)	Typhaceae	Taboa, bucha, capim-de-esteira, erva-de-esteira, espadana, paina, paineira-de-flecha, paineira-do-brejo, partasana, pau-de-lagoa, tabebuia, tabuca
Pterocaulon alopecuroides (sin. *Pterocaulon latifolium*)	Asteraceae	–
Pterocaulon balansae (sin. *Pterocaulon paniculatum*)	Asteraceae	–
Pterocaulon polystachyum (sin. *Pterocaulon polystachyum*)	Asteraceae	Quitoco
Polygala sabulosa	Polygalaceae	Timutu-pinheirinho
Conchocarpus fontanesianus (sin. *Angostura fontanesiana*)	Rutaceae	Pitaguará
Helietta apiculata (sin. *Helietta cuspidata*)	Rutaceae	Canela-de-veado
Fridericia platyphylla (sin. *Arrabidaea brachypoda*)	Bignoniaceae	Cervejinha-do-campo
Bryophyllum pinnatum (sin. *Kalanchoe pinnata*)	Crassulaceae	Flores-da-fortuna, folha-da-costa, erva-da-costa, folha-grossa, folha-da-vida, coirama, coirama-branca, coirama-brava, roda-da-fortuna, saião, saião-roxo, amor-verde, paratudo, planta-do-amor, sempre-viva,
Passiflora quadrangularis (sin. *Passiflora macrocarpa*)	Passifloraceae	–
Croton betulaster (sin. *Oxydectes betulaster*)	Euphorbiaceae	–
Dimorphandra mollis	Fabaceae	Barbatimão-falso
Amburana cearensis (sin. *Amburana claudii*)	Fabaceae	Cerejeira, cumaru-do-cerá-, cumaré, cumaru-das-caatingas, imburana-de-cheiro, umburana, amburana-de-cheiro, imburana, cerejeira-rajada, cumaru-de-cheiro
Neoraputia alba	Rutaceae	Arapoca, arapoca branca, arapoca verdadeira, banha-de-galinha, mucanga, sucanga.
Neoraputia paraensis (sin. *Raputia paraensis*)	Rutaceae	–
Cenostigma macrophyllum	Fabaceae	Caneleiro, canela de Velho, maraximbé, fava do Campo
Garcinia brasiliensis (sin. *Garcinia gardneriana*)	Clusiaceae	Bacupari, bacupari do achado, bacupari mirim
Phyllanthus amarus (sin. *Diasperus nanus*)	Euphorbiaceae	Quebra-pedra, arrebenta-pedra, erva-pombinha
Piper solmsianum	Piperaceae	–
Ocotea fasciculata (sin. *Ocotea duckei*)	Lauraceae	Louro-de-cheiro
Zanthoxylum petiolare (sin. *Zanthoxylum naranjillo*)	Rutaceae	Juva, tembetari, espinho

(Continued)

TABLE 3.1 *(Continued)*
Summary of the Names of All Plant Species, Their Respective Family and Common Names Presented in This Chapter

Species	Family	Common Names
Combretum fruticosum (sin. *Combretum loeflingii*)	Combretaceae	Escova-de-macaco-alaranjada, escovinha, flor-de-fogo, limpa-garrafa-laranja, bugio, escovinha-raspadeira, escovinha-flor-de-fogo
Aniba burchellii	Lauraceae	Abacaterana
Ocotea cymbarum (sin. *Licaria cymbarum*)	Lauraceae	Louro inamuí
Nectandra hihua (sin. *Nectandra glabrescens*)	Lauraceae	–
Styrax ferrugineus (sin. *Styrax ferrugineus* var. *grandifolius*)	Styracaceae	Laranjinha do cerrado

alkaloids, 14.16% flavonoids, 12.3% aromatic derivates, 9.6% lignoids, 5.2% coumarins, 3.5% phenylpropanoids and others (Figure 3.10).

In the chosen set of topics below, the known distributions of the above class of compounds in Brazilian plants are revised and the data were obtained from Chemical Abstracts (SciFinder) and Web of Science from 1998 to 2018.

3.2.1 ALKALOIDS

The most practical classification of alkaloids, due to their structural diversity, is in accordance with their known or hypothetical biogenesis. Thus, the reported alkaloids from the Brazilian plants are derivatives of anthranilic acid, tryptophan, phenylalanine and/or tyrosine, histidine, nicotinic acid, ornithine, and lysine, or in some cases, they are formed from two amino acid precursors (Dewick P. M., 2002; Wink M., 2016) (Figure 3.11).

From Brazilian plants, 34 alkaloids were isolated from 1998 to 2018: these were isolated from 15 species belonging to 14 genera and 9 families (Table 3.2, Figure 3.12). Alkaloids from phenylalanine and/or tyrosine are the most widespread in the Brazilian plants studied, such as isoquinolines (**42, 43**); tetrahydroisoquinoline (**45**), aporphines (**46, 47, 48, 49, 50**) and protoberberine (**51**). They are distributed in Annonaceae, Fabaceae and Amaranthaceae.

The alkaloids derived from anthranilic acid are the second group most isolated, for example, a simple 2-quinolone (**44**) and acridones (**52–59**). They were found in Amaranthaceae (simple 2-quinolone) and Rutaceae (acridones). The tryptophan derivative alkaloids, as terpene indole alkaloids (**60–65**) occur in Apocynaceae and Rubiaceae. Lysine derivatives piperidine (**66–69**) and pyridine (**70**) alkaloids were reported in Fabaceae and Piperaceae.

Amaryllidaceae alkaloids are also derived from phenylalanine and tyrosine; however, they are classified according to their main skeleton structure, as galanthamine-type (**71**), montanine-type (**72**), homolycorine-type (**73**) and tazettine-type (**74**).

Alkaloids derived from ornithine, such as pyrrolizidine alkaloid (**75**), were reported only in Fabaceae. This family is characterized by a diversity of alkaloids.

The interest in alkaloids has always been very great because these compounds exhibit marked biological activities such as leishmanicidal, antimalarial, antiprotozoal, antimicrobial and anticancer activities. However, a wide spectrum of other biological properties for the above alkaloids has been discovered (Table 3.2). One example is they act as inhibitors of acetylcholinesterase (AChE), which are currently one of the few therapies approved for the treatment of Alzheimer's disease (Table 3.2). The enzyme AChE acts in the central nervous system and rapidly hydrolyzes the active neurotransmitter acetylcholine into the inactive compounds choline and acetic acid. Low levels of acetylcholine in the synaptic cleft are associated with a decrease in cholinergic function characterizing Alzheimer's disease, which is the most common cause of dementia among the elderly.

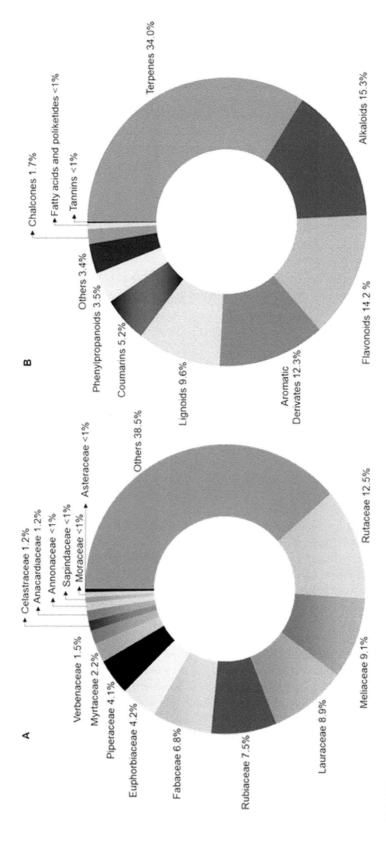

FIGURE 3.10 Statistics of the distribution of isolated and/or identified metabolites of Brazilian biodiversity according to (A) the families of plants and (B) metabolic classes. Data available in NuBBE database.

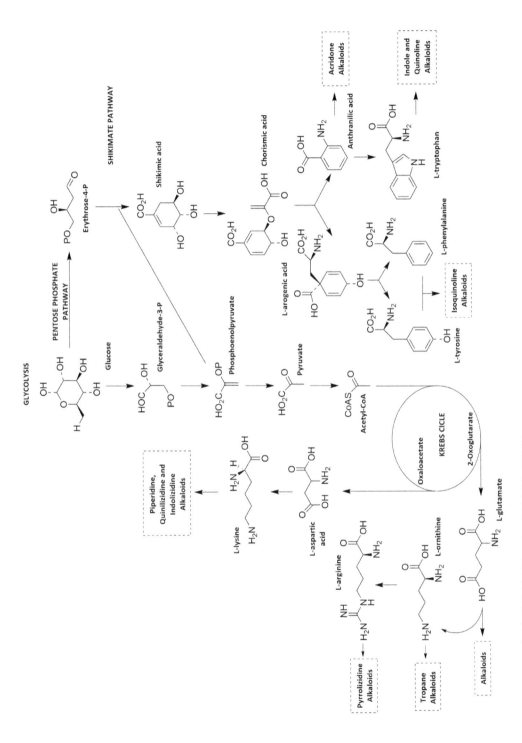

FIGURE 3.11 Biosynthesis of alkaloids. (Based on Wink M. (2016).)

TABLE 3.2

Occurrence of Alkaloids Identified from Brazilian Plants over the Last 20 Years

Code	Alkaloid	Substituents				Activity	Occurrence (plant part) (references)
42	(*R*)-1-(3,4-dihydroxyphenyl)-1,2,3,4-tetrahydroisoquinoline-6,7-diol					Antiprotozoal and antioxidant	*Alternanthera littoralis* Amaranthacea (Aerial parts)
43	6,7-dihydroxy-3,4-dihydroquinoline-1-one						(Koolen H. H. F. et al., 2016).
44	7,8-dihydroxy-1,2,4,5-tetrahydro-3*H*-1,5-ethano[c]azepin-3-one						
45	(+)-11α-Hydroxyerythravine	**R₁**	**R₂**			Anxiolytic	*Erythrina mulungu*
		H	OH				Fabaceae (flowers)
	(+)-Erythravine	H	H				(Flausino O. et al., 2007).
	(+)-α-Hydroxyerysotrine	CH₃	OH				
46		**R₁**	**R₂**	**R₃**	**R₄**	Antimicrobial	*Annona hypoglauca*
	Isoboldine	CH₃	OH	OCH₃	OH		Annonaceae (bark)
	Nornuciferine	H	H	H	OCH₃		(Rinaldi et al., 2017).
47	Stephalagine					Citotoxic	*Annona crassiflora* Annonaceae (fruits) (Pereira M. N. et al., 2017).
48	Duguetine					Antitrypanocidal and antileishmanial	*Duguetia furfuracea* Annonaceae (bark)
49	Dicentrinone						(da Silva B. D. et al., 2009).
50	*N*-methylglaucine						
51	*N*-methyltetrahydropalmatine						
52	Citrusinine-II	**R₁**	**R₂**	**R₃**		Inhibition of cathepsin V	*Swinglea glutinosa*
		H	OH	OCH₃			Rutaceae (roots)
	Citrusinine-I	H	OCH₃	OCH₃			(Severino R. P. et al., 2011).
	Citibrasine	OCH₃	OCH₃	OCH₃			
53	3,4-dihydro-3,5,8-trihydroxy-6-ethoxy-2,2,7-trimethyl-2*H*-pyrano[2,3-]acridin-12(7*H*)-one						
54	Corydine	**R₁**	**R₂**	**R₃**		Anti-HIV	*Croton echinocarpus*
		CH₃	OCH₃	H			Euphorbiaceae (leaves)
	Norisoboldine	H	H	OH			Ravanelli N. et al., 2016.
55	Glycocitrine-IV					Cytotoxic	*Swinglea glutinosa*
56	5-Dihydroxyacronycine						Rutaceae (fruits)
57	*bis*-5-Hydroxyacronycine						(Braga P. A. C. et al., 2007).
58	2,3-dihydro-4,9-dihydroxy-2-(2-hydroxypropan-2-yl)-11-methoxy10-methylfuro[3,2-*b*]acridin-5(10*H*)-one					Antimalarial	*Swinglea glutinosa* Rutaceae (stem bark) (dos Santos D. A. P. et al., 2009).
59	5-hydroxynoracronycine						
60	Strictosidine	**R₁**				Antioxidant	*Chimarrhis turbinata*
		H					Rubiaceae (bark)
	5α-Carboxystrictosidine	CO₂H					(Cardoso C. L. et al., 2004, Cardoso C. L. et al., 2008).

(Continued)

TABLE 3.2 *(Continued)*

Occurrence of Alkaloids Identified from Brazilian Plants over the Last 20 Years

Code	Alkaloid	Substituents			Activity	Occurrence (plant part) (references)
61	Coronaridine	R_1	R_2	R_3	Inhibition of acetylcholinesterase	*Tabernaemontana australis*
		H	$COOCH_3$	H		
	Voacangine	OCH_3	$COOCH_3$	H		Apocynaceae (bark)
62	Voacangine hydroxyindolenine					(Andrade M. T. et al.,
63	Rupicoline					2005).
64	Prunifoleine				Inhibition of	*Psychotria laciniata*
65	14-Oxoprunifoleine				cholinesterases and monoamine	Rubiaceae (leaves) (Passos C. S. et al., 2013).
66	(Z)-piplartine				Antifungal and antitrypanossomal	*Piper tuberculatum*
67	Piperine					Piperaceae (seeds and leaves) (Cotinguiba F. et al., 2009, Navickiene H. M. D. et al., 2000).
68		R_1			Antioxidant and anti-inflamatory	*Cassia spectabilis*
	(-)-Spectaline	H				Fabaceae (fruits)
	(-)-3-O-acetylspectaline	CH_3				(Viegas JR. C. et al., 2004).
69	(+)-3-O-feruloylcassine					Viegas JR. C. et al., 2007.
70	7′-multijuguinone	R_1			Inhibition of acetylcholinesterase	*Senna multijuga*
		CH_3				Fabaceae (leaves)
	12′-hydroxy-7′-multijuguinone	OH				(Serrano M. A. R. et al., 2010).
71	Galanthamine				Inhibition of acetylcholinesterase	*Hippeastrum psittacinum*
72	Montanine					Amaryllidaceae (bulbs)
73	Hippeastrine					(Pagliosa L. B. et al.,
74	Pretazettine					2010).
75	Usaramine				Antibacterial	*Crotalaria retusa* Fabaceae (seeds) (Neto T. S. N. et al., 2016).

The low enzymatic stability that is associated with the high cost of purification makes offline, in-solution enzymes assays impractical for high-throughput screening. However, the achievement of online assays using immobilised enzymes is a valuable alternative. Immobilized enzyme reactors (IMERs) are highly stabile in the presence of organic solvents and temperature variations; furthermore, these reactors enable the use of small amounts of enzyme and enzyme reuse. The use of IMERs for online screening using different formats and settings is extended because a high number of different immobilization approaches can be allied with a wide range of available chromatographic media. IMERs have been efficiently used in selective affinity chromatography methods and have been exploited through frontal and zonal (linear and nonlinear) chromatography. The optimized preparation of capillary enzyme reactors (ICERs) based on AChE for the screening of selective inhibitors was developed recently by a Brazilian group (Silva et al., 2013). The AChE-ICERs were prepared using the homo-bifunctional linker glutaraldehyde through a Schiff base linkage. The enzyme was anchored onto a modified fused silica capillary and used as an LC bio-chromatography column for online studies with UV-vis detection. Not only did the tailored AChE-ICER maintain the activity of the immobilized enzyme, but it also significantly improved the stability of the enzyme in the presence of organic solvents. In addition, kinetic studies demonstrated that the enzyme retained its

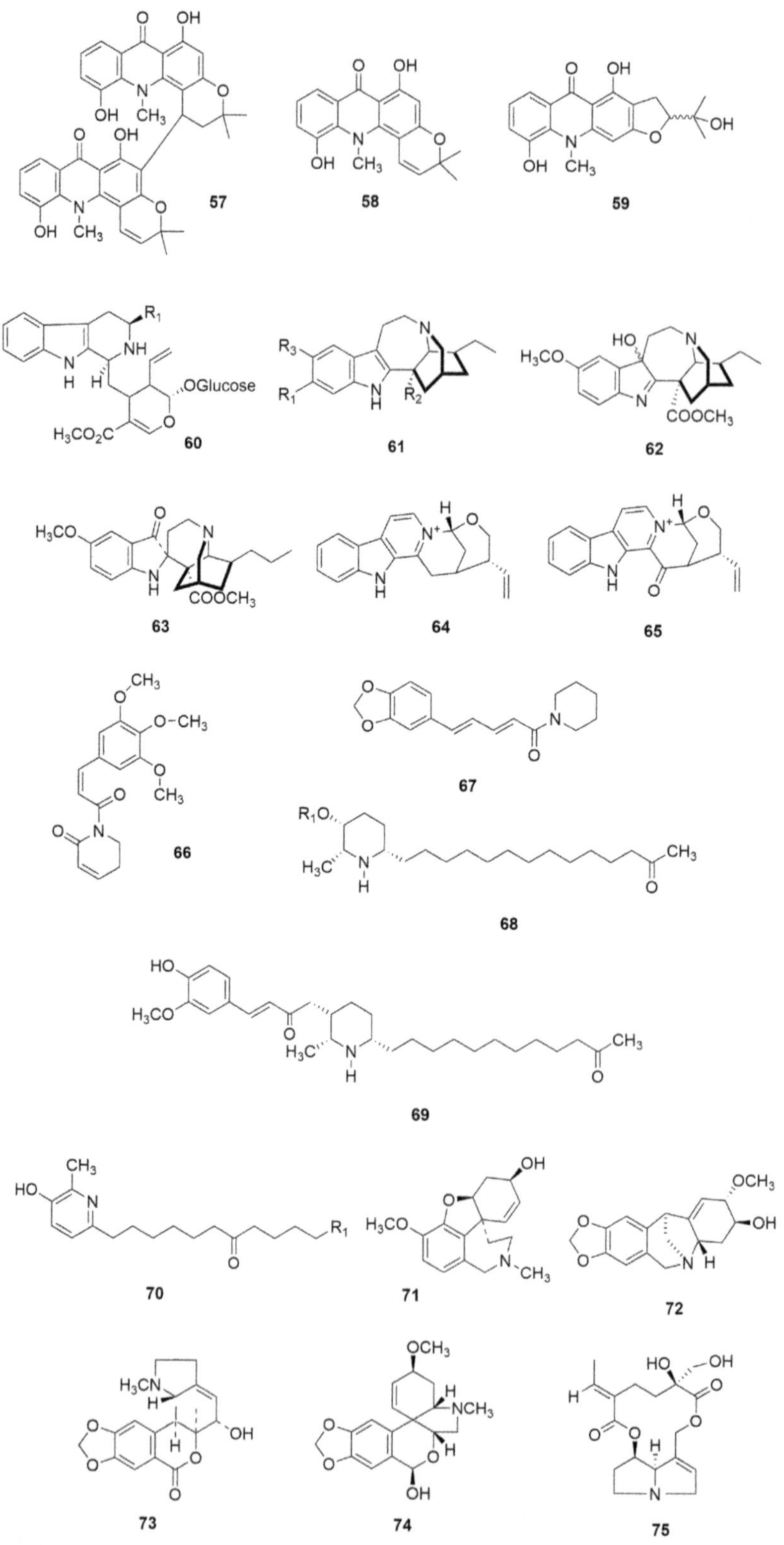

FIGURE 3.12 Structure of alkaloids identified from Brazilian plants over the last 20 years, based on information outlined in Table 3.2.

activity with high stability, preserving its initial activity over ten months. The absence of nonspecific matrix interactions, the immediate recovery of the enzymatic activity and the short analysis time were the main advantages of this AChE-ICER. The use of AChE-ICER in the ligand recognition assay was validated by the evaluation of four known reversible inhibitors (galantamine, tacrine, propidium and rivastigmine), and the same order of inhibitory potencies as described in the literature was found.

The immobilized enzyme was used to screen 21 synthetic coumarin derivatives. In this library, two new potent inhibitors were identified: [3-ethylcarboxylate-7-(2-piperidine-ethoxyl)coumarin] (IC_{50} 17.14 ± 3.50 µM) and [3-ethylcarboxylate-7-hydroxy-8-(1-piperidine-methoxyl)coumarin] (IC_{50} 6.35 ± 1.20 µM), or piperidine coumarins, which were compared to the standard galantamine (IC_{50} 12.68 ± 2.40 µM). Considering the high inhibitory activities of these compounds with respect to the AChE-ICER, the mechanism of action was investigated. Both coumarins exhibited a competitive mechanism of action, producing K_i values of 8.04 ± 0.18 and 2.67 ± 0.18 µM, respectively. The results revealed that the AChE-ICER is useful for the biological screening of inhibitor candidates and evaluating the mechanism of action.

A second example of alkaloids acting as enzyme inhibitors is the acridones, which have been shown to be potent inhibitors of cathepsin V. Several natural products have been investigated for their inhibitory effects on the catalytic activity of cathepsins, and Brazilian researchers developed bioassay methodologies with these enzymes (Severino et al., 2011). Cathepsins, also known as lysosomal cysteine peptidases, are members of the papain-like peptidase family, which are implicated in many pathological conditions. These enzymes have been intensively studied as valuable targets for drug discovery and development. Although the major role of cathepsins is related to the terminal protein degradation in lysosomes, it has been shown that these enzymes are also involved in other relevant biochemical pathways, acting at selective and controlled processes with specific functions associated to their restricted tissue localization. In addition, evidence has indicated that cysteine cathepsins have specific intra- and extracellular functions, being involved in a number of diseases including cancer, osteoarthritis, osteoporosis, autoimmune disorders and viral infections. Cathepsin V was identified as a lysosomal cysteine protease specifically expressed in thymus, testis and corneal epithelium. It is believed that cathepsin V plays a role in cancer progression, thus becoming a valuable drug target for oncology.

Eleven acridone alkaloids were isolated from *Swinglea glutinosa* (Bl.) Merr. (Rutaceae), with eight of them being identified as potent and reversible inhibitors of cathepsin V (IC_{50} values ranging from 1.2 to 3.9 µM). Detailed mechanistic characterization of the effects of these compounds on the cathepsin V-catalyzed reaction showed clear competitive inhibition with respect to substrate, with dissociation constants (Ki) in the low micromolar range. The most potent inhibitor citibrasine (**52**) (Table 3.2) has a Ki value of 200 nM (Severino et al., 2011).

3.2.2 TERPENES

The sesqui-, di- and triterpenes appear to be common in Brazilian plants. During the last few years several accounts have been published on the biosynthesis of terpenes. The literature clearly has shown that all compounds of the family are derived from the two building blocks isopentenyl diphosphate (IPP) and dimethylallyl diphosphate (DMAPP) and are biosynthesized either by the mevalonate (MVA) or the 2-C-methyl-D-erythritol 4-phosphate (MEP) pathway (Figure 3.13). Nevertheless, no studies of the biosynthesis of this class have apparently been reported in Brazilian plants. Until recently the only work in this series is due to Maysa Furlan and her colleagues, who showed that enzymatic extracts obtained from leaves and/or root bark of *Maytenus aquifolium* (Celastraceae) and *Salacia campestris* (Hippocrateaceae) displayed cyclase activity with conversion of the substrate oxidosqualene to the triterpenes, 3-friedelanol and friedelin. In addition, administration of (±)5-³H mevalonolactone in leaves of *M. aquifolium* seedlings produced radio labeled friedelin in the leaves, twigs and stems, while the root bark accumulated labeled maytenin and pristimerin (**89**).

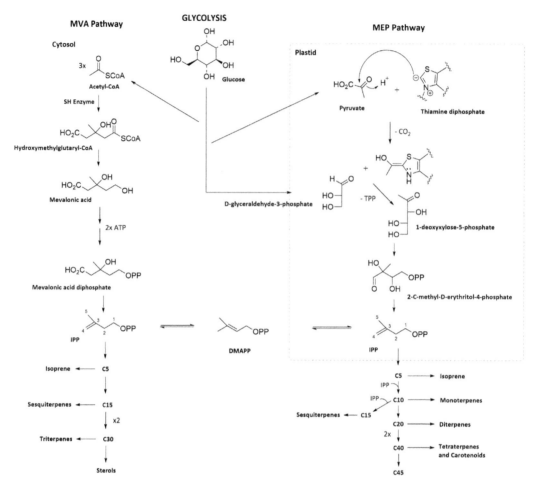

FIGURE 3.13 Biosynthesis of terpenes through the mevalonate pathways (MVA) and 2-C-methyl-D-erythritol 4-phosphate (MEP). Based on Lange B. M., 2015.

These experiments indicated that the triterpenes once biosynthesized in the leaves are translocated to the root bark and further transformed to the quinonemethide triterpenoids (Corsino et al., 2000).

Three secoiridoids were found in Brazilian Rubiaceae (skeleton **76**), which are common in this family. A sesquiterpene complex, named triquinane type (**77**), was mentioned in Asteraceae, whose class is characteristic of the family. The diterpenes were found in greater numbers: five labdanes (skeletons **78**, **79**), nine clerodanes (skeletons **80–83**), two isopiparane (skeleton **86**) and three kaurene (skeleton **87**). These were isolated from Brazilian plants of families Asteraceae, Euphorbiaceae, Flacourtiaceae, Meliaceae, Scrophulariaceae and Velloziaceae (Table 3.3). The triterpenes were the most abundant in the Brazilian plants mentioned in Table 3.3: a total of 13. Among these the tirucallanes/euphanes (**88**, **89**) and their derivatives, the limonoids (**90–92**), were found in Meliaceae, which is well characterized by such compounds. The cycloartane series is very common in higher plant families, and they (**84**, **85**) were also found in Meliaceae. Friedelanes (**93**) are also common in higher plant families, and one was isolated from Malpighiaceae, however, their biogenetic derivatives quinonemethides (**94**) are restricted to Celastraceae and Hippocrateaceae. Two quinonemethides are cited for Brazilian plants of the latter family (Table 3.3).

The above terpenes (Figure 3.14) exhibit a variety of biological activities, such as antitumoral, antibiotic, antimalarial, antileishmanial and trypanocidal activities (Table 3.3).

TABLE 3.3

Occurrence of Terpenes Identified from Brazilian Plants over the Last 20 Years

Code	Terpene	Substituents			Activity	Occurrence (plant part) (references)
		R_1	R_2	R_3		
76	7-Methoxydiderroside	CH_3	CH_3	H	Antitrypanosomal	*Calycophyllum spruceanum* Rubiaceae (wood bark) (Zuleta L. M. C. et al., 2003).
	6′-Acetyl-β-D-glucopyranosyldiderroside	H	CH_3	OCH_3		
	8-*O*-Tigloyldiderroside	H	$(CH_3)C=CH(CH_3)_3$	H		
77	Triquinane				Antileishmanial	*Laurencia dendroidea* Rhodomelaceae (Machado F. L. S. et al., 2011).
		R_1	R_2			*Vellozia graminifolia*
78	ent-3b-Hydroxylabd-8(17)-en-15-oic acid	βOH,H	CO_2H			Velloziaceae (roots, steam and leaves)
	3-*oxo*-Labd-8(17)-en-15-oate	O	CO_2CH_3			(Branco A. et al., 2004).
	ent-3b-Hydroxylabd-8(17)-ene-15-ol	βOH,H	CH_2OH			
		R_1				*Stemodia foliosa*
79	6α-Acetoxymanoyl oxide	$COCH_3$			Antibacterial	Scrophulariaceae (aerial parts)
	6α-Malonyloxymanoyl oxide	$COCH_2CO_2H$				(da Silva L. L. D. et al., 2008).
80	*trans*-Dehydrocrotonin				Antileishmanial	*Croton cajucara*
81	Crotonin					Euphorbiaceae (bark steam) (Lima G. S. et al., 2015).
		R_1	R_2	R_3		*Casearia sylvestris*
82	Casearin L	CH_3O	CH_3CO_2	OH	Cytotoxic and Anticancer	Flacourtiaceae (leaves) (Ferreira P. M. P. et al., 2010).
	Casearin O	CH_3O	CH_3CO_2	n-$C_3H_7CO_2H$		
	Casearin X	n-$C_3H_7CO_2H$	OH	H		
		R_1	R_2	R_3		*Casearia rupestres*
83	Casearupestrin A	OAc	a	OH	Citotoxic	Flacourtiaceae (leaves) (Vieira-Júnior G. M. et al., 2011).
	Casearupestrin B	OAc	OH	A		
	Casearupestrin C	OCH_3	OH	A		
	Casearupestrin D	OAc	OAc	A		

(Continued)

TABLE 3.3 (Continued)

Occurrence of Terpenes Identified from Brazilian Plants over the Last 20 Years

Code	Terpene	Substituents				Activity	Occurrence (plant part) (references)
84	Cycloart-23E-ene-3β,25-diol					Antitumoral	Guarea macrophylla Meliaceae (leaves) (Conserva G. A. A. et al., 2017).
85	(23S*,24S*)-Dihydroxycicloart-25-en-3-one						
86	Isopimara-7,15-diene-2α,3β-diol	H					
	Isopimara-7,15-dien-3β	OH					
87	ent-15β-Senecioyloxy-kaur-16-en-19-oic acid	R_1	R_2			Antitrypanosomal	Baccharis retusa Asteraceae (aereal parts) (Ueno A. K. et al., 2018).
		CH_2	$O_2CH=C(CH_3)_2$				
	ent-Kaur-16-en-19-oic acid	CH_2	H				
	ent-16-oxo-17-nor-Kauran-19-oic acid	O	H				
88	Hispidol A	R_1				Trypanocidal	Cedrela fissilis Meliaceae (steam) (Leite A. C. et al., 2008).
		A					
	Pentaol	B					
	Iso-odoratol	C					
	Odoratone	D					
89	Odoratol					Antitumor	Cabralea canjerana Meliaceae (fruits) (Cazal C. M. et al., 2010).
90	Gedunin						
91	6α-acetoxy-14β,15β-epoxyazadirone						
92	Cedrelona						
93	6α, 7α, 15β, 16β, 24-Pentacetoxy-22α-carbometoxy-21β,22β-epoxy-18β-hydroxy-27,30-bis nor-3,4-secofriedela-1,20 (29)-dien-3,4 R-olide					Antileishmanial	Lophanthera lactescens Malpighiaceas (stems) (Danelli M. G. M. et al., 2009).
94	Salacin	R_1	R_2	R_3	R_4	Antioxidant	Salacia campestres Hippocrateaceae (roots) (Carvalho P. R. F. et al., 2005).
		H	CH_3	OH	OH		
	Pristimerin	CH_3	CO_2CH_3	H	H		

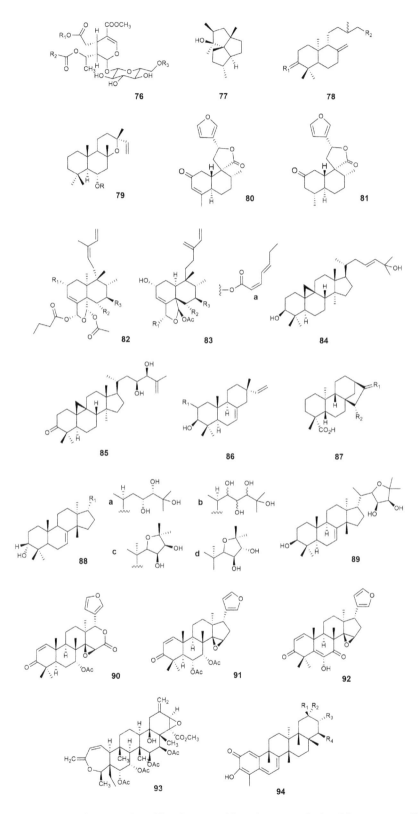

FIGURE 3.14 Structure of terpenes identified from Brazilian plants over the last 20 years, according to code of Table 3.3.

3.2.3 PHENYLPROPANOIDS

Several phenylpropanoids have recently been described from Brazilian plants; these are coumarins, flavonoids, lignans and neolignans (Table 3.4). Simple coumarins (**95–98**; see Figure 3.16) occur widely in the plant kingdom, but the proliferation of a wide range of complex furo- and pyrano-coumarins is a feature that is largely confined to Rutaceae and to Apiaceae. Trans-*p*-coumaric acid is the general precursor of 7-oxygen coumarins synthesized in plants (Figure 3.15). Prenylation of umbelliferone at C-6 leads to linear pyrano and furocoumarins and at C-8 to pyrano- and furo-angular coumarins (**99–101**; Figure 3.16). This class of compounds showed anti-inflammatory, anti-microbial and anticholinesterase properties, in Brazilian studies.

Flavonoids **102–111** (Figure 3.17) are formed by a mixed pathway, the C_6C_3 moiety being derived from the shikimic acid and the A ring from malonate (Figure 3.15). Flavonoids are widely occurring in nature and may suggest they are present in all angiosperms (Table 3.4). They showed antibacterial, antifungal, anticancer and other properties in Brazilian bioassays (Table 3.4).

Lignans and neolignans are a diverse group of compounds formed by the coupling of two phenylpropanoid (C_6C_3) units (Figure 3.15). The nomenclature recommended by International Union of Pure and Applied Chemistry (IUPAC) in 2000 respects the first definition of lignan introduced by Haworth, for whom the term "lignan" was introduced for the structures where the two units C_6C_3 are linked by a bond between positions 8 and 8′-linked (or β,β′) (Moss, 2000). For nomenclature purposes the C_6C_3 unit is treated as propylbenzene and numbered from 1 to 6 in the ring, starting from the propyl group, and with the propyl group numbered from 7 to 9, starting from the benzene ring. When the two are coupled in other ways they are called neolignans. This group is also considered to include examples where the two units are joined by an ether oxygen atom which for nomenclature purposes is treated as linking oxygen of an assembly. In addition, the class names lignan and neolignan are spelled in the conventional way without a terminal "e". The structures are spelled with a terminal "e" to indicate a saturated side chain unless modified to show unsaturation.

The compounds (Figure 3.18) encountered may be classified into lignane: 3′,4′,5,9,9′-pentamethoxy-3,4-methylenedioxy-lignane (**112**), 3,4,5,5′-tetramethoxy-3′,4′-methylenedioxy-7,7′-epoxy-lignane (**113**), 5,5′-dimethoxy-3,4,3′,4′-dimethylenedioxy-7,7′-epoxy-lignane (**114**), 3,4,5,3′,4′,5′-hexamethoxy-7,9′:7′,9-diepoxy-lignane (**115**), 9-hydroxy-3,4,3′,4′-dimethylenedioxy-9,9′-diepoxy-lignane (**116**), 4′,8-dihydroxy-3,3′,4′-trimethoxy-lignano-9,9′-lactone (**117**); neolignane: 3′-methoxy-3,4-methylenedioxy-2′,7-epoxy-4′*H*-8,1′-neolign-8′-en-4′-one (**118**), 4-hydroxy-3,3′-dimethoxy-4′,7-epoxy-8,3′-neolign-7′-ene (**119**), 9-hydroxy-3′,4′,5-trimethoxy-4,7′-epoxy-3,8′-9′norneolignan (**120**), 9-hydroxy-5-methoxy-3′,4′-methylenedioxy-4,7′-epoxy-3,8′-9′norneolignan (**120**). In Table 3.3 the name given to these compounds in the cited reference was considered, and in several of them they considered the Gottlieb nomenclature.

According to Gottlieb the difference between both these subclasses, rather than structural features, is the presence (lignans) or absence (neolignans) of oxygen functions at terminal carbon of the C_3 side chains. This difference is due to different biosynthetic pathways; cinnamyl alcohol or less commonly cinnamic acid is the precursors of lignans, while propenylphenol or allyphenol are the precursors of neolignans. However, the tendency is for the IUPAC proposal to be universally accepted as it has been discussed with representatives of numerous countries and respects the first definition.

Many lignans show physiological activity as tumor-inhibiting, some exhibit antibiotic, antimalarial, antileishmanial and trypanocidal activities (Table 3.4).

3.3 IMPACT ON BRAZILIAN ECOSYSTEMS

Finally, this survey indicates that the families most studied in Brazil are, in order of their decreasing total metabolites reported: Rutaceae, Meliaceae, Lauraceae, Rubiaceae, Fabaceae, Euphorbiaceae, Piperaceae, Myrtaceae, Verbenaceae, Anarcadiaceae, Celastracea, Annonaceae, Asteraceae, Moraceae and Sapindacea, and they offer considerable potential for the discovery of new or known compounds with significant and possibly valuable biological activities.

TABLE 3.4
Occurrence of Phenylpropanoids Identified from Brazilian Plants over the Last 20 Years

Code	Phenylpropanoids	Substituents	Activity	Occurrence (plant part) (references)
	Coumarins			
95	Umbelliferone		Anti inflamatory	*Typha domingensis* Typhaceae (aerial parts) (Vasconcelos J. F. et al., 2009).
96	5-Methoxy-6,7-ethylenedioxycoumarin	OCH_3	Antifungal	*Pterocaulon alopecuroides, Pterocaulon balansae, Pterocaulon polystachyum* Asteraceae (leaves) (Stein A. C. et al., 2006).
	Ayapin	H		
97		R_1 / R_2		
	Prenylatin	OH / $CH_2CHC(CH_3)_2$		
	Prenylatin-methyl-ether	OCH_3 / $CH_2CHC(CH_3)_2$		
	7-(2′,3′-Epoxy-3′-methylbutyloxy)-6-methoxycoumarin	OCH_3 / $CH_2CHOC(CH_3)_2$		
98		R_1	Antinociceptive	*Polygala subulosa* Polygalaceae (whole plant) (Meotti F. C. et al., 2006).
	7-Prenyloxy-6-methoxycoumarin	$CH_2CHC(CH_3)_2$		
	Scopoletin	H		
	Acetylscopoletin	$COCH_3$		
	Benzoylscopoletin	C_6H_6		
99	Marmesin		Anticholinesterase	*Conchocarpus fontanesianus* Rutaceae (stems) (Cabral R. S. et al., 2011).
100	(+)-3-(1′-Dimethylallyl)-decursinol		Antileishmanial	*Helietta apiculate* Rtaceae (stem bark) (Ferreira M. E. et al., 2010).
101	(-)-Heliettin			
102	*Flavonoids* 3′,4′-Dihydroxy-5,6,7-trimethoxyflavone	R_1 CH$_3$ / R_2 CH$_3$ / R_3 OH	Antimicrobial	*Arrabidaea brachypoda* Bignoniaceae (leaves) (Alcerito T. et al., 2002).
	Cirsiliol	H / CH_3 / OH		
	Cirsimaritin	H / CH_3 / H		
	Hispidulin	H / H / H		

(Continued)

TABLE 3.4 (Continued)

Occurrence of Phenylpropanoids Identified from Brazilian Plants over the Last 20 Years

Code	Phenylpropanoids	Substituents			Activity	Occurrence (plant part) (references)
103					Citotoxic	*Kalanchoe pinnata* Crassulaceae (leaves) (Muzitano M. F. et al., 2006).
	Kaempferol 3-*O*-α-L-arabinopyranosyl (1 → 2) α-L-rhamnopyranoside	R_1 H	R_2 3-*O*-α-L-arabinopyranosyl (1→2) α-L-rhamnopyranoside			
	Quercetin 3-*O*-α-L-arabinopyranosyl (1 → 2) α-L-rhamnopyranoside	OH	3-*O*-α-L-arabinopyranosyl (1→2) α-L-rhamnopyranoside			
	Quercitrin	OH	α-L-rhamnopyranose			
	Apigenin	R_1 H	R_2 H		Sedative	*Passiflora quadrangulares* Passifloraceae (fruits) (Gazola A. C. et al., 2015).
		H	H		Antitumoral	*Croton betulaster* Euphorbiaceae (Santos B. L. et al., 2015).
104	Rutin	R_1 3-*O*-rutinoside			Antitumoral	*Dimorphandra mollis* Fabaceae (Santos B. L. et al., 2015).
	Quercetin	OH				
105	Odoratin	R_1 H	R_2 OH	R_3 OCH$_3$	Inhibition of DNA topoisomerase II	*Amburana cearenses* Fabaceae (resin) (de Oliveira G. P. et al., 2017).
	Calycosin	OH	H	H		
106	Erycibenin D					
107	Penduletin	R_1 H	R_2 OH		Antitumoral	*Croton betulaster* Euphorbiaceae (aerial parts) (Coelho P. L. C et al., 2016).
	Casticin	OH	OCH$_3$			

(Continued)

TABLE 3.4 (Continued)

Occurrence of Phenylpropanoids Identified from Brazilian Plants over the Last 20 Years

Code	Phenylpropanoids	Substituents	Activity	Occurrence (plant part) (references)
108	3',4',5,7,8-Pentamethoxyflavone	R_1 OCH_3	Antitrypanossomal	*Neoraputia alba Neoraputia paraensis* Rutaceae (leaves) (Moraes V. R. S., et al., 2003).
109	3',4',5',5,7-Pentamethoxyflavone	H		
	3',4',7,8-Tetramethoxy-5,6-(2'',2''-dimethylpyrano)-flavone			
110	Agathisflavone		Antiviral	*Cenostigma macrophyllum* Fabaceae (stem and leaves) (de Sousa R. L. F. et al., 2015).
111	Amentoflavone Podocarpusflavone	R_1 OH OCH_3	Antioxidant	*Garcinia brasiliensis* Clusiaceae (leaves) (Arwa P. S. et al., 2015).
112	*Lignans and neolignans* Niranthin		Anti-inflammatory and antiallodynic	*Phyllanthus amarus* Euphorbiaceae (aerial parts) (Kassuya C. A. L. et al., 2006).
113	*rel-(7R,8R,7'R,8'R)*-3',4'-Methylenedioxy -3,4,5,5'-tetramethoxy-7,7'-epoxylignan		Antitrypanossomal	*Piper solmsianum* Piperaceae (inflorescence) Martins R. C. C. et al., 2003.
114	*rel-(7R,8R,7'R,8'R)*-3,4,3',4'-dimethylenedioxy-5,5'-dimethoxy-7,7'-epoxylignan			
115	Yangambin		Antileishmanial	*Ocotea duckei* Lauraceae (leaves) (Neto R. L. M. et al., 2011).
116	Cubebin		Anti-inflammatory	*Zanthoxyllum naranjillo* Rutaceae (leaves) (Bastos J. K. et al., 2001).

(Continued)

TABLE 3.4 *(Continued)*

Occurrence of Phenylpropanoids Identified from Brazilian Plants over the Last 20 Years

Code	Phenylpropanoids	Substituents		Activity	Occurrence (plant part) (references)
		R_1	R_2		
117	(−)-Trachelogenin			Antitumoral	*Combretum fruticosum* Combretaceae (stalks) (Moura A. F. et al., 2018).
118	Burchellin			Antidiuretic and antitrypanossomal	*Aniba burchelli* Lauraceae (leaves), *Ocotea cymbarum* Lauraceae (bark) (Cabral M. M. O. et al., 2000, Cabral M. M. O. et al., 2010).
119	Licarin A			Antitrypanossomal	*Nectandra glabrescens* Lauraceae (fruits) (Cabral M. M. O. et al., 2010).
120				Antibacterial and antifungal	*Styrax ferrugineus* Styracaceae (leaves) (Pauletti P. M. et al., 2000).
	5-(3″-Hydroxypropyl)-7-methoxy-2-(3′,4′-ethylenedioxyphenyl) benzofuran	-CH₂-			
	5-(3″-Hydroxypropyl)-7-methoxy-2-(3′,4′-dimethoxyphenyl)benzofuran	CH₃	CH₃		

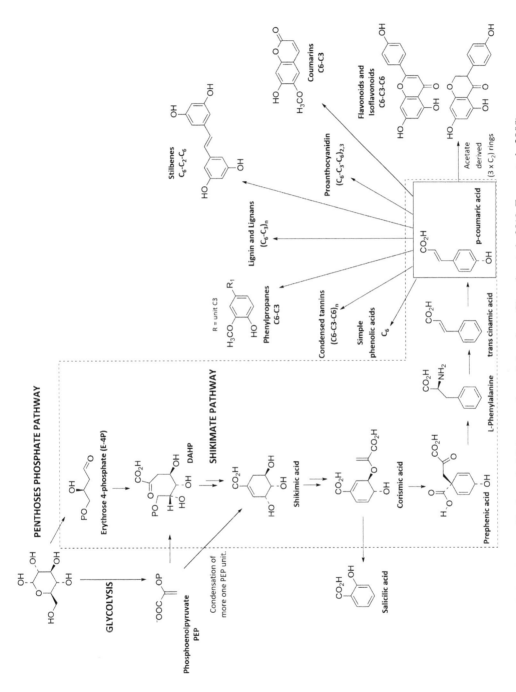

FIGURE 3.15   Biosynthesis of phenylpropanoids through shikimate pathway (taken from Cheynier et al., 2013; Ferrer et al., 2008)

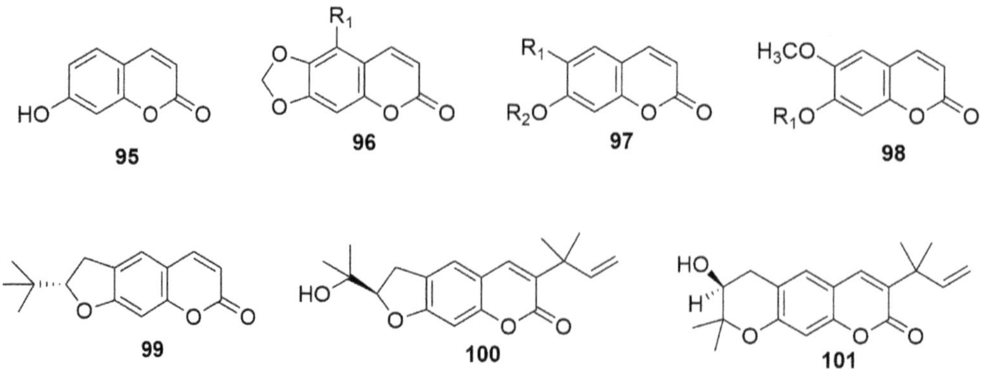

FIGURE 3.16 Structure of coumarins identified from Brazilian plants over the last 20 years, based on information in Table 3.4.

For over 20 years, Brazil has become the largest consumer of pesticides worldwide. The increased farming of biofuel crops and the use of genetically modified seeds caused the pesticide sales in Brazil to more than double recently, with the country surpassing the United States to become the largest market in the world. In spite of intensive research on plant natural products over the past three decades, only one type of botanical insecticides has been commercialized with any success in the past 15 years. It is based on neem seed extracts (limonoid azadirachtin, Meliacee) (da Silva et al., 2010). In Brazil, few groups are working to search for agricultural pesticides; this review showed that the vast majority have the objective of research to search for potential drugs. The data discussed above show that it is urgent to stimulate the scientific community to use scientific knowledge and expertise to improve pest management practices for the benefit of Brazil and the environment.

Brazil recently had an initiative to address this problem. The National Institutes of Science and Technology Program, which was launched in July 2008 by the Ministry of Science and Technology – National Council for Scientific and Technological Development (CNPq), with the collaboration of São Paulo Research Foundation (FAPESP), recruited scientists to work in networks in research areas that are strategic to the sustainable development of the country. Thus, the São Carlos Federal University (UFSCar, Brazil) aggregated in networks the best research groups in chemical ecological areas from five states and seven institutions to transform Brazil into the model country for the control of insects with a low environmental impact and created the National Institute of Science and Technology for the Biorational Control of Pest-Insect (NIST-BCPI). The NIST-BCPI is involved in teaching, research and extension oriented for the development of skilled researchers and the generation of knowledge and agrochemical products through the following areas: (i) Natural products as sources for new pesticides; (ii) Semisynthetic modifications; (iii) The mode of action of natural and synthetic pesticides via the inhibition of enzymes: immobilized enzymes reactors; (iv) Nanotechnology to improve activity, solubility and stability and (v) Citrus diseases and resistance mechanisms.

The example below should serve as a stimulus for the young Brazilian scientific community to start their research line with a focus on agriculture.

Citrus trees can exhibit a host of symptoms reflecting various disorders that can affect their health, vigour and productivity to varying degrees. Correctly identifying symptoms is an important aspect of management, as inappropriate remedial applications or actions can be costly and sometimes detrimental. Many diseases are difficult to distinguish from one another. Early disease detection and management are essential to ensuring the continued viability of the citrus industry. The rapid communication of new diseases, significant outbreaks and accurate information are vital. One of the major biotic diseases in Citrus is citrus variegated chlorosis (CVC). *Xylella fastidiosa*, a Gram-negative bacterium, colonizes plants xylem, thereby causing CVC in sweet orange. Flavonoids are the most bioactive secondary metabolites of citrus; however, only a few references to the role of

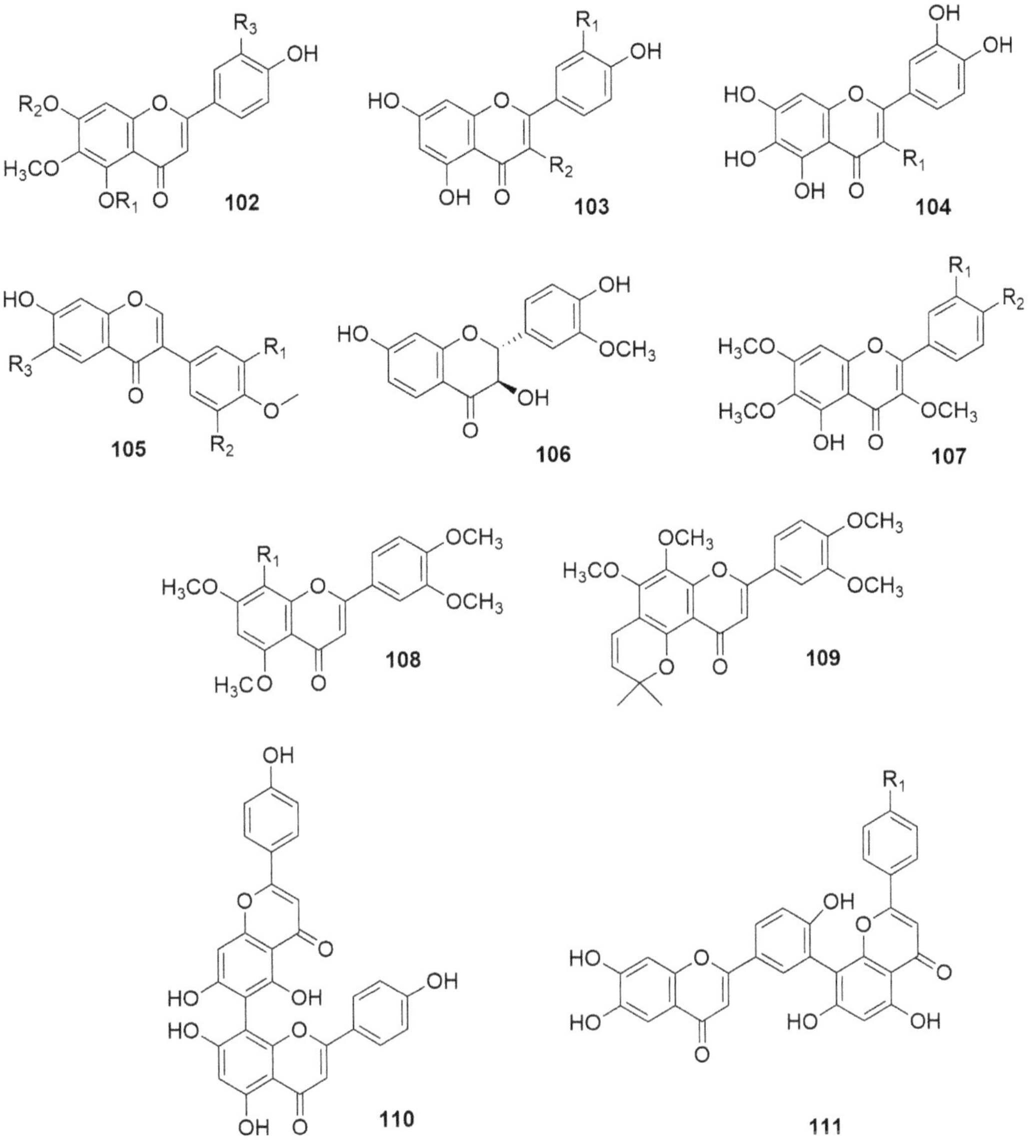

FIGURE 3.17 Structure of flavonoids identified from Brazilian plants over the last 20 years, based on information in Table 3.4.

these compounds in citrus tolerance to the CVC bacterium could be found. Alves and collaborators (2009) studied how *X. fastidiosa* colonizes and spreads within the xylem vessels of the sweet orange *C. sinensis* cultivar (cv.) Pêra (Alves et al., 2009). The authors reported that *X. fastidiosa* initially attached to the cell wall followed by an increase in the number of bacteria, the production of strand-like material and the formation of biofilm. Needle-like crystallized material was often present in xylem vessels of *C. sinensis* that were infected with *X. fastidiosa*. One hypothesis was that the needle-like crystal was hesperidin (**121**; Figure 3.19). These crystals were not observed in healthy plants. Hesperidin (**121**) is most likely involved in natural defence or in resistance mechanisms against *X. fastidiosa* in sweet orange varieties. A HPLC-UV method was developed to quantify hesperidin in the leaves and stems of *C. limonia*. This quantification method was applied to *C. sinensis* grafted onto *C. limonia* with and without CVC symptoms after *X. fastidiosa* infection. The total

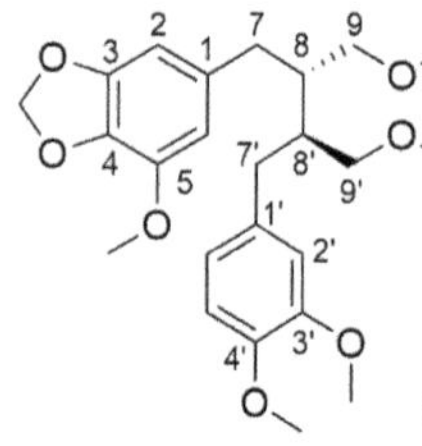
112

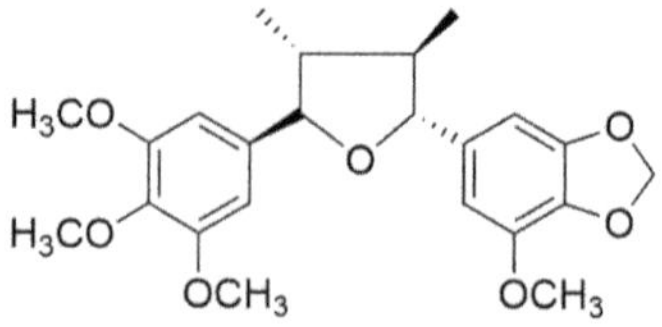
113

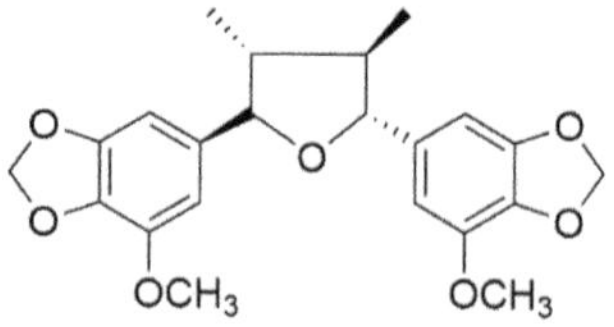
114

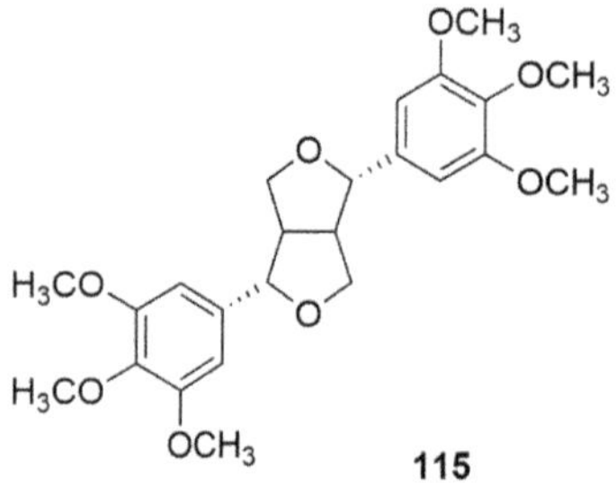
115

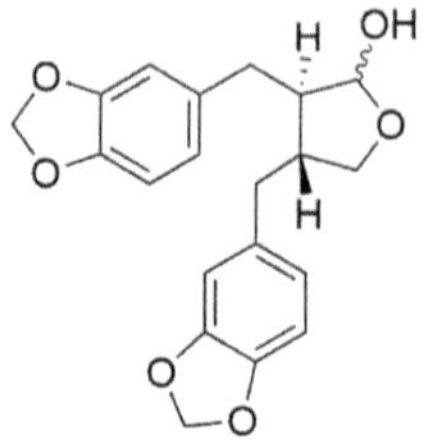
116

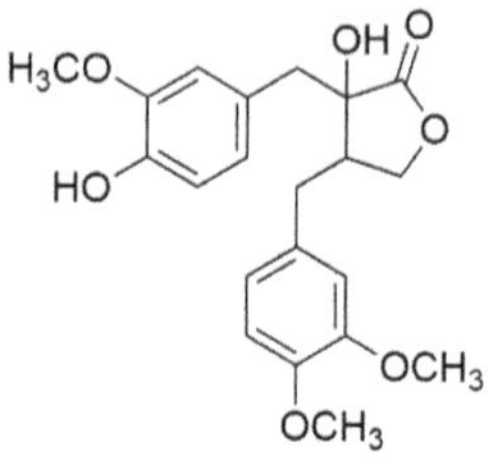
117

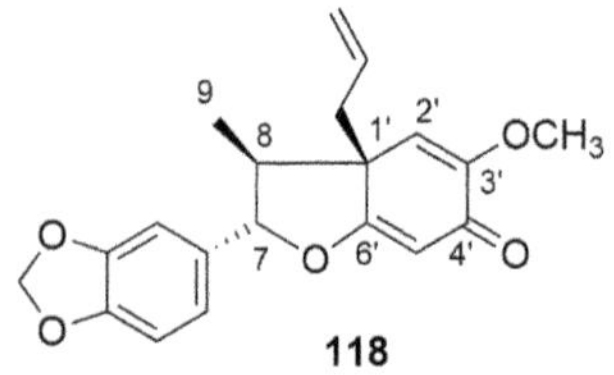
118

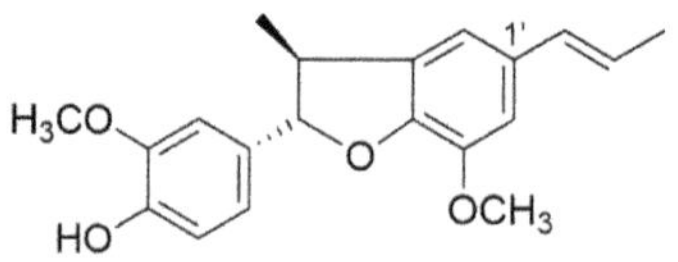
119

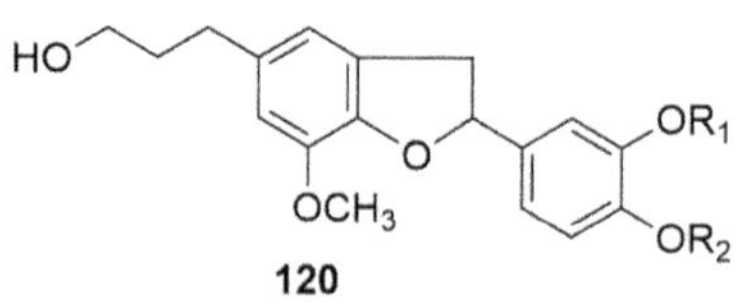
120

FIGURE 3.18 Structure of lignans and neolignans identified from Brazilian plants over the last 20 years, based on information in Table 3.4.

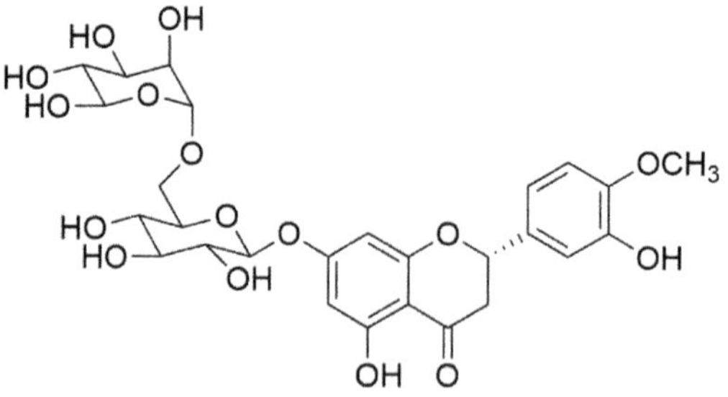
121

FIGURE 3.19 Structure of the flavonoid hesperidin (**121**).

content of hesperidin significantly increases in symptomatic leaves. Scanning electron microscopy studies on leaves with CVC symptoms showed vessel occlusion by biofilm and revealed crystallized material. Considering the impossibility of isolating these crystals for analysis, tissue sections were analyzed by Matrix-Assisted Laser Desorption/Ionization Mass Spectrometric Imaging (MALDI-MSI) to confirm the presence of hesperidin at the site of infection. The images that were constructed from Mass Spectrometry/Mass Spectrometry (MS/MS) data with a specific diagnostic fragment ion also showed higher concentrations of hesperidin in infected plants than in healthy ones, mainly in the vessel regions. These data suggest that hesperidin plays a role in plant-pathogen interactions, most likely as a phytoanticipin. This HPLC-UV method is simple and accurate for the determination of hesperidin in *C. sinensis*, *C. limonia* and their respective grafts. Citrus producers in Brazil have undertaken several control measures, including the eradication of diseased plants to remove the inoculum, spraying insecticides to reduce the population of transmission vectors (sharpshooters), and producing seedlings in greenhouses that are covered by plastic and laterally protected by screens. These measures have increased production costs that could reach US$ 286 to 322 million per year. The HPLC-UV method that was developed and applied to 60 citrus plants showed a 22% increase in the hesperidin content in asymptomatic plants. This increase may indicate the presence of the bacteria. Plants without CVC symptoms, also known as asymptomatic, are the plants which showed test positive by polymerase reaction chain (PCR) for *X. fastidiosa*, and the negative controls are the plants that were not inoculated with the bacteria. Therefore, the HPLC-UV method has become a powerful tool for detecting CVC in citrus before symptoms appear, thereby informing the citrus producers in advance when the plant should be removed from the orchard. This method could prevent the disease from being transmitted to other plants by insects and represents significant savings in pesticide application costs. In addition, it is less expensive to detect CVC disease in asymptomatic sweet orange trees using HPLC-UV than using other methods, such as PCR, and many samples can be screened per hour using approximately 1 mg of leaves (Soares et al., 2015).

Several flavonoids were tested for *in vitro* activity on the growth of *X. fastidiosa*, with hesperidin showing moderate activity (MIC 3.3 μM), suggesting that it can act as a good barrier for small-sized colonies from *X. fastidiosa* (Ribeiro et al., 2008). The main problem with natural products is generally low solubility in aqueous media when performing the tests, which has been and will continue to be a challenge. Therefore, the search for novel semisynthetic modifications of some flavonoids has attracted attention. One strategy would be to promote the metal chelation of hesperidin and study the potential biological relevance of these interactions. Magnesium is a metal of interest because of its biological importance as an essential metal for life, participating in a variety of metabolic and physiological functions. Moreover, compounds with magnesium (II) are good models for semisynthetic modifications of hesperidin because these species readily react with O-heterocyclic ligands, yielding stable compounds, and may provide a water-soluble compound. Because hesperidin activity is attributed to the generation of reduced metabolites that are involved in its antioxidant activity, it is expected that the Mg-hesperidin complex, in which the hesperidin ligand would be more accessible to oxidation, exhibits greater antioxidant activity than free hesperidin. Using this strategy and aiming to develop a more water-soluble compound that is bioactive and has reduced toxicity and luminescent diagnostic properties, the [Mg(hesp)2(phen)] complex, where hesp is hesperidin and phen is 1,10′-phenanthroline were prepared (Oliveira et al., 2013).

The complex [Mg(hesp)2(phen)] is more hydrosoluble (S = 472 ± 3.05 μg mL^{-1}) and liposoluble (log P = −0.15 ± 0.01) than free hesperidin (S = 5.92 ± 0.49 μg mL^{-1}, log P = 0.30). This complex is a better radical scavenger for superoxide radical (IC$_{50}$ = 68.3 μM at pH 7.8) than free hesperidin (IC$_{50}$ = 116.68 μmol L^{-1}) and vitamin C (IC$_{50}$ = 852 μmol L^{-1}). Hesperidin and its complex were assayed on the growth of *X. fastidiosa in vitro*, and the complex showed a better MIC than hesperidin, 0.34 and 3.3 μM, respectively (*in vivo* bioassay is in development).

Finally, the study of these interactions has led to new potential models for pesticides, suggesting that this line of research addresses the needs of Brazilian agriculture and should continue.

REFERENCES

Alcerito, T.; Barbo, F. E.; Negri, G.; Santos, Y. A. C. D.; Meda, C. I.; Young, M. C. M.; Chávez, D.; Blatt, C. T. T. 2002. Foliar epicuticular wax of Arrabidaea brachypoda: flavonoids and antifungal activity. Biochemical Systematics and Ecology, 30(7), 677–683.

Almeida, J. P.; Kowalski, L. P. 2010. Pilocarpine used to treat xerostomia in patients submitted to radioactive iodine therapy: a pilot study. Brazilian Journal of Otorhinolaryngology, 76(5), 659–662.

Almeida, M. R.; Martinez, S. T.; Pinto, A. C. 2017. Química de produtos naturais: plantas que testemunham histórias. Revista Virtual de Química, 9(3), 1117–1153.

Alves, E.; Leite, B.; Pascholati, S. F.; Ishida, M. L.; Andersen, P. C. 2009. *Citrus sinensis* leaf petiole and blade colonization by *Xylella fastidiosa*: details of xylem vessel occlusion. Scientia Agricola, 66(2), 218–224.

Andrade, M. T.; Lima, J. A.; Pinto, A. C.; Rezende, C. M.; Carvalho, M. P.; Epifanio, R. A. 2005. Indole alkaloids from *Tabernaemontana australis* (Müell. Arg) Miers that inhibit acetylcholinesterase enzyme. Bioorganic & Medicinal Chemistry, 13(12), 4092–4095.

Angolini, C. F. F.; Gonçalves, A. B.; Sigrist, R.; Paulo, B. S.; Samborskyy, M.; Cruz, P. L. R.; Vivian, A. F.; et al. 2016. Genome mining of endophytic *Streptomyces wadayamensis* reveals high antibiotic production capability. Journal of the Brazilian Chemical Society, 27(8), 1465–1475.

Arwa, P. S.; Zeraik, M. L.; Ximenes, V. F.; da Fonseca, L. M.; da Silva Bolzani, V.; Silva, D. H. S. 2015. Redox-active biflavonoids from *Garcinia brasiliensis* as inhibitors of neutrophil oxidative burst and human erythrocyte membrane damage. Journal of Ethnopharmacology, 174, 410–418.

Bastos, J. K.; Carvalho, J. C.; de Souza, G. H.; Pedrazzi, A. H.; Sarti, S. J. 2001. Anti-inflammatory activity of cubebin, a lignan from the leaves of *Zanthoxyllum naranjillo* Griseb. Journal of Ethnopharmacology, 75(2–3), 279–282.

Batista, A. N. L.; dos Santos, F. M.; Batista Jr, J. M.; Cass, Q. B. 2018. Enantiomeric mixtures in natural product chemistry: separation and absolute configuration assignment. Molecules, 23, 492.

Bellete, B. S.; Ramin, L. Z.; Porto, D.; Ribeiro, A. I. ; Forim, M. R.; Zuin, V. G.; Fernandes, J. B.; Silva, M. F. G. F. 2018. An environmentally friendly procedure to obtain flavonoids from Brazilian citrus waste. Journal of the Brazilian Chemical Society, 29(5), 111–115.

Bolzani, V. S.; Valli, M.; Pivatto, M.; Viegas, C. 2012. Natural products from Brazilian biodiversity as a source of new models for medicinal chemistry. Pure and Applied Chemistry, 84(9), 1837–1846.

Braga, P. A. C.; dos Santos, D. A. P.; da Silva, M. F. D. G. F.; Vieira, P. C.; Fernandes, J. B.; Houghton, P. J.; Fang, R. 2007. In vitro cytotoxicity activity on several cancer cell lines of acridone alkaloids and N-phenylethyl-benzamide derivatives from *Swinglea glutinosa* (Bl.) Merr. Natural Product Research, 21(1), 47–55.

Branco, A.; Pinto, A. C.; Braz Filho, R. 2004. Chemical constituents from *Vellozia graminifolia* (Velloziaceae). Anais da Academia Brasileira de Ciências, 76(3), 505–518.

Cabral, M. M. O.; Barbosa-Filho, J. M.; Maia, G. L. A.; Chaves, M. C. O.; Braga, M. V.; de Souza, W.; Soares, R. O. A. 2010. Neolignans from plants in northeastern Brazil (Lauraceae) with activity against *Trypanosoma cruzi*. Experimental Parasitology, 124(3), 319–324.

Cabral, M. M. O.; Kollien, A. H.; Kleffman, T.; Azambuja, P.; Gottlieb, O. R.; Garcia, E. S.; Schaub, G. A. 2000. *Rhodnius prolixus*: effects of the neolignan burchellin on in vivo and in vitro diuresis. Parasitology Research, 86(9), 710–716.

Cabral, R. S.; Sartori, M. C.; Cordeiro, I.; Queiroga, C. L.; Eberlin, M. N.; Lago, J. H. G.; Moreno, P. R. H.; Young, M. C. M. 2011. Anticholinesterase activity evaluation of alkaloids and coumarin from stems of *Conchocarpus fontanesianus*. Revista Brasileira de Farmacognosia, 22(2), 374–380.

Cao, G.; Sofic, E.; Prior, R. L. 1997. Antioxidant and prooxidant behavior of flavonoids: structure-activity relationships. Free Radical Biology & Medicine, 22(5), 749–760.

Caparica, C. 2012. Otto Gottlieb impulsionou a química de produtos naturais. Ciência e Cultura, 64(1), 10–11.

Cardoso, C. L.; Castro-Gamboa, I.; Silva, D. H. S.; Furlan, M.; Epifanio, R. A.; Pinto, A. C.; Rezende, C. M.; Lima, J. A.; Bolzani, V. S. 2004. Indole glucoalkaloids from *Chimarrhis turbinata* and their evaluation as antioxidant agents and acetylcholinesterase inhibitors. Journal of Natural Products, 67(11), 1882–1885.

Cardoso, C. L.; Silva, D. H. S.; Young, M. C. M.; Castro-Gamboa, I.; Bolzani, V. S. 2008. Indole monoterpene alkaloids from *Chimarrhis turbinata* DC Prodr.: a contribution to the chemotaxonomic studies of the rubiaceae family. Brazilian Journal of Pharmacognosy, 18(1), 26–29.

Carvalho, P. R.; Silva, D. H.; Bolzani, V. S.; Furlan, M. 2005. Antioxidant quinonemethide triterpenes from *Salacia campestris*. Chemistry & Biodiversity, 2(3), 367–372.

Cazal, C. M.; Choosang, K.; Severino, V. G. P.; Soares, M. S.; Sarria, A. L. F.; Fernandes, J. B.; Silva, M. F. G. F.; et al. 2010. Evaluation of effect of triterpenes and limonoids on cell growth, cell cycle and apoptosis in human tumor cell lines. Anti-Cancer Agents in Medicinal Chemistry, 10(10), 769–776.

Cheynier, V.; Comte, G.; Davies, K. M.; Lattanzio, V.; Martens, S. 2013. Plant phenolics: recent advances on their biosynthesis, genetics, and ecophysiology. Plant Physiology and Biochemistry, 72, 1–20.

Coelho, P. L. C.; de Freitas, S. R. V. B.; Pitanga, B. P. S.; da Silva, V. D. A.; Oliveira, M. N.; Grangeiro, M. S.; Souza, C. S.; et al. 2016. Flavonoids from the Brazilian plant *Croton betulaster* inhibit the growth of human glioblastoma cells and induce apoptosis. Revista Brasileira de Farmacognosia, 26(1), 34–43.

Conserva, G. A. A.; Girola, N.; Figueiredo, C. R.; Azevedo, R. A.; Mousdell, S.; Sartorelli, P.; Soares, M. G.; Antar, G. M.; Lago, J. H. G. 2017. Terpenoids from leaves of *Guarea macrophylla* display *in vitro* cytotoxic activity and induce apoptosis in melanoma cells. Planta Medica, 83(16), 1289–1296.

Corsino, J.; Carvalho, P. R. F.; Kato, M. J.; Latorre, L. R.; Oliveira, O. M. M. F.; Araujo, A. R.; Bolzani, V. S.; França, S. C.; Pereira, A. M. S.; Furlan, M. 2000. Biosynthesis of friedelane and quinonemethide triterpenoids is compartmentalized in Maytenus aquifolium and salacia campestris. Phytochemistry, 55, 741–748.

Cotinguiba, F.; Regasini, L. O.; Bolzani, V. S.; Debonsi, H. M.; Passerini, G. D.; Cicarelli, R. M. B.; Kato, M. J.; Furlan, M. 2009. Piperamides and their derivatives as potential anti-trypanosomal agents. Medicinal Chemistry Research, 18(9), 703.

Cragg, G. M.; Newman, D. J. 2005. Biodiversity: a continuing source of novel drug leads. Pure and Applied Chemistry, 77(1), 7–24.

Cragg, G. M.; Newman, D. J. 2013. Natural products: a continuing source of novel drug leads. Biochimica et Biophysica Acta, 1830, 3670–3695.

Cushnie, T. P. T.; Lamb, A. J. 2005. Antimicrobial activity of flavonoids. International Journal of Antimicrobial Agents, 26, 343–356.

da Silva, D. B.; Tulli, E. C. O.; Militão, G. C. G.; Costa-Lotufo, L. V.; Pessoa, C.; de Moraes, M. O.; Albuquerque, S.; de Siqueira, J. M. 2009. The antitumoral, trypanocidal and antileishmanial activities of extract and alkaloids isolated from *Duguetia furfuracea*. Phytomedicine, 16(11), 1059–1063.

da Silva, L. L. D.; Nascimento, M. S.; Cavalheiro, A. J.; Silva, D. H. S.; Castro-Gamboa, I.; Furlan, M.; Bolzani, V. S. 2008. Antibacterial activity of labdane diterpenoids from *Stemodia foliosa*. Journal of Natural Products, 71(7), 1291–1293.

da Silva, M. F. G. F.; Vieira, P. C.; Fernandes, J. B.; Oliva, G. 2010. A diversidade molecular dos metabólitos especiais da ordem rutales e sua importância na química medicinal. In: Química Medicinal. Métodos e Fundamentos em Planejamento de Fármacos. Editora da Universidade de São Paulo.

Danelli, M. G. M.; Soares, D. C.; Abreu, H. S.; Peçanha, L. M. T.; Saraiva, E. M. 2009. Leishmanicidal effect of LLD-3 (1), a nor-triterpene isolated from *Lophanthera lactescens*. Phytochemistry, 70(5), 608–614.

de Oliveira, G. P.; da Silva, T. M. G.; Camara, C. A.; Santana, A. L. B. D.; Moreira, M. S. A.; Silva, T. M. S. 2017. Isolation and structure elucidation of flavonoids *from Amburana cearensis* resin and identification of human DNA topoisomerase II-α inhibitors. Phytochemistry Letters, 22, 61–70.

de Sousa, L. R. F.; Wu, H.; Nebo, L.; Fernandes, J. B.; Silva, M. F. G. F.; Kiefer, W.; Kanitz, M.; et al. 2015. Flavonoids as noncompetitive inhibitors of dengue virus NS2B-NS3 protease: inhibition kinetics and docking studies. Bioorganic & Medicinal Chemistry, 23(3), 466–470.

Dewick, P. M. 2002 Medicinal Natural Products: A Biosynthetic Approach, 3rd ed. John Wiley & Sons.

Dias, D. A.; Urban, S.; Roessner, U. 2012. A historical overview of natural products in drug discovery. Metabolites, 2(2), 303–336.

dos Santos, D. A.; Vieira, P. C.; Silva, M. F. G. F.; Fernandes, J. B.; Rattray, L.; Croft, S. L. 2009. Antiparasitic activities of acridone alkaloids from *Swinglea glutinosa* (Bl.) Merr. Journal of the Brazilian Chemical Society, 20(4), 644–651.

Engler, A. 1931. In: Die Naturlichen Pflanzenfamilien. 3, part 4. Engelmann, Leipzig.

Ferreira, M. E.; de Arias, A. R.; Yaluff, G.; de Bilbao, N. V.; Nakayama, H.; Torres, S.; Schinini, A.; Guy, I.; Heinzen, H.; Fournet, A. 2010. Antileishmanial activity of furoquinolines and coumarins from *Helietta apiculata*. Phytomedicine, 17(5), 375–378.

Ferreira, P. M. P.; Santos, A. G.; Tininis, A. G.; Costa, P. M.; Cavalheiro, A. J.; Bolzani, V. S.; Moraes, M. O.; Costa-Lotufo, L. V.; Montenegro, R. C.; Pessoa, C. 2010. Casearin X exhibits cytotoxic effects in leukemia cells triggered by apoptosis. Chemico-Biological Interactions, 188(3), 497–504.

Ferrer, J. L.; Austin, M. B.; Stewart Jr, C.; Noel, J. P. 2008. Structure and function of enzymes involved in the biosynthesis of phenylpropanoids. Plant Physiology and Biochemistry, 46(3), 356–370.

Flausino, O.; de Ávila Santos, L.; Verli, H.; Pereira, A. M.; Bolzani, V. D. S.; Nunes-de-Souza, R. L. 2007. Anxiolytic effects of erythrinian alkaloids from *Erythrina mulungu*. Journal of Natural Products, 70(1), 48–53.

Gazola, A. C.; Costa, G. M.; Castellanos, L.; Ramos, F. A.; Reginatto, F. H.; Lima, T.; Schenkel, E. P. 2015. Involvement of GABAergic pathway in the sedative activity of apigenin, the main flavonoid from *Passiflora quadrangularis* pericarp. Revista Brasileira de Farmacognosia, 25(2), 158–163.

Gottlieb, O. R. 1988. Lignóides de plantas amazónicas: investigações biológicas e químicas. Acta Amazonica, 18(1–2), 333–344.

Gottlieb, O. R. 1992. Biodiversidade: uma teoria molecular. Química Nova, 15(2), 167–172.

Gottlieb, O. R.; Kaplan, M. A. C.; Borin, M. R. M. B. 1996. Biodiversidade: um enfoque químico-biológico. Ed. UFRJ.

Gottlieb, O. R.; Mors, W. B. 1980. Potential utilization of brazilian wood extractives. Journal of Agricultural and Food Chemistry, 28, 196–215.

Gottlieb, O. R.; Yoshida, M. 1989. Lignans. In: Natural Products of Wood Plants I. New York: Springer Series in Wood Science. Springer-Verlag.

Groppo, M.; Pirani, J. R.; Salatino, M. L. F.; Blanco, S. R.; Kallunki, J. A. 2008. Phylogeny of Rutaceae based on twononcoding regions from cpDNA. American Journal of Botany, 95(8), 985–1005.

Halpern, O.; Schmid, H. 1958. Zur Kenntnis des plumierids. Helvetica Chimica Acta, 41, 1109–1154.

Joly, C. A.; Casatti, L.; de Brito, M. C. W.; Menezes, N. A.; Rodrigues, R. R.; Bolzani, V. S. 2008. Histórico do programa Biota-FAPESP – O instituto virtual da biodiversidade. In Diretrizes Para a Conservação e Restauração da Biodiversidade no Estado de São Paulo. São Paulo: Editora Secretaria do Meio Ambiente.

Joly, C. A.; Rodrigues, R. R.; Metzger J. P.; Haddad, C. F. B.; Verdade, L. M.; Oliveira, M. C.; Bolzani, V. S. 2010. Biodiversity conservation research, training, and policy in São Paulo. Science, 328, 1358–1359.

Kassuya, C. A.; Silvestre, A.; Menezes-de-Lima, O.; Marotta, D. M.; Rehder, V. L. G.; Calixto, J. B. 2006. Antiinflammatory and antiallodynic actions of the lignan niranthin isolated from *Phyllanthus amarus*: evidence for interaction with platelet activating factor receptor. European Journal of Pharmacology, 546(1), 182–188.

Koolen, H. H. F.; Pral, E. M. F.; Alfieri, S. C.; Marinho, J. V. N.; Serin, A. F.; Hernández-Tasco, A. J.; Andreazza, N. L.; Salvador, M. J. 2016. Antiprotozoal and antioxidant alkaloids from *Alternanthera littoralis*. Phytochemistry, 134, 106–113.

Lange, B. M. 2015. The evolution of plant secretory structures and emergence of terpenoid chemical diversity. Annual Review of Plant Biology, 66, 139–159.

Lani, R.; Hassandarvish, P.; Shu, M.; Phoon, W. H.; Chu, J. J. H.; Higgs, S.; Vanlandingham, D.; Bakar, S. A.; Zandi, K. 2016. Antiviral activity of selected flavonoids against Chikungunya vírus. Antiviral Research, 133, 50–61.

Leite, A. C.; Ambrozin, A. R. P.; Fernandes, J. B.; Vieira, P. C.; Silva, M. F. G. F.; de Albuquerque, S. 2008. Trypanocidal activity of limonoids and triterpenes from *Cedrela fissilis*. Planta Medica, 74(15), 1795–1799.

Leme, G. M.; Coutinho, I. D.; Creste, S.; Hojo, O.; Carneiro, R. L.; Bolzani, V. S.; Cavalheiro, A. J. 2014. HPLC-DAD method for metabolic fingerprinting of the phenotyping of sugarcane genotypes. Analytica Methods, 6, 7781–7788.

Lima, G. S.; Castro-Pinto, D. B.; Machado, G. C.; Maciel, M. A.; Echevarria, A. 2015. Antileishmanial activity and trypanothione reductase effects of terpenes from the Amazonian species *Croton cajucara* Benth (Euphorbiaceae). Phytomedicine, 22(12), 1133–1137.

Machado, F. L. S.; Pacienza-Lima, W.; Rossi-Bergmann, B.; Gestinari, L. M. S.; Fujii, M. T.; de Paula, J. C.; Costa, S. S.; Lopes, N. P.; Kaiser, C. R.; Soares, A. R. 2011. Antileishmanial sesquiterpenes from the Brazilian red alga *Laurencia dendroidea*. Planta Medica, 77(7), 733–735.

Martins, R. C.; Lago, J. H. G.; Albuquerque, S.; Kato, M. J. 2003. Trypanocidal tetrahydrofuran lignans from inflorescences of *Piper solmsianum*. Phytochemistry, 64(2), 667–670.

Meotti, F. C.; Ardenghi, J. V.; Pretto, J. B.; Souza, M. M.; Moura, J. D.; Junior, A. C.; Soldi, C.; Pizzolatti, M. G.; Santos, A. R. S. 2006. Antinociceptive properties of coumarins, steroid and dihydrostyryl-2-pyrones from polygala sabulosa (Polygalaceae) in mice. Journal of Pharmacy and Pharmacology, 58(1), 107–112.

Moraes, V. R. S.; Tomazela, D. M.; Ferracin, R. J.; Garcia, C. F.; Sannomiya, M.; Soriano, M. P. C.; Silva, M. F. G. F.; et al. 2003. Enzymatic inhibition studies of selected flavonoids and chemosystematic significance of polymethoxylated flavonoids and quinoline alkaloids in neoraputia (Rutaceae). Journal of the Brazilian Chemical Society, 14(3), 380–387.

Moss, G. P. 2000. Nomenclature of lignans and neolignans (IUPAC Recommendations 2000). Pure and Applied Chemistry, 72, 1493–1523.

Moura, A. F.; Lima, K. S. B.; Sousa, T. S.; Marinho-Filho, J. D. B.; Pessoa, C.; Silveira, E. R.; Pessoa, O. D. L.; Costa-Lotufo, L. V.; Moraes, M. O.; Araújo, A. J. 2018. In vitro antitumor effect of a lignan isolated from *Combretum fruticosum*, trachelogenin, in HCT-116 human colon cancer cells. Toxicology in Vitro, 47, 129–136.

Muzitano, M. F.; Tinoco, L. W.; Guette, C.; Kaiser, C. R.; Rossi-Bergmann, B.; Costa, S. S. 2006. The antileishmanial activity assessment of unusual flavonoids from *Kalanchoe pinnata*. Phytochemistry, 67(18), 2071–2077.

Navickiene, H. M. D.; Alécio, A. C.; Kato, M. J.; Bolzani, V. S.; Young, M. C. M.; Cavalheiro, A. J.; Furlan, M. 2000. Antifungal amides from *Piper hispidum* and *Piper tuberculatum*. Phytochemistry, 55(6), 621–626.

Neto, R. L. M.; Sousa, L. M.; Dias, C. S.; Barbosa Filho, J. M.; Oliveira, M. R.; Figueiredo, R. C. 2011. Morphological and physiological changes in *Leishmania* promastigotes induced by yangambin, a lignan obtained from *Ocotea duckei*. Experimental Parasitology, 127(1), 215–221.

Neto, T. S. N.; Gardner, D.; Hallwass, F.; Leite, A. J. M.; de Almeida, C. G.; Silva, L. N.; Roque, A. A.; et al. 2016. Activity of pyrrolizidine alkaloids against biofilm formation and *Trichomonas vaginalis*. Biomedicine & Pharmacotherapy, 83, 323–329.

Nicacio, K. J.; Ióca, L. P.; Fróes, A. M.; Leomil, L.; Appolinario, L. R.; Thompson, C. C.; Thompson, F. L.; et al. 2017. Cultures of the marine bacterium *Pseudovibrio denitrificans* Ab134 produce bromotyrosine-derived alkaloids previously only isolated from marine sponges. Journal of Natural Products 80(2), 235–240.

Oliveira, G. G.; Neto, F. C.; Demarque, D. F.; Pereira-Junior, J. A. S.; Filho, R. C. S. P.; de Melo, S. J.; Almeida, J. R. G. S.; Lopes, J. L. C.; Lopes, N. P. 2017. Dereplication of flavonoid glycoconjugates from *Adenocalymma imperatoris-maximilianii* by untargeted tándem mass spectrometry-based molecular networking. Planta Medica, 83(7), 636–646.

Oliveira, R. M. M.; Daniel, J. F. S.; Aguiar, I.; Silva, M. F. G. F.; Fernandes, J. B.; Carlos, R. M. 2013. Structural effects on the hesperidin properties obtained by chelation to magnesium complexes. Journal of Inorganic Biochemistry, 129, 35–42.

Pagliosa, L. B.; Monteiro, S. C.; Silva, K. B.; de Andrade, J. P.; Dutilh, J.; Batista, J.; Cammarota, M.; Zuanazzi, J. A. S. 2010. Effect of isoquinoline alkaloids from two *Hippeastrum* species on in vitro acetylcholinesterase activity. Phytomedicine, 17(8–9), 698–701.

Passos, C. S.; Simões-Pires, C. A.; Nurisso, A.; Soldi, T. C.; Kato, L.; de Oliveira, C. M. A.; de Faria, E. O.; et al. 2013. Indole alkaloids of *Psychotria* as multifunctional cholinesterases and monoamine oxidases inhibitors. Phytochemistry, 86, 8–20.

Pauletti, P. M.; Araújo, A. R.; Young, M. C. M.; Giesbrecht, A. M.; Bolzani, V. S. 2000. *nor*-Lignans from the leaves of *Styrax ferrugineus* (Styracaceae) with antibacterial and antifungal activity. Phytochemistry, 55(6), 597–601.

Pereira, M. N.; Justino, A. B.; Martins, M. M.; Peixoto, L. G.; Vilela, D. D.; Santos, P. S.; Teixeira, T. L.; et al. 2017. Stephalagine, an alkaloid with pancreatic lipase inhibitory activity isolated from the fruit peel of *Annona crassiflora* mart. Industrial Crops and Products, 97, 324–329.

Pilon, A. C.; Valli, M.; Dametto, A. C. 2017. NuBBE$_{DB}$: na updated database to uncover chemical and biological information from Brazilian biodiversity. Scientific Reports, 7(1), 7215.

Pinto, A. C. 1995. O Brasil dos viajantes e dos exploradores e a química de produtos naturais brasileira. Química Nova, 18(6), 608–615.

Pinto, A. C.; Silva, D. H. S.; Bolzani, V. S.; Lopes, N. P.; Epifanio, R. D. A. 2002. Produtos naturais: atualidade, desafios e perspectivas. Química Nova, 25(1), 45–61.

Ravanelli, N.; Santos, K. P.; Motta, L. B.; Lago, J. H. G.; Furlan, C. M. 2016. Alkaloids from *Croton echinocarpus* baill.: anti-HIV potential. South African Journal of Botany, 102, 153–156.

Ribeiro, A. B.; Abdelnur, P. V.; Garcia, C. F.; Belini, A.; Severino, V. G. P.; Silva, M. F. G. F.; Fernandes, J. B.; et al. 2008. Chemical characterization of *Citrus sinensis* grafted on *C. limonia* and the effect of some isolated compounds on the growth of *Xylella fastidiosa*. Journal of Agricultural and Food Chemistry, 56(17), 7815–7822.

Rinaldi, M. V.; Díaz, I. E.; Suffredini, I. B.; Moreno, P. R. 2017. Alkaloids and biological activity of beribá (*Annona hypoglauca*). Revista Brasileira de Farmacognosia, 27(1), 77–83.

Santos, B. L.; Oliveira, M. N.; Coelho, P. L. C.; Pitanga, B. P. S.; da Silva, A. B.; Adelita, T.; Silva, V. D. A.; et al. 2015. Flavonoids suppress human glioblastoma cell growth by inhibiting cell metabolism, migration, and by regulating extracellular matrix proteins and metalloproteinases expression. Chemico-Biological Interactions, 242 (5), 123–138.

Santos, N. P.; Pinto, A. C.; Alencastro, R. B. 1998. Theodoro peckolt: naturalista e farmacêutico do Brasil imperial. Química Nova, 21(5), 666–670.

Serrano, M. A.; Pivatto, M.; Francisco, W.; Danuello, A.; Regasini, L. O.; Lopes, E. M. C.; Lopes, M. N.; Young, M. C. M.; Bolzani, V. S. 2010. Acetylcholinesterase inhibitory pyridine alkaloids of the leaves of *Senna multijuga*. Journal of Natural Products, 73(3), 482–484.

Severino, R. P.; Guido, R. V. C.; Marques, E. F.; Brömme, D.; Siva, M. F. G. F.; Fernandes, J. B.; Andricopulo, A. D.; Vieira, P. C. 2011. Acridone alkaloids as potent inhibitors of cathepsin V. Bioorganic & Medicinal Chemistry, 19(4), 1477–1481.

Severino, V. G. P.; Braga, P. A. C.; Silva, M. F. G. F.; Fernandes, J. B.; Vieira, P. C.; Theodoro, J. E.; Ellena, J. A. 2012. Cyclopropane- and spirolimonoids and related compounds from *Hortia oreadica*. Phytochemistry, 76, 52–59.

Severino, V. G. P.; de Freitas, S. D. L.; Braga, P. A. C.; Forim, M. R.; Silva, M. F. G. F.; Fernandes, J. B.; Vieira, P. C.; Venâncio, T. 2014. New limonoids from *Hortia oreadica* and unexpected coumarin from *H. superba* using chromatography over cleaning sephadex with sodium hypoclorite. Molecules, 19(8), 12031–12047.

Silva, D. H. S.; Castro-Gamboa, I.; Bolzani, V. S. 2010. Plant diversity from Brazilian cerrado and Atlantic forest as a tool for prospecting potential therapeutic drugs. In Comprehensive Natural Products II Chemistry and Biology. Oxford: Elsevier.

Silva, J. I.; Moraes, M. C.; Vieira, L. C. C.; Correa, A. G.; Cass, Q. B.; Cardoso, C. L. 2013. Acetylcholinesterase capillary enzyme reactor for screening and characterization of selective inhibitors. Journal of Pharmaceutical and Biomedical Analalysis, 73, 44–52.

Silva, M. F. G. F.; Bolzani, V. S. 2012. Otto Gottlieb, um cientista à frente de seu tempo. Química Nova, 35(11), 2103.

Silva, V. C.; Rodrigues, C. M. 2014 Natural products: an extraordinary source of value-added compounds from diverse biomasses in Brazil. Chemical and Biological Technologies in Agriculture, 1, 14.

Silva Filho, A. A.; Costa, E. S.; Cunha, W. R.; Silva, M. L. A.; Nanayakkara, N. P. D.; Bastos, J. K. 2008. *In vitro* antileishmanial and antimalarial activities of tetrahydrofuran lignans isolated from *Nectrandra megapotamica* (Lauraceae). Phytotherapy Research, 22, 1307–1310.

Silva-Junior, E. A.; Ruzzini, A. C.; Paludo, C. R.; Nascimento, F. S.; Currie, C. R.; Clardy, J.; Pupo, M. T. 2018. Pyrazines from bacteria and ants: convergente chemistry within an ecological niche. Scientific Reports, 8, 2595.

Soares, M. S.; da Silva, D. F.; Forim, M. R.; Silva, M. F. G. F.; Fernandes, J. B.; Vieira, P. C.; Silva, D. B.; et al. 2015. Quantification and localization of hesperidin and rutin in citrus sinensis grafted on C. limonia after Xylella fastidiosa infection by HPLC-UV and MALDI imaging mass spectrometry. Phytochemistry, 115, 161–170.

Souza, V. A.; Nakamura, C. V.; Corrêa, A. G. 2012. Atividade antichagásica de lignanas e neolignanas. Revista Virtual de Química, 4(3), 197–207.

Stein, A. C.; Alvarez, S.; Avancini, C.; Zacchino, S.; Von Poser, G. 2006. Antifungal activity of some coumarins obtained from species of pterocaulon (Asteraceae). Journal of Ethnopharmacology, 107(1), 95–98.

Ueno, A. K.; Barcellos, A. F.; Costa-Silva, T. A.; Mesquita, J. T.; Ferreira, D. D.; Tempone, A. G.; Romoff, P.; Antar, G. M.; Lago, J. H. G. 2018. Antitrypanosomal activity and evaluation of the mechanism of action of diterpenes from aerial parts of *Baccharis retusa* (Asteraceae). Fitoterapia, 125, 55–58.

Valli, M.; dos Santos, R. N.; Figueira, L. D.; Nakajima, C. H.; Castro-Gamboa, I.; Andricopulo, A. D.; Bolzani, V. S. 2013. Development of a natural products database from the biodiversity of Brasil. Journal of Natural Products, 76(3), 439–444.

Valli, M.; Russo, H. M.; Bolzani, V. S. 2018. The potential contribution of the natural products from Brazilian biodiversity to economy. Annals of the Brazilian Academy of Sciences, 90(1), 763–778.

Vasconcelos, J. F.; Teixeira, M. M.; Barbosa-Filho, J. M.; Agra, M. F.; Nunes, X. P.; Giulietti, A. M.; Ribeiro-dos-Santos, R.; Soares, M. B. P. 2009. Effects of umbelliferone in a murine model of allergic airway inflammation. European Journal of Pharmacology, 609(1–3), 126–131.

Viegas, C.; Bolzani, V. S.; Furlan, M.; Barreiro, E. J.; Young, M. C. M.; Tomazela, D.; Eberlin, M. N. 2004. Further bioactive piperidine alkaloids from the flowers and green fruits of *Cassia s pectabilis*. Journal of Natural Products, 67(5), 908–910.

Viegas Jr, C.; Bolzani, V. S. 2006. Os produtos naturais e a química medicinal moderna. Química Nova, 29(2), 326–337.

Viegas, Jr. C.; Silva, D. H. S.; Pivatto, M.; Rezende, A.; Castro-Gamboa, I.; Bolzani, V. S.; Nair, M. G. 2007. Lipoperoxidation and cyclooxygenase enzyme inhibitory piperidine alkaloids from *Cassia spectabilis* green fruits. Journal of Natural Products, 70(12), 2026–2028.

Vieira-Junior, G. M.; Dutra, L. A.; Ferreira, P. M. P.; de Moraes, M. O.; Costa-Lotufo, L. V.; Ó Pessoa, C.; Torres, R. B.; Boralle, N.; Bolzani, V. S.; Cavalheiro, A. J. 2011. Cytotoxic clerodane diterpenes from *Casearia rupestris*. Journal of Natural Products, 74(4), 776–781.

Wang, M.; Carver, J. J.; Phelan, V. V.; Sanchez, L. M.; Garg, N.; Peng, Y.; Nguyen, D. D.; et al. 2016. Sharing and community curation of mass spectrometry data with global natural products social molecular networking. Nature Biotechnology, 34, 828–837.

Wink, M. 2016. Alkaloids: properties and determination. In Encyclopedia of Food Sciences and Nutrition. Heidelberg: Heidelberg University.

Zuleta, L. M. C.; Cavalheiro, A. J.; Silva, D. H. S.; Furlan, M.; Marx, Y. M. C.; Albuquerque, S.; Castro-Gamboa, I.; Bolzani, V. S. 2003. *seco*-Iridoids from *Calycophyllum spruceanum* (Rubiaceae). Phytochemistry, 64(2), 549–553.

4 Molecular Biology Tools to Boost the Production of Natural Products
Potential Applications for Brazilian Medicinal Plants

Luzia Valentina Modolo, Samuel Chaves-Silva,
Thamara Ferreira da Silva, and Cristiane Jovelina da-Silva
Departamento de Botânica, Instituto de Ciências Biológicas,
Universidade Federal de Minas Gerais, Belo Horizonte, Brazil

CONTENTS

4.1 INTRODUCTION

The ability to synthesize a variety of chemically diverse metabolites makes plants a potential source of inspiration to produce valuable compounds. Natural products have been used since ancient times in fragrances as pigments, as pesticides and as therapeutics (Facchini et al., 2012). Even though, potentially, many more secondary metabolites are yet to be discovered, of 200,000–1,000,000 bioactive substances estimated to occur in the plant kingdom (Afendi et al., 2012; Dixon and Strack, 2003).

A known fact is that many plants synthesize valuable compounds in amounts that are usually not enough to meet the various commercial demands, thus, molecular biology is an interesting approach for the large-scale production of structurally complex substances. In fact, the advance in sequencing technologies has boosted research on plant genomics in recent years. The development of bioinformatic tools for the analysis of transcriptional data has been of considerable help in studies on evolution, organization and gene expression regulation. The generation of metabolite libraries and elucidation of species-specific biosynthetic pathways beyond those reported for model plants such as *Arabidopsis thaliana* (Brassicaceae), *Medicago truncatula* (Fabaceae) or

Oryza sativa (Poaceae) serve to expand the knowledge on the biosynthesis of natural products in medicinal plants (Unamba et al., 2015). Once identified, the genes involved in the biosynthesis of a valuable substance, new biotechnological approaches such as synthetic biology, genome editing and reverse genetics can then assist the manipulation of biosynthetic pathways *in planta* or in a host (e.g. bacteria, yeast, insect cells, cell culture, hairy root culture, etc.). This approach can lead to improve the production of desired compounds. The following topics provide details on some technological approaches based on molecular biology that are useful to enrich the accumulation of natural products in living cells.

4.2   FORWARD GENETICS

The biosynthesis of secondary metabolites in plants is a complex process controlled by genetics and environment (Kessler and Kalske, 2018). The identification of genes involved in a pathway is the very first step to rationally modulate plant secondary metabolism (Yang et al., 2014). Prior to the advent of whole genome sequencing, forward genetics was predominantly used to identify the molecular basis associated with a trait of interest (Tierney and Lamour, 2005). From the observation of a particular phenotype originating from a naturally occurring mutation, one can identify the genes involved in that trait. This approach is based on "phenotype to gene" determinations. The observable variation (phenotype) may also be purposely induced in cells or an organism using a DNA mutagen. The investigator eventually ends up sequencing the gene or genes thought to be involved in a certain biosynthetic pathway related to the observable trait (Figure 4.1). The main strategies used in forward genetics are map-based cloning and mutational breeding. The former consists of mapping of a biparental population based on recombination frequency during meiosis and identifies the underlying genetic cause of a mutant phenotype. Mutational breeding, on the other hand, induces mutation to generate various phenotypes to identify candidate genes. This technique is used to obtain new genetic combinations, without changing the major total genetic setup of an organism (Abbai et al., 2017). Both techniques are widely used in the study of medicinal plants: recognizing there is the disadvantage that only one trait can be analyzed at a time, some genes may be missed in the screening and the procedure is feasible only in plants that are amenable to *Agrobacterium* transformation. Examples of mutagens used in forward genetics include X-rays and ethylmethanesulfonate (EMS), in which the former may generate large deletions in the chromosomes or chromosomes rearrangements (it induces point mutation while the latter causes point mutation – changes at a single nucleotide position).

4.3   GENOME EDITING AND REVERSE GENETICS

In the early 2000s, a large repository of sequence data was created with the disclosure of *A. thaliana* and *O. sativa* (rice) genomes (Arabidopsis Genome Initiative, 2000; Goff et al., 2002; Yu et al., 2002), among others. Nowadays, at the post-genomic stage, technology has advanced, the cost to generate omics data has been reduced and a set of computational tools for data analysis has been developed, leading to the annotation and identification of thousands of genes (Abbai et al., 2017). Although nucleotide sequences of several plant genes have been revealed, the function of most of them and the involvement in specific networks remains to be elucidated. Thus, the current challenge in plant research is the analysis of gene function. Therefore, genome editing techniques is a useful tool to achieve this goal.

Genome editing is a resource used to make changes in specific regions of a genome (e.g. insertion, substitution or deletion of DNA fragments) of a cell or organism for several purposes. This technique is based on the cleavage of the DNA double strand in targeted regions followed by the use of the own cell repair system to introduce precise mutations (Tan et al., 2018). In this scenario, reverse genetics (from gene to mutant phenotype) arises as a powerful tool to unravel gene function (Alonso and Ecker, 2006). In reverse genetics, a specific gene or gene product is disrupted or

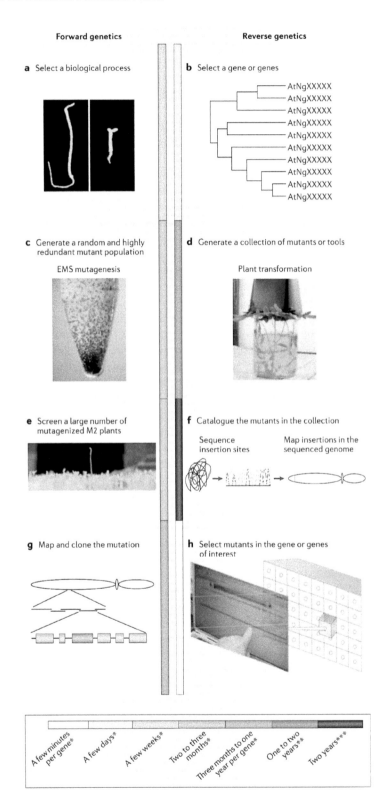

FIGURE 4.1 Some differences between forward and reverse genetics. Reproduced from Alonso and Ecker (2006) with permission by Springer Nature.

modified, and the plant phenotype is consequently determined (Figure 4.1; Tierney and Lamour, 2005). Some strategies of reverse genetics are described in the literature. Among them are included those that rely on the use of zinc-finger nucleases (ZFNs), transcription activator-like effector nucleases (TALENs), homologous recombination, RNAi, T-DNA insertional mutagenesis, targeting induced local lesions in genome (TILLING) and clustered, regularly interspaced, short palindromic repeat-Cas9 (CRISPR-Cas9) (Abbai et al., 2017).

The first technologies developed to break down DNA focused on the use of ZFNs and TALENs, artificial enzymes generated by the fusion of DNA-binding domains to a nonspecific cleavage domain of the bacterial endonuclease FokI produced by *Flavobacterium okeanokoites* (Li et al., 2011). Such enzymes, however, were shown to introduce additional mutations in regions of the DNA different from the target one (Zych et al., 2018). Furthermore, the process of generating customized enzymes by specialized enterprises is time consuming, which makes the first-generation technology expensive. TALEN was first applied to plants to generate *Solanum tuberosum* lines with lower levels of both cholesterol and the toxic glycoalkaloids, such as α-solanine and α-chaconine (Sawai et al., 2014).

Homologous recombination is a technique based on the similarity of sequence between the host gene and the gene to be replaced. This method is simple and site-specific, but nevertheless works better for less complex organisms (Abbai et al., 2017). Few studies highlighted the success of homologous recombination in crops (Iida and Terada, 2004), but no work using this technique on medicinal plants has been reported up to date. Nevertheless, homologous recombination can be considered an important tool for synthetic biology in the coming years (Alonso and Ecker, 2006).

The RNA interference (RNAi) or posttranscriptional gene silencing technology has become an important tool to speed up the breeding of medicinal plants, from which a conventional mutation breeding approach was shown to fail (Allen et al., 2004). The RNAi works by knocking down the expression of the target gene (Abbai et al., 2017). This provides an alternative to block the activity of enzymes that are encoded by a multigene family and are expressed in different plant tissues at distinct developmental stages. This technique has been employed to modulate the biosynthesis of morphine-like alkaloids (psychoactive drugs) by interfering with the activity of codeinone reductase (Allen et al., 2004). The gene that encodes for codeinone reductase was knocked down in *Papaver somniferum* plants (opium poppy) through DNA-directed RNAi, which resulted in the accumulation of (*S*)-reticuline, the precursor of isoquinoline alkaloid biosynthesis, at the expense of morphine, codeine, oripavine and thebaine (Allen et al., 2004). This same technique was used to block the activity of the berberine bridge enzyme in *California poppy* culture cells, also resulting in the accumulation of (*S*)-reticuline (Fujii et al., 2007). Recently, RNAi technology was used to elucidate the role of cytochrome CYP76AH1 in the metabolism of hairy roots of *Salvia miltiorrhiza*. The silencing of the *CYP76AH1* gene affected the production of tanshinones (Ma et al., 2016). Therefore, this gene is a potential target for metabolic engineering in medicinal plants.

The T-DNA vector is used to insert DNA fragments, in a completely random way, into the target genome. During transformation, there is a differential loss of several T-DNA genes (Abbai et al., 2017). This loss can affect the growth and morphological patterns of hairy roots, expression of biosynthetic pathway genes and accumulation of specific metabolites. In addition, the variability in different insertion lines can be used to select the lines that are better producers of a desired metabolite. One disadvantage of this technique is that it can be applied only for transformation and tissue culture friendly plant species (Abbai et al., 2017). However, the T-DNA insertional mutagenesis technique is most commonly used for reverse genetics to produce secondary metabolites. This approach increased the production of alkaloids in Solanaceae (Moyano et al., 1999), ginkgolides in Ginkgoaceae (Ayadi and Trémouillaux-Guiller, 2003), isoflavones in Fabaceae (Shinde et al., 2009), plumbagin in Droseraceae (Putalun et al., 2010) and phenols in Lamiaceae (Sitarek et al., 2018) plants.

Targeting induced local lesions in genome (TILLING) or chemical mutagenesis is a transformation-free functional genomic technique. This technique comprises an alternative to apply on plants that are not conducive to transformation or production of tissue cultures (Abbai et al., 2017). Two of the most widely used mutagens in this technique are EMS (also used in forward genetics) and ethylnitrosourea (ENU). The EMS is the most common mutagen applied to plants that functions by alkylating guanine bases. The alkylated guanine will then pair with thymine instead of the preferred cytosine base, ultimately resulting in a G/C to A/T transition (Tierney and Lamour, 2005). The ENU is a more potent mutagen than SEM and induces point mutations. It is also an alkylating agent that transfers an ethyl group to oxygen or nitrogen atoms present in the DNA structure. Such alkylation leads to mispairing, base pair substitutions and even base pair losses (Tierney and Lamour, 2005). A variant of TILLING is the EcoTILLING, in which the natural population is the starting material (Abbai et al., 2017). The TILLING technology has already been used to increase production of triterpenoid saponins in soybean (*Glycine max*) by a loss-of-function mutation in a gene that encodes for the cytochrome P450 CY72A69 (Yano et al., 2017). The production of cyanogenic glucosides decreased in *Sorghum bicolor* by using this same technique (Blomstedt et al., 2012). Similarly, the EMS TILLING was useful to generate *Catharanthus roseus* plants with new alkaloid profiles, accumulating mainly intermediates of vindoline biosynthesis (Edge et al., 2018).

The main current disadvantage of using reverse genetics in medicinal plant research is that not all techniques related to this approach can be applied to all organisms. This is because many valuable secondary metabolites are produced by exotic plant species or in some cases by woody species. Hence, the available methods of reverse genetics may not be suitable for such species due to unavailability of transgenesis protocols, besides unreasonable time scales (Gandhi et al., 2015). For organisms that are difficult to be transformed, TILLING appears to be an interesting alterative, unless the genome of the target plant is riddled with mutations that make hard the detection of mutational phenotypes (Tierney and Lamour, 2005). Despite the drawback, reverse genetics can be improved since it is promising for boosting the production of metabolites of pharmacological interest. It may be used to activate naturally silenced routes: the "silence part" of a pathway to allow for the accumulation of intermediate valuable substances and/or silence routes to obtain cleaner and contaminant-free desired metabolites.

A new generation genome editing technology called clustered, regularly interspaced, short palindromic repeat-Cas9 (CRISPR-Cas9) has been recently developed and does not rely on the use of customized restriction enzymes. More details about this technology will be given in the next topic of this chapter since this know-how can be used for both forward and reverse genetics.

The current challenge for the broad application of available genome editing technologies to boost natural products production is the establishment of protocols for transformation and regeneration of plant species other than the model ones.

4.4 GENE EDITION BASED ON CRISPR/CAS9 TECHNOLOGY

In CRISPR/Cas9 technology, the cleavage of double-strand DNA is assisted by a single nuclease (Cas9) that is guided to the target by a small RNA through Watson–Crick base pairing (Gasiunas et al., 2012). This method is of relatively low cost, versatile, easy to perform, highly specific and efficient that can be used for both forward and reverse genetics. This gene editing system enables the generation of an organism with a "clean" modified genome, which would essentially make it a non-genetically modified organism (Abbai et al., 2017). CRISPR-Cas9 technique was developed from the observation of the defense system, naturally occurring in bacteria. One of the defense mechanisms exhibited by prokaryotes is the regularly interspaced short palindromic repeats (CRISPR). The locus received this name because it is constituted of sequence repeats of 29 base pairs separated by variable sequences of 32 nucleotides referred to as spaces (Ishino et al., 1987). The sequencing of some bacteria and virus genomes shed light on the compatibility of the CRISPR spaces with phages and plasmids. Currently, it is known that CRISPR is involved in a protection system highly conserved among prokaryotes: 45% of bacteria and 90% of Archaea possess CRISPR locus (Nemudryi et al., 2014). For instance, bacteria capture DNA fragments from an invading virus to generate small

DNA sequences called CRISPR arrays that will function as a "memory" of the pathogen attack for future self-defense, in case the virus and related pathogens try to infect the bacteria again. The bacterial enzyme Cas9, breaks down the viral DNA to prevent its action (Barrangou et al., 2007). The system CRISPR-Cas9, for the sake of genome editing, works in a similar way. First, it is synthesized as a short RNA sequence (about 20 base pair long) containing a guiding sequence that is complementary to a specific sequence of the DNA to be edited. A complex formed between Cas9 and the short RNA sequence then "search" for the complementary region in the target DNA and intercalates the DNA double strand at that point to indicate the region where Cas9 is required to break down (Hsu et al., 2014). The host cell repairing machinery can take care of the damaged DNA, adding or suppressing DNA fragments or even substituting DNA fragments for customized DNA. The DNA repair may occur by two mechanisms: (1) nonhomologous end-joining or (2) homology-directed repair (Zych et al., 2018). In the first mechanism, the DNA ends become adjacent to recombine without a template, which in turn can lead to insertions/deletions, altering the gene open reading frame. As for the homology-directed repair strategy, disrupted sequences are resynthesized, using as a template a homologous sequence throughout the genome (Wyman and Kanaar, 2006). Thus, homology-directed repair is useful to promote site-directed genome editions, while nonhomologous end-joining is used to rearrange chromosomes or generate functional knockouts (Montano et al., 2018; Tan et al., 2018).

The most common employed CRISPR/Cas9 system is the one that uses the endonuclease Cas9 from *Streptococcus pyogenes* and a chimeric single guide RNA (sgRNA) to direct the endonuclease to the target. The cleavage sites need to be near a sequence termed PAM (Protospacer Adjacent Motif; 5′-NGG-3′). This can be achieved by customizing the 5′ region of sgRNA so that it can reach any genome sequence near PAM (Cong et al., 2013). In addition, multiples genes can be edited simultaneously by using several sgRNAs.

Since the beginning of the 2010s, the CRISPR/Cas9 system has efficiently edited specific genes in bacteria (Jiang et al., 2013), mice (Yin et al., 2014), human cells (Cong et al., 2013) and plants (Ito et al., 2015). In the scope of plants, CRISPR/Cas9 has been used to edit genes in crops such as *O. sativa, Solanum lycopersicum, Zea mays* and *Triticum aestivum* (Mishra and Zhao, 2018). This technology was employed to increase the levels of γ-aminobutyric acid (GABA) in *S. lycopersicum* (Nonaka et al., 2017). GABA is a non-proteinogenic amino acid that possesses hypotensive properties.

Such success opens a window for the application of CRISPR/Cas9 on medicinal plants as well. The efficiency of CRISPR/Cas9 system on medicinal plants was first demonstrated in a study with opium poppy. CRISPR-Cas9 system was used to knock out an *O*-methyltransferase gene (*4′OMT2*) involved in the biosynthesis of benzylisoquinoline alkaloids. This significantly decreased the production of alkaloids, such as thebaine, codeine, noscapine and papaverine, in the transgenic plants. Furthermore, a novel uncharacterized alkaloid was observed only in CRISPR/Cas9 edited plants, demonstrating how this technique is useful for metabolic engineering and the discovery of new compounds in genome-edited plants (Alagoz et al., 2016). Furthermore, the CRISPR-Cas9 technology efficiently knocked out a diterpene synthase gene (*SmCPS1*) involved in a committed step of tanshinones biosynthesis in *S. miltiorrhiza* (Chinese medicinal plant), such as cryptotanshinone, tanshinone IIA and tanshinone I. The use of this technique decreased the amount of the target metabolites, without interfering with the biosynthesis of other phenolic acid metabolites (Li et al., 2017). Likewise, the gene *SmRAS*, which encodes for rosmarinic acid synthase, was silenced in the same species using the CRISPR/Cas9 system, causing disruption in the production of rosmarinic acid and lithospermic acid B (Zhou et al., 2018).

4.5   SYNTHETIC BIOLOGY

Synthetic biology is a transdisciplinary science based on the knowledge on biology, chemistry, engineering, physics and informatics used to further understand how living systems function to redesign them for specific purposes (Moses and Goossens, 2017). A milestone article on synthetic biology was published in 2010, in which scientists successfully created the bacteria *Mycoplasma mycoides* JCVI-syn1.0 controlled by a synthetic genome (Gibson et al., 2010). This was the beginning of a most ambitious goal – to transform bacteria, yeast, algae and virus in synthetic organisms (bearing synthetic genomes) to perform specific functions such as the sustainable production of valuable molecules and biomaterials. Researchers who work on this branch of science use and/or modify techniques related to genetic engineering, microbiology and bioinformatics to design, synthesize and transfer DNAs to microorganisms (van der Helm et al., 2018). In a manner analogous to a computer, the microorganisms would be the hardware, while the synthetic DNA, the software. Idealized in a virtual environment, the synthetic DNA contains "scripts", a series of programmable commands, that once integrated, will result in several responses by the microorganism that host the synthetic DNA (Wohlsen, 2011). Synthetic biology has the potential to revolutionize the production of plant-derived natural products from unicellular organisms (Moses and Goossens, 2017). Once a biosynthetic pathway is well characterized from the genetic point of view, synthetic biology can be used to introduce such pathways in heterologous expression systems such as *Saccharomyces cerevisiae* or *Escherichia coli*. Such organisms are of relatively easy maintenance in the laboratory, thus the shorter life cycle when compared, for instance, with whole medicinal plants. In this context, microorganisms are more advantageous when taking into account the production of natural products on a large scale (Moses and Goossens, 2017).

One of the most notable examples of a successful application of synthetic biology to produce plant metabolites in microorganisms was the production of artemisinic acid (precursor of the antimalarial agent, artemisinin) in *S. cerevisiae* (Paddon et al., 2013). Briefly, the genes that encode for the enzymes of the mevalonate pathway were over-expressed in yeast cells together with the gene of *Artemisia annua* that encodes for amorphadiene synthase. In *A. annua*, amorphadiene formed from the activity of amorphadiene synthase is further converted to artemisinic acid through an oxidative process that involves three steps. Based on this work, the following genes were inserted in the yeast: *CYP71AV1*, *CPR1* and *CYB5* (they encode for enzymes involved in the production of artemisinic alcohol from amorphadiene); *ADH1* (it encodes for an enzyme that oxidizes artemisinic alcohol to artemisinic aldehyde) and *ALDH1* (it encodes for an enzyme that oxidizes artemisinic aldehyde to artemisinic acid).

By using this yeast strain, engineered via synthetic biology, the global biopharmaceutical company Sanofi started the large scale production of artemisinic acid in 2013/2014 (Paddon and Keasling, 2014; Peplow, 2013). Notably, the use of such a biotechnological approach can increase by over 30% the production of artemisinic acid to meet global demand for the antimalarial artemisinin.

The pharmacological proprieties of plant natural products belonging to the class of benzylisoquinoline alkaloids (BIAs) have caught the attention of synthetic biologists. For instance, BIAs such as oxycodone, hydrocodone and hydromorphine are opioid analgesics supplied by pharmaceutical companies using semi-synthesis approaches. It was recently estimated that *P. somniferum* (opium poppy) was cultivated in approximately 100,000 hectares to obtain 800 tons of thebaine or morphine (natural precursors of semisynthetic BIAs) to meet the medical demand of analgesic opiods) (Galanie et al., 2015). Although some synthetic routes to provide morphine and derivatives are disclosed, none of them are commercially competitive or viable in large scale compared to the semisynthetic approach (Reed and Hudlicky, 2015). Efforts in synthetic biology have been made since the end of the 2000s to produce BIAs in microorganisms. *S. cerevisiae* was genetically modified to produce reticuline, a key intermediate of BIA's biosynthesis, from the commercially available *(R,S)*-norlaudanosoline (Hawkins and Smolke, 2008). A few years later, a fermentation system constituted from *E. coli* was developed to produce reticuline from simpler and cheaper carbon sources (Nakagawa et al., 2012). In 2014, researchers achieved the introduction of ten plant genes in *S. cerevisiae*, which in turn, resulted in the production of dihydrosanguinarine and sanguinarine, BIAs of notable antimicrobe and antineoplasic activities (Fossati et al., 2014). Additionally, 21 and 23 genes (of plants, mammalians and bacteria origin) were introduced to yeast strains to make them competent to produce thebaine and hydrocodone, respectively, from sugar (Galanie et al. 2015).

Another natural product group of medicinal interest contains the monoterpene indole alkaloids (MIAs), represented, but not limited to, the anticancer agents vinblastine, vincristine and vinflunine (Leggans et al., 2013). Strictosidine is the common intermediate to produce the structurally diverse MIAs. The insertion of 21 genes (from which 14 are of known to participate in the biosynthesis of MIAs) in *S. cerevisiae* and deletion of three genes from the yeast genome resulted in the production of strictosidine by the altered yeast cells (Brown et al., 2015).

Other plant natural products such as glycyrrhetinic acid (Seki et al., 2011), taxadien (precursor of the anticancer agent paclitaxel; Ajikumar et al., 2010), sapogenins and saponins (Moses et al., 2014) were successfully produced in genetically engineered yeast or *E. coli*.

The upcoming challenges will be the improvement of tools to make viable the synchronized expression of multiple genes, reduction of metabolic loads on the host and the development of microorganisms with increased efficiency to produce xenobiotics (Moses and Goossens, 2017). Although *E. coli* and *S. cerevisiae* have been used as the host to produce valuable phytochemicals, other organisms might prove to be more efficient for large scale production purposes. Furthermore, one might consider that a great microbial system is the one that transforms substrates of renewable sources or industrial byproducts to metabolites of multiple interests (Eisenstein, 2016; Rai et al., 2017).

4.6 MOLECULAR BIOLOGY IN THE CONTEXT OF BRAZILIAN MEDICINAL PLANTS

According to the Convention on Biological Diversity, Brazil hosts the greatest number of endemic species on the planet, sheltering an estimated biota of over 170 thousand species, which corresponds to about 13.1% of the world's known wealth. Endemism rates as high as 55% of plant biodiversity are reported from Brazilian biomes (Stehmann and Sobral, 2017). Many natural products originating from Brazilian medicinal plants have been disclosed. They are either used as phytoterapics or as inspiration source for the design of valuable substances of pharmacological, cosmetic, agronomic and supplemental food interests (Fougat et al., 2015). Indeed, over 60% of the anticancer drugs are derived from natural products (Cragg et al., 1997). Among them, β-lapachone and lapachol, used for the treatment of various neoplasms, are extracted from the bark of *Handroanthus impetiginosus*, native to Brazil (Melo et al., 2011). Many other commercial phytopharmaceuticals are known to be extracted from the Brazilian native flora. Quercetin, used in treatment of heart disease, is extracted from *Dimorphandra mollis*, a native tree of Cerrado. One of the main drugs used for the treatment of chronic glaucoma is pilocarpine, a phytopharmaceutical used in treatment of glaucoma and extracted from *Pilocarpos spp.* Pilocarpine is also used to treat xerostomia in patients undergoing radiotherapy for head and neck cancer, since it stimulates the secretion of large amounts of saliva and sweat (Nogueira et al., 2010). Some of these compounds, such as D-tubocurarine (extracted from *Chondrodendron tomentosum*) and emetine (isolated from *Carapichea ipecacuanha*) are supplied by foreign pharmaceutical companies as anesthetics and vomiting inducer, respectively (Figure 4.2; Nogueira et al., 2010). Other examples include, but are not limited to, emetine, pilocarpine, rupununine, D-tubocuranine and vitexin (Figure 4.2).

Despite the great biodiversity, only 8% of Brazilian plant species were so far investigated for bioactive compounds and relatively few of them had their medicinal properties evaluated (Simões et al., 2003). Additionally, most medicinal plants used in the preparation of medicines are exotic species brought to Brazil during the country's colonization (Brandão et al., 2009). Nevertheless, many efforts have been made to voucher native species for the purpose of valuing and prioritizing the research with Brazilian flora wealth. Indeed, a database and samples of aromatic, medicinal and toxic plants namely DATAPLAMT was created in the beginning of the 2000s, in which a lot of information about native medicinal plants was compiled, mainly, for species located in the Minas Gerais (Southeast Brazil). Some examples of bioactive compounds identified in plants

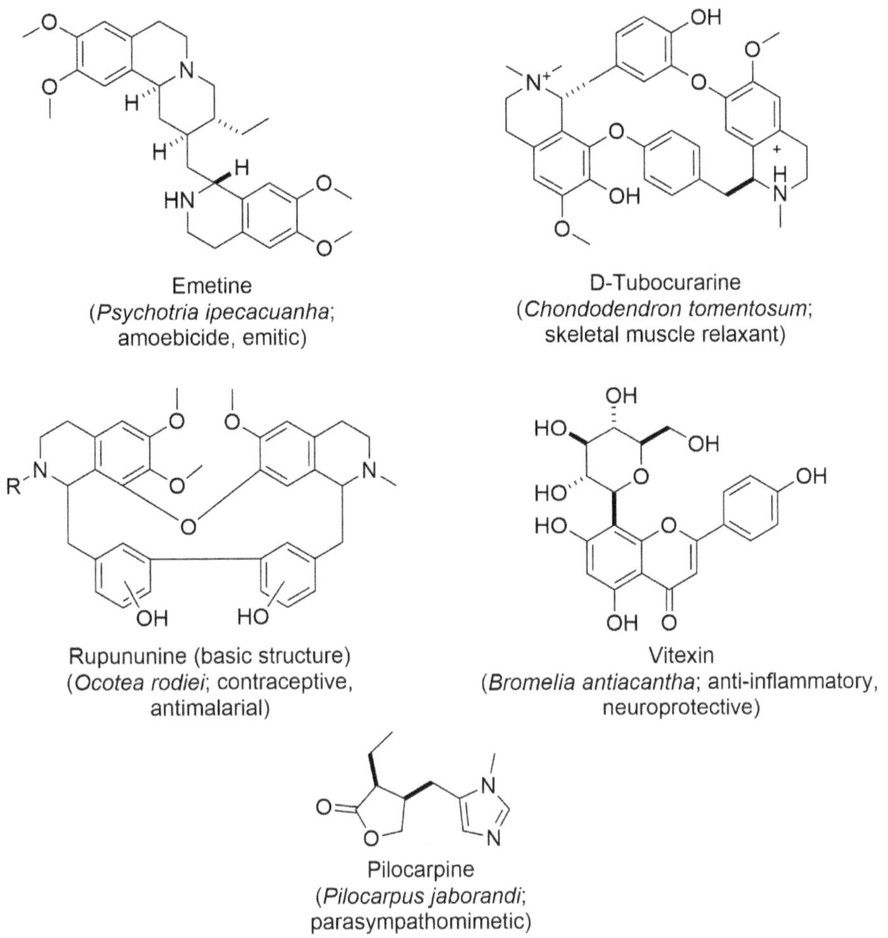

Emetine
(*Psychotria ipecacuanha*;
amoebicide, emitic)

D-Tubocurarine
(*Chondodendron tomentosum*;
skeletal muscle relaxant)

Rupununine (basic structure)
(*Ocotea rodiei*; contraceptive,
antimalarial)

Vitexin
(*Bromelia antiacantha*; anti-inflammatory,
neuroprotective)

Pilocarpine
(*Pilocarpus jaborandi*;
parasympathomimetic)

FIGURE 4.2 Structure of some notable pharmaceuticals produced by plant species native to Brazil. The indicated pharmacological properties are described elsewhere (He et al., 2016; Nogueira et al., 2010; www.thoughtco.com/drugs-and-medicine-made-from-plants-608413).

species belonging to the different Brazilian biomes are shown in Table 4.1, while Figure 4.3 shows representative images of Brazilian medicinal plants that occurs in Amazon Forest, Atlantic Forest, Caatinga, Cerrado, Pampa and Pantanal.

4.7 USE OF MOLECULAR BIOLOGY TO IMPROVE THE PRODUCTION OF VALUABLE METABOLITES IN MEDICINAL PLANTS

The great advance in genomics integrated with high-resolution metabolomics became a powerful tool to investigate structural aspects and the regulation of secondary metabolism in organisms (Kim and Buell, 2015; Scossa et al., 2018). These approaches allow increasing the number of cells that produce a valuable natural product, increasing the carbon flux toward a particular biosynthetic pathway by overexpressing key genes, blocking feedback inhibition, among others (Karuppusamy, 2009). This can be better exemplified by the improvement of the accumulation of the alkaloid chemotherapeutics, vinblastine and vincristine in *C. roseus* hairy roots (Wang et al., 2010). Both vinblastine and vincristine are produced in very low levels in *C. roseus* leaves. However, the overexpression of *G10H* and *ORCA3* that encodes for geraniol 10-hydroxylase and octadecanoid-derivative responsive

TABLE 4.1

Some Bioactive Compounds Isolated from Plant Species Native to the Brazilian Amazon Forest, Atlantic Forest, Caatinga, Cerrado, Pampa and Pantanal

Family	Scientific Name	Common Names	Bioactive Compounds	Brazilian Phytogeographical Domains	Reference
Adoxaceae	*Sambucus australis* (sin. *Sambucus pentagynia*)	Sabugueiro, acapora	Ursolic acid	Atlantic Forest	Rao et al. (2011)
Anacardiaceae	*Anacardium occidentale* (sin. *Anacardium microcarpum*)	Cajueiro	Catechin, epicatechin	Amazon and Atlantic Forests, Cerrado, Caatinga	Trox et al. (2011)
Anacardiaceae	*Myracrodruon urundeuva*	Aroeira	Artemiseole, bergamotene, terpinolene	Atlantic Forest, Caatinga, Cerrado	Figueredo et al. (2014)
Annonaceae	*Annona crassiflora* (sin. *Annona macrocarpa*)	Araticum	Epicatechin, peltatoside, quercetin	Cerrado	Lage et al. (2014)
Aquifoliaceae	*Ilex paraguariensis* (sin. *Ilex curitibensis*)	Erva-mate	Theobromine, theophylline	Atlantic Forest, Caatinga, Cerrado	Heck and de Mejia (2007)
Asteraceae	*Achyrocline satureioides* (sin. *Gnaphalium saturejaefolium*)	Marcela	Quercitrin, isoquercitrin, luteolin	Atlantic Forest, Cerrado, Pampa	de souza et al. (2007)
Asteraceae	*Mikania laevigata*	Guaco	Campestero, taraxasterol	Atlantic Forest, Cerrado, Pampa	Ferreira and Oliveira (2010)
Asteraceae	*Baccharis genistelloides* var. *trimera* (sin. *Baccharis trimera*)	Carqueja	Bicyclogermacrene, caryophyllene, germacrene D	Atlantic Forest, Pampa, Cerrado	de Oliveira et al. (2012)
Asteraceae	*Eremanthus arboreus* (sin. *Vanillosmopsis arborea*)	Candeeiro	Bisabolol	Caatinga	Matos et al. (1988)
Bignoniaceae	*Handroanthus impetiginosus* (sin. *Tabebuia impetiginosa*)	Ipê-roxo	Lapachol, β-lapachona	Cerrado	Gupta et al. (2002)
Boraginaceae	*Cordia curassavica* (sin. *Varronia curassavica*)	Erva-baleeira	Sabinene, δ-elemene, α-gurjunene, hellandrene	Amazon Forest, Atlantic Forest, Caatinga, Cerrado	Nizio et al. (2015)
Bromeliaceae	*Bromelia antiacantha* (sin. *Hechtia longifolia*)	Gravatá	Chrisin, hesperidin, hyperoside, orientin, quercetin, rutin, vitexin	Atlantic Forest Pampa	Santos et al. (2009)
Celastraceae	*Maytenus ilicifolia* (sin. *Monteverdia truncata*)	Espinheira-santa	Erythrodiol, oxotingenol, pristimerin	Caatinga	Ohsaki et al. (2004)
Cucurbitaceae	*Wilbrandia ebracteata* (sin. *Wilbrandia ebracteata var. ebracteata*)	Taiuiá	Cucurbitacin	Atlantic Forest, Cerrado	Peters et al. (1999)

(Continued)

TABLE 4.1 *(Continued)*

Some Bioactive Compounds Isolated from Plant Species Native to the Brazilian Amazon Forest, Atlantic Forest, Caatinga, Cerrado, Pampa and Pantanal

Family	Scientific Name	Common Names	Bioactive Compounds	Brazilian Phytogeographical Domains	Reference
Equisetaceae	*Equisetum giganteum* (sin. *Equisetum bolivianum*)	Cavalinha	Caffeic acid, kaempferol	Atlantic Forest, Cerrado	Jabeur et al. (2017)
Euphorbiaceae	*Manihot esculenta* (sin. *Manihot flexuosa*)	Mandioca	Esculentoic acids, linamarin	Amazon and Atlantic Forests, Caatinga, Cerrado, Pantanal	Idibie et al. (2007), Chaturvedula et al. (2003)
Euphorbiaceae	*Croton grewioides* (sin. *Croton zehntneri*)	Cunha	Anethole, caryophyllene, estragole, myrcene	Caatinga	Donati et al. (2015)
Fabaceae	*Copaifera langsdorffii* (sin. *Copaifera nitida*)	Copaíba, copaibeira pau-de-óleo	Kaurenoic acid	Atlantic Forests, Cerrado, Caatinga	Costa-Lotufo et al. (2002)
Fabaceae	*Hymenaea courbaril* (sin. *Hymenaea multiflora*)	Jatobá	Halimane, diterpenoids	Amazon and Atlantic Forests, Cerrado, Caatinga, Pantanal	Abdel-Kader et al. (2002)
Fabaceae	*Stryphnodendron adstringens* (sin. *Stryphnodendron barbatimam*)	Barbatimão	Epigallocatechin, gallocatechin,	Caatinga Cerrado	de Mello et al. (1996)
Fabaceae	*Dimorphandra mollis* (sin. *Ocotea rodiei*)	Faveiro	Rutin	Amazon Forest, Cerrado, Pantanal	Lucci and Mazzafera (2009)
Lauraceae	*Aniba rosaeodora (sin. Aniba duckei)*	Pau-rosa	Linalool	Amazon Forest	d'Acampora-Zellner et al. (2006)
Lauraceae	*Chlorocardium rodiei* (sin. *Ocotea rodiei*)	Bibiri	Rupununine	Amazon Forest	Gorinsky (1996)
Lauraceae	*Ocotea odorifera* (sin. *Ocotea pretiosa*)	Canela-sassafraz	Camphor, safrole	Atlantic Forest, Cerrado	Mossi et al. (2013)
Meliaceae	*Trichilia catigua* (sin. *Trichilia alba*)	Catuaba	Flavalignans	Amazon and Atlantic Forests, Caatinga, Cerrado, Pantanal	Pizzolatti et al. (2002)
Menispermaceae	*Chondrodendron tomentosum* (sin. *Chondrodendron hypoleucum*)	Curare, parreira-brava, uva-da-serra, uva-do-mato	D-Tubocurarin	Amazon Forest	King (1940)
Menispermaceae	*Cissampelos sympodialis*	Jarrinha, orelha-de-onça, abuteira, milona	Desmethylroraime, roraimine	Amazon and Atlantic Forests, Caatinga	de Lira et al. (2002)
Moraceae	*Brosimum glaziovii* (sin. *Alicastrum glaziovii*)	Leiteira	Campesterol, lupenone, stigmasterol	Atlantic Forest	Coqueiro et al. (2014)

(Continued)

TABLE 4.1 *(Continued)*

Some Bioactive Compounds Isolated from Plant Species Native to the Brazilian Amazon Forest, Atlantic Forest, Caatinga, Cerrado, Pampa and Pantanal

Family	Scientific Name	Common Names	Bioactive Compounds	Brazilian Phytogeographical Domains	Reference
Nyctaginaceae	*Boerhavia diffusa* (sin. *Boerhavia caespitosa*)	Pega-pinto	Boeravinones, punarnavine	Amazon and Atlantic Forests, Caatinga, Cerrado	Ahmed-Belkacem et al. (2007), Manu and Kuttan (2009)
Piperaceae	*Piper hispidum* (sin. *Piper hispidinervum*)	Pimenta-longa	Safrole	Amazon and Atlantic Forests, Caatinga, Cerrado	Estrela et al. (2006)
Plumbaginaceae	*Plumbago zeylanica* (sin. *Plumbago scandens*)	Caataia, folha-de-louro, queimadura	Plumbagin	Amazon and Atlantic Forests, Caatinga	Paiva et al. (2004)
Rubiaceae	*Carapichea ipecacuanha* (sin. *Psychotria ipecacuanha*)	Ipecacuanha	Emetine, ephaeline	Amazon and Atlantic Forests, Caatinga, Cerrado	Alves et al. (2005)
Rubiaceae	*Genipa americana* (sin. *Genipa venosa*)	Genipapo	Iridoids, genipin, gardendiol, shanzhiside	Amazon and Atlantic Forests, Caatinga, Cerrado, Pantanal	Ono et al. (2007)
Rubiaceae	*Uncaria tomentosa* (sin. *Ourouparia tomentosa*)	Unha-de-gato	Isopteropodine, speciophylline, Akuammigine, hirsuteine	Amazon Forest	Laus et al. (1997)
Rutaceae	*Pilocarpus microphyllus*	Jaborandi	Pilocarpine	Amazon Forest	Sawaya et al. (2011)
Salicaceae	*Casearia sylvestris* (sin. *Casearia subsessiliflora*)	Guaçatonga	Casearvestrins	Amazon and Atlantic Forests, Caatinga, Cerrado, Pampa, Pantanal	Oberlies et al. (2002)
Sapindaceae	*Paullinia cupana* (sin. *Paullinia sorbilis*)	Guaraná	Caffeine, catechin, epicatechin	Amazon Forest	Marques et al. (2016)
Solanaceae	*Solanum mauritianum* (sin. *Solanum tabacifolium*)	Fumo-bravo	Caulophyllumine	Atlantic Forest	Jayakumar et al. (2016)
Urticaceae	*Cecropia pachystachya* (sin. *Cecropia catarinensis*)	Embaúba	Orientin, apigenin, luteolin	Amazon and Atlantic Forests, Caatinga, Cerrado, Pantanal	Cruz et al. (2013)
Verbenaceae	*Lippia origanoides* (sin. *Lippia schomburgkiana*)	Alecrim-pimenta	Thymol, carvacrol	Amazon and Atlantic Forests, Caatinga, Cerrado	Botelho et al. (2007)
Winteraceae	*Drimys brasiliensis* (sin. *Drimys retorta*)	Casca-d'anta	Polygodial, drimanial	Atlantic Forest, Caatinga, Cerrado	Malheiros et al. (2005)

Anacardium occidentale   *Bromelia antiacantha*   *Hymenaea courbaril*

FIGURE 4.3 Brazilian medicinal plant species *Anacardium occidentale, Bromelia antiacantha* and *Hymenaea courbaril*. *A. occidentale* is reported in Caatinga biome, *B. antiacantha* occurs in the Pampa and Atlantic Forest biomes, whereas *H. courbaril* is found in Amazon and Atlantic Forests, Cerrado and Pantanal. *A. occidentale* and *H. courbaril* images were kindly provided by Dr. João Renato Stehmann (Federal University of Minas Gerais, Brazil). Dr. Mara Rejane Ritter (Federal University of Rio Grande do Sul, Brazil) kindly provided the *B. antiacantha* image. Pictures are reproduced with permission by the owners.

Catharanthus AP2-domain, respectively, yielded higher amounts of catharanthine in hairy roots, a precursor of vinblastine and vincristine biosynthesis (Wang et al., 2010). As a result, much larger amounts of these anticancer compounds accumulated in *C. roseus* hairy roots. The biosynthesis of the terpene artemisinin, a frontline antimalarial drug, was increased by roughly 80% in *A. annua* due to overexpression of the transcription factor gene AaNAC1 that, in turn, led to the increased expression of the genes that encode amorphous-4,11-diene synthase (ADS), artemisinic aldehyde Δ11 (13) reductase (DBR2) and aldehyde dehydrogenase 1 (ALDH1) (Lv et al., 2016). The transgenic plants were also found to exhibit tolerance to drought and the fungus *Botrytis cinerea* (Lv et al., 2016). Similarly, the simultaneous overexpression of the genes that encode for putrescine *N*-methyltransferase (PMT) and hyoscyamine 6β-hydroxylase (H6H) in *Hyoscyamus niger* hairy roots increased the accumulation of the alkaloid anticholinergic agent scopolamine by nine-fold in comparison with the wild-type hairy roots (Zhang et al., 2004). The biosynthesis of resveratrol, anti-inflammatory and antioxidant stilbene, was considerably stimulated in cell cultures of *Vitis amurensis* overexpressing *VaCPK29*, a member of a multigene family of calcium-dependent protein kinases (Aleynova et al., 2015).

Some efforts have been made to identify genes involved in synthesis of bixin, an apocarotenoid produced by the South American native species *Bixa orellana*. Bixin is widely used in pharmaceutical, food, cosmetic and dye industries (Teixeira da Silva et al., 2018). Therefore, genetic engineering could be an interesting approach for the heterologous expression of *B. orellana* genes in hairy roots system, for instance, as an alternative for the large-scale production of such an important pigment.

Although the establishment of protocols to produce cell or hairy root cultures of some plant species is quite challenging, there are still some approaches to be considered for the development of bioreactor systems based on medicinal plants. Additionally, one can consider the heterologous expression of genes from species native to Brazil in plant systems that are already used as models in the scope of molecular biology. In this sense, the Nucleus of Bioassays, Biosynthesis and Ecophysiology of Natural Products (NuBBE) created a database that became a source of eligible molecules that can be targeted for the improvement of their production in heterologous systems. Conceived in 1998, the NuBBE database (nubbe.iq.unesp.br/portal/nubbe-search.html) comprises a catalog of bioactive molecules, in which the chemical, pharmacological and toxicological features of metabolites and derivatives are disclosed and identified in species from the Brazilian biodiversity. The NuBBE database serves as a useful tool for studies on multidisciplinary interfaces related to chemistry and biology, including virtual screening, dereplication, metabolomics and medicinal chemistry. Such a database certainly is contributing to a more sustainable development of the Cerrado and Atlantic Forest.

REFERENCES

Abbai, R.; Subramaniyam, S.; Mathiyalagan, R.; Yang, D. C. 2017. Functional genomic approaches in plant research. In K. Hakeem; A. Malik; F. Vardar-Sukan; M. Ozturk, eds., Plant Bioinformatics, pp. 215–239. Cham: Springer.

Abdel-Kader, M.; Berger, J. M.; Slebodnick, C.; Hoch, J.; Malone, S.; Wisse, J. H.; Werkhoven, M. C. M.; Mamber, S.; Kingston, D. G. 2002. Isolation and absolute configuration of *ent*-Halimane diterpenoids from *Hymenaea courbaril* from the Suriname rain forest. Journal of Natural Products, 65, 11–15.

Afendi, F. M.; Okada, T.; Yamazaki, M.; Hirai-Morita, A.; Nakamura, Y.; Nakamura, K.; Ikeda, S.; et al. 2012. KNApSAcK family databases: integrated metabolite-plant species databases for multifaceted plant research. Plant Cell Physiology, 53, 1–12.

Ahmed-Belkacem, A.; Macalou, S.; Borrell, F.; Capasso, R.; Fattorusso, E.; Taglialatela-Scafati, O.; Di Pietro, A. 2007. Nonprenylated rotenoids, a new class of potent breast cancer resistance protein inhibitors. Journal of Medicinal Chemistry, 50, 1933–1938.

Ajikumar, P. K.; Xiao, W. -H.; Tyo, K. E. J.; Wang, Y.; Simeon, F.; Leonard, E.; Mucha, O.; Phon, T. H.; Pfeifer, B.; Stephanopoulos, G. 2010. Isoprenoid pathway optimization for taxol precursor overproduction in *Escherichia coli*. Science, 330, 70–74.

Alagoz, Y.; Gurkok, T.; Zhang, B.; Unver, T. 2016. Manipulating the biosynthesis of bioactive compound alkaloids for next-generation metabolic engineering in opium poppy using CRISPR-Cas 9 genome editing technology. Scientific Reports, 6, 309–310.

Aleynova, O. A.; Dubrovina, A. S.; Manyakhin, A. Y.; Karetin, Y. A.; Kiselev, K. V. 2015. Regulation of resveratrol production in *Vitis amurensis* cell cultures by calcium-dependent protein kinases. Applied Biochemistry and Biotechnology, 175, 1460–1476.

Allen, R. S.; Millgate, A. G.; Chitty, J. A.; Thisleton, J.; Miller, J. A.; Fist, A. J.; Gerlac, W. L.; Larkin, P. J. 2004. RNAi-mediated replacement of morphine with the nonnarcotic alkaloid reticuline in opium poppy. Nature Biotechnology, 22, 1559–1566.

Alonso, J. M.; Ecker, J. R. 2006. Moving forward in reverse: genetic technologies to enable genome-wide phenomic screens in *Arabidopsis*. Nature Reviews Genetics, 7, 524–536.

Alves, G. R. M.; Oliveira, L. O.; Alves, M. M.; Silva, B. W. 2005. Variation in emetine and cephaeline contents in roots of wild ipecac (*Psychotria ipecacuanha*). Biochemical Systematics and Ecology, 33, 233–243.

Arabidopsis Genome Initiative. 2000. Analysis of the genome sequence of the flowering plant *Arabidopsis thaliana*. Nature, 408, 796–815.

Ayadi, R.; Trémouillaux-Guiller, J. 2003. Root formation from transgenic calli of *Ginkgo biloba*. Tree Physiology, 23, 713–718.

Barrangou, R.; Fremaux, C.; Deveau, H.; Richards, M.; Boyaval, P.; Moineau, S.; Romero, D. A.; Horvath, P. 2007. CRISPR provides acquired resistance against viruses in prokaryotes. Science, 315, 1709–1712.

Blomstedt, C. K.; Gleadow, R. M.; O'Donnell, N.; Naur, P.; Jensen, K.; Laursen, T.; Olsen, C. E.; et al. 2012. A combined biochemical screen and TILLING approach identifies mutations in *Sorghum bicolor* L. Moench resulting in acyanogenic forage production. Plant Biotechnology Journal, 10, 54–66.

Botelho, M. A.; Nogueira, N. A.; Bastos, G. M.; Fonseca, S. G.; Lemos, T. L.; Matos, F. J.; Montenegro, D.; Heukelbach, J.; Rao, V. S.; Brito, G. A. 2007. Antimicrobial activity of the essential oil from *Lippia sidoides*, carvacrol and thymol against oral pathogens. Brazilian Journal of Medical and Biological Research, 40, 349–56.

Brandão, M. G. L.; Cosenza, G. P.; Grael, C. F. F.; Netto Junior, N. L.; Monte-Mór, R. L. M. 2009. Traditional uses of American plant species from the 1st edition of Brazilian official pharmacopoeia. Revista Brasileira de Farmacognosia, 19, 478–487.

Brown, S.; Clastre, M.; Courdavault, V.; O'Connor, S. E. 2015. *De novo* production of the plant-derived alkaloid strictosidine in yeast. Proceedings of the National Academy of Sciences of the USA, 112, 3205–3210.

Chaturvedula, V. S.; Schilling, J. K.; Malone, S.; Wisse, J. H.; Werkhoven, M. C.; Kingston, D. G. 2003. New cytotoxic triterpene acids from aboveground parts of *Manihot esculenta* from the Suriname rainforest. Planta Medica, 69, 271–274.

Cong, L.; Ran, F. A.; Cox, D.; Lin, S. L.; Barretto, R.; Habib, N.; Hsu, P. D.; et al. 2013. Multiplex genome engineering using CRISPR/Cas systems. Science, 339, 819–823.

Coqueiro, A.; Regasini, L. O.; Leme, G. M.; Polese, L.; Nogueira, C. T.; Del Cistia, M. L.; Graminha, M. A. S.; Bolzani, V. S. 2014. Leishmanicidal activity of *Brosimum glaziovii* (Moraceae) and chemical composition of the bioactive fractions by using high-resolution gas chromatography and GC-MS. Journal of the Brazilian Chemical Society, 25, 1839–1847.

Costa-Lotufo, L. V.; Cunha, G. M.; Farias, P. A.; Viana, G. S.; Cunha, K. M.; Pessoa, C.; Moraes, M. O.; Silveira, E. R.; Gramosa, N. V.; Rao, V. S. 2002. The cytotoxic and embryotoxic effects of kaurenoic acid, a diterpene isolated from *Copaifera langsdorffii* oleo-resin. Toxicon, 40, 1231–1234.

Cragg, M.; Newman, D. J.; Snader, K. M. 1997. Natural products in drug discovery and development. Journal of Natural Products, 60, 52–60.

Cruz, E. M.; Silva, E. R.; Maquiaveli, C. C.; Alves, E. S. S.; Lucon, J. F.; Reis, M. B. G.; Vannier-Santos, M. A. 2013. Leishmanicidal activity of *Cecropia pachystachya* flavonoids: arginase inhibition and altered mitochondrial DNA arrangement. Phytochemistry, 89, 71–77.

d'Acampora-Zellner, B.; Lo Presti, M.; Barata, L. E.; Dugo, P.; Dugo, G.; Mondello, L. 2006. Evaluation of leaf-derived extracts as an environmentally sustainable source of essential oils by using gas chromatography-mass spectrometry and enantioselective gas chromatography-olfactometry. Analytical Chemistry, 7, 883–890.

de Lira, G. A.; De Andrade L. M.; Florencio, K. C.; Da Silva M. S.; Barbosa-Filho, J. M.; Leitão da-Cunha, E. V. 2002. Roraimine: a bisbenzylisoquinoline alkaloid from *Cissampelos sympodialis* roots. Fitoterapia, 73, 356–358.

de Mello, J. P.; Petereit, F.; Nahrstedt, A. 1996. Flavan-3-ols and prodelphinidins from *Stryphnodendron adstringens*. Phytochemistry, 41, 807–813.

de Oliveira, R. N.; Rehder, V. L. G.; Santos Oliveira, A. S.; Júnior, I. M.; de Carvalho, J. E.; de Ruiz, A. L. T. G.; Allegretti, S. M. 2012. *Schistosoma mansoni: in vitro* schistosomicidal activity of essential oil of *Baccharis trimera* (less) DC. Experimental Parasitology, 132, 135–143.

de Souza, K. C. B.; Bassani, V. L.; Schapoval, E. E. S. 2007. LC determination of flavonoids: separation of quercetin, luteolin and 3-*O*-methylquercetin in *Achyrocline satureioides* preparations. Phytomedicine, 14, 102–108.

Dixon, R. A.; Strack, D. 2003. Phytochemistry meets genome analysis, and beyond. Phytochemistry, 62, 815–816.

Donati, M.; Mondin, A.; Chen, Z.; Miranda, F. M.; Nascimento Junior, B. B.; Schirato, G.; Pastore, P.; Froldi, G. 2015. Radical scavenging and antimicrobial activities of *Croton zehntneri*, Pterodon emarginatus and *Schinopsis brasiliensis* essential oils and their major constituents: estragole, trans-anethole, β-caryophyllene and myrcene. Natural Product Research, 29, 939–946.

Edge, A.; Qu, Y.; Easson, M. L.; Thamm, A. M.; Kim, K. H.; De Luca, V. 2018. A tabersonine 3-reductase *Catharanthus roseus* mutant accumulates vindoline pathway intermediates. Planta, 247, 155–169.

Eisenstein, M. 2016. Living factories of the future. Nature, 531, 401–403.

Estrela, J. L. V.; Murilo Fazolin, M.; Catani, V.; Alécio, M. R.; de Lima, M. S. 2006. Toxicidade de óleos essenciais de *Piper aduncum* e *Piper hispidinervum* em *Sitophilus zeamais*. Pesquisa Agropecuária Brasileira, 41, 217–222.

Facchini, P. J.; Bohlmann, J.; Covello, P. S.; De Luca, V.; Mahadevan, R.; Page, J. E.; Ro, D. K.; Sensen, C. W.; Storms, R.; Martin, V. J. J. 2012. Synthetic biosystems for the production of high-value plant metabolites. Trends in Biotechnology, 30, 127–131.

Ferreira, F. P.; Oliveira, D. C. R. 2010. New constituents from *Mikania laevigata* shultz bip. ex baker. Tetrahedron Letters, 51, 6856–6859.

Figueredo, F. G.; Lucena, B. F. F.; Tintino, S. R.; Matias, E. F. F.; Leite, N. F.; Andrade, J. C.; Nogueira, L. F. B.; et al. 2014. Chemical composition and evaluation of modulatory of the antibiotic activity from extract and essential oil of *Myracrodruon urundeuva*. Pharmaceutical Biology, 2, 560–565.

Fossati, E.; Ekins, A.; Narcross, L.; Zhu, Y.; Falgueyret, J. P.; Beaudoin, G. A. W.; Facchini, P. J.; Martin, V. J. J. 2014. Reconstitution of a 10-gene pathway for synthesis of the plant alkaloid dihydrosanguinarine in *Saccharomyces cerevisiae*. Nature Communications, 5, 1–11.

Fougat, R. S.; Kumar, S.; Sakure, A. A. 2015. Advances in molecular biology of medicinal plants. In G. R. Smitha, ed., Compendium-Advances in Medicinal & Aromatic Plants Research, pp. 62–65. Boriyavi, Gujrat: ICAR-DMAPR.

Fujii, N.; Inui, T.; Iwasa, K.; Morishige, T.; Sato, F. 2007. Knockdown of berberine bridge enzyme by RNAi accumulates (S)-reticuline and activates a silent pathway in cultured *California poppy* cells. Transgenic Research, 16, 363–375.

Galanie, S.; Thodey, K.; Trenchard, I.; Interrante, M. F.; Smolke, C. 2015. Complete biosynthesis of opioids in yeast. Science, 349, 1095–1100.

Gandhi, S. G.; Mahajan, V.; Bedi, Y. S. 2015. Changing trends in biotechnology of secondary metabolism in medicinal and aromatic plants. Planta, 241, 303–317.

Gasiunas, G.; Barrangou, R.; Horvath, P.; Siksnys, V. 2012. Cas9-crRNA ribonucleoprotein complex mediates specific DNA cleavage for adaptive immunity in bacteria. Proceedings of the National Academy of Sciences of the USA, 109, E2579–E2586.

Gibson, D. G.; Glass, J. I.; Lartigue, C.; Noskov, V. N.; Chuang, R. Y.; Algire, M. A.; Benders, G. A.; et al. 2010. Creation of a bacterial cell controlled by a chemically synthesized genome. Science, 329, 52–56.

Goff, S. A.; Ricke, D.; Lan, T. H.; Presting, G.; Wang, R.; Dunn, M.; Glazebrook, J.; et al. 2002. A draft sequence of the rice genome (*Oryza sativa* L. ssp. japonica). Science, 296, 92–100.

Gorinsky, C. 1996. Biologically active rupununines. U.S. Patent No. 5569456, October 29, 1996.

Gupta, D.; Podar, K.; Tai, Y. T.; Lin, B.; Hideshima, T.; Akiyama, M.; LeBlanc, R.; Anderson, K. C. 2002. β-Lapachone, a novel plant product, overcomes drug resistance in human multiple myeloma cells. Experimental Hematology, 30, 711–720.

Hawkins, K. M.; Smolke, C. D. 2008. Production of benzylisoquinoline alkaloids in *Saccharomyces cerevisiae*. Nature Chemical Biology, 4, 564–573.

He, M.; Min, J. -W.; Kong, W. -L.; He, X. -H.; Li, J. -X.; Peng, B. -W. 2016. A review on the pharmacological effects of vitexin and isovitexin. Fitoterapia, 115, 74–85.

Heck, C. I.; de Mejia E. G. 2007. Yerba Mate Tea (*Ilex paraguariensis*): a comprehensive review on chemistry, health implications, and technological considerations. Journal of Food Science, 72, 138–151.

Hsu, P. D.; Lander, E. S.; Zhang, F. 2014. Development and applications of CRISPR-Cas9 for genome engineering. Cell, 157, 1262–1278.

Idibie, C. A.; Davids, H.; Iyuke, S. E. 2007. Cytotoxicity of purified cassava linamarin to a selected cancer cell lines. Bioprocess and Biosystems Engineering, 30, 261–269.

Iida, S.; Terada, R. 2004. A tale of two integrations, transgene and T-DNA: gene targeting by homologous recombination in rice. Current Opinion in Biotechnology, 15, 132–138.

Ishino, Y.; Shinagawa, H.; Makino, K.; Amemura, M.; Nakata, A. 1987. Nucleotide-sequence of the IAP gene, responsible for alkaline-phosphatase isozyme conversion in *Escherichia-coli*, and identification of the gene-product. Journal of Bacteriology, 169, 5429–5433.

Ito, Y.; Nishizawa-Yokoi, A.; Endo, M.; Mikami, M.; Toki, S. 2015. CRISPR/Cas9-mediated mutagenesis of the *RIN* locus that regulates tomato fruit ripening. Biochemical and Biophysical Research Communications, 467, 76–82.

Jabeur, I.; Martins, N.; Barros, L.; Calhelha, R. C.; Vaz, J.; Achour, L.; Ferreira, I. C. F. R. 2017. Contribution of the phenolic composition to the antioxidant, anti-inflammatory and antitumor potential of *Equisetum giganteum* L. and *Tilia platyphyllos* Scop. Food & Function, 8, 975–984.

Jayakumar, K.; Meenu Krishnan, V. G.; Murugan, K. 2016. Evaluation of antioxidant and antihemolytic activities of purified caulophyllumine-A from *Solanum mauritianum* Scop. Journal of Pharmacognosy and Phytochemistry, 5, 195–199.

Jiang, W. Y.; Bikard, D.; Cox, D.; Zhang, F.; Marraffini, L. A. 2013. RNA-guided editing of bacterial genomes using CRISPR-Cas systems. Nature Biotechnology, 31, 233–239.

Karuppusamy, S. 2009. A review on trends in production of secondary metabolites from higher plants by in vitro tissue, organ and cell cultures. Journal of Medicinal Plants Research, 3, 1222–1239.

Kessler, A.; Kalske, A. 2018. Plant secondary metabolite diversity and species interactions. Annual Review of Ecology, Evolution, and Systematics, 49, 115–138.

Kim, J.; Buell, C. R. 2015. A revolution in plant metabolism: genome-enabled pathway discovery. Plant Physiology, 169, 1532–1539.

King, H. 1940. Curare alkaloids. Part V. Alkaloids of some *Chondrodendron* species and the origin of radix pareirae bravae. Journal of Chemical Society, 137, 737–46.

Lage, G. A.; Medeiros, F. D. S.; Furtado, W. D. L.; Takahashi, J. A.; Filho, J. D. D. S.; Pimenta, L. P. S. 2014. The first report on flavonoid isolation from *Annona crassiflora* mart. Natural Products Research, 28, 808–811.

Laus, G.; Brössner, D.; Keplinger, K. 1997. Alkaloids of Peruvian *Uncaria tomentosa*. Phytochemistry, 45, 855–860.

Leggans, E. K.; Duncan, K. K.; Barker, T. J.; Schleicher, K. D.; Boger, D. L. 2013. A remarkable series of vinblastine analogues displaying enhanced activity and an unprecedented tubulin binding steric tolerance: C20' urea derivatives. Journal of Medicinal Chemistry, 56, 628–639.

Li, B.; Cui, G.; Shen, G.; Zhan, Z.; Huang, L.; Chen, J.; Qi, X. 2017. Targeted mutagenesis in the medicinal plant *Salvia miltiorrhiza*. Scientific Reports, 7, 43320.

Li, T.; Huang, S.; Jiang, W. Z.; Wright, D.; Spalding, M. H.; Weeks, D. P.; Yang, B. 2011. TAL nucleases (TALNs): hybrid proteins composed of TAL effectors and FokI DNA-cleavage domain. Nucleic Acids Research, 39, 359–372.

Lucci, N.; Mazzafera, P. 2009. Rutin synthase in fava d'anta: purification and influence of stressors. Canadian Journal of Plant Science, 89, 895–902.

Lv, Z.; Wang, S.; Zhang, F.; Chen, L.; Hao, X.; Pan, Q.; Fu, X.; Li, L.; Sun, X.; Tang, K. 2016. Overexpression of a novel NAC domain-containing transcription factor gene (*AaNAC1*) enhances the content of artemisinin and increases tolerance to drought and Botrytis cinerea in *Artemisia annua*. Plant Cell Physiology, 57, 1961–1971.

Ma, Y.; Ma, X. H.; Meng, F. Y.; Zhan, Z. L.; Guo, J.; Huang, L. Q. 2016. RNA interference targeting CYP76AH1 in hairy roots of *Salvia miltiorrhiza* reveals its key role in the biosynthetic pathway of tanshinones. Biochemical and Biophysical Research Communications, 477, 155–160.

Malheiros, A.; Cechinel-Filho, V.; Schmitt, C. B.; Yunes, R. A.; Escalante, A.; Svetaz, L.; Zacchino, S.; Delle-Monache, F. 2005. Antifungal activity of drimane sesquiterpenes from *Drimys brasiliensis* using bioassay-guided fractionation. Journal of Pharmacy & Pharmaceutical Sciences, 15, 335–339.

Manu, K. A.; Kuttan, G. 2009. Anti-metastatic potential of Punarnavine, an alkaloid from *Boerhaavia diffusa* Linn. Immunobiology, 14, 245–255.

Marques, L. L.; Panizzon, G. P.; Aguiar, B. A.; Simionato, A. S.; Cardozo-Filho, L.; Andrade, G.; de Oliveira, A. G.; Guedes, T. A.; Mello, J. C. 2016. Guaraná (*Paullinia cupana*) seeds: selective supercritical extraction of phenolic compounds. Food Chemistry, 212, 703–711.

Matos, M. E. O.; De Sousa, M. P.; Matos, F. J. A.; Craveiro, A. A. 1988. Sesquiterpenes from *Vanillosmopsis arborea*. Journal of Natural Products, 51, 780–782.

Melo, J. G.; Santos, A. G.; Amorim, E. L. C.; Nascimento, S. C.; Albuquerque, U. P. 2011. Medicinal plants used as antitumor agents in Brazil: an ethnobotanical approach. Evidence-Based Complementary and Alternative Medicine, 2011, 1–14.

Mishra, R.; Zhao, K. J. 2018. Genome editing technologies and their applications in crop improvement. Plant Biotechnology Reports, 12, 57–68.

Montano, A.; Forero-Castro, M.; Hernandez-Rivas, J. M.; Garcia-Tunon, I.; Benito, R. 2018. Targeted genome editing in acute lymphoblastic leukemia: a review. BMC Biotechnology, 18, 1–10.

Moses, T.; Goossens, A. 2017. Plants for human health: greening biotechnology and synthetic biology. Journal of Experimental Botany, 68, 4009–4011.

Moses, T.; Pollier, J.; Almagro, L.; Buyst, D.; Van Montagu, M.; Pedreño, M. A.; Martins, J. C.; Thevelein, J. M.; Goossens, A. 2014. Combinatorial biosynthesis of sapogenins and saponins in *Saccharomyces cerevisiae* using a C-16α hydroxylase from *Bupleurum falcatum*. Proceedings of the National Academy of Sciences of the USA, 111, 1634–1639.

Mossi, A. J.; Zanella, C. A.; Kubiak, G.; Lerin, L. A.; Cansian, R. L.; Frandoloso, F. S.; Treichel, H. 2013. Essential oil of *Ocotea odorifera*: an alternative against *Sitophilus zeamais*. Renewable Agriculture and Food Systems, 29, 161–166.

Moyano, E.; Fornalé, S.; Palazón, J.; Cusidó, R. M.; Bonfill, M.; Morales, C.; Piñol, M. T. 1999. Effect of *Agrobacterium rhizogenes* T-DNA on alkaloid production in Solanaceae plants. Phytochemistry, 52, 1287–1292.

Nakagawa, A.; Minami, H.; Kim, J. S.; Koyanagi, T.; Katayama, T.; Sato, F.; Kumagai, H. 2012. Bench-top fermentative production of plant benzylisoquinoline alkaloids using a bacterial platform. Bioengineered, 3, 49–53.

Nemudryi, A. A.; Valetdinova, K. R.; Medvedev, S. P.; Zakian, S. M. 2014. TALEN and CRISPR/Cas genome editing systems: tools of discovery. Acta Naturae, 6, 19–40.

Nizio, D. A. C.; Brito, F. A.; Sampaio, T. S.; Melo, J. O.; Silva, F. L. S.; Gagliardi, P. R.; Arrigoni-Blank, M. F.; Anjos, C. S.; Alves, P. B.; Wisniewski Junior, A.; Blank, A. F.; 2015. Chemical diversity of native populations of *Varronia curassavica* Jacq. and antifungal activity against *Lasiodiplodia theobromae*. Industrial Crops and Products, 76, 437–448.

Nogueira, R. C.; de Cerqueira, H. F.; Soares, M. B. 2010. Patenting bioactive molecules biodiversity: the Brazilian experience. Expert Opinion on Therapeutic Patents, 20, 1–13.

Nonaka, S.; Arai, C.; Takayama, M.; Matsukura, C.; Ezura, H. 2017. Efficient increase of gamma-aminobutyric acid (GABA) content in tomato fruits by targeted mutagenesis. Scientific Reports, 7.

Oberlies, N. H.; Burgess, J. P.; Navarro, H. A.; Pinos, R. E.; Fairchild, C. R.; Peterson, R. W.; Soejarto, D. D.; Farnsworth, N. R.; Kinghorn, A. D.; Wani, M. C.; et al. 2002. Novel bioactive clerodane diterpenoids from the leaves and twigs of *Casearia sylvestris*. Journal of Natural Products, 65, 95–99.

Ohsaki, A.; Imai, Y.; Naruse, M.; Ayabe, S.; Komiyama, K.; Takashima J. 2004. Four new triterpenoids from *Maytenus ilicifolia*. Journal of Natural Products, 67, 469–471.

Ono, M.; Ishimatsu, N.; Masuoka, C.; Yoshimitsu, H.; Tsuchihashi, R.; Okawa, M.; Kinjo, J.; Ikeda, T.; Nohara, T. 2007. Three new monoterpenoids from the fruit of *Genipa americana*. Chemical and Pharmaceutical Bulletin, 55, 632–634.

Paddon, C. J.; Keasling, J. D. 2014. Semi-synthetic artemisinin: a model for the use of synthetic biology in pharmaceutical development. Nature Reviews Microbiology, 12, 355–367.

Paddon, C. J.; Westfall, P. J.; Pitera, D. J.; Benjamin, K.; Fisher, K.; McPhee, D.; Leavell, M. D.; et al. 2013. High-level semi-synthetic production of the potent antimalarial artemisinin. Nature, 496, 528–536.

Paiva, S. R.; Lima, L. A.; Figueiredo, M. R.; Kaplan, M. A. C. 2004. Plumbagin quantification in roots of *Plumbago scandens* L. obtained by different extraction techniques. Anais da Academia Brasileira de Ciências, 76, 499–504.

Peplow, M. 2013. Malaria drug made in yeast causes market ferment. Nature, 494, 160–161.

Peters, R. R.; Saleh, T. F.; Lora, M.; Patry, C.; Brum-Fernandes, A. J.; Farias, M. R.; Ribeiro-do-Valle, R. M. 1999. Anti-inflammatory effects of the products from *Wilbrandia ebracteata* on carrageenan-induced pleurisy in mice. Life Science, 64, 2429–2437.

Pizzolatti, M. G.; Venson, A. F.; Smânia Junior, A.; Smânia, E. F. A.; Braz-Filho R. 2002. Two epimeric flavalignans from *Trichilia catigua* (Meliaceae) with antimicrobial activity. Zeitschrift Für Naturforschung C, 57, 483–488.

Putalun, W.; Udomsin, O.; Yusakul, G.; Juengwatanatrakul, T.; Sakamoto, S.; Tanaka, H. 2010. Enhanced plumbagin production from *in vitro* cultures of *Drosera burmanii* using elicitation. Biotechnology Letters, 32, 721–724.

Rai, A.; Saito, K.; Yamazaki, M. 2017. Integrated omics analysis of specialized metabolism in medicinal plants. Plant Journal, 90, 764–787.

Rao, V. S.; de Melo, C. L.; Queiroz, M. G. R.; Lemos, T. L. G.; Menezes, D. B.; Melo, T. S.; Santos, F. A. 2011. Ursolic acid, a pentacyclic triterpene from *Sambucus australis*, prevents abdominal adiposity in mice fed a high-fat diet. Journal of Medicinal Food, 14, 1375–1382.

Reed, J. W.; Hudlicky, T. 2015. The quest for a practical synthesis of morphine alkaloids and their derivatives by chemoenzymatic methods. Accounts of Chemical Research, 48, 674–687.

Santos, Vanessa N. C.; Freitas, Rilton A. de; Deschamps, Francisco C.; Biavatti, Maique W. 2009. Ripe fruits of *Bromelia antiacantha*: investigations on the chemical and bioactivity profile. Revista Brasileira de Farmacognosia, 19, 358–365.

Sawai, S.; Ohyama, K.; Yasumoto, S.; Seki, H.; Sakuma, T.; Yamamoto, T.; Takebayashi, Y.; et al. 2014. Sterol side chain reductase 2 is a key enzyme in the biosynthesis of cholesterol, the common precursor of toxic steroidal glycoalkaloids in potato. Plant Cell, 26, 3763–3774.

Sawaya, A. C. H. F.; Vaz, B. G.; Eberlin, M. N.; Mazzafera, P. 2011. Screening species of *Pilocarpus* (Rutaceae) as sources of pilocarpine and other imidazole alkaloids. Genetic Resources and Crop Evolution, 58, 471–480.

Scossa, F.; Benina, M.; Alseekh, S.; Zhang, Y.; Fernie, A. R. 2018. The integration of metabolomics and next-generation sequencing data to elucidate the pathways of natural product metabolism in medicinal plants. Planta Medica, 84, 855–873.

Seki, H.; Sawai, S.; Ohyama, K.; Mizutani, M.; Ohnishi, T.; Sudo, H.; Fukushima, E. O.; et al. 2011. Triterpene functional genomics in licorice for identification of CYP72A154 involved in the biosynthesis of glycyrrhizin. Plant Cell, 23, 4112–4123.

Shinde, A. N.; Malpathak, N.; Fulzele, D. P. 2009. Enhanced production of phytoestrogenic isoflavones from hairy root cultures of *Psoralea corylifolia* L. using elicitation and precursor feeding. Biotechnology and Bioprocess Engineering, 14, 288–294.

Simões, C. M. O.; Schenkel, E. P.; Gosman, G.; Mello, J. C. P.; Mentz, L. A.; Petrovick, P. R. 2003. Farmacognosia – da planta ao medicamento, 5.ed. Porto Alegre/Florianópolis: UFGRS/UFSC.

Sitarek, P.; Kowalczyk, T.; Rijo, P.; Białas, A. J.; Wielanek, M.; Wysokińska, H.; Garcia, C.; Toma, M.; Śliwiński, T.; Skała, E. 2018. Over-expression of AtPAP1 transcriptional factor enhances phenolic acid production in transgenic roots of *Leonurus sibiricus* L. and their biological activities. Molecular Biotechnology, 60, 74–82.

Stehmann; R. R.; Sobral, M. 2017. Biodiversidade no Brasil. In.: C. M. O. Simões; E. P. Schenkel; G. Mello; L. A. Mentz; P. R. Petrovick Eds. Farmacognosia: do Produto Natural Planta ao Medicamento. 1–10. São Paulo: Artmed.

Tan, Z. Y.; Huang, T. S.; Ngeow, J. 2018. 65 Years of the double helix. the advancements of gene editing and potential application to hereditary cancer. Endocrine-Related Cancer, 25, T141–T158.

Teixeira da Silva, J. A.; Dobránszki, J.; Rivera-Madrid, R. 2018. The biotechnology (genetic transformation and molecular biology) of *Bixa orellana* L. (achiote). Planta, 248, 267–277.

Tierney, M. B.; Lamour, K. H. 2005. An introduction to reverse genetic tools for investigating gene function. The Plant Health Instructor. DOI: 10.1094/PHI-A-2005-1025-01.

Trox, J.; Vadivel, V.; Vetter, V.; Stuetz, W.; Kammerer, D. R.; Carle, R. B. 2011. Catechin and epicatechin in testa and their association with bioactive compounds in kernels of cashew nut (*Anacardium occidentale* L.). Food Chemistry, 128, 1094–1099.

Unamba, C. I. N.; Nag, A.; Sharma, R. K. 2015. Next generation sequencing technologies: the doorway to the unexplored genomics of non-model plants. Frontiers in Plant Science, 6, 1–16.

van der Helm, E.; Genee, H. J.; Sommer, M. O. A. 2018. The evolving interface between synthetic biology and functional metagenomics. Nature Chemical Biology, 14, 752–759.

Wang, C. -T.; Liu, H.; Gao, X. -S.; Zhang, H. -X. 2010. Overexpression of *G10H* and *ORCA3* in the hairy roots of *Catharanthus roseus* improves catharanthine production. Plant Cell Reports, 29, 887–894.

Wohlsen, M., 2011. Biopunk: DIY Scientists Hack the Software of Life.

Wyman, C.; Kanaar, R. 2006. DNA double-strand break repair: all's well that ends well. Annual Review of Genetics, 40, 363–383.

Yang, D.; Du, X.; Yang, Z.; Liang, Z.; Guo, Z.; Liu, Y. 2014. Transcriptomics, proteomics, and metabolomics to reveal mechanisms underlying plant secondary metabolism. Engineering in Life Sciences, 14, 456–466.

Yano, R.; Takagi, K.; Takada, Y.; Mukaiyama, K.; Tsukamoto, C.; Sayama, T.; Saito, K. 2017. Metabolic switching of astringent and beneficial triterpenoid saponins in soybean is achieved by a loss-of-function mutation in cytochrome P450 72A69. The Plant Journal, 89, 527–539.

Yin, H.; Xue, W.; Chen, S. D.; Bogorad, R. L.; Benedetti, E.; Grompe, M.; Koteliansky, V.; Sharp, P. A.; Jacks, T.; Anderson, D. G. 2014. Genome editing with Cas9 in adult mice corrects a disease mutation and phenotype, Nature Biotechnology, 32, 551–553.

Yu, J.; Hu, S.; Wang, J.; Wong, G. K. S.; Li, S.; Liu, B.; Deng, Y.; et al. 2002. A draft sequence of the rice genome (*Oryza sativa* L. ssp. indica). Science, 29, 79–92.

Zhang, L.; Ding, R.; Chai, Y.; Bonfill, M.; Moyano, E.; Oksman-Caldentey, K.-M.; Tang, K. 2004. Engineering tropane biosynthetic pathway in *Hyoscyamus niger* hairy root cultures. Proceedings of the National Academy of Sciences of the USA, 101, 6786–6791.

Zhou, Z.; Tan, H. X.; Li, Q.; Chen, J. F.; Gao, S. H.; Wang, Y.; Chen, W. S.; Zhang, L. 2018. CRISPR/Cas9-mediated efficient targeted mutagenesis of RAS in *Salvia miltiorrhiza*. Phytochemistry, 148, 63–70.

Zych, A. O.; Bajor, M.; Zagozdzon, R. 2018. Application of genome editing techniques in immunology. Archivum Immunologiae et Therapiae Experimentalis, 66, 289–298.

5 Diversity of Endophytes and Biotechnological Potential

Daiani Cristina Savi and Chirlei Glienke
Federal University of Paraná, Department of Genetics,
Curitiba, Brazil

CONTENTS

5.1 ENDOPHYTES – GENERAL ASPECTS

The term "endophyte" is derived from the Greek whereby "endon" means within, and "phyton" means plant. Thus, endophytes comprise microorganisms that for part of their life colonize plants tissues without causing any damage to the host or disease symptoms (Hardoim et al., 2015). However, the interaction between endophytes and their hosts depends on several aspects, such as environmental, biological, chemical and physiological characteristics (Jia et al., 2016; Rajamanikyam et al., 2017). Some studies have shown that the host plant and the interactions between other microorganisms can balance this symbiotic relationship (Jalgaonwala et al., 2012), which can be disrupted, and the symptoms of the disease can occur due to endophytic virulence factors or due to mechanisms of host defense (Kusari et al., 2012).

Endophytes colonize the intercellular space of various parts of the plant including roots, leaves, stems, flowers and seeds (Liu et al., 2017). However, fungi are more commonly isolated from leaves and stems, and bacteria in the roots (Roy and Sharma, 2015; Savi et al., 2015; Wang et al., 2016), colonizing intercellular space in the plant, mainly because these areas have an abundance of carbohydrates, amino acids and inorganic nutrients (Kandel et al., 2017). In view of the specificity of chemical compounds found in the host, one or two endophytic species are frequently predominant in a specific host, while other isolates are considered rare (Bernardi-Wenzel et al., 2010). Thus, studies of endophytes isolated from hotspot biomes can contribute to a better understanding of Brazilian diversity, as well as the effect of the anthropological and environmental conditions in this ecosystem.

5.2 EXPLORING ENDOPHYTES WITH BIOTECHNOLOGICAL POTENTIAL

Our key interest is to ask what makes medicinal plants found in Brazil a unique source for exploring endophytes with biotechnological potential? Brazil has an inestimable biodiversity of plants, being at the top of 17 megadiversity countries in the world (Mittermeier et al., 1997), with more than 55,000 species of plants (http://www.unesco.org/new/en/brasilia/natural-ciences/environment/biodiversity/). These plants are distributed in six biomes, Amazonian rainforest, Caatinga, Cerrado (Savanna), Atlantic Rainforest, Pampa and Pantanal (Swampland) (Myers et al., 2000), based on the adaptation to biotic and abiotic factors (Antonelli and Sanmartín, 2011; BFG, 2015; Forzza et al., 2010; 2012).

Two Brazilian biomes are recognized as biodiversity hotspots: Cerrado and Atlantic Rainforest (Mittermeier et al., 1998; Myers et al., 2000). However, the diversity of some other biomes, such as the Pantanal, remains relatively unexplored (Alho, 2011). Considering that some endophytes may be host specific, the plant diversity found in Brazil represents an extraordinary variety of habitats, life forms and biological associations confined to particular environments at different geographical scales (Forzza et al., 2010).

Medicinal plants in Brazil have been widely documented and investigated for their unique biological and chemical diversity (Atanasov et al., 2015). The great diversity of the Brazilian flora has been used for pharmaceutical purposes and is known in the global market as the origin of many products (Castro et al., 2014). We have been studying the biological potential of the endophytic community of four medicinal plants in Brazil (Table 5.1), *Vochysia divergens* from the Pantanal (Gomes et al., 2013; Gos et al., 2017; Hokama et al., 2017; Noriler et al., 2018; Savi et al., 2015), *Stryphnodendron adstringens* from the Cerrado (Noriler et al., 2018), *Maytenus ilicifolia* and *Schinus terebinthifolius* from the Atlantic forest (Figueiredo et al., 2018; Gomes-Figueiredo et al., 2007; Lima et al., 2012; Tonial et al., 2016).

The Pantanal is the largest wetland in the world, and it is subject to different climatic conditions during the year, experiencing flooding and a dry season (Alho, 2008). As a consequence, the Pantanal harbors its own flora (Arieira et al., 2006), especially the medicinal plant *Vochysia divergens* used in the treatment of colds, coughs, fever, pneumonia and other diseases (Pott et al., 2004). Extracts of stem and leaves of *V. divergens* have been reported to possess considerable anti-inflammatory, antibacterial and molluscicidal activities (Corrêa et al., 2018; Dos Santos and Sant'Ana, 2000; Hess et al., 1995). Chemical analysis revealed the presence of 3-sitosterol, betulinic acid, sericic acid (Hess et al., 1995), 5-methoxyluteolin-7-O-β-glucopyranoside, rutin, galloyl-HHDP-glucopyranoside, 3',5-dimethoxyluteolin-7-O-β-glucopyranoside (Corrêa et al., 2018), divergioic acid (Hess et al., 1999) and tormentic acid (Bortalanza et al., 2002). Among the compounds identified, tormentic acid has been reported as a promising metabolite due to its strong anticancer, anti-inflammatory and antiatherogenic properties (Bortalanza et al., 2002; Ma et al., 2015; Yang et al., 2018).

In contrast to the Pantanal, which is subject to flooding, the Cerrado is characterized by periods of natural burning due to fires affecting the vegetation in this biome (Felfili and Fagg, 2007). Due to this peculiar condition, the Cerrado is considered one of the most diverse places in the world (Myers et al., 2000), with approximately 6000 species of plants (Felfili and Fagg, 2007; Oliveira-Filho and Ratter, 2002). For example, *S. adstringens* is a medicinal plant commonly found in the Cerrado, and has been used in the treatment of cutaneous and mucosal lesions, diseases of genitourinary system, such as STDs, and as anti-inflammatory and antiseptic (Ferrão et al., 2014; Morey et al., 2016; Pinto et al., 2015; Rodrigues and Andrade, 2014). Several studies have evaluated the toxicity of the *S. adstringens* extract using rats and *Drosophila melanogaster* as models, and these evaluations

TABLE 5.1

Taxonomic Classification of the Medicinal Plants Used for the Isolation of Endophytes in the *BIOGEMM* Laboratory – UFPR

Scientific Name	Common Name	Family
Maytenus ilicifolia (sin. *Celastrus spinifolius*)	Espinheira Santa	Celastraceae
Schinus terebinthifolia (sin. *Schinus mellisii*)	Peppertree, Aroeira	Anacardiaceae
Vochysia divergens	Cambará	Vochysiaceae
Stryphnodendron adstringens (sin. *Stryphnodendron barbatimam*)	Barbatimão	Fabaceae

have shown that the plant extract has no genotoxicity or mutagenic activity (Costa et al., 2010; De Sousa et al., 2003). In addition, a clinical study has demonstrated the efficacy of *S. adstringens* in the healing of decubitus ulcers in 51 patients (Ricardo et al., 2018).

The Atlantic Forest is considered a hotspot biome, but most of this biodiversity has been reduced to less than 8% of its original coverage, being replaced primarily by sugarcane, coffee, cocoa and *Eucalyptus* forest for cellulose and pulp production (Colombo et al., 2010). *M. ilicifolia* is widely distributed in the Atlantic Forest and is commonly used in folk medicine in the treatment of gastric diseases (Sá et al., 2017). In addition, foliar extracts of *M. ilicifolia* showed antinociceptive, antioxidant, anti-inflammatory and antiulcerogenic activities (Cipriani et al., 2009; Jorge et al., 2004; Sá et al., 2017), with active compounds: polygalacturonic acid, catechin, friedelan-3β-ol, friedelin, and several phenolic and flavonoids (Cipriani et al., 2009; Queiroga et al., 2000; Sá et al., 2017; Tiberti et al., 2007).

Among the four plants studied by our group, *S. terebinthifolius* has been more exploited due to its biotechnological potential and chemical proprieties (Bernardes et al., 2014; Fedel-Miyasato et al., 2014; Richter et al., 2010; Rosas et al., 2015; Sereniki et al., 2016; Silva et al., 2017; Salem et al., 2018). It has been reported that the extract of *S. terebinthifolius* acts in Parkinson's disease, as an anti-inflammatory, immunomodulatory, chemopreventive, wound healing, antioxidant, anti-mycobacterial and antiproliferative (Bernardes et al., 2014; Fedel-Miyasato et al., 2014; Richter et al., 2010; Rosas et al., 2015; Sereniki et al., 2016; Silva et al., 2017; Salem et al., 2018).

Based on the specificity of the biomes, in the use of these plants in folk medicine, biotechnological potential and in the absence of extensive studies on the endophytic community, we have tried to catalog and explore the diversity of endophytes, as well as to understand the endophyte-host association in these four medicinal plants (Figueiredo et al., 2018; Gomes-Figueiredo et al., 2007; Gomes et al., 2013; Gos et al., 2017; Hokama et al., 2017; Lima et al., 2012; Medeiros et al., 2018; Noriler et al., 2018; Savi et al., 2015; Savi et al., 2016; Savi et al., 2018; Tonial et al., 2016; Tonial et al., 2017).

5.3 ISOLATION OF ENDOPHYTES

Upon selection of the medicinal plant, choosing the methodology for the isolation of endophytes is necessary. A convenient and common method accepted by many researchers for the isolation of endophytes is the adaptation of the method developed by Petrini (1982). The method consists of superficially disinfecting the plant samples (Figure 5.1A) using 70% alcohol for about 30 seconds, followed by immersion in 0.5–3.5% sodium hypochlorite for 1-10 minutes, 70% alcohol for about 30 seconds again and finally the sample is rinsed in sterile doubly distilled water (Figure 5.1B). Exposure to alcohol eliminates bacteria living on the tissue surface (Kampf et al., 2008), and exposure to sodium hypochlorite acts on fungi (Amirabadi and Sasannejad, 2016), so the remaining microorganisms will be those that live inside the plant tissue. Variations in the exposure time to sodium hypochlorite can be applied, depending on the composition of plant leaves. Some plants have more sensitive tissues and long exposure may cause tissue decomposition allowing sodium hypochlorite to enter the tissue, which may inhibit the isolation of fungal endophytes. After disinfection, the plant tissue is fragmented and deposited in culture medium for the isolation of endophytes (Figures 5.1C and D).

Selection of suitable culture media is one of the prerequisites for studying microorganisms. Different endophytic microorganisms live in different environments and have a variety of growth requirements, such as nutrients, pH, osmotic conditions and temperature. Due to the lack of sufficient variability in the composition of the culture medium, replication of the exact environmental conditions in the laboratory is almost impossible (Basu et al., 2015). However, some media have been considered more suitable to isolate most fungi, bacteria and actinomycetes, such as potato dextrose agar (VanderMolen et al., 2013), Luria Bertani agar (Ramalashmi et al., 2018) and AAC (Akemi et al., 2013), respectively. These media have proven effective in providing the nutrients necessary for the growth of the microorganism, such as vitamins, amino acids and sugar sources.

The addition of antibiotics (Figure 5.1C), such as 50 mg/L chloramphenicol or ampicillin, may be used to suppress bacterial growth (Savi et al., 2015) and cycloheximide (50 μg/mL) may be used

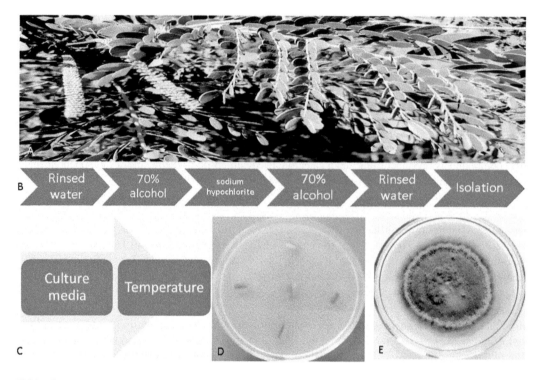

FIGURE 5.1 Work-up scheme for isolation of endophytes: (A) plant species; (B) disinfection process; (C) selection of culture conditions; (D) Petri dish with fragments of leaves; (E) isolated microorganism.

to inhibit the development of fungi (Gos et al., 2017). To stimulate the isolation of actinomycetes, nalidixic acid (50 µg/mL) and cycloheximide (50 µg/mL) were used to inhibit bacterial and fungal development (Savi et al., 2016).

A temperature range of 25–36°C is suitable for the growth of most endophytic microorganisms. After the incubation of the leaf fragments, the growth of endophytes needs to be verified daily, for about 30 days, since different microorganisms or species have different growth rates. As an example, some fungal species grow in only 2 days, while others require more than 10 days (Romão-Dumaresq et al., 2016). The emerging mycelia are transferred to a new plate (Figure 5.1E) and stored at 4°C in a suitable medium for further identification (Noriler et al., 2018).

Using these criteria, our group has isolated fungi and actinomycetes from the above-mentioned medicinal plants (*M. ilicifolia, S. terebinthifolius, V. divergens* and *S. adstringens*) to assess the diversity of endophytes. Interestingly, leaves and stems of the studied plants were colonized by fungi with higher frequency than by actinomycetes (Gomes-Figueiredo et al., 2007; Gomes et al., 2013; Gos et al., 2017; Hokama et al., 2017; Noriler et al., 2018; Savi et al., 2015; Savi et al., 2016; Tonial et al., 2016). This is probably related to the chemical and biological composition of the host leaves. Thus, the selection of plant tissues has a considerable effect on the observed biodiversity.

5.4 DIVERSITY OF ENDOPHYTES

Once the endophytes have been isolated, the first question is how to exploit such diversity? Second, how to work with such a large number of isolates? The first step is to group the microorganisms based on their macro- and micromorphology, considering the growth rate, colony color, hyphae aspects, presence/absence and spores' morphology. Based on morphological characteristics, one isolate from each morphotype is randomly selected for identification and bioprospecting.

The pure culture needs to be obtained through a single spore culture approach (Gilchrist-Saavedra et al., 2006). The single spore culture ensures that the isolates are pure and not contaminated with a close morphological species. The technique consists in making a spore solution 10^2 spores/mL, which is spread on a plate containing the appropriate culture medium, and after 2 or 3 days a colony of a single spore is transferred to a new plate and used for the next steps. In our group, the single spore colonies are deposited in the Culture Collection "Centro de Coleções de Culturas Biológicas do Estado do Paraná" of the Taxonline (http://taxonline.bio.br/index.php), at the Federal University of Paraná, Brazil (http://taxonline.bio.br/colecoes/index.php?id=2-coleções-microbiológicas). The extraction of DNA is performed using standard techniques (Noriler et al., 2018), as described by Raeder and Broda (1985), or using a commercial kit (Savi et al., 2016).

The identification of microorganisms is based on morphological, phylogenetic and ecological aspects, and the correct identification is a critical step to ensure biotechnological reproducibility (Raja et al., 2017). For many years, morphological characters were used as the single criterion for species identification (Militão et al., 2014). However, the classification of microorganisms isolated from the environment based on morphological analysis is complicated, since it is a highly variable group, which does not always produce spores under laboratory conditions (Rodriguez et al., 2009). In addition, morphological analyses are time-consuming and not compatible with the identification of several isolates in a short time. In view of these limitations, we first identify isolates based on phylogenetic analyses and use the complete morphological analysis only to describe new species (Noriler et al., 2018; Savi et al., 2015; Savi et al., 2016; Savi et al., 2018).

The internally transcribed spacer (ITS) and 16S rRNA regions remain the first choice for identifying fungi and bacteria, respectively, at a lower level, such as genus or species. In an analysis performed to select the barcode sequence for fungal identification the ITS region was selected in view of its easy amplification in different groups, and among the ribosomal regions analyzed, ITS region presents the highest probability of successful identification, with the most clearly defined barcode gap between inter- and intraspecific variation (Schoch et al., 2012). However, in some cryptic genera, such as *Diaporthe* (Gomes et al., 2013) and *Fusarium* (Chitrampalam at al., 2016), a multigene sequence analysis using protein-coding genes is required for species identification.

While ITS sequences are fungal barcodes, the 18S nuclear ribosomal small subunit rRNA gene (SSU) is commonly used in phylogenetic analysis at the family level, because it has fewer hypervariable domains. The 28S nuclear ribosomal large subunit rRNA gene (LSU) sometimes discriminates species on its own or combined with ITS. As an example, LSU can be used to confirm that an isolate can represent a new genus or new family, or even to point out an inconsistency observed in the ITS sequence to classify species in close related genera. According to Schoch et al. (2012) for yeasts, the D1/D2 region of LSU is useful for a long time for species identification.

Thus, we used the strategy of initially identifying the endophytes based on ITS or 16S rRNA sequence analysis. The amplification of the ITS region can be performed using different primers, the most used being ITS1 and ITS4, but if there is a problem in the amplification or sequencing, other primers may be used (Table 5.2). For 16S rRNA, several primers are described in the literature, and the most used are listed in Table 5.3. Normally, the names of primers used for 16S rRNA are numbers that designate their position in that gene in *Escherichia coli*.

The ribosomal sequence is compared based on the similarity to the available sequences in the GenBank database, using the BLAST tool (Figure 5.2). The GenBank was selected to compare the sequence since it is the largest sequence database with approximately 210 million sequences (www.ncbi. nlm.nih.gov/genbank/statistics/), corresponding to sequences of approximately 95,000 species (www. nature.com/nature/debates/e-access/Articles/lipman.html). Of these sequences about 172,000 represent fungal ITS sequences, in 2500 genera and 15,500 species (Schoch et al., 2012).

However, the blast result is not a conclusive identification, since approximately 20% of the fungal ITS sequences in this database were incorrectly annotated (Federhen, 2015; Nilson et al., 2006). An interesting alternative is to use the filter "sequence-from-type" in the blast searches, or to perform the searches on the RefSeq Targeted Loci project (http://www.ncbi.nlm.nih.gov/refseq/targetedloci/),

TABLE 5.2
Primers Used to Amplify ITS Region in Fungi

Primer	F/R	Sequence	Reference
ITS5	F	GGAAGTAAAAGTCGTAACAAGG	White et al. (1990)
ITS1	F	TCCGTAGGTGAACCTGCGG	White et al. (1990)
ITS3	F	GCATCGATGAAGAACGCAGC	White et al. (1990)
ITS2	R	GCTGCGTTCTTCATCGATGC	White et al. (1990)
ITS4	R	TCCTCCGCTTATTGATATGC	White et al. (1990)
LR1	R	GGTTGGTTTCTTTTCCT	Vilgalys and Hester (1990)

Note: F. forward; R. reverse.

TABLE 5.3
Primers Used to Amplify 16 rDNA Gene in Bacteria

Primers Pair	Sequence (5′-3′)	References
68f	TNANACATGCAAGTCGRRCG	McAllister et al. (2011)
518r	WTTACCGCGGCTGCTGG	Lee et al. (2010)
341f	CCTACGGGNGGCWGCAG	Klindworth et al. (2013)
785r	GACTACHVGGGTATCTAATCC	Klindworth et al. (2013)
799f	AACMGGATTAGATACCCKG	Chelius and Triplett (2001)
1193r	ACGTCATCCCCACCTTCC	Bodenhausen et al. (2013)
967f	CAACGCGAAGAACCTTACC	Sogin et al. (2006)
1391r	GACGGGCGGTGWGTRCA	Walker and Pace (2007)

Note: f. forward; r. reverse.

FIGURE 5.2   Work-up scheme for identification of the endophytes.

which maintains curated sets of full-length reference sequences from type for ribosomal RNAs (Federhen, 2015).

First, we use the blast tool to identify the possible fungal or bacterial genus to which the isolate belongs. For a final or more precise identification, a dataset containing all the sequences of type species of the valid fungal species belonging to the respective genus is obtained through a search on the Mycoback database (www.mycobank.org/) and for bacteria and actinomycetes using search on the List of Prokaryotic Names With Standing in Nomenclature (www.bacterio.net/). After selecting the sequences of valid species, the identification is based on an evolutionary framework using a phylogenetic approach (Figure 5.2). Phylogeny reconstructs the tree-like pattern that describes the evolutionary relationships between species with a predictive value (Pace et al., 2012), different from a similarity analysis via Blast. Many approaches to phylogenetic inference have been used and the relative merits of these methods have been an important consideration for phylogenetic analysis (Holder et al., 2008). The topology of the phylogenetic tree, as well as the order of branching events, is determined from the sequences of the analyzed region and, despite the fact some methods use distance-matrix to perform the phylogeny analysis, the most valuable methods are based on standard statistical techniques, such as maximum likelihood and Bayesian inference (Bogusz and Whelan, 2017).

In some genera, such as *Phaeophleospora*, we were able to identify an isolate as a new species using only the ITS sequence, because, for this genus the ITS sequence has enough information to differentiate species (Savi et al., 2018). However, it is not true for other critical genera, such as *Diaporthe*, in which five genes are required for accurate species identification (Gomes et al., 2013). In other cases, such as the *Colletotrichum* genus, the ITS sequence has enough information to identify which species' complex the isolate belongs to (Damn et al., 2012). Within each of the *Colletotrichum* species complexes, different protein-coding genes are recommended for species identification, such as the GAPDH intron region for identification of species within the *Colletotrichum acutatum* complex (Silva et al., 2017). Thus, identification of the isolate should be performed carefully, using sequences of the type species, and the analysis should begin with the ITS sequence and, if necessary, other genes need to be sequenced for identification at the species level.

Using phylogenetic analysis, we identified 46 genera (including 6 possible new genera of the Pleomassariaceae and Xylariaceae families) and 49 species as endophytes of *M. ilicifolia, S. terebinthifolius, V. divergens* and *S. adstringens* (Table 5.4). Among these isolates, 9 strains were described as new species, and the description of other 15 species is in progress (Table 5.4). This data reinforce the biodiversity found in the Cerrado, Pantanal and Atlantic Rainforest biomes and suggest the medicinal plants found in Brazil as a repository for fungi and actinomycetes.

5.5 EXPLORING THE BIOTECHNOLOGICAL POTENTIAL OF THE ISOLATED ENDOPHYTES

The search for interesting biological activity from microorganisms has been the basis for the development of several biotechnological applications, mainly in the pharmaceutical and agricultural industries (Vitorino and Bessa, 2017). The most promising compounds in the clinic for treatment of bacterial infections were isolated from microorganisms, such as penicillin isolated from *Penicillium digitatum* (Laich et al., 2002); vancomycin produced by *Streptomyces orientalis* (Levine, 2006); streptomycin isolated from *Streptomyces griseus* and erythromycin produced by *Saccharopolyspora erythraea* (Donadio et al., 1996), among several others. Besides the high exploration of microorganisms for active compounds in the past, studies have shown that unknown species and genetically different strains are still abundant in nature, and natural products remain the most promising source for new compounds (Monciardini et al., 2014).

One of the strategies for screening of endophyte producers of antibacterial and antifungal compounds, in our laboratory, is performed by dual culture (Hokama et al., 2017). In the dual cultures,

TABLE 5.4

Endophytic Microorganisms of the Medicinal Plants *Maytenus Ilicifolia*, *Schinus Terebinthifolius*, *Vochysia Divergens* and *Stryphnodendron Adstringens* Isolated and Studied in the *BIOGEMM* Laboratory – UFPR

Genus	Species	Reference	Medicinal Plant
Acrocalymma	*A. medicaginis*	Noriler et al., 2018	Sa
Actinomadura	*Actinomadura* sp.	Gos et al., 2017	Vd
Aeromicrobium	*A. ponti*	Gos et al., 2017	Vd
Alternaria	*Alternaria* section alternate	Tonial et al., 2016; Noriler et al., 2018	Vd/St
Annellosympodiella	*Annellosympodiella* sp.	Hokama et al., 2017	Vd
Antrodia	*Antrodia* sp.	Hokama et al., 2017	Vd
Bjerkandera	*Bjerkandera* sp.	Tonial et al., 2016; Noriler et al., 2018	Vd/Sa/St
Cladosporium	*Cladosporium* sp.	Tonial et al., 2016; Hokama et al., 2017	Vd/St
Colletotrichum	*Colletotrichum* sp.	Hokama et al., 2017; Noriler et al., 2018	Vd/Sa
	C. boninense sensu lato	Noriler et al., 2018	Vd/Sa
	C. gloeosporioides sensu lato	Noriler et al., 2018	Vd/Sa
	C. siamense	Noriler et al., 2018	Vd/Sa
	C. simmondsii	Lima et al., 2012	St
	C. acutatum	Lima et al., 2012	St
	C. fioriniae	Lima et al., 2012	St
Coniochaeta	*C. nepalica*	Noriler et al., 2018	Vd
Corynespora	*C. cambrensis*	Noriler et al., 2018	Vd/Sa
Curvularia	*Curvularia* sp.	Noriler et al., 2018	Vd/Sa
Daldinia	*Daldinia* sp.	Noriler et al., 2018	Vd
Diaporthe	*Diaporthe* sp. **1, 2, 3, 4**	Gomes et al., 2013; Noriler et al., 2018	Vd/Sa/Mi/St
	Diaporthe cf. *heveae* 1,	Noriler et al., 2018	Vd/Sa
	D. endophytica	Gomes et al., 2013	Mi/St
	D. inconspícua	Gomes et al., 2013	Mi
	D. infecunda	Gomes et al., 2013	Mi/St
	D. mayteni	Gomes et al., 2013	Mi
	D. novem	Gomes et al., 2013	Mi
	D. oxe	Gomes et al., 2013	Mi/St
	D. paranensis	Gomes et al., 2013	Mi
	D. phaseolorum	Gomes et al., 2013	Mi
	D. schini	Gomes et al., 2013	Vd/Sa/St
	D. terebinthifolii	Gomes et al., 2013	St
Hypoxylon	*Hypoxylon* sp.1	Noriler et al., 2018	Vd/Sa
Irpex	*I. lacteus*	Hokama et al., 2017	Vd
Lanceispora	*Lanceispora* sp.	Hokama et al., 2017	Vd
Lasiodiplodia	*Lasiodiplodia* sp.	Noriler et al., 2018	Vd/Sa
Microbacterium	*Microbacterium* sp.	Gos et al., 2017	Vd
Microbispora	*Microbispora* sp. **1, 2, 3**	Savi et al., 2016	Vd
Micrococcus	*Micrococcus* sp.	Gos et al., 2017	Vd
Micromonospora	*Micromonospora* sp.	Savi et al., 2015	Vd
Neofusicoccum	*N. brasiliense*	Noriler et al., 2018	Vd/Sa
	N. grevilleae	Hokama et al., 2017	Vd

(Continued)

TABLE 5.4 *(Continued)*

Endophytic Microorganisms of the Medicinal Plants *Maytenus Ilicifolia, Schinus Terebinthifolius, Vochysia Divergens* and *Stryphnodendron Adstringens* Isolated and Studied in the *BIOGEMM* Laboratory – UFPR

Genus	Species	Reference	Medicinal Plant
Neopestalotiopsis	*Neopestalotiopsis* sp.	Noriler et al., 2018	Vd
Nigrospora	*Nigrospora* sp.	Noriler et al., 2018	Vd/Sa
	N. hainanensis	Hokama et al., 2017	Vd
Penicillium	*Penicillium* sp.	Tonial et al., 2016	St
Phaeosphaeria	*Phaeosphaeria* sp.	Hokama et al., 2017	Vd
Paraphaeosphaeria	*Paraphaeosphaeria* sp.	Noriler et al., 2018	Vd/Sa
Peniophora	*Peniophora laxitexta*	Hokama et al., 2017	Vd
Pestalotiopsis	*Pestalotiopsis* sp.	Gomes-Figueiredo et al., 2007; Noriler et al., 2018	Vd/Sa/Mi
	P. microspora	Gomes-Figueiredo et al., 2007	Mi
Phaeophleospora	*Phaeophleospora* sp.**1, 2**	Noriler et al., 2018	Vd/Sa
	P. vochysiae	Savi et al., 2018	Vd
Phyllosticta	*Phyllosticta* sp.	Noriler et al., 2018	Vd/Sa
	Phyllosticta capitalensis	Hokama et al., 2017	Vd
Pleomassariaceae	Pleomassariaceae sp.	Noriler et al., 2018	Vd/Sa
Polyporus	*Polyporus* sp.	Hokama et al., 2017	Vd
Pseudofusicoccum	*Pseudofusicoccum* sp.	Noriler et al., 2018	Vd/Sa
	P. stromaticcum	Noriler et al., 2018	Vd/Sa
Roussoella	*Roussoella* sp.	Noriler et al., 2018	Vd
Sphaerisporangium	*Sphaerisporangium* sp.	Gos et al., 2017	Vd
	S. thermocarboxydus	Gos et al., 2017	Vd
Streptomyces	*S. sampsonii*	Savi et al., 2015a, 2015b	Vd
Xylaria	*Xylaria* sp.	Tonial et al., 2016	St
	X. cubensis	Figueiredo et al., 2018	Mi
Xylariaceae	Xylariaceae sp.**1, 2, 3, 4, 5**	Hokama et al., 2017; Noriler et al., 2018	Vd/Sa
Williamsia	*Williamsia* sp.	Gos et al., 2017	Vd

Note: Mi: *M. ilicifolia*, St: *S. terebinthifolius*, Vd: *V. divergens*, Sa: *S. adstringens*.
The names in **bold** represent new species or genera.

the ability of the endophyte to produce active compounds in the presence of a pathogenic strain is assessed (Fierro-Cruz et al., 2017). First, the endophytic and pathogenic strains are cultured on a specific medium for the growth of fungi or actinomycetes, during the time necessary for the development of the microorganism (Savi et al., 2015a, 2015b). One disc (6 mm) of the endophyte and one of the phytopathogen are inoculated on opposite sides of the Petri dish and incubated under favorable conditions for the development of endophytes and pathogens. The inhibition percentage is calculated by comparing the diameter of the mycelial growth of the pathogen in dual culture with the diameter of the mycelial growth of the pathogen in the individual inoculation. The antimicrobial activity is classified as low (50–59%), moderate (60–69%) and high (≥70%) according to the percentage of inhibition (Noriler et al., 2018).

Once any promising isolates have been selected, a small-scale culture is performed using culture media containing different concentrations containing nitrogen and carbon sources at different temperatures to select the best condition to produce active secondary metabolites (Gos et al., 2017). The most commonly used media to produce metabolites of actinomycetes are M2, R5A and SG (Savi et al., 2015b), and for fungi the most used are PD (potato dextrose), Czapeck and ME (malt extract)

(Savi et al., 2018). Generally, the culture is filtered over celite, the biomass is extracted with MeOH and the supernatant is mixed with the XAD-16 resin and the metabolites which are retained on the resin are extracted with MeOH. The recovered organics are evaporated under vacuum at 40°C to yield the crude extracts (Savi et al., 2015a). The crude extracts produced in different culture media are evaluated for their antibacterial or antifungal activities, using the disc diffusion evaluation following the instructions of the Clinical and Laboratory Standards Institute (CLSI, 2015). The culture condition giving the most active extract is used for any subsequent large-scale cultivation of 8-10 liters and the metabolites are purified using various chromatographic techniques such as HPLC, Sephadex LH-20 and TLC (Savi et al., 2018). Pure compounds are then evaluated for their antimicrobial activity, and the compound identification is performed generally by mass spectra and NMR analysis (Savi et al., 2015b). The Work-up scheme for production, purification and identification of secondary metabolites produced by endophytes is shown in Figure 5.3.

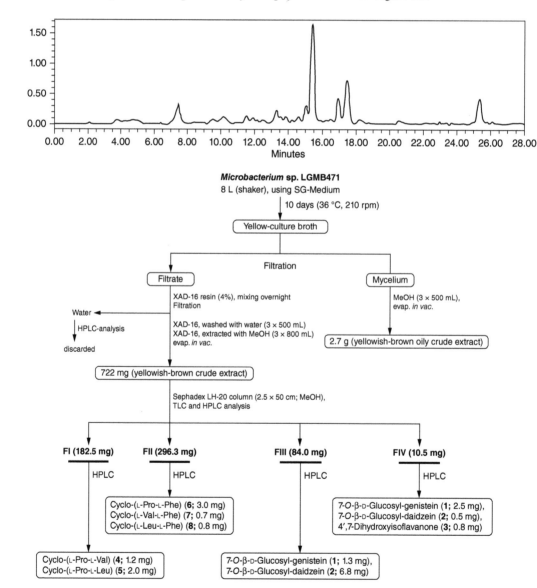

FIGURE 5.3 HPLC analysis of crude extract and work-up scheme for production, purification and identification of secondary metabolites produced by endophytes, using the data produced by Savi et al. (2015b).

In our screening program for metabolites produced by endophytes of different biomes in Brazil, we identified β-carbolines and indoles produced by a strain of actinomycetes, *Microbispora* sp. LGMB259, isolated from the medicinal plant *V. divergens* in the Pantanal – Brazil (Savi et al., 2015b). Among the isolated metabolites, the compound 1-vinil-β-carboline-3-carboxylic acid showed high antibacterial and cytotoxic activities. In addition to the identification of metabolites, the manuscript also highlighted the importance of the chemical group bound to carbon 1 in the biological activity of β-carbolines.

We also described a new species, *Phaeophleospora vochysiae* (LGMF1215b), isolated from *V. divergens* that produced secondary metabolites with considerable antifungal activity. Although the strain LGMF1215 was isolated as endophyte, it produced phytotoxic perylen-equinones as major compounds, cercosporin and isocercosporin, two toxic metabolites commonly produced by *Cercospora* species (Savi et al., 2018a). The resistance to cercosporins by *P. vochysiae* and by the host *V. divergens* may be due to cercosporin being produced associated with fungal hyphae, in this way the compound is present in reduced form, which makes the compound nontoxic or photoactive. In addition, strain LGMF1215 produced a new compound having antibacterial activity, 3-(*sec*-butyl)-6-ethyl-4,5-dihydroxy-2-methoxy-6-methylcyclohex-2-enone and absence of cytotoxic activity for human cell lines (Savi et al., 2018), suggesting the possibility to using this compound to treat clinical infections caused by bacteria.

In addition to the previously reported compounds, the endophytes isolated by our group also produced alkaloids (Gos et al., 2017; Tonial et al., 2016), diketopiperazines (Gos et al., 2017; Savi et al., 2018b), isocoumarins (Medeiros et al., 2018; Savi et al., 2019), perylenequinones (Savi et al., 2018a), dioxolanones (Savi et al., 2019), isoflavones (Savi et al., 2018b), tyrosols, phenolic acids (Savi et al., 2019) and two new compounds: 5-hydroxy-orthosporin and phenguignardic acid butyl ester. These new compounds exhibited antibacterial and cytotoxic activity against tumor cells (Savi et al., 2019). The structure of secondary metabolites belonging to different chemical classes are represented in Figure 5.4.

5.6   *DIAPORTHE TEREBINTHIFOLII*: A PROMISING SPECIES TO CONTROL THE CITRUS PHYTOPATHOGEN *PHYLLOSTICTA CITRICARPA*

In 2013, we described the new species *Diaporthe terebinthifolii* (Gomes et al., 2013), and since then, we have focused on exploring its biotechnological potential to act against the phytopathogen *Phyllosticta citricarpa* (Medeiros et al., 2018; Santos et al., 2016; Tonial et al., 2016). We demonstrated that the strain LGMF907 produces secondary metabolites in ME medium that completely inhibit the development of *P. citricarpa* (Figure 5.5) (Santos et al., 2016). Endophytes can be used to produce secondary metabolites used in the pharmaceutical or agricultural industries, or they can be used as a biological controller (Kandel et al., 2017). In order to study the possibility of using *D. terebinthifolii* in the biological control of *P. citricarpa*, we evaluated the ability of this endophyte to colonize citrus plants using a strain of *D. terebinthifolii* expressing the DsRed fluorescent protein (Santos et al., 2016). Microscopic analysis demonstrated that *D. terebinthifolii* colonizes the intercellular region and oil glands of citrus plants without causing any negative damage (Figure 5.5), suggesting the potential to be used in the biological control of Citrus Black Spot disease caused by *P. citricarpa*.

In a chemical analysis of *D. terebinthifolii* extracts, two major compounds were identified: orthosporin and diaporthin (Medeiros et al., 2018). However, neither of these compounds has activity against *P. citricarpa*, which suggests that the compounds responsible for the biological activity are produced in small amounts.

An alternative way to explore the metabolic potential of endophytes is to assess the long evolutionary coexistence of endophytes, epiphytes and phytopathogens with the host, resulting in different scenarios of interactions that need to be better understood and explored (Liu et al., 2017). The

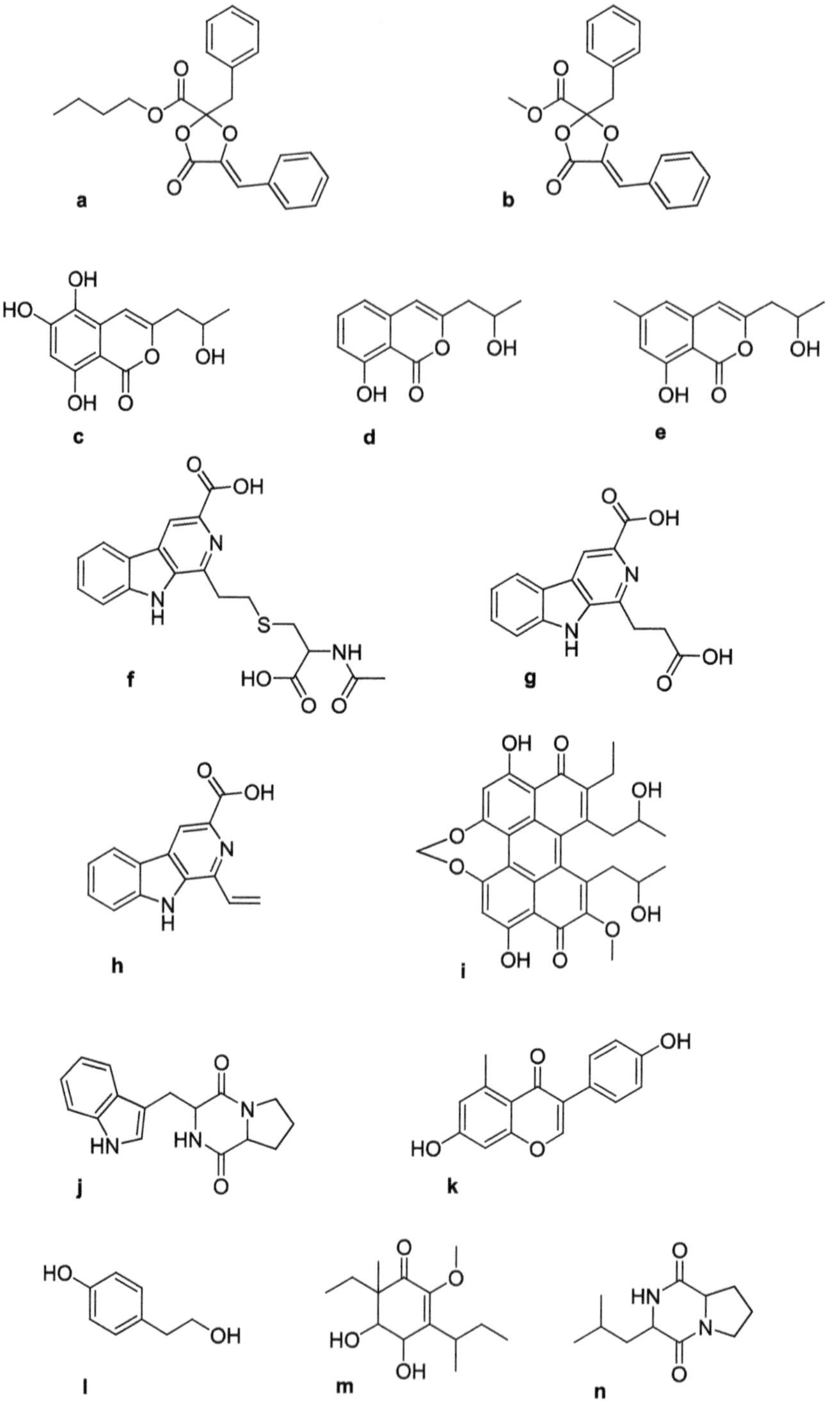

FIGURE 5.4 Representative chemical diversity of secondary metabolites produced by endophytic isolates of different Brazilian biomes: (A) phenguignardic acid butyl ester; (B) phenguignardic acid methyl ester; (C) 5-hydroxy-orthosporin; (D) orthosporin; (E) diaporthin; (F) kitasetaline; (G) methyl 1-(propionicacid)-β-carboline-3-carboxylic acid; (H) 1-vinil-β-carboline-3caboxylic acid; (I) cercosporin; (J) brevianamide; (K) 4′,7-dihydroxyisoflavanone; (L) tyrosol; (M) 3-(*sec*-butyl)-6-ethyl-4,5-dihydroxy-2-methoxy-6-methylcyclohex-2-enone; (N) Cyclo-(L-Pro-L-Leu). The chemical structures were obtained using the Chemdraw software (https://chemistry.com.pk/software/chemdraw-free/).

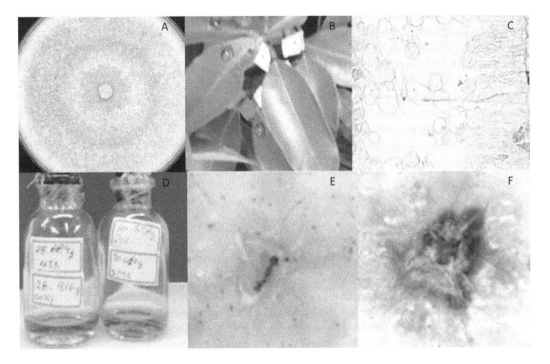

FIGURE 5.5 Potential of *Diaporthe terebinthifolii* to be used in the biological control of *Phyllosticta citricarpa*: (A) *D. terebinthifolii*; (B) inoculation of *D. terebinthifolii* in citrus plant; (C) *D. terebinthifolii* colonizing citrus leaves without causing any damage; (D) extract of *D. terebinthifolii*; (E) extract of *D. terebinthifolii* (10 μL) inhibiting citrus black spot lesions in citrus fruits; (F) induced citrus black spot lesion.

classical approach used to select endophytes with biotechnological potential is based on the screening of a single strain in culture, as presented above (Tonial et al., 2016). However, this excludes the interaction between different microorganisms that occupy the same environment. Thus, co-cultivation of different microorganisms has been used to understand the chemical ecological interaction between different organisms and to induce the expression of inactive metabolic pathways, or even increase the production of active compounds (Reen et al., 2015). As an example, we have the HPLC profile of an endophytic strain, *D. terebinthifolii* LGMF907, cultivated in the presence and absence of the *P. citricarpa*, a pathogen of citrus plants. The co-culture of both endophytic and phytopathogenic microorganisms was performed under the same condition as the single culture. Preliminary HPLC analysis of the extract obtained from the co-cultivation showed a 10-fold increase of a compound eluting after 9:0 min, which is produced by *D. terebinthifolii* in the presence of *P. citricarpa*. This result suggests that the eluting compound at 9 min may be responsible for the biological activity of *D. terebinthifolii* against *P. citricarpa*. The chemical characterization of this compound is in progress.

5.7 CONCLUSION

Brazil has a high biodiversity and herein the diversity of endophytic microorganisms is highlighted. The endophytes that we have isolated from medicinal plants found in the Brazilian biomes of the Pantanal and Cerrado (Table 5.1) have represented a great source of bioactive molecules and represented in Figure 5.4 with some of them possessing a new molecular framework. The biotechnological advances contribute to increasing the importance of Brazilian diversity, and new species and bioactive compounds are waiting to be reported.

REFERENCES

Alho, C. J. 2008. Biodiversity of the Pantanal: response to seasonal flooding regime and to environmental degradation. Brazilian Journal of Biology, 68, 957–966.

Alho, C. J. 2011. Biodiversity of the Pantanal: its magnitude, human occupation, environmental threats and challenges for conservation. Brazilian Journal of Biology, 71, 229–232.

Amirabadi, F.; Sasannejad, S. 2016. Evaluation of the antimicrobial effects of various methods to disinfect toothbrushes contaminated with *Streptococcus mutans*. International Journal of Medical Research & Health Sciences, 5, 536–540.

Antonelli, A.; Sanmartín, I. 2011. Why are there so many plant species in the Neotropics? Taxon, 60, 403–414.

Arieira, J.; Cunha, C. N. 2006. Fitossociologia de uma floresta inundável monodominante de *Vochysia divergens* Pohl (Vochysiaceae), no Pantanal Norte, MT, Brasil. Acta Botanica Brasilica, 20, 569–580.

Atanasov, A. G.; Waltenberger, B.; Pferschy-Wenzig, E. M.; Linder, T.; Wawrosch, C.; Uhrin, P.; Temml, V.; et al. 2015. Discovery and resupply of pharmacologically active plant-derived natural products: a review. Biotechnology Advances, 33, 1582–1614.

Basu, S.; Bose, C.; Ojha, N.; Das, N.; Das, J.; Pal, M.; Khurana, S. 2015. Evolution of bacterial and fungal growth media. Bioinformation, 11, 182–184.

Bernardesa, N. R.; Heggdorne-Araújo, M.; Borges, I. F. J.; Almeida, F. M.; Amaral, E. P.; Lasunskai, E. B.; Muzitano, M. F.; Oliveira, D. B. 2014. Nitric oxide production, inhibitory, antioxidant and antimycobacterial activities of the fruits extract and flavonoid content of *Schinus terebinthifolius*. Revista Brasileira de Farmacognosia, 24, 644–650.

Bernardi-Wenzel, J.; García, A.; Filho, C. J. R.; Prioli, A. J.; Pamphile, J. A. 2010. Evaluation of foliar fungal endophyte diversity and colonization of medicinal plant *Luehea divaricata* Martius et Zuccarini. Biological Research, 43, 375–384.

BFG. 2015. Growing knowledge: an overview of seed plant diversity in Brazil. Rodriguésia, 66, 1085–1113.

Bodenhausen, N.; Horton, M. W.; Bergelson, J. 2013. Bacterial communities associated with the leaves and the roots of *Arabidopsis thaliana*. PLoS ONE, 8:e56329.

Bogusz, M.; Whelan, S. 2017. Tree estimation with and without alignment: new distance methods and benchmarking. Systematic Biology, 66, 218–231.

Bortalanza, L. B.; Ferreira, J.; Hess, S. C.; Delle Monache, F.; Yunes, R. A.; Calixto, J. B. 2002. Anti-allodynic action of the tormentic acid, a triterpene isolated from plant, against neuropathic and inflammatory persistent pain in mice. European Journal of Pharmacology, 453, 203–208.

Castro, R. D.; Oliveira, J. A.; Vasconcelos, L. C.; Maciel, P. P.; Brasil, V. L. M. 2014. Brazilian scientific production on herbal medicines used in dentistry. Revista Brasileira de Plantas Medicinais, 16, 618–627.

Cipriani, T. R.; Mellinger, C. G.; de Souza, L. M.; Baggio, C. H.; Freitas, C. S.; Marques, M. C.; Gorin, P. A. J. et al. 2009. Polygalacturonic acid: another anti-ulcer polysaccharide from the medicinal plant *Maytenus ilicifolia*. Carbohydrate Polymers, 78, 361–363.

Chelius, M. K.; Triplett, E. W. 2001. The diversity of archaea and bacteria in association with the roots of *Zea mays* L. Microbiology and Ecology, 41, 252–263.

Chitrampalam, P.; Nelson, B. D. 2014. Effect of *Fusarium tricinctum* on growth of soybean and molecular-based method of identification. Plant Health Progress, doi:10.1094/PHP-RS-14-0014.

Colombo, A. F.; Joly, C. A. 2010. Brazilian Atlantic Forest *lato sensu*: the most ancient Brazilian forest, and a biodiversity hotspot, is highly threatened by climate change. Brazilian Journal of Biology, 70, 697–708.

Corrêa, M. F. P.; Ventura, T. L. B.; Muzitano, M. F.; dos Anjos da Cruz, E.; Bergonzi, M. C.; Bilia, A. R.; Rossi-Bergmann, B.; Soares Costa, S. 2018. Suppressive effects of *Vochysia divergens* aqueous leaf extract and its 5-methoxyflavone on murine macrophages and lymphocytes. Journal of Ethnopharmacology, 221, 77–85.

Costa, M. A.; Ishida, K.; Kaplum, V.; Koslyk, E. D. A.; Mello, J. C. P.; Ueda-Nakamura, T.; Dias Filho, B. P.; Nakamura, V. 2010. Safety evaluation of proanthocyanidin polymer-rich fraction obtained from stem bark of *Stryphnodendron adstringens* Barbatimão for use as a pharmacological agent. Regulatory Toxicology and Pharmacology, 58, 330–335.

de Sousa, N. C.; de Carvalho, S.; Spanó, M. A.; Graf, U. 2003. Absence of genotoxicity of a phytotherapeutic extract from *Stryphnodendron adstringens* mart. coville in somatic and germ cells of *Drosophila melanogaster*. Environmental and Molecular Mutagenesis, 41, 293–299.

Donadio, S.; Staver, M. J.; Katz, L. 1996. Erythromycin production in *Saccharopolyspora erythraea* does not require a functional propionyl-CoA carboxylasel. Molecular Microbiology, 19, 977–984.

dos Santos, A. F.; Sant'Ana, A. E. G. 2000. The molluscicidal activity of plants used in Brazilian folk medicine. Phytomedicine, 6, 431–438.

Fedel-Miyasato, E. S.; Kassuya, C. A. L.; Auharek, S. A.; Formagio, A. S. N.; Cardoso, C. A. L.; Mauro, M. O.; Cunha-Laura, A. L.; Monreal, A. C. D.; Vieira, M. C.; Oliveira, R. J. 2014. Evaluation of anti-inflammatory, immunomodulatory, chemopreventive and wound healing potentials from *Schinus terebinthifolius* methanolic extract. Revista Brasileira de Farmacognosia, 24, 565–575.

Federhen, S. 2015. Type material in the NCBI taxonomy database. Nucleic Acids Research, 28, (43), D1086–D1098.

Felfili, J. M.; Fagg, C. W. 2007. Floristic composition, diversity and structure of the "cerrado" *sensu stricto* on rocky soils in northern Goiás and Southern Tocantins, Brazil. Revista Brasileira de Botânica, 30, 375–385.

Ferrão, B. H.; Oliveira, H. B.; Molinari, R. F.; Teixeira, M.; Fontes, G. G.; Amaros, M. O. F.; Rosa, M. B.; Carvalho, C. A. 2014. Importância do conhecimento tradicional no uso de plantas medicinais em Buritis, MG, Brasil. Ciência e Natura, 36. https://www.redalyc.org/articulo.oa.

Fierro-Cruz, J. E.; Jiménez, P.; Coy-Barrera, E. 2017. Fungal endophytes isolated from *Protium heptaphyllum* and *Trattinnickia rhoifolia* as antagonistic of *Fusarium oxysporum*. Revista Argentina de Microbiologia, 49, 255–263.

Figueiredo, J. A. G.; Savi, D. C.; Goulin, E. H.; Tonial, F.; Stringari, D.; Kava, V.; Terasawa, L. V. G.; Glienke, C. 2018. Antagonistic activity and agrotransformation of *Xylaria cubensis*, isolated from the medicinal plant *Maytenus ilicifolia*, against *Phyllosticta citricarpa*. Current Biotechnology, 7, 59–64.

Forzza, R. C.; Baumgratz, J. F. A.; Bicudo, C. E. M.; Carvalho Jr, A. A.; Costa, A.; Costa, D. P.; Hopkins, M. et al. 2010. Catálogo de plantas e fungos do Brasil. Andrea Jakobsson Estúdio Editorial, Jardim Botânico do Rio de Janeiro, Rio de Janeiro. 870p.; 830p.

Forzza, R. C.; Baumgratz, J. F. A.; Bicudo, C. E. M.; Canhos, D. A. L.; Carvalho Jr, A. A.; Costa, A.; Costa, D. P. et al. 2012. New Brazilian floristic list highlights conservation challenges. BioScience, 62, 39–45.

Gilchrist-Saavedra, L.; Fuentes-Dávila, G.; Martínez-Cano, C.; López Atilano, R. M.; Duveiller, E.; Singh, R. P.; Henry, M. et al. 2006. Practical guide to the identification of selected diseases of wheat and barley (2nd ed.). Mexico, DF: CIMMYT.

Gomes, R. R.; Glienke, C.; Videira, S. I.; Lombard, L.; Groenewald, J. Z.; Crous, P. 2013. *Diaporthe*: a genus of endophytic, saprobic and plant pathogenic fungi. Persoonia, 31, 1–41.

Gomes-Figueiredo, J.; Pimentel, I. C.; Vicente, V. A.; Pie, M. R.; Kava-Cordeiro, V.; Galli-Terasawa, L.; Pereira, J. O.; de Souza, A. Q.; Glienke, C. 2007. Bioprospecting highly diverse endophytic *Pestalotiopsis* spp. with antibacterial properties from *Maytenus ilicifolia*, a medicinal plant from Brazil. Canadian Journal of Microbiology, 53, 1123–1132.

Gos, F. M. W. R.; Savi, D. C.; Shaaban, K. A.; Thorson, J. S.; Aluizio, R.; Possiede, Y. M.; Rohr, J.; Glienke, C. 2017. Antibacterial activity of endophytic actinomycetes isolated from the medicinal plant *Vochysia divergens* Pantanal, Brazil. Frontiers in Microbiology, 6, 1642.

Hardoim, P. R.; van Overbeek, L. S.; Berg, G.; Pirttila, A. M.; Company, S.; Campisano, A.; Doring, M.; Sessitsch, A. 2015. The hidden world within plants: ecological and evolutionary considerations for defining functioning of microbial endophytes. Microbiology and Molecular Biology Review, 79, 293–320.

Hess, S. C.; Brum, R. L.; Honda, N. K.; Cruz, A. B.; Moretto, E.; Cruz, R. B.; Messana, I.; Ferrari, F.; Cechinel Filho, V.; Yunes, R. A. 1995. Antibacterial activity and phytochemical analysis of *Vochysia divergens* Vochysiaceae. Journal of Ethnopharmacology, 47, 97–100.

Hess, S. C.; Monache, F. D. 1999. Divergioic acid, a triterpene from *Vochysia divergens*. Journal of Chemistry Society, 10, 104–106.

Hokama, Y.; Savi, D. C.; Assad, B.; Aluizio, R.; Gomes-Figueiredo, J.; Adamoski, D.; Possiede, Y. M. et al. 2017. Endophytic fungi isolated from *Vochysia divergens* in the pantanal, Mato Grosso do Sul: diversity, phylogeny and biocontrol of *Phyllosticta citricarpa* in Endophytic Fungi: Diversity, Characterization and Biocontrol, 4th Edn, ed E. Hughes (Hauppauge, NY: Nova), 1–25.

Jalgaonwala, R. E.; Mohite, B. V.; Mahajan, R. T. 2012 A review: Natural products from plant associated endophytic fungi. Journal of Microbiology and Biotechnology Research, 1, 21–32.

Jia, M.; Chen, L.; Xin, H. L.; Zheng, C. J.; Rahman. K.; Han, T.; Qin, L. P. 2016. A friendly relationship between endophytic fungi and medicinal plants: a systematic review. Frontiers in Microbiology, 7, 906.

Jorge, R. M.; Leite, J. P. V.; Oliveira, A. B.; Tagliati, C. A. 2004. Evaluation of anticonceptive, anti-inflammatory and antiulcerogenic activities of *Maytenus ilicifolia*. Journal of Ethnopharmacology, 94, 93–100.

Kandel, S. L.; Joubert, P. M.; Doty, S. L. 2017. Bacterial endophyte colonization and distribution within plants. Microorganisms. 5 doi: 10.3390/microorganisms5040077.

Klindworth, A.; Pruesse, E.; Schweer, T.; Peplies, J.; Quast, C.; Horn, M. 2013. Evaluation of general 16S ribosomal RNA gene PCR primers for classical and next-generation sequencing-based diversity studies. Nucleic Acids Research, 41:e1. doi: 10.1093/nar/gks808.

Kusari, S.; Hertweck, C.; Spiteller, M. 2012. Chemical ecology of endophytic fungi: origins of secondary metabolites. Chemical Biology, 19, 792–798.

Laich, F.; Fierro, F.; Martin, J. F. 2002. Production of penicillin by fungi growing on food products: identification of a complete penicillin gene cluster in *Penicillium griseofulvum* and truncated cluster in *Penicillium verrucosum*. Applied Environmental Microbiology, 68, 1211–1219.

Lee, T. K.; Van Doan, T.; Yoo, K.; Choi, S.; Kim, C.; and Park, J. 2010. Discovery of commonly existing anode biofilm microbes in two different wastewater treatment MFCs using FLX titanium pyrosequencing. Applied Microbiology and Biotechnology, 87, 2335–2343.

Levine, D. P. 2006. Vancomycin: a history. Clinical Infection Diseases, 42, 5–12.

Lima, J. S.; Figueiredo, J. G.; Gomes, R. G.; Stringari, D.; Goulin, E. H.; Adamoski, D.; Kava-Cordeiro, V.; Galli-Terasawa, L. V.; Glienke. C. 2012. Genetic diversity of *Colletotrichum* spp. an endophytic fungi in a medicinal plant Brazilian pepper tree. ISRN Microbiology, 2012, 215716.

Liu, H.; Carvalhais, L. C.; Crawford, M.; Singh, E.; Dennis, P. G.; Peterse, C. M. J.; Schenk, P. M. 2017. Inner plant values: diversity colonization and benefits from endophytic bacteria. Frontiers in Microbiology, 8, 2552.

Ma, C. X.; Sun, Y. H.; Wang, H. Y. 2015. ABCB1 polymorphisms correlate with susceptibility to adult acute leukemia and response to high-dose methotrexate. Tumour Biology, 36, 7599–7606.

Matsubara, A.; Hurtado, J. E. 2013. Isolation and characterization of actinomycetes from acidic cultures of ores and concentrates. Advanced Materials Research, 825, 406–409.

McAllister, S. M.; Davis, R. E.; McBeth, J. M.; Tebo, B. M.; Emerson, D.; Moyer, C. L. 2011. Biodiversity and emerging biogeography of the neutrophilic iron-oxidizing Zetaproteobacteria. Applied. Environmental Microbiology, 77, 5445–5457.

Medeiros, A. G.; Savi, D. C.; Mitra, P.; Shaaban, K. A.; Jha, A. K.; Thorson, J. S.; Rohr, J. et al. 2018. Bioprospecting of *Diaporthe terebinthifolii* LGMF907 for antimicrobial compounds. Folia Microbiologica, 63, 499–505.

Militão, T.; Gómez-Díaz, E.; Kaliontzopoulou, A.; González-Solís, J. 2014. Comparing multiple criteria for species identification in two recently diverged Seabirds. PLoS One, 9, e115650.

Mittermeier, R. A.; Robles Gil, P.; Mittermeier, C. G. 1997. Megadiversity: Earth's Biologically Wealthiest Nations. CEMEX and Agrupación Sierra Madre.

Monciardini, P.; Iorio, M.; Maffioli, S.; Sosio, M.; Donadio, S. 2014. Discovering new bioactive molecules from microbial sources. Microbiology and Biotechnology, 7, 209–220.

Morey, A. T.; de Souza, F. C.; Santos, J. P.; Pereira, C. A.; Cardoso, J. D.; de Almeida, R. S.; Costa, M. A.; et al. 2016. Antifungal activity of condensed tannins from *Stryphnodendron adstringens*: effect on *Candida tropicalis* growth and adhesion properties. Current Pharmacology and Biotechnology, 17, 365–375.

Myers, N.; Mittermeier, R. A.; Mittermeier, C. G.; Da Fonseca, G. A.; Kent, J. 2000. Biodiversity hotspots for conservation priorities. Nature, 403, 853–858.

Noriler, S. A.; Savi, D. C.; Aluizio, R.; Palácio-Cortes, A. M.; Possiede, Y. M.; Glienke, C. 2018. Bioprospecting and structure of fungal endophyte communities found in the Brazilian biomes Pantanal and Cerrado. Frontiers in Microbiology, 9, https://doi.org/10.3389/fmicb.2018.01526.

Oliveira-Filho, A. T.; Ratter, J. A. 1995. A study of the origin of central Brazilian forests by the analysis of plant species distribution patterns. Edinburgh Journal of Botany, 52, 141–194.

Pinto, S. C. G.; Bueno, F. G.; Panizzon, G. P.; Morais, G.; dos Santos, P. V. P.; Baesso, M. L.; de Souza Leite-Mello, E. V.; de Mello, J. C. P. 2015. *Stryphnodendron adstringens*: clarifying wound healing in Streptozotocin-induced diabetic rats. Planta Medica, 81, 1090–1096.

Pott, A.; Pott, V. J.; Sobrinho, A. A. B. 2004. Plantas úteis à sobrevivência no Pantanal. In Anais do IV Simpósio sobre recursos Naturais e Sócio econômicos do Pantanal. Empresa Brasileira de Pesquisa Agropecuária, Corumbá, 81–92.

Queiroga, C. L.; Silva, G. F.; Dias, P. C.; Possenti, A.; Carvalho, J. E. 2000. Evaluation of the antiulcerogenic activity of friedelan-3β-ol and friedelin isolated from *Maytenus ilicifolia* (Celastraceae). Journal of Ethnopharmacology, 72, 465–468.

Raja, H. A.; Miller, A. N.; Pearce, C. J.; Oberlies, N. H. 2017. Fungal identification using molecular tools: a primer for the natural products research community. Journal of Natural Products, 80, 759–770.

Rajamanikyam, M.; Vadlapudi, V.; Ramars, A.; Upadhyayula, S. M. 2017. Endophytic fungi novel resources of natural therapeutics. Brazilian Archives of Biology and Technology, 60, e17160542.

Ramalashmi, K.; Prasanna Vengatesh, K.; Magesh, K.; Sanjana, R.; Siril, J. S.; Ravibalan, K. 2018. A potential surface sterilization technique and culture media for the isolation of endophytic bacteria from *Acalypha indica* and its antibacterial activity. Journal of Medicinal Plants Studies, 6, 181–184.

Reen, F. J.; Romano, S.; Dobson, A. D. W.; O'Gara, F. 2015. The sound of silence: activating silent biosynthetic gene clusters in marine microorganism. Marine Drugs, 13, 4754–4783.

Ricardo, L. M.; Dias, B. M.; Muqqe, F. L. B.; Leite, V. V.; Brandão, M. G. L. 2018. Evidence of traditionality of Brazilian medicinal plants: the case studies of *Stryphnodendron adstringens* mart. coville barbatimão barks and *Copaifera* spp. copaiba oleoresin in wound healing. Journal of Ethnopharmacology, 219, 319–336.

Richter, R.; von Reuss, S. H.; Köning, W. A. 2010. Spirocyclopropane-type sesquiterpene hydrocarbons from *Schinus terebinthifolius* Raddi. Phytochemistry, 71, 1371–1374.

Rodrigues, A. P.; Andrade, L. H. C. 2014. An ethnobotanical survey of medicinal plants used by the rural community of Inhamã state of Pernambuco Notheastern Brazil. Revista Brasileira de Plantas Medicinais, 16, 721–730.

Rodriguez, R. J.; White, J. F.; Arnold, A. E.; Redman, R. S. 2009. Fungal endophytes: diversity and functional roles. New Phytology, 182, https://doi.org/10.1111/j.1469-8137.2009.02773.x.

Romão-Dumaresq, A. S.; Dourado, M. N.; Fávaro, L. C. L.; Mendes, R.; Ferreira, A.; Araújo, W. L. 2016. Diversity of cultivated fungi associated with conventional and transgenic sugarcane and the interaction between endophytic *Trichodera virens* and the host plant. PLoS One, 11, e0158974.

Rosas, E. C.; Correa, L. B.; Pádua, T.; de A Costa, T. E.; Mazzei, J. L.; Heringer, A. P.; Bizarro, C. A.; Kaplan, M. A.; Figueiredo, M. R.; Henriques, M. G. 2015. Anti-inflammatory effect of *Schinus terebinthifolius* Raddi hydroalcoholic extract on neutrophil migration in Zymosan-induced arthritis. Journal of Ethnopharmacology, 175, 490–498.

Roy, S.; Sharma, S. 2015. Isolation and identification of a novel endophyte from a plant *Amaranthus spinosus*. International Journal of Current Microbiology and Applied Science, 4, 785–798.

Sá, F. A. S.; De Paula, J. A. M.; Dos Santos, P. A.; Oliveira, L. A. R.; Oliveira, G. A. R.; Lião, L. M.; De Paula, J. R. et al. 2017. Phytochemical analysis and antimicrobial activity of *Myrcia tomentosa* (Aubl.) DC. leaves. Molecules, 22, 1100–1110.

Salem, M. Z. M.; El-Hefny, M.; Ali, H. M.; Elansary, H. O.; Nasser, R. A.; El-Settawy, A. A. A.; El Shanhorey, N.; Ashmawy, N. A.; Salem, A. Z. M. 2018. Antibacterial activity of extracted bioactive molecules of *Schinus terebinthifolius* ripened fruits against some pathogenic bacteria. Microbiology and Pathology, 120, 119–127.

Santos, P. J.; Savi, D. C.; Gomes, R. R.; Goulin, E. H.; Senkiv, C. C.; Tanaka, F. A. O.; Almeida, A. M. R. et al. 2016. *Diaporthe endophytica* and *D. terebinthifolii* from medicinal plants for biological control of *Phyllosticta citricarpa*. Microbiological Research, 186, 153–160.

Savi, D. C.; Haminiuk, C. W. I.; Sora, G. T. S.; Adamoski, D. M.; Kenski, J.; Winnischofer, S. M. B.; Glienke, C. 2015a. Antitumor antioxidant and antibacterial activities of secondary metabolites extracted by endophytic actinomycetes isolated from *Vochysia divergens*. International Journal of Pharmaceutical, Chemical and Biological Sciences, 5, 347–356.

Savi, D. C.; Aluizio, R.; Galli-Terasawa, L.; Kava, V.; Glienke, C. 2016. 16S-gyrB-rpoB multilocus sequence analysis for species identification in the genus *Microbispora*. Antonie Van Leeuwenhoek, 109, 801–815.

Savi, D. C.; Shaaban, K. A.; Gos, F. M. W. R.; Ponomareva, L. V.; Thorson, J. S.; Glienke, C.; Rohr, J. 2018a *Phaeophleospora vochysiae* Savi & Glienke sp. nov. isolated from *Vochysia divergens* found in the Pantanal Brazil produces bioactive secondary metabolites. Scientific Reports, 8, 3122, https://DOI:10.1038/s41598-018-21400-2.

Savi, D. C.; Shaaban, K. A.; Gos, F. M. W.; Thorson, J. S.; Glienke, C.; Rohr, J. 2018b. Secondary metabolites produced by *Microbacterium* sp. LGMB471 with antifungal activity against the phytopathogen *Phyllosticta citricarpa*. Folia Microbiologica, https://doi.org/10.1007/s12223-018-00668-x.

Savi, D. C.; Shaaban, K. A.; Mitra, P.; Ponomareva, L. V.; Thorson, J. S.; Glienke, C.; Rohr, J. 2019. Secondary metabolites produced by the citrus phytopathogen *Phyllosticta citricarpa*. Journal of Antibiotics, https://doi.org/10.1038/s41429-019-0154-3DO.

Savi, D. C.; Shaaban, K. A.; Vargas, N.; Ponomareva, L. V.; Possiede, Y. M.; Thorson, J. S.; Glienke, C.; Rohr, J. 2015b. *Microbispora* sp. LGMB259 endophytic actinomycete isolated from *Vochysia divergens* Pantanal Brazil producing β-carbolines and indoles with biological activity. Current Microbiology, 70, 345–354.

Schoch, C. L.; Seifert, K. A.; Huhndorf, S.; Robert, V.; Spouge, J. L.; Levesque, C. A.; Chen, W.; Fungal Barcoding Consortium. 2012. Nuclear ribosomal internal transcribed spacer (ITS) region as a universal DNA barcode marker for fungi. Proceedings of the National Academy of Sciences, 109, 6241–6246.

Silva, A. O.; Savi, D. C.; Gomes, F. B.; Gos, F. M. W. R.; Silva, G. J.; Glienke, C. 2017. Identification of *Colletotrichum* species associated with postbloom fruit drop in Brazil through GAPDH sequencing analysis and multiplex PCR. European Journal of Plant Pathology, 147, 731–748.

Silva, M. M.; Iriguchi, E. K. K.; Kassuya, C. A. L.; Vieira, M. C.; Foglio, M. A.; Carvalho, J. E.; Ruiz, A. L. T. G.; Souza, K. P.; Formagio, A. S. N. 2017. *Schinus terebinthifolius*: phenolic constituents and in vitro antioxidant antiproliferative and *in vivo* anti-inflammatory activities. Revista Brasileira de Farmacologia, 27, 445–452.

Sogin, M. L.; Morrison, H. G.; Huber, J. A.; Welch, D. M.; Huse, S. M.; Neal, P. R. 2006. Microbial diversity in the deep sea and the underexplored "rare biosphere". Proceedings of the National Academy of Sciences, 103, 12115–12120.

Tiberti, L. A.; Yariwake, J. H.; Ndjoko, K.; Hostettmann, K. 2007. Identification of flavonols in leaves of *Maytenus ilicifolia* and *M. aquifolium* (Celastraceae) by LC/UV/MS analysis. Journal of Chromatography, 846, 378–384.

Tonial, F.; Maia, B. H.; Gomes-Figueiredo, J. A.; Sobottka, A. M.; Bertol, C. D.; Nepel, A.; Savi, D. C.; Vicente, V. A.; Gomes, R. R.; Glienke, C. 2016. Influence of culturing conditions on bioprospecting and the antimicrobial potential of endophytic fungi from *Schinus terebinthifolius*. Current Microbiology, 72, 173–183.

Tonial, F.; Maia, B. H. L. N. S.; Sobottka, A. M.; Savi, D. C.; Vicente, V. A.; Gomes, R. R.; Glienke, C. 2017. Biological activity of *Diaporthe terebinthifolii* extracts against *Phyllosticta citricarpa*. FEMS Microbiology Letters, 364, https://doi: 10.1093/femsle/fnx026.

VanderMolen, K. M.; Raja, H. A.; El-Elimat, T.; Oberlies, N. H. 2013. Evaluation of culture media for the production of secondary metabolites in a natural products screening program. AMB Express, 3, 71, https://doi: 10.1186/2191-0855-3-71.

Vilgalys, R.; Hester, M. 1990. Rapid genetic identification and mapping of enzymatically amplified ribosomal DNA from several *Cryptococcus* species. Journal of Bacteriology, 172, 4238–4246.

Vitorino, L. C.; Bessa, L. A. 2017. Technological microbiology: development and applications. Frontiers in Microbiology, 8, 827, https://doi.org/10.3389/fmicb.2017.00827.

Walker, J. J.; Pace, N. R. 2007. Phylogenetic composition of rocky mountain endolithic microbial ecosystems. Applied Environmental and Microbiology, 73, 3497–3504.

Wang, W.; Zhai, Y.; Cao, L.; Tan, H.; Zhang, R. 2016. Endophytic bacterial and fungal microbiota in sprouts roots and stems of rice *Oryza sativa* L. Microbiological Research, 188–189, https://doi.org/10.1016/j.micres.2016.04.009.

Yang, Y.; Wang, Y.; Zhao, M.; Jia, H.; Li, B.; Xing, D. 2018. Tormentic acid inhibits IL-1B-induced chondrocyte apoptosis by activating the Pl3k/Akt signaling pathway. Molecular Medicine Reports, 17, 4753–4758.

6 Environmental Factors Impacting Bioactive Metabolite Accumulation in Brazilian Medicinal Plants

Camila Fernanda de Oliveira Junkes,
Franciele Antonia Neis, Fernanda de Costa,
Anna Carolina Alves Yendo,
and Arthur Germano Fett-Neto
Plant Physiology Laboratory, Center for Biotechnology
and Department of Botany, Federal University of Rio Grande
do Sul (UFRGS), Campus do Vale, Porto Alegre, Brazil

CONTENTS

6.1 INTRODUCTION

Secondary metabolites help protect plants against attacks by insects, herbivores and pathogens, improve survival under abiotic stresses, besides participating in allelopathy and interactions with pollinators, dispersers and symbionts. Strategies for production of metabolites in culture have been developed based on these properties to increase the yield of metabolites of interest. It is well-known that plant treatments with elicitors, or exposure to pathogens and herbivores, cause a series of defense reactions, including the accumulation of a variety of metabolites such as phytoalexins in intact plants or in plant cell cultures (Matsuura et al., 2018). Elicitors can be abiotic, such as metal ions and inorganic compounds, and biotic, such as fungi, bacteria, viruses or herbivores, plant cell wall components, as well as chemicals derived from pathogens or herbivores and plant defense signaling molecules.

The richness of the Brazilian flora is remarkable. Knowledge on the complexity and medicinal potential of its chemical components remains only partial, despite massive research efforts. The eco-chemical roles of secondary metabolites, which are often the pharmacologically active

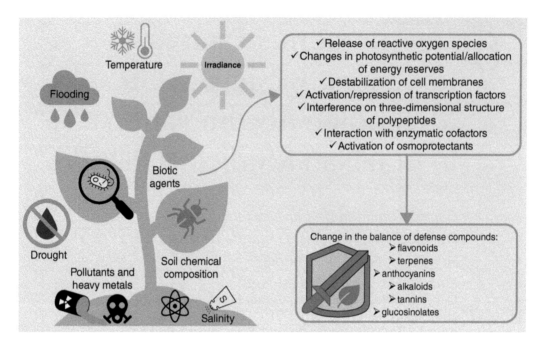

FIGURE 6.1   Scheme of some key steps and players involved in environmental or external modulation of medicinal plant secondary metabolite profile.

molecules in plants, provide a working platform for improving yields of target products. Plant metabolic profile responses to environmental stimuli can vary with taxon, genetic makeup and chemical nature of metabolites, and, as such, should be addressed on a case-by-case basis whenever possible. Nonetheless, some general trends are identifiable and can serve as useful guidelines to modulate metabolic pathways toward compounds of interest through environmental variables. First, some of the key players and steps involved in environmental or external modulation of medicinal plant secondary metabolite profile (summarized in Figure 6.1) are explored.

Second, the Brazilian territory encompasses a wide variety of climates and regional characteristics, such as precipitation and temperature regimes. The seasonality in various regions is explored in depth in a section on the topic and a list of all plant species exemplified in this chapter is shown in Table 6.1.

6.2   IRRADIANCE

Both irradiance intensity and quality are important environmental factors for plant growth and development, affecting plant's morphological, biochemical and physiological parameters (Gobbo-Neto and Lopes, 2007). Compounds such as flavonoids, anthocyanins, tannins and carotenes are known to respond to irradiance stress because of their capacity of absorbing UV radiation and antioxidant activity (Bian et al., 2015). The molecular structure of phenolic compounds contributes to convert short-wave, high-energy-destructive radiation into longer wavelengths, which are less destructive to the photosynthetic apparatus and other cellular structures (Teixeira et al., 2013). The effects of irradiance intensity and quality on the production of secondary metabolites in plants are often species-specific, but in general increasing their concentration can protect plants' structures

TABLE 6.1

Summary of the Names of All Plant Species, Their Respective Family and Common Names Discussed in This Chapter

Scientific Name	Family	Common Names
Aloe arborescens (sin. *Aloe principis*)	Asphodelaceae	Babosa, babosa-de-arbusto (aloe)
Aloysia citriodora (sin. *Aloysia triphylla*)	Verbenaceae	Limonete, cedrina, cidró (lemon beebrush)
Alternanthera philoxeroides (sin. *Alternanthera philoxerina*)	Amaranthaceae	Mata-bicho, perna-de-saracura (alligator weed)
Amburana cearensis (sin. *Amburana claudii*)	Fabaceae	Amburana, imburana, cumaru, Cerejeira
Anadenanthera colubrina (sin. *Acacia colubrina*)	Fabaceae	Angico-branco
Araucaria angustifolia (sin. *Araucaria dioica*)	Araucariaceae	Araucaria (Brazilian pine)
Baccharis dentata (sin. *Baccharis macrodonta*)	Asteraceae	Vassourinha
Baccharis dracunculifolia (sin. *Baccharis pulverulenta*)	Asteraceae	Vassourinha, alecrim-do-campo
Baccharis trimera (sin. *Baccharis genistelloides* var. *trimera*)	Asteraceae	Carqueja
Bauhinia cheilantha (sin. *Pauletia cheilantha*)	Fabaceae	Pata-de-vaca, mororó
Cordia curassavica (sin. *Cordia verbenacea*)	Boraginaceae	Erva baleeira, Maria milagrosa
Cunila galioides (sin. *Hedeoma glaziovii*)	Lamiaceae	Poejo
Dimorphandra mollis	Fabaceae	Fava-de-anta
Elionurus muticus (sin. *Elionurus marunguensis*)	Poaceae	Capim cheiroso (lemon grass)
Eugenia uniflora (sin. *Stenocalyx uniflorus*)	Myrtaceae	Pitanga (Surinam cherry)
Hevea brasiliensis (sin. *Hevea camargoana*)	Euphorbiaceae	Seringueira (rubber tree)
Hydrocotyle umbellata (sin. *Hydrocotyle caffra*)	Araliaceae	Acariçoba
Hypericum brasiliense (sin. *Sarothra brasiliensis*)	Hypericaceae	Milfacadas, alecrim-bravo, erva-da-vida
Hypericum polyanthemum (sin. *Hypericum rivulare*)	Hypericaceae	Hipérico
Hyptis carpinifolia (sin. *Mesosphaerum carpinifolium*)	Lamiaceae	Rosmaninho, mata-pasto
Ilex paraguariensis (sin. *Ilex curitibensis*)	Aquifoliaceae	Mate, erva mate
Bryophyllum pinnatum (sin. *Kalanchoe pinnata*)	Crassulaceae	Folha-da-fortuna
Lafoensia pacari	Lythraceae	Dedaleiro, mangaba brava, pacari
Lippia origanoides (sin. *Lippia schomburgkiana*)	Verbenaceae	Alecrim-do-campo
Lychnophora ericoides (sin. *Lychnophora rosmarinus*)	Asteraceae	Arnica-do-campo, arnica-falsa, candeia
Martianthus leucocephalus (sin. *Hyptis leucocephala*)	Lamiaceae	Bamburral
Maytenus ilicifolia (sin. *Celastrus spinifolius*)	Celastraceae	Espinheira santa
Mikania glomerata (sin. *Mikania glomerata* var. *montana*)	Asteraceae	Guaco
Mikania laevigata	Asteraceae	Guaco
Myrcia tomentosa (sin. *Aguava tomentosa*)	Myrtaceae	Araçazinho
Ocimum gratissimum (sin. *Ocimum dalabaense*)	Myrtaceae	Goiaba brava
Ocimum carnosum (sin. *Ocimum selloi*)	Lamiaceae	Alfavaca, manjericão
Achetaria azurea (sin. *Otacanthus azureus*)	Lamiaceae	Aniseto, alfavaca anis, anis-do-campo
Palicourea rigida (sin. *Uragoga rigida*)	Scrophulariaceae	Erva-copaíba
Passiflora alata (sin. *Passiflora phoenicia*)	Rubiaceae	Chapéu-de-couro, bate-caixa
Passiflora edulis (sin. *Passiflora vernicosa*)	Passifloraceae	Maracujá (passion fruit)
Passiflora incarnata (sin. *Passiflora edulis* var. *kerii*)	Passifloraceae	Maracujá azedo
Passiflora ligularis (sin. *Passiflora lowei*)	Passifloraceae	Maracujá-guaçu (purple passion flower)

(Continued)

TABLE 6.1 *(Continued)*
Summary of the Names of All Plant Species, Their Respective Family and Common Names Discussed in This Chapter

Scientific Name	Family	Common Names
Paullinia cupana var. *sorbilis* (sin. *Paullinia sorbilis*)	Passifloraceae	Granadilla
Pfaffia glomerata (sin. *Pfaffia divergens*)	Sapindaceae	Guaraná
Physalis angulata (sin. *Physalis esquirolii*)	Amaranthaceae	Ginseng brasileiro (Brazilian ginseng)
Pilocarpus jaborandi (sin. *Pilocarpus cearensis*)	Solanaceae	Camapu, joá-de-capote, buch-de-rã
Pilocarpus microphyllus	Rutaceae	Jaborandi
Piper aduncum (sin. *Piper reciprocum*)	Rutaceae	Jaborandi
Pisum sativum (sin. *Pisum vulgare*)	Piperaceae	Pimenta-de-macaco
Psidium guajava (sin. *Psidium igatemyense*)	Fabaceae	Ervilha (pea)
Psychotria brachyceras (sin. *Uragoga brachyceras*)	Myrtaceae	Goiaba (guava)
Psychotria leiocarpa (sin. *Psychotria tenella*)	Rubiaceae	Cafezinho-do-mato
Quillaja brasiliensis (sin. *Fontenellea brasiliensis*)	Rubiaceae	Cafeeiro-do-mato
Quillaja saponaria	Quillajaceae	Saboneteira, pau-sabão
Schinus terebinthifolia (sin. *Schinus mellisii*)	Quillajaceae	Soapbark
Theobroma cacao (sin. *Theobroma sapidum*)	Anacardiaceae	Aroeira vermelha, roeira mansa (Brazilian pepper tree)
Tithonia diversifolia (sin. *Helianthus quinquelobus*)	Malvaceae	Cacau (cocoa tree)
Valeriana glechomifolia	Asteraceae	Margaridão, girassol mexicano (Mexican sunflower)
Cordia curassavica (sin. *Varronia curassavica*)	Caprifoliaceae	Valeriana serrana
Aloe arborescens (sin. *Aloe principis*)	Boraginaceae	Erva-baleeira, Maria-preta

from photodamage. Some examples of impacts of irradiance quality and intensity on secondary metabolite profiles are listed in Table 6.2.

UV radiation is an effective elicitor for boosting the biosynthesis of various plant secondary metabolites. UV is divided into three classes: UV-C, UV-B and UV-A. Although UV-C is completely absorbed by atmospheric gases, this highly energetic radiation (200-280 nm) can severely affect metabolism, cell structure and DNA, depending on its intensity. For example, the concentration of the monoterpene indole alkaloid brachycerine, (Figure 6.2), which has antioxidant and antimutagenic properties, can be increased by approximately an order of magnitude upon UV-C exposure of *Psychotria brachyceras* Müll Arg. leaves (Matsuura et al., 2013). UV-C treatment provides a means for obtaining plant biomass with higher yields for pharmacological applications. Lewinski et al. (2015) used UV light to accelerate maturation of processed mate (*Ilex paraguariensis* A. St.-Hil.), resulting in increase of some dicaffeoylquinic acid isomers and decrease of methylxanthines, rutin and isomers of chlorogenic acids.

Quillaja brasiliensis Mart., a native species from Southern Brazil, is known as soldier's soap due to the fact that its leaves and bark yield persistent foam in water (Yendo et al., 2010). Red light or UV-C treatment of young plants and detached leaves of *Q. brasiliensis* promoted QB-90 accumulation, an immunoadjuvant triterpene saponin fraction with strong capacity to induce cellular and humoral immune response pathways with low toxicity to mammals (Yendo et al., 2015). These saponins showed remarkable structural similarities with saponins from *Quillaja saponaria* Molina bark, a related Chilean species and one of the major sources of industrial saponins that are used as adjuvants in vaccine formulations (Kauffmann et al., 2004). Irradiation-based treatments may be important in improving yields of immunoadjuvant saponins of soap tree, both before and after leaf harvest.

TABLE 6.2
Examples of Abiotic Stress and Its Effect on Production and/or Accumulation of Secondary Metabolites

Species	Affected Secondary Metabolite(s)	Tissue/Organ	Condition	Effect	References
Irradiance					
Quillaja brasiliensis	Triterpene saponin QB-90	Leaves (in stem and post-harvest)	Higher irradiance, red light and UV-C	Increase	De Costa et al. (2013), Yendo et al. (2015)
Kalanchoe pinnata	Quercetin 3-rhamnopyranoside	Leaves	UV-B	Higher diversity of phenolic compounds and increased quantity of quercitrin	Nascimento et al. (2015)
Piper aduncum	e-Nerolidol, linalol, α-humulene, cis-cadin-4-en-7-ol and caryophyllene	Leaves	Blue light	Increase	Pacheco et al. (2016)
Psychotria leiocarpa	GPV (N, β-D-glucopyranosylvincosamide)	Whole seedlings	Far red and blue light enrichment	Increase	Matsuura et al. (2016)
Otacanthus azureus	Essential oil	Shoots	Higher irradiance	Increase	Silva et al. (2006)
Temperature					
Psychotria brachyceras	Brachycerine	Leaves	Increased temperature	Brachycerine accumulation and protection of heat-sensitive species under severe heat stress	Magedans et al. (2017)
Ocimum selloi	Elimicin, trans-caryophyllene, germacrene D and bicyclogermacrene	Leaf essential oil	Temperature above 40°C	Reduction	David et al. (2006)
Lafoensia pacari	Phenolic metabolites	Leaves	Warmer months	Reduction	Sampaio et al. (2011)
Ilex paraguariensis	Total polyphenol content	Leaves	Drying at 45°C versus natural aging	Increase	Holowaty et al. (2015)
Water Availability and Osmotic Stress					
Bauhinia cheilantha	Flavonoid content	Leaves	Elevated rainfall	Mild reduction	Sobrinho et al. (2009)
Hypericum brasiliense	Betulinic acid; isouliginosin B; 1,5-dihydroxyxanthone and rutin	Shoots and roots	15 days of drought or waterlogging	Increase	Abreu and Mazzafera (2005)
Amburana cearenses	Phenolic compounds	Cotyledons	Water restriction	Reduction	Pereira et al. (2014)
Dimorphandra mollis	Quercetin, isoquercitrin and rutin	Plants and seedlings	Drought, flooding or 75 mM NaCl	Increase	Lucci and Mazzafera (2009a, 2009b)
Physalis angulata	Physalin	Leaves	13 days exposure to saline solution 0.9%	Increase	de Souza et al. (2013)

(Continued)

TABLE 6.2 (Continued)

Examples of Abiotic Stress and Its Effect on Production and/or Accumulation of Secondary Metabolites

Species	Affected Secondary Metabolite(s)	Tissue/Organ	Condition	Effect	References
Quillaja brasiliensis	Triterpene saponin (QB-90)	Leaves	150 mM NaCl or isosmotic concentrations of sorbitol and polyethylene glycol	Increase	de Costa et al. (2013)
Pilocarpus jaborandi	Pilocarpine	Leaves	Salt stress and hypoxia	Reduction	Avancini et al. (2003)
Alternanthera philoxeroides	Betacyanins	In vitro plants	400 mM salt stress	Increase	Ribeiro et al. (2014)
Mineral Composition and Fertilization					
Lafoensia pacari	Phenolic metabolites	Leaves	Foliar micronutrients Cu, Fe, Mn and Zn	Positive correlation	Sampaio et al. (2011)
Eugenia uniflora	Spathulenol and caryophyllene oxide	Leaves	S, Ca and Fe balance	Positive correlation	Costa et al. (2009)
	Selina-1,3,7 (11)-trien-8-one epoxide	Leaves	K, Cu and Mn	Positive correlation	
	Flavonoids and tannins	Leaves	Mn and Cu	Negative correlation	
Myrcia tomentosa	Oxygenated sesquiterpenes	Flowers and leaves	Foliar concentration of Cu and P	Positive correlation	Sá et al. (2012)
Palicourea rigida	Sesquiterpene hydrocarbons	Flowers and leaves	Foliar concentration of Cu and P	Negative correlation	Morel et al. (2011)
	Iridoids	Aerial parts	Low fertility and acidic soil	Increase	
	Loganin	Aerial parts	All the soil macro and micronutrients except for Mg	Negative correlation	
Tithonia diversifolia	Esters of trans-cinnamic acid	Root	Soil nutrients Ca, Mg, P, K and Cu	Negative correlation	Sampaio et al. (2016)
Lychnophora ericoides	Monoterpenes and sesquiterpene	Leaf essential oil	Organic matter and P in soils	Positive correlation	Curado et al. (2006)
Passiflora incarnata	Total flavonoids	Leaves	Soil elements Fe, B and Cu	Negative correlation	Reimberg et al. (2009)
Passiflora alata and Passiflora ligularis	Total phenols	Primary branch leaves	Substrate with organic N (relatively low-leaf N), rich in P and K and pH 6	Increase	Sousa et al. (2013)
Baccharisdracunculifolia	γ-Muroleno, valenceno, δ-cadineno, E-nerolidol and espatulenol	Shoot essential oil	Organic compost fertilization at 30–40 t.ha^{-1}	Increase	Santos et al. (2012)
Schinus terebinthifolius	Flavonoid and phenol	Leaf essential oil	Growth on poultry litter	Increase	Tabaldi et al. (2016)
Cunila galioides	Flavonoids	Leaves	Soil with high level of aluminum	Increase	Mossi et al. (2011)
Baccharis trimera	Phenolic acids and flavonoids	Shoots	Coal burning area	Reduction	Menezes et al. (2016)
Alternanthera philoxeroides	Betacyanin	Shoots	175 μM $CuSO_4$ in vitro	60% increase	Perotti et al. (2010)

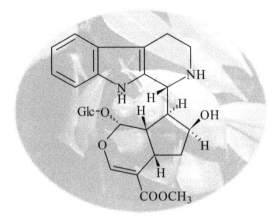

FIGURE 6.2   Structure of brachycerine.

UV-B radiation is also an important abiotic factor to promote accumulation of secondary metabolites such as polyphenolic compounds. *Kalanchoe pinnata* Pers. is a widespread species of Crassulaceae used in popular medicine in Brazil and around the world to treat several diseases, mainly inflammatory processes. Flavonoids isolated from the leaves are mainly quercetins, such as quercetin 3-rhamnopyranoside. Analyses of the effects of supplemental UV-B radiation on phenolic profile, antioxidant activity and total flavonoid content of leaves of *K. pinnata* showed that extracts had higher diversity of phenolic compounds and increased quantity of quercitrin after UV-B irradiation (Nascimento et al., 2015).

Light conditions can contribute to increase the production of essential oil of commercial medicinal and aromatic species. *Ocimum* spp. are generally characterized as rich in essential oils used in pharmaceuticals, fragrances and cosmetics. The major component in oil of *Ocimum gratissimum* L. is eugenol. In *O. gratissimum*, the essential oil yield per plant increased with light intensity as a function of increased leaf biomass, although this was not related to variation in trichome morphology, density or oil content (Fernandes et al., 2013). In contrast, *Varronia curassavica* Jacq., known as erva-baleeira, had increased essential oil content under higher irradiance in association with higher frequency of glandular trichomes (Feijó et al., 2014). Oil terpenes were differentially affected; although the concentration of caryophyllene derivatives increased with irradiance, α-humulene amount was unchanged.

Psychotria leiocarpa Cham. and Schltdl., a highly abundant bush in the understory of the Atlantic Forest biome, produces GPV, an antioxidant glycosylated alkaloid that accounts for up to 2.5% of dry leaves and accumulates mainly in reproductive structures. Accumulation of GPV is stimulated by light, particularly under far red or blue wavelength enrichment (Matsuura and Fett-Neto, 2013; Matsuura et al., 2016).

Piper aduncum L., the monkey-pepper, is an essential oil source, with biological activities, which include insecticidal, antimicrobial and larvicidal, among others. Pacheco et al. (2016) evaluated the effect of different light conditions on production and profile of essential oil constituents of leaves and roots of the species. Major compounds in the oil increased under an environment rich in blue light. Leaves were rich in E-nerolidol, linalol, α-humulene, cis-cadin-4-en-7-ol and caryophyllene, whereas roots had mostly apiol. While oil concentration of the former organs was stimulated by exposure to 50% reduction in sunlight, the latter were unaffected.

Increasing irradiance promoted the total yield of essential oil per plant of *Baccharis trimera* (Less.) DC., popularly known as carqueja (Silva et al., 2006). The highest total essential oil yield of *Otacanthus azureus* (Linden) Ronse was also observed in treatments with 100% irradiance, with decreasing production as light levels decreased (Serudo et al., 2013).

6.3   TEMPERATURE

Daily temperature changes have considerable influence on cell physiology. Both cold and heat can cause severe impacts on vegetative and reproductive tissues. Chilling causes rigidification of cell membranes, that may lead to release of cell content, accumulation of reactive oxygen species, stabilization of RNA secondary structure, destabilization of protein complexes, reductions in enzymatic activity and impairment of photosynthesis. Freezing can kill cells due to the formation of ice crystals, which lead to the rupture of organelles and cell compartments. Heat causes unfolding of proteins, affects membrane fluidity, metabolism, enzymatic activity and cytoskeleton rearrangement. On the other hand, temperature changes can regulate movements, like the opening/closing of flower corolla, and may reset the internal clocks and diurnal synchronization. Some species require exposure to low temperature to trigger developmental processes, such as flowering or germination (Ruelland and Zachowski, 2010).

In order to tolerate temperature variations, plants have developed adaptive mechanisms. These organisms can coordinate specific responses to the different components of abiotic stress, that include accumulation of sugar or compatible solutes, changes in membrane composition, synthesis of dehydrin-like proteins, synthesis of chaperones and increase of antioxidant capacity (Bita and Gerats, 2013). The response of the different classes of secondary metabolites is variable, depending on the magnitude of temperature variation, degree of exposure and species analyzed. In general, the production of volatile oils seems to increase at higher temperatures, although very hot days can lead to an excessive loss of these metabolites (Gobbo-Neto and Lopes, 2007). Total phenolic content in leaves of *Lafoensia pacari* A. St.-Hil. diminished in warmer months of the cerrado (Brazilian savannah), in part possibly due to photosynthetic limitations (Sampaio et al., 2011).

P. brachyceras is a woody understory plant with bushy habit, reaching up to 3 m in height, and distributed between the states of Rio de Janeiro and Rio Grande do Sul (Porto et al., 2009). This species is characterized by the production of brachycerine, an antioxidant alkaloid, which occurs mostly in leaves and whose accumulation can be stimulated by several stressful environmental conditions (Gregianini et al., 2004; Nascimento et al., 2013). Magedans et al. (2017) showed high temperature increased brachycerine accumulation, suggesting that heat treatment represents a viable means to improve yields of brachycerine for use as an antioxidant. Brachycerine was able to protect heat-sensitive species against severe heat stress, suggesting its involvement in mitigating heat-promoted oxidative imbalance (Table 6.2).

I. paraguariensis originates from the subtropical region of South America and grows in Northeastern Argentina, Southern Brazil and Eastern Paraguay. Holowaty et al. (2015) studied the variation of different parameters related to mate quality, specifically levels of caffeine, total polyphenol content, color parameters and antioxidant activity in samples obtained by three different aging methods: natural aging, temperature-controlled aging, temperature- and humidity-controlled aging. It was noted that the different aging methods of mate-yielded products with slight differences in their physico-chemical properties. Considering the main interest on mate, that is the content of polyphenols and its potential benefits to human health, the temperature- and humidity-controlled aging method proved to be the best procedure.

Negri et al., (2009) investigated the influence of drying temperature on the content of secondary metabolites of *Maytenus ilicifolia* Mart. ex Reissek, the espinheira santa, which is widely used for treating gastritis and stomach ulcers, finding that their values decreased as temperature increased. Temperatures at 40°C to 50°C were more adequate for leaf drying, preserving higher levels of active principles. Similarly, David et al. (2006) observed a drop in main components of essential oil of *Ocimum selloi* Benth. (elimicin, trans-caryophyllene, germacrene D and bicyclogermacrene), with temperature above 40°C.

6.4   WATER AND SALINITY

Changes in water availability can severely interfere with a variety of biochemical processes, such as mobilization of reserves, photosynthetic and respiratory capacity, cellular turgor, opening and closing of stomata, structures of membranes and organelles, osmotic stress, enzyme activity,

ATP synthesis, hormonal balance, etc. Water stress can be caused both by drought or flooding, whose effects, although distinct, are detrimental to plants often resulting in compounded stresses. Prolonged periods of exposure to drought, for example, may cause stomata closure, decreased CO_2 uptake, reduced photosynthesis, increased photooxidative damage, decreased protein synthesis, inactivation of chloroplast enzymes, disturbance of electron transport chain, as well as perturbation of membrane integrity and function. All these factors directly or indirectly result in increased activity of enzymes involved in the quenching of reactive oxygen species, ROS, in order to mitigate the harmful effects of their excess in cells. Drought stress often also provokes salinity stress, since the lack of water increases the concentration of salts in soil and cells, which causes a difference in water potential favoring loss of water by the plant (Selmar and Kleinwächter, 2013). Flooding conditions, in turn, lead to a reduction in transpiration rates, difficulties in mineral nutrient uptake and accumulation of the gaseous hormone ethylene, which can trigger tissue senescence, foliar abscission and synthesis of defense compounds.

Secondary plant metabolites may respond positively or negatively to water stress (see Table 6.2). Some species increase the synthesis of metabolites that act in osmotic adjustment, whereas accumulation of other metabolites can significantly decrease as a result of nutrient deficiency, affecting enzymatic cofactors, photosynthetic pattern and energy metabolism (Verma and Shukla, 2015). There are several reports describing how water deficiency may lead to increased production of several types of secondary metabolites, such as cyanogenic glycosides, glucosinolates, terpenes, anthocyanins and alkaloids. This effect, however, depends on the degree of severity and the period of exposure to the drought conditions, which, in the short term, seem to lead to an increased production, whereas an opposite effect is observed in longer times of exposure. Excess water, on the other hand, can result in the loss of water-soluble substances from leaves and roots by leaching, which can lead to the loss of some alkaloids, glycosides and volatile oils (Gobbo-Neto and Lopes, 2007). Depending on the degree of flooding and time of exposure, ROS generation and ethylene production may trigger secondary metabolite accumulation.

Plant growth under drought conditions is reduced, which often leads to an increase in the concentration of secondary metabolites in relation to the dry mass, but not in their total content. However, the effect may be the reverse in some cases. In *Pisum sativum* L., in spite of the total biomass of plants grown under drought stress being about a third of that of plants cultivated under standard conditions, the concentration of anthocyanins practically doubled in stressed plants (Nogués et al., 1998). The active constituents of *Hypericum brasiliense* Choisy, a herb found in Southern and Southeastern Brazil, which include betulinic acid, isouliginosin B, 1,5-dihydroxyxanthone and rutin, increased in plants under water stress, both in hypoxic and dry conditions, although the fresh mass of the plants decreased in both cases (Abreu and Mazzafera, 2005). Drought also increased the amount of uliginosin B and total phenolic compounds in acclimatized *Hypericum polyanthemum* Klotzsch ex H. Reich. plants (Nunes et al., 2014). Likewise, higher essential oil content of *Aloysia triphylla* Britton., native from South America, was observed in the seasons of lower biomass production (Schwerz et al., 2015).

In *Bauhinia cheilantha* (Bong.) D. Dietr., a native species of the caatinga, leaf flavonoid contents responded negatively to higher rainfall rates, but without significant effects (Sobrinho et al., 2009). Water restriction resulted in decreased total phenolic compounds in *Amburana cearensis* (Allemão) A.C.Sm. seeds, an endemic tree from semiarid region of Northeastern Brazil (Pereira et al., 2014). *Hydrocotyle umbellata* L., found in all Brazilian states, on the other hand, did not suffer decreases either in biomass production or in flavonoid content when cultivated under two intensities of irrigation for three months (Alves et al., 2015). Rainfall did not correlate with phenolic compounds concentration in bark of *Anadenanthera colubrine* (Vell.) Brenan, a tree of the Brazilian caatinga and semideciduous forest (Araújo et al., 2015).

Salt stress negatively affects plant growth and influences development, impacting on the water potential difference between the plant and exterior environment. Water deficit as a result of salt stress affects the availability, competitive uptake and translocation of nutrients to aboveground plant parts (Park et al., 2016). Besides under salt stress, the excessive concentrations of Na$^+$ and Cl$^-$ hinder the absorption and/or assimilation of other elements, including boron, zinc, calcium, copper, magnesium, iron, nitrogen, phosphorus and potassium (Farooq et al., 2017). The imbalance in nutrient uptake and assimilation and ROS generation during salt stress interfere in photosynthetic reactions, cause stomatal limitation, reduction in intercellular CO$_2$ concentration and damage to photosystems (Zhu, 2016). Plants show several adaptations to deal with saline stress, such as production of osmoprotectant molecules (cell compatible solutes), hormonal regulation, activation of antioxidant defense systems and mechanisms of ion exclusion or compartmentalization (Acosta-Motos et al., 2017).

The effect of salinity on secondary metabolism is dual depending on severity and has been relatively less described for medicinal plants. Among Brazilian medicinal plants, there are several examples of the positive effect of moderate salinity on the accumulation of secondary metabolites. In plants and seedlings of *Dimorphandra mollis* Benth, the fava-de-anta, present in the cerrado vegetation, quercetin, isoquercitrin and rutin amounts generally increased under water stress by drought, flooding and salinity (NaCl 75 mM) (Lucci and Mazzafera, 2009a, 2009b). *Physalis angulata* L., common throughout Brazil, also accumulated more physalin in leaves after 13 days exposure to saline solution 0.9%, in spite of showing biomass reduction (de Souza et al., 2013). Saponins from leaf disks of *Q. brasiliensis* were significantly increased by application of osmotic stress agents, such as sodium chloride 150 mM or isosmotic concentrations of sorbitol and polyethylene glycol (de Costa et al., 2013).

Plants of *Alternanthera philoxeroides* Mart. (Griseb.), a native species from the temperate regions of South America, showed increased concentration of betacyanins in stems when exposed to 200 mM or 400 mM of salt (Ribeiro et al., 2014). *Pilocarpus microphyllus* Stapf. ex Wardlew. and *Pilocarpus jaborandi* Holmes, commonly known as jaborandi, are distributed exclusively in South America, mainly in Northern and Northeast Brazil (Abreu et al., 2005). The leaves of jaborandi are a source of the imidazolic alkaloid pilocarpine, used mainly in the first stages of glaucoma and xerostomia treatment and for stimulation of lacrimal and salivary glands (Caldeira et al., 2017). In contrast to *A. philoxeroides*, salt stress and hypoxia resulted in decreased pilocarpine amount in *P. jaborandi* leaves; reduction of alkaloid amounts in salt-treated plants was concentration and time of exposure dependent (Avancini et al., 2003). However, in callus cultures of jaborandi, salt and osmotic stress promoted pilocarpine release in medium (Abreu et al., 2005). Accumulation of brachycerine in *P. brachyceras* leaf disks was promoted by osmotic stress and abscisic acid, a key signaling hormone in drought stress (Nascimento et al., 2013).

6.5   MINERAL NUTRITION AND HEAVY METALS

Soil composition and mineral nutrient availability can greatly influence the synthesis of secondary metabolites. Changes in mineral nutrition impact on the availability of metallic ions that act as enzymatic cofactors for defense metabolites synthesis. Nitrogen and sulfur can be components of certain molecules, such as alkaloids, cyanogenic glycosides, glucosinolates, thiophenes and nonprotein amino acids. Nutrient-deficient soils often result in lower growth rate and higher secondary metabolite accumulation. High nitrogen supply frequently results in lower production of phenolics (Gobbo-Neto and Lopes, 2007). Phosphorus deficiency is known to promote anthocyanin accumulation (Treutter, 2010).

The metabolic effects of nutritional imbalance are not entirely predictable. Although patterns can be recognized, it is not possible to establish a consensus on the classes of metabolites that will be benefited or impaired in each case (see Table 6.2). This is due to the complexity of the biochemical pathways and factors that are involved in mineral nutrition. Soil pH, for example, alters the availability of some nutrients, so that uptake by plants can be disrupted and their metabolism affected. Likewise, the presence of microorganisms in the rhizosphere may interfere with the absorption and availability of nutrients. The positive effect of mycorrhizal fungi on acquisition of phosphorus by the roots is well established. However, despite the recognized influence on plant development, few studies show relationships between pH or soil microorganisms in secondary metabolism. Furthermore, environmental factors do not seem to exert a homogeneous effect on the metabolism of different parts of the plant, which may respond differently to such external signals (Sampaio et al., 2016).

Plants of *Palicourea rigida* Kunth, a Brazilian cerrado bush, grown on low-fertility and acidic soil displayed higher concentration of iridoids. Except for Mg, all soil macro- and micronutrients showed a negative correlation with loganin concentration, particularly Ca and K (Morel et al., 2011). Accumulation of the alkaloid pilocarpine in *P. microphyllus* seedlings was negatively affected by N and K omission, but not influenced by P removal from nutrient solution (Avancini et al., 2003). In partially immersed callus cultures, however, the omission or excess of N or P improved alkaloid accumulation (Abreu et al., 2005). This difference in metabolic response may be related to the presence of sucrose and partial submersion of the tissues in the axenic cultures. In *Tithonia diversifolia* (Hemsl.) A. Gray, found in tropical regions, the presence of sesquiterpene lactones, flavonoids and trans-cinnamic acid derivatives in inflorescences and roots was affected by the soil nutrients Ca, Mg, P, K and Cu (Sampaio et al., 2016). Monoterpenes and sesquiterpene hydrocarbons of the essential oil of *Lychnophora ericoides* Mart., a native plant from the cerrado, were strongly correlated with organic matter, P and base saturation in soils (Curado et al., 2006). Phenolic metabolites in the leaves of the tree *L. pacari* A. St.-Hil. are positively influenced by foliar micronutrients Cu, Fe, Mn and Zn, whereas macronutrients appear to have no effect on the production of these compounds (Sampaio et al., 2011).

In leaves of *Eugenia uniflora* L., widely distributed in Brazil, the accumulation of spathulenol and caryophyllene oxide was strongly correlated with phenolic content and S, Ca and Fe balance, whereas selina-1,3,7(11)-trien-8-one epoxide was affected by K, Cu, Mn and water availability during the wet season (Costa et al., 2009). Negative correlation of leaf phenolic concentration with Mn and Cu was reported for the same species in the cerrado, indicating that metabolism of flavonoids and tannins depends on leaf mineral nutrition (Santos et al., 2011). The same authors proposed that both Mn and Cu could affect phenolic metabolism by their action as cofactors of enzymes dedicated to phenolic catabolism and lignin synthesis. Data indicated a potential modulation of the amount of these metabolites by foliar or soil micronutrient application. Oxygenated sesquiterpenes in *Myrcia tomentosa* DC. flowers and leaves were positively correlated with foliar concentration of Cu and P, whereas sesquiterpene hydrocarbons showed the opposite behavior (Sá et al., 2012). *M. tomentosa* leaf oil composition was shown to be influenced by minerals in leaves (N, Fe, P, K, Ca, Mg, Mn, Al) and soil (Fe, Al, Cu), as well as rainfall (Borges et al., 2013a). Concentration of total phenolics in leaves of *M. tomentosa* was affected by foliar nutrients, such as Ca, Mg (both negatively) and Mn (positively). Rainfall and soil K were negatively correlated with the concentration of hydrolysable tannins (Borges et al., 2013b).

The genus *Passiflora* originates from South America and its main distribution center is in Northern Central Brazil. *Passiflora alata* Curtis cultivated in low-nutrient soil inoculated with the arbuscular mycorrhizal fungi *Gigaspora albida* yielded amounts of phenolics equivalent to those of nutrient-sufficient soil. This result indicated that mycorrhizae association can help overcome poor soil fertility, particularly phosphate deficiency (Riter Netto et al., 2014). Similar positive effects of *G. albida* inoculation on *P. alata* were reported depending on the level of soil fertilization, with the fungus generally promoting flavonoid and total phenolic yield (Oliveira et al., 2015). *P. alata* and *Passiflora ligularis* Juss. showed higher concentrations of total phenols under conditions of low leaf N (Sousa et al., 2013). *Passiflora edulis* Sims. displayed higher content of tocopherols and ascorbic acid in organic system cultivation, whereas conventionally grown passion fruit displayed more carotenoids (Pertuzatti et al., 2015). It has been shown that the soil elements Fe, B and Cu showed an inverse correlation with the concentration of total flavonoids in *Passiflora incarnata* L. (Reimberg et al., 2009).

Baccharis dracunculifolia DC. displayed higher yields of γ-muuroleno, valenceno, δ-cadineno, E-nerolidol and espatulenol in essential oil when submitted to organic compost fertilization at

30 t.ha⁻¹ versus unfertilized soil (Santos et al., 2012). Yields of essential oil of *O. selloi*, a native plant of South and Southeast regions of Brazil, were increased in treatments of fertilization with bovine and avian manure (Costa et al., 2008), which also led to an increase in leaf content of the major component, methyl chavicol. However, unfertilized plants showed a higher diversity of compounds in essential oil. Phosphorus and nitrogen influenced the essential oil content of *Schinus terebinthifolius* Raddi (Pinto et al., 2016), known as Brazilian pepper tree and found in South America. Overall growth, flavonoid and total phenolics concentrations of the same species increased by application of poultry litter at 20 t.ha⁻¹ (Tabaldi et al., 2016). Roots of *Pfaffia glomerata* (Spreng.) Pedersen, the Brazilian ginseng, showed higher total amounts of β-ecdysone at 360 days after seedling emergence, but was little affected by different levels of organic fertilization (Guerreiro et al., 2009)

Many anthropogenic activities can release heavy metals in the environment. Soil and sediments tend to accumulate these elements, becoming a large reservoir to which plants are exposed. Plants are generally sensitive and vulnerable to varying concentrations of these elements. Although some metals are essential for plants, involved in synthesis, structure or activity of enzymes and proteins, these same elements can become toxic when in high concentrations. Other metals, however, have no known function in plant metabolism, so that they induce toxicity symptoms at minimum concentrations due to a variety of interactions at the molecular level that may disrupt cellular homeostasis.

The plasma membrane is one of the first structures whose function is affected by heavy metals, due to rupture or interference in protein sulfhydryl groups, increased lipid peroxidation, inactivation of proteins, decrease in phosphorylation of target molecules and consequent changes in composition and fluidity of lipids, frequently leading to structural damage and leakage of cellular contents (Lara Lanza de Sá e Melo Marques et al., 2011). Inside the cells, one of the first physiological damages is the inactivation of several cytoplasmic enzymes, due to the modification of their three-dimensional structure, subunit binding or displacement of essential elements through chemical competition, leading to the interruption of normal functions and the establishment of deficiency symptoms. In addition, heavy metals can cause oxidative stress, which results in changes in nuclear proteins and DNA, degradation of biological macromolecules and lipid membrane peroxidation (Rodrigues et al., 2016). Photosynthesis is also compromised in plants exposed to heavy metal contamination, since these elements can reduce the levels of chlorophyll and carotenoids, via inactivation of enzymes of pigment biosynthesis, damage to electron transport chain, inactivation of enzymes of the Calvin cycle and reduction of stomatal conductance. Some elements may interfere with the absorption and metabolism of nutrients, which compromises the development of plants in a global way (Küpper et al., 2007).

Plants can show tolerance to heavy metals based on mechanisms modulating regulation of their absorption in the rhizosphere and accumulation in roots. These strategies may preserve plant cell integrity and primary functions, which, in association with the low translocation to the aerial part, may avoid overload on the photosynthetic apparatus and damage to the vascular bundles. The concentration of free heavy metals in the cytosol is reduced especially via compartmentalization in subcellular structures, exclusion and/or decrease in membrane transport. In addition, production and formation of cysteine rich peptides, known as phytochelatins and metallothioneins, can complex several metals. In conjunction, the actions of antioxidant defense systems, both enzymatic and non-enzymatic, are capable of removing, neutralizing or cleaning free radicals (Emamverdian et al., 2015). All these mechanisms can also impact the synthesis and extrusion of several plant metabolites, both from primary and secondary metabolism.

The oil of *Cunila galioides* Benth plants grown on soil with a high level of aluminum showed no difference in the content of oil major components, although the concentration of flavonoids increased significantly under the same condition (Mossi et al., 2011). In contrast, extracts of *B. trimera* (Less.) DC. grown in coal-burning area had lower levels of phenolic acids and flavonoids compared with plants grown in nearby regions without contamination by these pollutants (Menezes et al., 2016). *In vitro* culture of *A. philoxeroides* plants supplemented with 175 µM CuSO₄ had approximately 60% increase in betacyanin yield in relation to control, although growth reduction was also observed (Perotti et al., 2010). Aluminum or silver exposure caused a two- to threefold increase in brachycerine accumulation in leaf disks of *P. brachyceras* (Nascimento et al., 2013).

6.6 BIOTIC ELICITATION OF SECONDARY METABOLITES

Biotic responses are normally mediated by signaling compounds, such as jasmonic acid (JA), jasmonoyl isoleucine (Ile-JA) and methyl jasmonate (MeJA), synthesized from linolenic or hexadecatrienoic acids starting with lipoxygenase (LOX) activity. JA and related compounds have a key role in regulation of herbivory and wounding responses by modulating global changes in gene expression. Another example of a major biotic signaling compound is salicylic acid (SA) and its methyl analog, methyl salicylate (MeSA), which have been shown to take part in defense signaling against pathogens, leading to systemic acquired resistance (SAR) and providing long-term defense (Heil and Ton, 2008). Both JA and SA may co-participate and cross talk in herbivory and pathogen responses. Ethylene (ET) also has an important role in plant protection, acting as virulence factor of pathogens and signaling compound in disease resistance, depending on the situation (van Loon et al., 2006). The simulation of herbivory by applying mechanical damage can induce the formation of JA and ET (Bailey et al., 2005). Besides, SA or JA exposure triggers events such as production of ROS and increased cytoplasmic Ca^{2+}, which can stimulate certain biochemical reactions and production of secondary metabolites (Lin et al., 2001) (see Table 6.3).

Q. brasiliensis leaf saponins show structural and functional similarities to those of *Q. saponaria* barks, which are currently used as adjuvants in vaccine formulations (Fleck et al., 2006, 2012; Kauffmann et al., 2004). Previous studies with saponins of *Q. brasiliensis* showed a pronounced immunoadjuvant activity in experimental vaccines against bovine herpesvirus type 1 and 5, poliovirus, rabies and bovine viral diarrhea virus in mice (Cibulski et al., 2016; Fleck et al., 2006; Yendo et al., 2016). The accumulation patterns of the immunoadjuvant fraction of leaf triterpene saponins QB-90 in response to stress factors were examined. Higher yields of bioactive saponins were observed upon exposure to SA, JA and by mechanical damage, as well as by applying ultrasound to leaves (de Costa et al., 2013). A significant increase in QB-90 content was observed when leaf disks were submitted to 1 mM SA or 40 μM JA, supporting a general defense role for these metabolites in plants.

The shoots of *A. philoxeroides* (alligator weed) contain betacyanins. Addition of 100 μM MeJA to standard *in vitro* culture medium resulted in a fourfold increase in the pigment amaranthine after 35 days, in spite of having a negative impact on development (Perotti et al., 2016).

Wounding seedling leaves of *P. microphyllus* with a hemostat resulted in a negative effect on pilocarpine content, changing from 0.7 to 0.1 μg.mg of leaf dry weight, after 1 week of treatment (Avancini et al., 2003). Nevertheless, exposure of seedlings to MeJa and SA led to a significant increase in the alkaloid concentration in leaves after 5 and 9 days of treatment, respectively. In a follow-up study performed with aseptic cultures of *P. microphyllus*, the amount of quantified pilocarpine was similar between MeJa-treated and control calluses kept for 4 and 8 days under agitation in the dark. However, control calluses released most of the alkaloid to the medium, while MeJa-treated calluses retained most of the metabolite (Abreu et al., 2005).

Psidium guajava L., commonly known as guava, is an evergreen tree native to South America, including Brazil. In a study of González-Aguilar et al. (2004), the application of MeJa to mature green guava fruits stored at 5°C for 10 days reduced deterioration and the development of chilling injury symptoms. MeJa treatment resulted in an increased activity of LOX and phenylalanine-ammonia lyase, enzymes, which have an important role in the activation of mechanisms against different stresses, including the biosynthesis of defense secondary metabolites.

The phloroglucinol derivative uliginosin B is found in the herbaceous plant *H. polyanthemum* native to South Brazil. This molecule is a promising antidepressant, which activates the monoaminergic neurotransmitter system (Nunes et al., 2014; Stein et al., 2012). Furthermore, *H. polyanthemum* accumulates benzopyrans, chlorogenic acid and flavonoids (Nunes et al., 2010). A weekly application of 2 mM

TABLE 6.3

Examples of Biotic Stress Effectors to Induce Production and/or Accumulation of Secondary Metabolites

Species	Treatment	Induced Secondary Metabolite(s)	Tissue/Organ	Dose	Time for Maximum Enhancement	Maximum Increase	References
A. philoxeroides	Methyl Jasmonate	Amaranthine	Shoots	100 µM	35 days	Fourfold	Perotti et al. (2016)
H. polyanthemum	Salicylic acid	Uliginosin B and TPC	Vegetative tissues	2 mM (weekly application)	18 weeks	Twofold	Nunes et al. (2014)
H. polyanthemum	Salicylic acid or mechanical damage or both	Uliginosin B	Vegetative tissues	10 mM/wounded with fine sterile needle	2 days for mechanical damage and SA alone and 7 days for both	2 days (fivefold) and 7 days (threefold)	
H. polyanthemum	Fungal elicitation with Nomuraea rileyi	Benzopyrans (HP1, HP2, HP3) and total phenolic compounds	Whole plants	10 ml of solution of dried autoclaved cell powder (containing the equivalent of 1.5 9×10^6 spores ml^{-1})	48 h and 72 h (benzopyrans) and 24 h (total phenolic compound)	Benzopyrans (approx. 1.7-fold), TPC (twofold)	Meirelles et al. (2013)
P. microphyllus	Salicylic acid	Pilocarpine	Leaves	5 mM	9 days	Fourfold	Avancini et al. (2003)
	Methyl Jasmonate	Pilocarpine	Leaves	Plant incubation in plastic bag with cotton piece impregnated with 50 µL of pure MeJA	5 days	Fourfold	
P. brachyceras	Mechanical damage	Brachycerine	Leaves	Wounded 75% of the total leaf area of cuttings with scissor	2 days	Twofold	Gregianini et al. (2004)
P. brachyceras	Jasmonic acid	Brachycerine	Leaves	40 µM and 400 µM	6 days (40 µM), 4 days (400 µM)	2.7-fold (40 µM) and 3.3-fold (400 µM)	

(Continued)

TABLE 6.3 (*Continued*)

Examples of Biotic Stress Effectors to Induce Production and/or Accumulation of Secondary Metabolites

Species	Treatment	Induced Secondary Metabolite(s)	Tissue/ Organ	Dose	Time for Maximum Enhancement	Maximum Increase	References
Q. brasiliensis	Jasmonic acid	QB-90 (triterpenic saponin fraction)	Leaf disks	40 μM	2 days	2.6-fold	de Costa et al. (2013)
	Salicylic acid	QB-90	Leaf disks	1 mM	2 days	Threefold	
	Ultrasound treatment (40 kHz and 135 W)	QB-90	Leaf disks	1 and 2 min	2 days after initial 1 min SA exposure	Twofold	
	Mechanical damage	QB-90	Seedlings	50% of leaves were wounded with scissors	2 days	1.5-fold	
V. glechomifolia	Salicylic acid	Valepotriates	Whole-plants cultured in MS liquid media	0.1 mM SA and 1 mM SA	72 and 96 h (0.1 mM SA), 48 and 72 h (1 mM SA)	1.7- to 1.8-fold (0.1 mM SA), approx. 1.5-fold (1 mM SA)	Russowski et al. (2013)
	Ultrasound (40 kHz and 135 W)	Valepotriates	Whole-plants cultured in MS liquid media	2.5 and 5 min	2.5 min for cultures in the 7th day and/or 14th day of growth cycle – harvested in the 21st day of growth cycle	1.4- to 1.8-fold	

SA increased the amount of uliginosin B and total phenolic compounds in acclimatized *H. polyanthemum* plants after 18 weeks (Nunes et al., 2014). Mechanical damage, 10 mM SA and the combination of both treatments were able to induce higher levels of uliginosin B in leaves after 1 or 2 days. The entomopathogenic fungus *Nomuraea rileyi* induced production of three benzopyrans and total phenolic compounds in *H. polyanthemum* plantlets (Meirelles et al., 2013). The former compounds have been studied for their antinociceptive and antitumoral activities (Ferraz et al., 2005; Haas et al., 2010). After 48 and 72 h of exposure of acclimatized plants to a dried autoclaved cell powder of *N. rileyi*, increased levels of the three benzopyrans were observed in vegetative parts. In addition, a twofold increase in total phenolic compounds was promoted after 24 h of fungal exposure (Meirelles et al., 2013).

P. edulis is cultivated commercially in tropical and subtropical areas of Southern Brazil. Jardim et al. (2010) observed an increase in LOX transcripts and higher enzyme activity in response to a specialist (*Agraulis vanillae vanilla*) and a generalist (*Spodoptera frugiperda*) caterpillar attack, suggesting that the herbivore response in passion fruit is mediated by JA signaling pathway, which may also impact on its secondary metabolite profile. LOX activity was previously shown to be modulated in *P. edulis* in response to wounding and exogenous MeJa application (Rangel et al., 2002).

The seeded-fruit transcriptome of guarana, *Paullinia cupana* var. *sorbilis* (Mart.) Ducke, a stimulant plant native to the central Amazon basin, was performed with a focus in finding Expressed Sequence Tags (ESTs) related to secondary metabolism. Several key genes related to flavonoid biosynthesis, plant biotic defense pathways and purine alkaloid metabolism were well represented in the analyzes (Angelo et al., 2008).

The effects of mechanical wounding, ethylene and 0.2 mM MeJa application on the expression of genes associated with stress responses on leaves of *Theobroma cacao* L. (cocoa tree) were analyzed. A higher expression of the transcripts of type III peroxidase, a class VII chitinase, and a caffeine synthase was found on treated-seedlings. Gene expression profile proved to be dependent on the ontogeny of the associated tissue (Bailey et al., 2005).

Monoterpene derived valepotriates are accumulated in both shoots and roots of *Valeriana glechomifolia* Meyer, a small herb which grows in rocky fields of Southern Brazil (Salles et al., 2002). A semi-purified valepotriate fraction from shoots and roots showed sedative effects and affected behavioral parameters related to recognition memory (Maurmann et al., 2011). A supercritical CO_2 *V. glechomifolia* extract displayed antidepressant potential, which is mediated by the dopaminergic and noradrenergic neurotransmission systems (Muller et al., 2015). The exposure of whole plants cultivated in liquid medium to SA and ultrasound increased by twofold the amount of the metabolite, best results being recorded with the latter treatment, which has not diminished biomass accumulation (Russowski et al., 2013).

Biomembranes derived from *Hevea brasiliensis* (Willd. ex A.Juss.) Müll.Arg. latex (rubber tree) have shown wound healing properties (Frade et al., 2012). The effect of ET, JA and wounding was analyzed in the barks of 3-month-old shoots of *H. brasiliensis*. Several defense and latex exudation-related genes had increased expression upon treatment application (Duan et al., 2010). A member of AP2/ERF transcription factors from laticifers induced by JA was fully characterized, further supporting the involvement of this hormone in latex-based defense responses (Chen et al., 2011).

Araucaria angustifolia (Bertol.) Kuntze, the Brazilian pine, is well-known for its medicinal properties, which include antioxidant, antibacterial, antiviral, anti-inflammatory and antiproliferative (Branco et al., 2016). Several of its active pharmacological constituents are present in the bark resin, including various polyphenolics and terpenes. A study examining the regulation of bark resin exudation in young *Araucaria* plants showed that ET, SA and JA stimulated resin yield and, in some cases, modified monoterpene relative concentration (Perotti et al., 2015).

6.7 SEASONAL VARIATION

The climate of the Northern and Northeastern regions is characterized by small annual variation of elevated temperatures and long photoperiod, with different average rainfall. Whereas the North presents a rainy equatorial climate, the Northeastern is characterized by a semiarid climate. On the other hand, the Southeast and Center-West regions are influenced by a dry season in winter and a rainy season in the summer. In the south of Brazil, due to its latitudinal location, cold air masses contribute to the predominance of lower winter temperatures and shorter day length, with generally well-defined four seasons.

Several environmental factors, such as seasonality, photoperiod, circadian rhythm, temperature, irradiance, altitude, humidity and water availability may affect plant secondary metabolism in an integrated way (Yao et al., 2004). These factors can influence secondary compound production throughout the year, highlighting the importance of harvest time for optimizing yield. In fact, there are several studies describing seasonal influence on the content of diverse classes of secondary metabolites, such as terpenes and phenolic compounds (Gobbo-Neto and Lopes, 2007).

M. tomentosa (Aubl.) DC., commonly known as guava brava, is used in folk medicine against gastrointestinal disorders, infectious diseases and hemorrhagic conditions. Sá et al. (2012) showed that its terpene content may vary according to the time of year. In general, the major component in samples collected in August, October, December and February was (2*E*,6*E*)-methyl farnesoate; epi-α-bisabolol was the main compound in April, whereas germacrene D, (2*E*,6*E*)-methyl farnesoate and bicyclogermacrene were the main components in June. The component γ-muurolene was absent in August and December but present in the total essential oil in February and April. Regarding the variation of percentage in the total content of essential oil, the sesquiterpene (*E*)-β-farnesene had a small variation only in August and October, but bicyclogermacrene amounts varied from 4.73% of the total essential oil content in October to 14.71% in June. Content of (2*E*,6*E*)-methyl farnesoate ranged from only 5.33% of the total essential oil content in April to more than 47% in October (Sá et al., 2012).

Elionurus muticus (Spreng.) Kuntze leaves essential oil was shown to vary throughout the year. Also known as lemon grass, it is one of the most abundant grass species in the mid-southern portion of the Pantanal biome, characterized by subtropical climate, with a short cold season and low occurrence of frost. The period of plant harvest affected the percentage of the sesquiterpenoids (*E*)-caryophyllene, bicyclogermacrene, spathulenol and caryophyllene oxide in essential oils. In winter and spring, (*E*)-caryophyllene was the main component in the oils, whereas in summer and autumn bicyclogermacrene became the main component. In spring, the antibacterial caryophyllene oxide and spathulenol displayed higher yields than in the other seasons (Hess et al., 2007).

Martianthus leucocephalus (Mart. ex Benth.) J.F.B.Pastore, formerly known as *Hyptis leucocephala*, is an aromatic herb whose leaves, flowers and branches produce antimicrobial essential oils which showed activity against *Bacillus cereus*, *Staphylococcus aureus* and *Candida albicans*. The chemical profile of this species was strongly affected by climatic factors during the year, and its oil content varied from 0.1% to 0.31%. Essential oil content was positively correlated with irradiance but showed a negative correlation with precipitation and relative humidity. Its highest production was between September and March, months of low rainfall and elevated level of solar radiation. However, in May and June the essential oil production was low, but vegetative growth was higher, suggesting that an optimized cultivation protocol can be established, so that planting could take place in May and June whereas harvesting would be done between September and March for improved yields (Azevedo et al., 2015). In contrast, in the related species *Hyptis carpinifolia* Benth., commonly known as rosmaninho and mata-pasto, the higher the humidity the greater the amount of α-copaene and pinonic acid (Sá et al., 2012).

The antileishmanial *K. pinnata* (Lam.) Pers., leaf extract traditionally used in Brazil to treat skin diseases and wounds, has active flavonoids that accumulate at higher levels in summer. This seasonal peak of accumulation was coincident with higher solar irradiances. A relationship between higher accumulation of flavonoids and irradiance availability was corroborated by comparing plant cultivation under direct sunlight or shade. Leaves under higher irradiance had an increment of sevenfold in the quercetin yield (Muzitano et al., 2011).

Plants of *Baccharis* sp. are widely used in Brazilian folk medicine, mainly to treat gastrointestinal disorders. Phenolic acids such as flavones, methylated-flavones and some flavanols, mainly aglycones, are the major compounds described for this genus, and have several biological activities, including antimicrobial and anti-inflammatory properties (Martinez et al., 2005). Sartor et al. (2013) showed that the total content of phenolic compounds and the flavonoid fraction of *Baccharis dentata* (Vell.) G.M. Barroso undergoes seasonal variations. The highest concentrations of total phenolics were recorded in autumn and winter. Rutin was the most abundant flavonoid, showing a concentration peak in winter. Phenolic compounds such as caffeic acid were also most abundant in winter and summer, possibly by activation of the phenylpropanoid pathway at more extreme temperatures. In *B. dracunculifolia*, caffeic acid contents were higher mostly in summer, but also in some of the spring and early autumn months (approximately between 3% and 6% of dry weight), further confirming that seasonality affects the profile of phenolics (Sousa et al., 2009). Quercetin, kaempferol and apigenin concentrations were higher in spring and summer than in winter, perhaps as part of an adaptive response against photooxidative stress caused by excess of light (Sartor et al., 2013).

Lima et al. (2017) observed that pilocarpine contents varied seasonally in three populations of *P. microphyllus*. Overall, pilocarpine content varied throughout the year in all samples, with the lowest levels being recorded in the rainy season and the highest in the dry season, suggesting a negative influence of rainfall on alkaloid content (Lima et al., 2017). The higher accumulation of alkaloid in the dry season may reflect its role in adaptive responses against drought, photooxidative damage and temperature stresses.

Mikania laevigata Sch.Bip. ex Baker and *Mikania glomerata* Spreng. are medicinal plants popularly known as guaco in Brazil. Both species are broadly used to treat inflammatory and allergic conditions, particularly disorders of the respiratory system. The leaves of *M. laevigata* and *M. glomerata* have been reported to have similar chemical compositions, and their major bioactive constituents are kaurene-type diterpenes and derivatives of cinnamic acid. The content of coumarin in *M. laevigata* reached the highest yields in summer and was significantly increased in plants cultivated under high shading levels, whereas this phenolic was absent in *M. glomerata*. The accumulation of kaurene-type diterpenes in both species was favored by growth under full sunlight, and *M. glomerata* had the highest seasonal accumulation of these metabolites in winter (Bertolucci et al., 2013).

Known as aloe, *Aloe arborescens* Mill. contain approximately 2.0% of dry weight of compounds with antimicrobial potential, such as the quinones barbaloin, aloe-emodin, aloin A and B and isobarbaloin. Cardoso et al. (2010) found that the quinone levels on leaves were higher in summer and autumn, whereas the flavonoid contents were similar for all seasons. Winter, spring and summer chloroform extracts presented higher antimicrobial activity than their autumn counterpart; winter extract had the lowest Minimum Inhibitory Concentration (MIC, 128 µg.mL^{-1}) on *Bacillus subtilis*. Ethanolic extracts of summer and autumn showed low antimicrobial activity, while winter ethanolic extract had again the lowest MIC (256 µg.mL^{-1}) on *Klebsiela pneumoniae*. The antimicrobial effects may involve other metabolites and/or a combined action of active compounds in the extracts. In fact, the correlation between seasonal variation, metabolite content and biological activity is not always apparent. In *Lippia origanoides* Kunth., antibacterial activity against *S. aureus* and *Escherichia coli*, as well as the average yield and overall oil composition, were little influenced by seasonal variation (Sarrazin et al., 2015).

Cordia verbenacea DC. is a native Brazilian medicinal plant, widely distributed along the Southeast coast of Brazil. This bushy plant is popularly known as cordia, blacksage or erva baleeira and has been known for its properties as antiulcer, antimicrobial and anti-inflammatory (Falcão et al., 2008; Michielin et al., 2009). Several compounds are found in the aerial parts of *C. verbenacea* including α-pinene, *trans*-caryophyllene, aloaromadendrene, cordialin A, cordialin B, rosmarinic acid and flavanols (Thirupathi et al., 2008). The phenolic rosmarinic acid is regarded as a phytochemical marker of *C. verbenaceae* due to its abundance in the species. A relatively strong positive correlation between the rainy season at a Central Brazil locality and rosmarinic acid content was reported, suggesting the wet period is the best choice for harvesting this medicinal plant (Matos et al., 2015). According to Queiroz et al. (2016), the harvest time during the day did not influence the content of essential oil, but it could modify its chemical profile. Even though the concentrations of the major compounds β-caryophyllene, xylene and γ-muurolene displayed no differences among collection times, sabinene was found only during early morning (6 am) harvest.

6.8 FINAL REMARKS

Both mild abiotic and biotic stresses often result in increased secondary metabolite accumulation, and this response is usually associated with some degree of growth rate reduction. This is expected since the flux of C, N and S toward secondary metabolism drains the pools of primary metabolite precursors supporting growth. The degree of stress exposure must be such that it finds a balance between maximum defense metabolism stimulation and minimal damage. This fine balance can be achieved by appropriate combinations of time and intensity of exposure to stimuli. At the center of several stresses signaling pathways is redox balance, which is a major regulator of secondary metabolism acting as a network hub. Developmental stage is frequently a determinant of response capacity, as is the type of nutrition, heterotrophic (most cell cultures), semi-autotrophic or fully autotrophic. Seasonal and circadian variations of secondary metabolic profiles may also be considered as a factor for optimizing target product yields.

Signaling pathways mediating environmental stress or regular signals involve key phytohormones, notably JA, SA and their derivatives. Such molecules and their precursors may be used to trigger metabolic profiles of interest without the need of the environmental signal itself, sparing significant energy and time for responses to arise. Similarly, adequate modulation of mineral nutrition to sustain biochemical activity, including enzyme components and cofactors, or even essential and nonessential metal levels that trigger ROS production, are also important tools in driving secondary metabolism.

To sum up, environmental signals and mild stresses are useful tools to control secondary metabolism and the yield of target bioactive metabolites in Brazilian medicinal plants. There is an array of relatively simple and low-cost crop management strategies to sustainably produce pharmacologically active phytomedicines and isolated compounds from this valuable and still relatively untapped biological resource.

REFERENCES

Abreu, I. N.; Sawaya, A. C. H. F.; Eberlin, M. N.; Mazzafera, P. 2005. Production of pilocarpine in callus of jaborandi (*Pilocarpus microphyllus* Stapf). In Vitro Cellular & Developmental Biology – Plant, 41, 806–811.

Abreu, I. N.; Mazzafera, P. 2005. Effect of water and temperature stress on the content of active constituents of *Hypericum brasiliense* choisy. Plant Physiology and Biochemistry, 43, 241–248.

Acosta-Motos, J.; Ortuño, M.; Bernal-Vicente, A.; Diaz-Vivancos, P.; Sanchez-Blanco, M.; Hernandez, J. 2017. Plant responses to salt stress: adaptive mechanisms. Agronomy, 7, 18.

Alves, N. M.; Lima, M. D. B.; Paula, J. R.; Simon, G. A. 2015. Lâminas de irrigação e sombreamento na produção de biomassa de Acariçoba (*Hydrocotyle umbellata* L.). Revista Brasileira de Plantas Medicinais, 17, 210–214.

Angelo, P. C.; Nunes-Silva, C. G.; Brigido, M. M.; Azevedo, J. S.; Assuncao, E. N.; Sousa, A. R.; Patricio, F. J.; et al. 2008. Guarana (*Paullinia cupana* var. *sorbilis*), an anciently consumed stimulant from the amazon rain forest: the seeded-fruit transcriptome. Plant Cell Reports, 27, 117–124.

Araújo, T. A. S.; Castro, V. T. N. A.; Solon, L. G. S.; da Silva, G. A.; Almeida, M. G.; da Costa, J. G. M.; de Amorim, E. L. C.; Albuquerque, U. P. 2015. Does rainfall affect the antioxidant capacity and production of phenolic compounds of an important medicinal species? Industrial Crops and Products, 76, 550–556.

Avancini, G.; Abreu, I. N.; Saldaña, M. D. A.; Mohamed, R. S.; Mazzafera, P. 2003. Induction of pilocarpine formation in jaborandi leaves by salicylic acid and methyljasmonate. Phytochemistry, 63, 171–175.

Azevedo, B. O.; Oliveira, L. M.; Lucchese, A. M.; Silva, D. J.; Ledo, C. A. S.; Nascimento, M. N. 2015. Growth and essential oil production by *Martianthus leucocephalus* grown under the edaphoclimatic conditions of Feira de Santana, Bahia, Brazil. Ciência Rural, 46, 593–598.

Bailey, B. A.; Strem, M. D.; Bae, H.; de Mayolo, G. A.; Guiltinan, M. J. 2005. Gene expression in leaves of *Theobroma cacao* in response to mechanical wounding, ethylene, and/or methyl jasmonate. Plant Science, 168, 1247–1258.

Bertolucci, S. K. V.; Pereira, A. B. D.; Pinto, J. E. B. P.; Braga, A. B. O. F. C. 2013. Seasonal variation on the contents of coumarin and kaurane-type diterpenes in *Mikania laevigata* and *M. glomerata* leaves under different shade levels. Chemistry & Biodiversity, 10, 288–295.

Bian, Z. H.; Yang, Q. C.; Liu, W. K. 2015. Effects of light quality on the accumulation of phytochemicals in vegetables produced in controlled environments: a review. Journal of the Science and Food Agriculture, 95, 869–877.

Bita, C. E.; Gerats, T. 2013. Plant tolerance to high temperature in a changing environment: scientific fundamentals and production of heat stress-tolerant crops. Frontiers in Plant Science, 4, 1–18.

Borges, L. L.; Alves, S. F.; Alves, M. T. F.; Conceição, E. C.; Ferri, P. H.; Paula, J. R. 2013a. Influence of environmental factors on the composition of essential oils from leaves of *Myrcia tomentosa* (Aubl.) DC. Boletín Latinoamericano y del Caribe de Plantas Medicinales y Aromáticas, 12, 572–580.

Borges, L. L.; Alves, S. F.; Sampaio, B. L.; Conceição, E. C. F.; Bara, M. T.; Paula, J. R. 2013b. Environmental factors affecting the concentration of phenolic compounds in *Myrcia tomentosa* leaves. Revista Brasileira de Farmacognosia, 23, 230–238.

Branco, C. S.; Rodrigues, T. S.; Lima, É. D.; Calloni, C.; Scola, G.; Salvador, M. 2016. Chemical constituents and biological activities of *Araucaria angustifolia* (Bertol.) O. Kuntze: a review. Journal of Organic & Inorganic Chemistry, 2, 1–10.

Caldeira, C. F.; Giannini, T. C.; Ramos, S. J.; Vasconcelos, S.; Mitre, S. K.; Pires, J. P. A.; Ferreira, G. C.; et al. 2017. Sustainability of Jaborandi in the eastern Brazilian amazon. Perspectives in Ecology and Conservation, 15, 161–171.

Cardoso, F. L.; Murakami, C.; Mayworm, M. A. S.; Marques, L. M. 2010. Análise sazonal do potencial antimicrobiano e teores de flavonoides e quinonas de extratos foliares de *Aloe arborescens* Mill.; Xanthorrhoeaceae. Brazilian Journal of Pharmacognosy, 20, 35–40.

Chen, Y. Y.; Wang, L. F.; Yang, S. G.; Tian, W. M. 2011. Molecular characterization of HbEREBP2, a jasmonate responsive transcription factor from Hevea brasiliensis Muell. Arg. African Journal of Biotechnology, 10, 9751–9759.

Cibulski, S. P.; Silveira, F.; Mourglia-Ettlin, G.; Teixeira, T. F.; dos Santos, H. F.; Yendo, A. C.; de Costa, F.; Fett-Neto, A. G.; Gosmann, G.; Roehe, P. M. 2016. *Quillaja brasiliensis* saponins induce robust humoral and cellular responses in a bovine viral diarrhea virus vaccine in mice. Comparative Immunology, Microbiology and Infectious Diseases, 45, 1–8.

Costa, D. P.; Santos, S. C.; Seraphin, J. C.; Ferri, P. H. 2009. Seasonal variability of essential oils of *Eugenia uniflora* leaves. Journal of the Brazilian Chemical Society, 20, 1287–1293.

Costa, L. C. B.; Pinto, J. E. B. P.; de Castro, E. M.; Bertolucci, S. K. V.; Corrêa, R. M.; Reis, É. S.; Alves, P. B.; Niculau, E. S. 2008. Tipos e doses de adubação orgânica no crescimento, no rendimento e na composição química do óleo essencial de elixir paregórico. Ciência Rural, 38, 2173–2180.

Curado, M. A.; Oliveira, C. B.; Jesus, J. G.; Santos, S. C.; Seraphin, J. C.; Ferri, P. H. 2006. Environmental factors influence on chemical polymorphism of the essential oils of *Lychnophora ericoides*. Phytochemistry, 67, 2363–2369.

David, E. F. S.; Pizzolato, M.; Facanali, R.; Morais, L. A. S.; Ferri, A. F.; Marques, M. O. M.; Ming, L. C. 2006. Influência da temperatura de secagem no rendimento e composição química do óleo essencial de *Ocimum selloi* Benth. Revista Brasileira de Plantas Medicinais, 8, 66–70.

de Costa, F.; Yendo, A. C.; Fleck, J. D.; Gosmann, G.; Fett-Neto, A. G. 2013. Accumulation of a bioactive triterpene saponin fraction of *Quillaja brasiliensis* leaves is associated with abiotic and biotic stresses. Plant Physiology and Biochemistry, 66, 56–62.

de Souza, M. O.; de Souza, C. L. M.; Pelacani, C. R.; Soares, M.; Mazzei, J. L.; Ribeiro, I. M.; Rodrigues, C. P.; Tomassini, T. C. B. 2013. Osmotic priming effects on emergence of *Physalis angulata* and the influence of abiotic stresses on physalin content. South African Journal of Botany, 88, 191–197.

Duan, C.; Rio, M.; Leclercq, J.; Bonnot, F.; Oliver, G.; Montoro, P. 2010. Gene expression pattern in response to wounding, methyl jasmonate and ethylene in the bark of *Hevea brasiliensis*. Tree Physiology, 30, 1349–1359.

Emamverdian, A.; Ding, Y.; Mokhberdoran, F.; Xie, Y. 2015. Heavy metal stress and some mechanisms of plant defense response. The Scientific World Journal, 2015, 1–18.

Falcão, H. S.; Mariath, I. R.; Diniz, M. F. F. M.; Batista, L. M.; Barbosa-Filho, J. M. 2008. Plants of the American continent with antiulcer activity. Phytomedicine, 15, 132–146.

Farooq, M.; Gogoi, N.; Hussain, M.; Barthakur, S.; Paul, S.; Bharadwaj, N.; Migdadi, H. M.; Alghamdi, S. S.; Siddique, K. H. M. 2017. Effects, tolerance mechanisms and management of salt stress in grain legumes. Plant Physiology and Biochemistry, 118, 199–217.

Feijó, E. V. R. S.; Oliveira, R. A.; Costa, L. C. B. 2014. Light affects *Varronia curassavica* essential oil yield by increasing trichomes frequency. Revista Brasileira de Farmacognosia, 24, 516–523.

Fernandes, V. F.; Almeida, L. B.; Feijó, E. V. R. S.; Silva, D. C.; Oliveira, R. A.; Mielke, M. S.; Costa, L. C. B. 2013. Light intensity on growth, leaf micromorphology and essential oil production of *Ocimum gratissimum*. Brazilian Journal of Pharmacognosy, 23, 419–424.

Ferraz, A. B.; Grivicich, I.; von Poser, G. L.; Faria, D. H.; Kayser, G. B.; Schwartsmann, G.; Henriques, A. T.; da Rocha, A. B. 2005. Antitumor activity of three benzopyrans isolated from *Hypericum polyanthemum*. Fitoterapia, 76, 210–215.

Fleck, J. D.; de Costa, F.; Yendo, A. C. A.; Segalin, J.; Dalla Costa, T. C. T.; Fett-Neto, A. G.; Gosmann, G. 2012. Determination of new immunoadjuvant saponin named QB-90, and analysis of its organ-specific distribution in *Quillaja brasiliensis* by HPLC. Natural Product Research, 27, 907–910.

Fleck, J. D.; Kauffmann, C.; Spilki, F.; Lencina, C. L.; Roehe, P. M.; Gosmann, G. 2006. Adjuvant activity of *Quillaja brasiliensis* saponins on the immune responses to bovine herpesvirus type 1 in mice. Vaccine, 24, 7129–7134.

Frade, M. A. C.; Assis, R. V. C.; Coutinho Netto, J.; Andrade, T. A. M.; Foss, N. T. 2012. The vegetal biomembrane in the healing of chronic venous ulcers. Anais Brasileiros de Dermatologia, 87, 45–51.

Gobbo-Neto, L.; Lopes, N. P. 2007. Plantas medicinais fatores de influência no conteúdo de metabólitos secundários. Química Nova, 30, 374–381.

González-Aguilar, G. A.; Tiznado-Hernández, M. E.; Zavaleta-Gatica, R.; Martínez-Téllez, M. A. 2004. Methyl jasmonate treatments reduce chilling injury and activate the defense response of guava fruits. Biochemical and Biophysical Research Communications, 313, 694–701.

Gregianini, T. S.; Porto, D. D.; Do Nascimento, N. C.; Fett, J. P.; Henriques, A. T.; Fett-Neto, A. G. 2004. Environmental and ontogenetic control of accumulation of brachycerine, a bioactive indole alkaloid from *Psychotria brachyceras*. Journal of Chemical Ecology, 30, 2023–2036.

Guerreiro, C. P. V.; Marques, M. O. M.; Ferracini, V. L.; Queiroz, S. C. N.; Ming, L. C. 2009. Produção de β-ecdisona em *Pfaffia glomerata* (Spreng.) Pedersen em função da adubação orgânica em 6 épocas de crescimento. Revista Brasileira de Plantas Medicinais, 11, 392–398.

Haas, J. S.; Viana, A. F.; Heckler, A. P.; von Poser, G. L.; Rates, S. M. 2010. The antinociceptive effect of a benzopyran (HP1) isolated from *Hypericum polyanthemum* in mice hot-plate test is blocked by naloxone. Planta Medica, 76, 1419–1423.

Heil, M.; Ton, J. 2008. Long-distance signaling in plant defense. Trends in Plant Science, 13, 264–272.

Hess, S. C.; Peres, M. T. L. P.; Batista, A. L.; Rodrigues, J. P.; Tiviroli, S. C.; Oliveira, L. G. L.; Santos, C. W. C.; Fedel, L. E. S. 2007. Evaluation of seasonal changes in chemical composition and antibacterial activity of *Elyonurus muticus* (sprengel) o. Kuntze (gramineae). Química Nova, 30, 370–373.

Holowaty, S. A.; Trela, V.; Thea, A. E.; Scipioni, G. P.; Schmalko, M. E. 2015. Yerba Maté (*Ilex paraguariensis* St. Hil.): chemical and physical changes under different aging conditions. Journal of Food Process Engineering, 39, 19–30.

Jardim, B. C.; Perdizio, V. A.; Berbert-Molina, M. A.; Rodrigues, D. C.; Botelho-Junior, S.; Vicente, A. C.; Hansen, E.; Otsuki, K.; Urmenyi, T. P.; Jacinto, T. 2010. Herbivore response in passion fruit (*Passiflora edulis* Sims) plants: induction of lipoxygenase activity in leaf tissue in response to generalist and specialist insect attack. Protein and Peptide Letters, 17, 480–484.

Kauffmann, C.; Machado, A. M.; Fleck, J. D.; Provensi, G.; Pires, V. S.; Guillaume, D.; Sonnet, P.; Reginatto, F. H.; Schenkel, E. P.; Gosmann, G. 2004. Constituents from leaves of *Quillaja brasiliensis*. Natural Product Research, 18, 153–157.

Küpper, H.; Parameswaran, A.; Leitenmaier, B.; Trtilek, M.; Setlik, I. 2007. Cadmium-induced inhibition of photosynthesis and long-term acclimation to cadmium stress in the hyperaccumulator *Thlaspi caerulescens*. New Phytologist, 175, 655–674.

Lara Lanza de Sá e Melo Marques, T.C.; Soares, A. M.; Gomes, M. P.; Martins, G. 2011. Respostas fisiológicas e anatômicas de plantas jovens de Eucalipto expostas ao cádmio. Revista Árvore, 35, 997–1006.

Lewinski, C. S.; Gonçalves, I. L.; Borges, A. C. P.; Dartora, N.; Souza, L. M.; Valduga, A. T. 2015. Effects of UV light on the physic-chemical properties of yerba-mate. Nutrition & Food Science, 45, 221–228.

Lima, D. F.; de Lima, L. I.; Rocha, J. A.; de Andrade, I. M.; Grazina, L. G.; Villa, C.; Meira, L.; et al. 2017. Seasonal change in main alkaloids of jaborandi (*Pilocarpus microphyllus* Stapf ex Wardleworth), an economically important species from the Brazilian flora. Plos One, 12(2).

Lin, L.; Wu, J.; Ho, K. P.; Qi, S. 2001. Ultrasound-induced physiological effects and secondary metabolite (saponin) production in *Panax ginseng* cell cultures. Ultrasound in Medicine & Biology, 27, 1147–1152.

Lucci, N.; Mazzafera, P. 2009a. Distribution of rutin in fava d'anta (*Dimorphandra mollis*) seedlings under stress. Journal of Plant Interactions, 4, 203–208.

Lucci, N.; Mazzafera, P. 2009b. Rutin synthase in fava d'anta: purification and influence of stressors. Canadian Journal of Plant Science, 89, 895–902.

Magedans, Y. V. S.; Matsuura, H. N.; Tasca, R. A. J. C.; Wairich, A.; Junkes, C. F. O.; de Costa, F.; Fett-Neto, A. G. 2017. Accumulation of the antioxidant alkaloid brachycerine from *Psychotria brachyceras* Müll. Arg. is increased by heat and contributes to oxidative stress mitigation. Environmental and Experimental Botany, 143, 185–193.

Martinez, M. J. A.; Bessa, A. L.; Benito, P. B. 2005. Biologically active substances from the genus *Baccharis* L. (compositae). Studies in Natural Products Chemistry, 30, 703–759.

Matos, D. O.; Tironi, F. L.; Martins, D. H. N.; Fagg, C. W.; Netto Júnior, N. L.; Simeoni, L. A.; Magalhães, P. O.; Silveira, D.; Fonseca-Bazzo, Y. M. 2015. Determinação de ácido rosmarínico em *Cordia verbenacea* por cromatografia líquida aplicabilidade em estudo sazonal. Revista Brasileira de Plantas Medicinais, 17, 857–864.

Matsuura, H. N.; de Costa, F.; Yendo, A. C. A.; Fett-Neto, A. G. 2013. Photoelicitation of bioactive secondary metabolites by ultraviolet radiation: mechanisms, strategies, and applications. In S. Chandra; H. Lata; A. Varma, eds., Biotechnology for Medicinal Plants: Micropropagation and Improvement, 1st ed., pp. 171–190. Berlin: Springer.

Matsuura, H. N.; Fett-Neto, A. G. 2013. The major indole alkaloid N,β-D-glucopyranosyl vincosamide from leaves of *Psychotria leiocarpa* Cham. & Schltdl. is not an antifeedant but shows broad antioxidant activity. Natural Product Research, 27, 402–411.

Matsuura, H. N.; Fragoso, V.; Paranhos, J. T.; Rau, M. R.; Fett-Neto, A. G. 2016. The bioactive monoterpene indole alkaloid N,β-d-glucopyranosyl vincosamide is regulated by irradiance quality and development in *Psychotria leiocarpa*. Industrial Crops and Products, 86, 210–218.

Matsuura, H. N.; Malik, S.; de Costa, F.; Yousefzadi, M.; Mirjalili, M. H.; Arroo, R.; Bhambra, A. S.; Strnad, M.; Bonfill, M.; Fett-Neto, A. G. 2018. Specialized plant metabolism characteristics and impact on target molecule biotechnological production. Molecular Biotechnology, 60,169–183.

Maurmann, N.; Reolon, G. K.; Rech, S. B.; Fett-Neto, A. G.; Roesler, R. 2011. A valepotriate fraction of *Valeriana glechomifolia* shows sedative and anxiolytic properties and impairs recognition but not aversive memory in mice. Evidence-Based Complementary and Alternative Medicine, 2011, 7. doi:10.1093/ecam/nep232.

Meirelles, G.; Pinhatti, A. V.; Sosa-Gomez, D.; Rosa, L. M. G.; Rech, S. B.; von Poser, G. L. 2013. Influence of fungal elicitation with *Nomuraea rileyi* (Farlow) samson in the metabolism of acclimatized plants of *Hypericum polyanthemum* Klotzsech ex Reichardt (Guttiferae). Plant Cell, Tissue and Organ Culture, 112, 379–385.

Menezes, A. P.; da Silva, J.; Fisher, C.; da Silva, F. R.; Reyes, J. M.; Picada, J. N.; Ferraz, A. G.; et al. 2016. Chemical and toxicological effects of medicinal baccharis trimera extract from coal burning area. Chemosphere, 146, 396–404.

Michielin, E. M. Z.; Salvador, A. A.; Riehl, C. A. S.; Smânia Jr., A.; Smânia, E. F. A.; Ferreira, S. R. S. 2009. Chemical composition and antibacterial activity of *Cordia verbenacea* extracts obtained by different methods. Bioresource Technology, 100, 6615–6623.

Morel, L. J. F.; Baratto, D. M.; Pereira, P. S.; Contini, S. H. T.; Momm, H. G.; Bertoni, B. W.; França, S. C.; Pereira, A. M. S. 2011. Loganin production in *Palicourea rigida* H.B.K. (Rubiaceae) from populations native to Brazilian Cerrado. Journal of Medicinal Plants Research, 5, 2559–2565.

Mossi, A.; Pauletti, G.; Rota, L.; Echeverrigaray, S.; Barros, I.; Oliveira, J.; Paroul, N.; Cansian, R. 2011. Effect of aluminum concentration on growth and secondary metabolites production in three chemotypes of *Cunila galioides* Benth. medicinal plant. Brazilian Journal of Biology, 71, 1003–1009.

Muller, L. G.; Borsoi, M.; Stolz, E. D.; Herzfeldt, V.; Viana, A. F.; Ravazzolo, A. P.; Rates, S. M. K. 2015 Diene valepotriates from *Valeriana glechomifolia* prevent lipopolysaccharide-induced sickness and depressive-like behavior in mice. Evidence-Based Complementary and Alternative Medicine, 2015, 12.

Muzitano, M. F.; Bergonzi, M. C.; de Melo, G. O.; Lage, C. L. S.; Bilia, A. R.; Vincieri, F. F.; Rossi-Bergmann, B.; Costa, S. S. 2011. Influence of cultivation conditions, season of collection and extraction method on the content of antileishmanial flavonoids from *Kalanchoe pinnata*. Journal of Ethnopharmacology, 133, 132–137.

Nascimento, L. B. S.; Leal-Costa, M. V.; Menezes, E. A.; Lopes, V. R.; Muzitano, M. F.; Costa, S. S.; Tavares, E. S. 2015. Ultraviolet-B radiation effects on phenolic profile and flavonoid content of *Kalanchoe pinnata*. Journal of Photochemistry and Photobiology B: Biology, 148, 73–81.

Nascimento, N. C.; Menguer, P. K.; Henriques, A. T.; Fett-Neto, A. G. 2013. Accumulation of brachycerine, an antioxidant glucosidic indole alkaloid, is induced by abscisic acid, heavy metal, and osmotic stress in leaves of *Psychotria brachyceras*. Plant Physiology and Biochemistry, 75, 33–40.

Negri, M. L. S.; Possamai, J. C.; Nakashima, T. 2009. Atividade antioxidante das folhas de espinheira-santa - *Maytenus ilicifolia* Mart. ex Reiss.; secas em diferentes temperaturas. Revista Brasileira de Farmacognosia, 19, 553–556.

Nogués, S.; Allen, D. J.; Morison, J. I. L.; Baker, N. R. 1998. Ultraviolet-B radiation effects on water relations, leaf development, and photosynthesis in droughted pea plants. Plant Physiology 117(1), 173–181.

Nunes, J. M.; Bertodo, L. O. O.; da Rosa, L. M. G.; Von Poser, G. L.; Rech, S. B. 2014. Stress induction of valuable secondary metabolites in *Hypericum polyanthemum* acclimatized plants. South African Journal of Botany, 94, 182–189.

Nunes, J. M.; Pinto, P. S.; Bordignon, S. A. L.; Rech, S. B.; von Poser, G. L. 2010. Phenolic compounds in *Hypericum* species from the *Trigynobrathys* section. Biochemical Systematics and Ecology, 387, 224–228.

Oliveira, M. S.; Campos, M. A.; Silva, F. S. 2015. Arbuscular mycorrhizal fungi and vermicompost to maximize the production of foliar biomolecules in *Passiflora alata* curtis seedlings. Journal of the Science of Food and Agriculture, 95, 522–528.

Pacheco, F. V.; Avelar, R. P.; Alvarenga, I. C. A.; Bertolucci, S. K. V.; de Alvarenga, A. A.; Pinto, J. E. B. P. 2016. Essential oil of monkey-pepper (*Piper aduncum* L.) cultivated under different light environments. Industrial Crops and Products, 85, 251–257.

Park, H. J.; Kim, W. Y.; Yun, D. J. 2016. A new insight of salt stress signaling in plant. Molecules Cells, 39, 447–459.

Pereira, E. P. L.; Ribeiro, P. R.; Loureiro, M. B.; de Castro, R. D.; Fernandez, L. G. 2014. Effect of water restriction on total phenolics and antioxidant properties of *Amburana cearensis* (Fr. Allem) A.C. Smith cotyledons during seed imbibition. Acta Physiologiae Plantarum, 36, 1293–1297.

Perotti, J. C.; Milech, C.; Kleinowski, A. M.; Lucho, S. R.; Soares, M. M.; Braga, E. J. B. 2016. Metil jasmonato na multiplicação in vitro e no incremento de betacianina em *Alternanthera philoxeroides*. Revista da jornada de pós-graduação e pesquisa Congrega URCAMP. http://trabalhos.congrega.urcamp.edu.br/index.php/jpgp/article/view/854.

Perotti, J. C.; Rodrigues, I. C. S.; Kleinowski, A. M.; Ribeiro, M. V.; Einhardt, A. M.; Peters, J. A.; Bacarin, M. A.; Braga, E. J. B. 2010. Produção de betacianina em erva-de-jacaré cultivada *in vitro* com diferentes concentrações de sulfato de cobre. Ciência Rural, 40, 1874–1880.

Perotti, J. C.; Rodrigues-Correa, K. C. S.; Fett-Neto, A. G. 2015. Control of resin production in *Araucaria angustifolia*, an ancient South American conifer. Plant Biology, 17, 852–859.

Pertuzatti, P. B.; Sganzerla, M.; Jacques, A. C.; Barcia, M. T.; Zambiazi, R. C. 2015. Carotenoids, tocopherols and ascorbic acid content in yellow passion fruit (*Passiflora edulis*) grown under different cultivation systems. LWT – Food Science and Technology, 64, 259–263.

Pinto, J. V. C.; Vieira, M. C.; Zárate, N. A. H.; Formagio, A. S. N.; Cardoso, C. A. L.; Carnevali, T. O.; Souza, P. H. N. 2016. Effect of soil nitrogen and phosphorus on early development and essential oil composition of *Schinus terebinthifolius* raddi. Journal of Essential Oil Bearing Plants, 19, 247–257.

Porto, D. D.; Henriques, A. T.; Fett-Neto, A. G. 2009. Bioactive alkaloids from south American *Psychotria* and related species. The Open Bioactive Compounds Journal, 2, 29–36.

Queiroz, T. B.; Mendes, A. D. R.; Silva, J. C. R. L.; Fonseca, F. S. A.; Martins, E. R. 2016. Teor e composição química do óleo essencial de erva-baleeira (*Varronia curassavica* Jaqc.) em função dos horários de coleta. Revista Brasileira de Plantas Medicinais, 18, 356–362.

Rangel, M.; Machado, O. L.; da Cunha, M.; Jacinto, T. 2002. Accumulation of chloroplast-targeted lipoxygenase in passion fruit leaves in response to methyl jasmonate. Phytochemistry, 60, 619–625.

Reimberg, M. C. H.; Colombo, R.; Yariwak, J. H. 2009. Multivariate analysis of the effects of soil parameters and environmental factors on the flavonoid content of leaves of *Passiflora incarnata* L.; Passifloraceae. Brazilian Journal of Pharmacognosy, 19, 853–859.

Ribeiro, M. V.; Deuner, S.; Benitez, L. C.; Einhardt, A. M.; Peters, J. A.; Braga, E. J. B. 2014. Betacyanin and antioxidant system in tolerance to salt stress in *Alternanthera philoxeroides*. Agrociencia, 48, 199–210.

Riter Netto, A. F.; Freitas, M. S. M.; Martins, M. A.; Carvalho, A. J. C.; Vitorazi Filho, J. A. 2014. Efeito de fungos micorrízicos arbusculares na bioprodução de fenóis totais e no crescimento de *Passiflora alata* Curtis. Revista Brasileira de Plantas Medicinais, 16, 1–9.

Rodrigues, A. C. D.; Santos, A. M.; Santos, F. S.; Pereira, A. C. C.; Sobrinho, N. M. B. A. 2016. Response mechanisms of plants to heavy metal pollution, possibility of using macrophytes for remediation of contaminated aquatic environments. Revista Virtual de Química, 8, 262–276.

Ruelland, E.; Zachowski, A. 2010. How plants sense temperature. Environmental and Experimental Botany, 69, 225–232.

Russowski, D.; Maurmann, N.; Rech, S. B.; Fett-Neto, A. G. 2013. Improved production of bioactive valepotriates in whole-plant liquid cultures of *Valeriana glechomifolia*. Industrial Crops and Products, 46, 253–257.

Sá, F. A. S.; Sampaio, B. L.; Borges, L. L.; Ferri, P. H.; Paula, J. R.; Paula, J. A. M. 2012. Essential oils in aerial parts of *Myrcia tomentosa*: composition and variability. Revista Brasileira de Plantas Medicinais, 22, 1233–1240.

Salles, L. A.; Silva, A. L.; Fett-Neto, A. G.; von Poser, G. L.; Rech, S. B. 2002. *Valeriana glechomifolia*: *in vitro* propagation and production of valepotriates. Plant Science, 163, 165–168.

Sampaio, B. L.; Bara, M. T. F.; Ferri, P. H.; Santos, S. C.; de Paula, J. R. 2011. Influence of environmental factors on the concentration of phenolic compounds in leaves of *Lafoensia pacari*. Revista Brasileira de Farmacognosia, 21, 1127–1137.

Sampaio, B. L.; Edrada-Ebel, R.; da Costa, F. B. 2016. Effect of the environment on the secondary metabolic profile of *Tithonia diversifolia*: a model for environmental metabolomics of plants. Scientific Reports, 6, 29265.

Santos, R. F.; Isobe, M. T. C.; Lalla, J. G.; Haber, L. L.; Marques, M. O. M.; Ming, L. C. 2012. Composição química e produtividade dos principais componentes do óleo essencial de *Baccharis dracunculifolia* DC. em função da adubação orgânica. Revista Brasileira de Plantas Medicinais, 14, 224–234.

Santos, R. M.; Fortes, G. A. C.; Ferri, P. H.; Santos, S. C. 2011. Influence of foliar nutrients on phenol levels in leaves of *Eugenia uniflora*. Revista Brasileira de Farmacognosia, 21, 575–580.

Sarrazin, S. L. F.; da Silva, L. A.; de Assunção, A. P. F.; Oliveira, R. B.; Calao, V. Y. P.; da Silva, R.; Stashenko, E. E.; Maia, J. G. S.; Mourão, R. H. V. 2015. Antimicrobial and seasonal evaluation of the carvacrol-chemotype oil from *Lippia origanoides* Kunth. Molecules, 20, 1860–1871.

Sartor, T.; Xavier, V. B.; Falcão, M. A.; Mondin, C. A.; Santos, M. A.; Cassel, E.; Astarita, L. V.; Santarém, E. R. 2013. Seasonal changes in phenolic compounds and in the biological activities of *Baccharis dentata* (Vell.) G.M. Barroso. Industrial Crops and Products, 51, 355–359.

Schwerz, L.; Caron, B. O.; Manfron, P. A.; Schmidt, D.; Elli, E. F. 2015. Biomassa e teor de óleo essencial em *Aloysia triphylla* (l'hérit) Britton submetida a diferentes níveis de reposição hídrica e à variação sazonal das condições ambientais. Revista Brasileira de Plantas Medicinais, 17, 631–641.

Selmar, D.; Kleinwächter, M. 2013. Influencing the product quality by deliberately applying drought stress during the cultivation of medicinal plants. Industrial Crops and Products, 42, 558–566.

Serudo, R. N.; Assis, I. M.; Klehm, C. S.; Silva, J. F.; Florêncio, V. 2013. Acúmulo de matéria seca e rendimento de óleo da planta *Otacanthus azureus* em função da luminosidade e adubação nitrogenada. Scientia Plena, 9, 1–5.

Silva, F. G.; Pinto, J. E. B. P.; Cardoso, M. G.; Nascimento, E. A.; Nelson, D. L.; Sales, J. F.; Mol, D. J. S. 2006. Influence of radiation level on plant growth, yield and quality of essential oil in carqueja. Ciência e Agrotecnologia, 30, 52–57.

Sobrinho, T. J. S. P.; Cardoso, K. C. M.; Gomes, T. L. B.; Albuquerque, U. P.; Amorim, E. L. C. 2009. Análise da pluviosidade e do efeito de borda sobre os teores de flavonóides em *Bauhinia cheilantha* (Bong.) Steud.; Fabaceae. Revista Brasileira de Farmacognosia, 19, 740–745.

Sousa, J. P. B.; Leite, M. F.; Jorge, R. F.; Resende, D. O.; Filho, A. A. S.; Furtado, N. A. J. C.; Soares, A. E. E.; Spadaro, A. C. C.; Magalhães, P. M.; Bastos, J. K. 2009. Seasonality role on the phenolics from cultivated *Baccharis dracunculifolia*. Evidence-Based Complementary and Alternative Medicine, 2011, 1–8.

Sousa, L. B.; Heitor, L. C.; Santos, P. C.; Freitas, J. A. A.; Freitas, M. S. M.; Freitas, S. J.; Carvalho, A. J. C. 2013. Crescimento, composição mineral e fenóis totais de espécies de Passiflora em função de fontes nitrogenadas. Bragantia, 72, 247–254.

Stein, A. C.; Viana, A. F.; Muller, L. G.; Nunes, J. M.; Stolz, E. D.; Do Rego, J. C.; Costentin, J.; von Poser, G. L.; Rates, S. M. 2012. Uliginosin B, a phloroglucinol derivative from *Hypericum polyanthemum*: a promising new molecular pattern for the development of antidepressant drugs. Behavioural Brain Research, 228, 66–73.

Tabaldi, L. A.; Vieira, M. C.; Zárate, N. A. H.; Formagio, A. S. N.; Pilecco, M.; da Silva, L. R.; dos Santos, K. P.; dos Santos, L. A. C.; Cardoso, C. A. L. 2016. Produção de biomassa e conteúdo de fenóis e flavonoides de *Schinus terebinthifolius* cultivada em fileira simples e dupla com cama de frango. Ciência Florestal, 26, 789–796.

Teixeira, A.; Eiras-Dias, J.; Castellarin, S. D.; Gerós, H. 2013. Berry phenolics of grapevine under challenging environments. International Journal of Molecular Sciences, 14, 18711–18739.

Thirupathi, K.; Kumar, S. S.; Raju, V. S.; Ravikumar, B.; Krishna, D. R.; Mohan, G. K. 2008. A review of medicinal plants of the *Genus Cordia*: Their chemistry and pharmacological uses. Journal of Natural Remedies, 8, 1–10.

Treutter, D. 2010. Managing phenol contents in crop plants by phytochemical farming and breeding-visions and constraints. International Journal of Molecular Sciences, 11, 807–857.

van Loon, L. C.; Geraats, B. P.; Linthorst, H. J. 2006. Ethylene as a modulator of disease resistance in plants. Trends Plant Science, 11, 184–191.

Verma, N.; Shukla, S. 2015. Impact of various factors responsible for fluctuation in plant secondary metabolites. Journal of Applied Research on Medicinal and Aromatic Plants, 2, 105–113.

Yao, L. H.; Caffin, N.; D'Arcy, B.; Jiang, Y. M.; Shi, J.; Singanusong, R.; Liu, X.; Datta, N.; Kakuda, Y.; Xu, Y. 2004. Seasonal variations of phenolic compounds in Australia-grown tea (*Camellia sinensis*). Journal of Agricultural and Food Chemistry, 53, 6477–6483.

Yendo, A. C. A.; de Costa, F.; Cibulski, S. P.; Teixeira, T. F.; Colling, L. C.; Mastrogiovanni, M.; Soulé, S.; et al. 2016. A rabies vaccine adjuvanted with saponins from leaves of the soap tree (*Quillaja brasiliensis*) induces specific immune responses and protects against lethal challenge. Vaccine, 34, 2305–2311.

Yendo, A. C. A.; de Costa, F.; Fleck, J. D.; Gosmann, G.; Fett-Neto, A. G. 2015. Irradiance-based treatments of *Quillaja brasiliensis* leaves (A. St.-Hil. & Tul.) Mart. as means to improve immunoadjuvant saponin yield. Industrial Crops and Products, 74, 228–233.

Yendo, A. C. A.; de Costa, F.; Gosmann, G.; Fett-Neto, A. G. 2010. Production of plant bioactive triterpenoid saponins: elicitation strategies and target genes to improve yields. Molecular Biotechnology, 46, 94–104.

Zhu, J. K. 2016. Abiotic stress signaling and responses in plants. Cell, 167, 313–324.

7 Brazilian Bryophytes and Pteridophytes as Rich Sources of Medicinal Compounds

Adaíses Simone Maciel-Silva and Lucas Vieira Lima
Universidade Federal de Minas Gerais, Laboratório de Sistemática Vegetal, Departamento de Botânica, Instituto de Ciências Biológicas, Belo Horizonte, Brazil

CONTENTS

7.1 INTRODUCTION

Bryophytes and pteridophytes are generalized names given to the five different plant lineages: "bryophytes" – Marchantiophyta (liverworts, ca. 9,000 spp.), Bryophyta (mosses, 12,700 spp.), and Anthocerotophyta (hornworts, 225 spp.) (Christenhusz and Byng, 2016; Crandall-Stotler et al., 2009; Goffinet et al., 2009; Renzaglia et al., 2009); "pteridophytes" include Lycopodiopsida (e.g. clubmosses, quillworts, and spikemosses, 1,290–1,338 spp.) and Polypodiopsida (e.g. ferns, horsetails, and whisk ferns, 10,560 spp.) (PPG I, 2016). Scientists have estimated that liverworts appeared ca. 470 million years ago, which is at least ~330 Ma before the angiosperms (flowering plants) (Magallón et al., 2015; Wellman et al., 2003). Because of their ancient origins, many taxa in these groups could produce unique and rare phytochemical compounds and rich sources of medicinal compounds. In this chapter, we shall henceforward use the term "bryophytes" and "pteridophytes" to refer to those plant groups.

Bryophytes and pteridophytes produce a wide diversity of chemical compounds such as terpenoids, steroids, flavonoids, alkaloids, and aromatic and phenolic compounds (Asakawa et al., 2013a; Chopra and Kumra 1988; Huneck 1983; Schofield 1985). Bibenzyl cannabinoids, pinguisane-type sesquiterpenoids, and sacculatane-type diterpenoids are examples of chemicals exclusive to bryophytes, while pteridophytes produce unique compounds such as the alkaloids huperzine, lycopodine, and the triterpenoid lycophlegmariol, among others (Asakawa et al., 2013a, 2013b; Cao et al., 2017).

Studies focusing on the phytochemistry of bryophytes and pteridophytes have increased in recent years (Asakawa, 2001; Asakawa et al., 2013a; Asakawa and Ludwiczuk, 2017; Cao et al., 2017) despite difficulties encountered in terms of the identification of plants and their collection in large quantities as pure samples (especially bryophytes). Countries such as China and United States stand out in terms of the numbers of papers published from 1999 to 2017 (Figure 7.1). The chemistries of bryophyte species are not yet very well-known, largely because of problems in obtaining pure samples of any species (Sabovljevic et al., 2009). The secondary metabolic compounds produced by bryophytes and ferns show antimicrobial, antifungal, anti-Alzheimer, cytotoxic, antitumor, vasopressin (VP) antagonist, cardiotonic, allergenic, irritant, tumor-affecting, insect antifeedants, insecticide, molluscicide, and piscicide effects; and plant growth regulation, superoxide anion radical release inhibition, 5-lipoxygenase, calmodulin, hyaluronidase, cyclooxygenase, and anti-HIV activities (Asakawa et al., 2013a; Liu et al., 1986a, 1986b; Sabovljevic et al., 2001; Sabovljevic et al., 2009; Santos et al., 2010).

Ethnobotanical studies have focused on the use of those plants by human societies (in places such as Africa, America, Europe, Poland, Argentina, Australia, New Zealand, Turkey, Japan, Taiwan, Pakistan, China, Nepal, and different parts of southern, northern, and eastern India; Beaujard (1998), Benjamin and Manickn (2007), Bonet and Valles (2007), Chandra et al. (2017), Glime (2017), Hammond et al. (1998), Harris (2008)) and their potential applications in the pharmaceutical and medicinal industries (Frahm, 2004; Harris 2008; Ho et al., 2011).

Approximately 5,718 species of bryophytes and 1,332 of pteridophytes are currently known from Brazil (Flora do Brasil, 2020). The Atlantic forest alone harbors more than 32% of all Brazilian species of bryophytes, lycophytes, and ferns. Although chemical studies of bryophytes and pteridophytes have indicated their great medicinal potential, these species remain poorly investigated in relation to their high taxonomic diversities (Pinheiro et al., 1989; Santos et al., 2010). That line of research has been very slow in Brazil despite its high biodiversity, with research on pteridophytes being slightly greater than on bryophytes (Figure 7.1).

The main aim of this review is to compile data that could help identify potential uses of the chemical compounds found in Brazilian species of bryophytes and pteridophytes. Here, we present accessible information on the morphology and taxonomy of those plants, techniques for their collection in the field and their taxonomic identification and herborization, the potential occurrence of secondary metabolites and their biological activities, and the geographical distributions of Brazilian taxa.

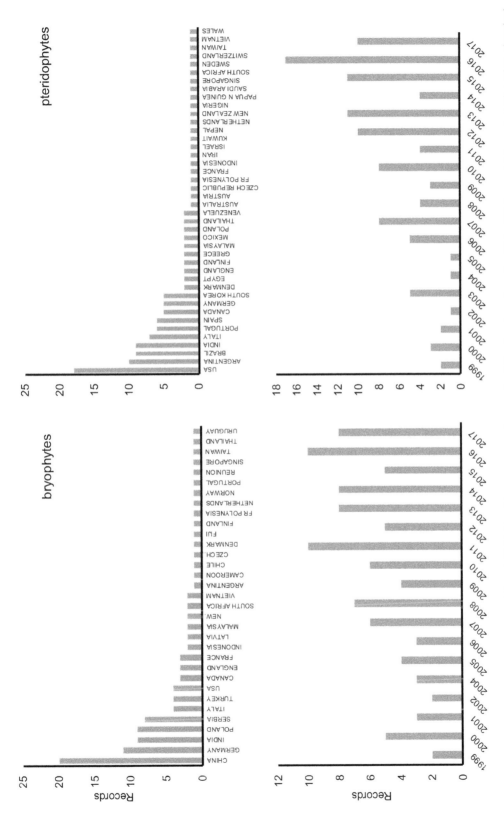

FIGURE 7.1 Studies published between 1999 and 2017, listed by country and year of publication; based on the Web of Science database (search topics: chemical compounds and bryophytes; chemical compounds and pteridophytes; chemical compounds and ferns).

7.2 MORPHOLOGY AND SYSTEMATICS OF BRYOPHYTES AND PTERIDOPHYTES

7.2.1 BRYOPHYTES

Bryophytes are probably the closest modern relatives of the earliest land plants, comprising plants, which have life cycles with alternating haploid and diploid generations with a dominant gametophyte (Figure 7.2; Gerrienne and Gonéz, 2011; Vanderpoorten and Goffinet, 2009). Bryophytes are the only land plants with a dominant, branched gametophyte, and they exhibit a large diversity of morphologies as compared to tracheophytes (Gerrienne and Gonéz, 2011; Vanderpoorten and Goffinet, 2009). In addition to the structural diversity of their gametophytes, bryophytes display physiological adaptations (e.g. poikilohydry, desiccation tolerance, efficient mechanisms for water and nutrient uptake, and specialized life cycles) that enable their successful colonization of many different biomes, from the tundra of the Northern hemisphere to Antarctica (Crandall-Stotler et al., 2009; Goffinet et al., 2009; Ligrone et al., 2000; Proctor et al., 2007; Renzaglia et al., 2007, 2009; Vanderpoorten and Goffinet, 2009).

Recent phylogenies, especially those based on molecular data, have resulted in different hypotheses concerning the evolution of bryophyte lineages and the relationships between them (Crandall-Stotler et al., 2009; Goffinet et al., 2009; Renzaglia et al., 2007, 2009). Among the current phylogenetic hypotheses of land plants or embryophytes is the proposal that Marchantiophyta would have diverged early in the evolution of the group, while Anthocerotophyta is the sister group to all vascular plants; Bryophyta would be an intermediate group. Many recent studies have focused on understanding the evolution of land plants using ever broader frameworks, and bryophyte lineages seem to mark the transition from the algal ancestors of land plants to vascular plants (Goffinet and Buck, 2012; Goffinet et al., 2009).

Liverworts (Marchantiophyta) have thallose or leafy gametophytes (Figure 7.3A–C) with leaves in two or three rows. Oil bodies (organelles rich in essential oils) are frequently found in the gametophytic cells of different liverwort taxa. Sporophytes produce one sporangium, elevated only at maturity by a hyaline stalk (seta) that extends by cell elongation. Sporangium dehiscence is typically along four vertical lines, and no stomata are present in the sporangial walls. Spores and elaters (elongated cells with spiral wall thickenings that facilitate spore dispersal) can be found inside the sporangia. After germination, the spores develop into a single, branched gametophyte (Crandall-Stotler et al., 2009; Vanderpoorten and Goffinet, 2009). Brazilian liverworts are represented by 135 genera and ca. 667 species (Flora do Brazil, 2020).

Mosses (Bryophyta) have leafy gametophytes with leaves arranged in spiral rows (Figure 7.3D). In sporophyte, complete seta development is prior to sporogenesis and elevates a terminal sporangium. Dehiscence occurs through an operculum in the majority of mosses. Stomata may occur on the sporangium wall. No elaters are found inside the sporangium, and spores generally germinate into filamentous sporelings called protonema, which can develop into several leafy gametophytes (Goffinet et al., 2009; Vanderpoorten and Goffinet, 2009). Brazilian mosses are currently represented by 276 genera and ca. 890 species (Flora do Brazil, 2020).

Hornworts (Anthocerotophyta) consist of thalloid gametophytes and linear sporophytes composed of a long sporangium with no seta (Figure 7.3E). Sporangia show nonsynchronous spore dispersal as the basal meristem adds new cells to its base. Dehiscence occurs by two longitudinal lines, exposing spore mass and multicellular pseudoelaters, unlike the elaters of liverworts. Stomata are present on the sporangia walls in some taxa. Endosymbiotic colonies of the cyanobacteria *Nostoc* are common among hornworts (Renzaglia et al., 2009; Vanderpoorten and Goffinet, 2009). Seven genera and 15 species of hornworts are currently known to Brazil (Flora do Brazil, 2020).

The life cycles of bryophytes (as in all land plants) are characterized by an alternation of two generations: gametophyte and sporophyte. After the fusion of two gametes (fertilization), the zygote develops into a sporophyte, which produces spores through meiosis. While the gametophyte has sex

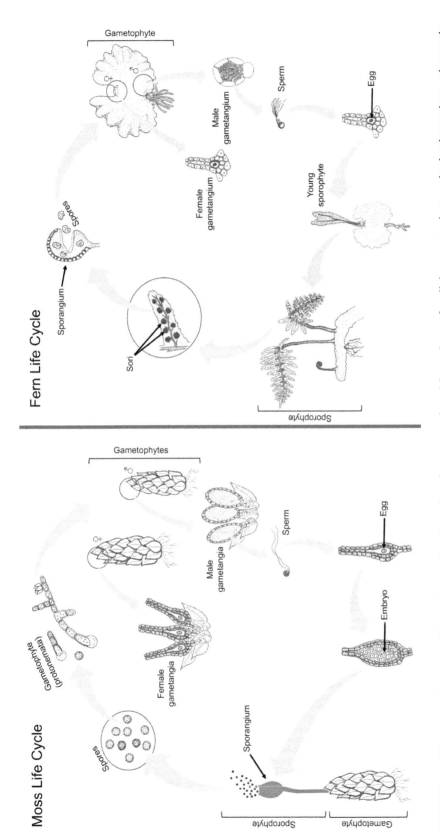

FIGURE 7.2 Life cycles of mosses and ferns, showing the alternation of two generations. Mosses have a free-living gametophyte as the dominant phase, whereas the sporophyte is dominant in ferns.

FIGURE 7.3   Examples of different bryophyte groups. (A and B) Ventral and dorsal views of a leafy liverwort (*Frullania brasiliensis*). (C) View of a thallose liverwort (*Marchantia polymorpha*). (D) Mosses (*Polytrichum* sp.) with sporophytes. (E) A hornwort (*Phaeoceros* sp.) containing several sporophytes. Photographs (A), (B), and (D) by Oliveira, M.F.; (C) by Oliveira, B.A.; and (E) by Araújo, C.A.T.

organs (female and male gametangia), the sporophyte develops a sporangium that produces only spores. The sporophyte is physiologically dependent on, and permanently attached to, the gametophyte (maternal plant) during its complete life cycle. Bryophytes are the only extant land plants with gametophyte as the dominant generation in their life cycles (Figure 7.2; Vanderpoorten and Goffinet, 2009; Maciel-Silva and Pôrto 2014).

7.2.2   Pteridophytes

There are many botanical texts that describe the morphologies of pteridophytes. Our main goal here is to provide a quick guide for nonspecialists who wish to become familiar with the subject. We recommend the following reports for more detailed approaches (Harris and Harris 1994; Lawrence, 1977; Lellinger, 2002; Stearn, 1998).

Pteridophytes are seedless vascular plants with a life cycle showing an alternation of generation, in which the gametophytic phase is independent from the sporophyte (Figure 7.2). Unlike bryophytes, the ephemeral and reduced phase of ferns is the gametophyte, while the sporophyte is complex, branched, and long-lived (Gifford and Foster, 1987).

Pteridophytes are widely distributed and inhabit nearly all tropical habitats, occurring in rain forests, high montane cloud forests, temperate forests, mangroves, and even floating or submerged in lakes. They are often pioneer species and weedy colonizers of disturbed landscapes and can be found scattered among rocks in semiarid landscapes, savannas, and coastal and high alpine mountains, resisting droughts, fires, and cold temperatures (Sharpe et al., 2010).

Ferns and lycophytes are two distinct and ancient phylogenetic lineages of seedless vascular plants traditionally addressed as pteridophytes (PPG I, 2016; Pryer et al., 2004). According to the most recent classification (PPG I, 2016), two monophyletic classes are recognized: Lycopodiopsida and Polypodiopsida.

FIGURE 7.4 (A) The quillworts, *Isoetes*, Isoetaceae. (B) The spike-moss *Selaginella*, Selaginellaceae. (C) The club-moss *Phlegmariurus*, Lycopodiaceae. (D) The horsetail *Equisetum*, Equisetaceae. (E) The moonwort *Botrychium*, Ophioglossaceae. (F) *Danea*, Maratticeae.

Lycopodiopsida is divided into three orders, three families, and 18 genera (PPG I, 2016), and is represented by spike mosses (Selaginellaceae), clubmosses (Lycopodiaceae), and quillworts (Isoetaceae). Those plants are principally characterized by the presence of microphylls, sporangia with transversal dehiscence, and by each fertile microphyll bearing only one sporangium on the adaxial surface, usually forming a strobilus at the branch apex (Øllgaard, 1990; Tryon and Tryon, 1982). Lycopodiopsida is represented by 11 genera and approximately 179 species in Brazil.

Selaginella (Figure 7.4B) is the only genus in the Selaginellaceae family (PPG I, 2016), and is characterized by the presence of rhizophores, leaves, a ligule, heterospores, and adaxial and reniform sporangia (Webster, 1992). In Brazil, the family is represented by 89 species, of which 30 are endemic (Flora do Brazil, 2020).

Lycopodiaceae (Figure 7.4C) is divided into three subfamilies, 16 genera, and an estimated 388 species (PPG I, 2016). Thedifference from the other lycophyte families is mainly by demonstrating homospory and by having eligulate microphylls. In Brazil, this family is represented by nine genera and 62 species, of which 31 are endemic (Flora do Brazil, 2020).

Isoetaceae (Figure 7.4A) comprises one genus and approximately 250 species. *Isoetes* L. is readily identified bythe species' microphylls having four air chambers in cross section, a single sunken sporangium at the base of the microphylls, ligulate microphylls, and heterospores (Gifford and Foster, 1987; Pigg, 1992). Isoetaceae is represented by 27 species in Brazil, of which 22 are endemic (Flora do Brazil, 2020).

The Polypodiopsida class is divided into four subclasses, 11 orders, 48 families, 319 genera, and an estimated 10,578 species (PPG I, 2016).

Equisetidae consists of one extant order, one family, one genus, and 15 species (PPG I 2016). Horsetails ferns (*Equisetum*) (Figure 7.4D-E) are principally characterized by the presence of a peltate sporangiophore, articulated rhizomes (usually hollow), and by reduced and verticillate leaves (Hauke, 1990). Only *Equisetum giganteum* L. occurs in Brazil (Nóbrega and Prado, 2018).

Ophioglossidae is divided into two orders, two families, 12 genera, and an estimated 129 species. This subclass comprises the whisk ferns (Psilotaceae), which are mainly characterized by rootless sporophytes, with dichotomous rhizomes, aerial branches, and scale-like or leaf-like leaves. Only *Psilotum nudum* (L.) P. Beauv. occurs in Brazil. The subclass also includes moonwort ferns (Ophiglossaceae) (Figure 7.4F). Those plants are readily recognized by having hemidimorphic fronds with eusporangia on the erect fertile portion of the frond, usually in a fertile spike or panicle-like sporangial cluster arising from the base of the sterile blades (Mickel and Smith, 2004). Ophioglossidae is represented by four genera and six species in Brazil (Prado et al., 2015).

Marattiidae has only one order, one family, six genera, and an estimated 111 species. Marattiaceae is mainly characterized by the presence of pairs of large, persistent photosynthetic stipules that protect the young croziers, and having free abaxial eusporangia, or eusporangia united to form a synangium (Camus, 1990). In Brazil, this subclass is represented by three genera and six species (Prado et al., 2015).

Polypodiidae (Figure 7.5) comprises the vast majority of extant fern diversity, with seven orders, 44 families, 300 genera, and an estimated 10,323 species (PPG I, 2016). Film ferns, tree ferns, maidenhair ferns, among other groups, demonstrate the wide morphological diversity of this subclass, which is characterized by leptosporangia. Polypodiidae is represented by 31 families, 134 genera, and 1,153 species in Brazil (Flora do Brasil, 2020).

7.3   COLLECTION TECHNIQUES AND PROCESSING

Scientific plant collecting is essential for many reasons, including specimen identification, herbarium collections, and the establishment of DNA banks (Vanderpoorten et al., 2010). Correct techniques for collecting and processing plant material should always be employed. The necessity of obtaining collecting permits must also be emphasized, as well as export licenses (if the material is to be taken out of the country) (Frahm, 2003; Gradstein et al., 2001; Vanderpoorten et al., 2010).

7.3.1   BRYOPHYTES

Collecting bryophytes is generally easier than collecting flowering plants, as they generally do not need to be pressed and can be held in simple paper bags (together with substrate samples; 1–3 cm) and allowed to air dry. A quick guide for collecting and processing of bryophytes is provided herein. For more details, see Gradstein et al. (2001), Frahm (2003), Vanderpoorten et al. (2010), and Glime and Wagner (2013).

During specimen collection, plants should be selected to include all organs needed for their identification. The sizes of plant samples will vary according the species and colony extension, although c. 4 × 4 cm samples are very common. However, some species grow intermingled (mainly tiny liverworts) and can be very difficult to identify in the field using just a hand lens. Sporophytes and perianths are useful for identifications and should be searched for and collected along with the gametophytes.

Useful tools for collecting bryophytes in the field include a 10–20× hand lens, different-sized paper bags, plastic bags (used for transporting fresh samples to the laboratory), a penknife, a chisel, waterproof markers, a field notebook, and a GPS for recording the geographic coordinates of the

FIGURE 7.5   Polypodiidae representatives. (A) *Asplenium*, Aspleniaceae. (B) Sticherus, Gleicheniaceae. (C) The film fern *Hymenophyllum*, Hymenophyllaceae. (D) The deer tongue fern *Elaphoglossum*, Dryopteridaceae. (E) *Pleopeltis*, Polypodiaceae. (F) The tree fern *Cyathea*, Cyatheaceae. (G) *Anemia*, Anemiaceae.

collection sites. Additionally, general data about the collect site should be noted, including the locality, collector names, the date, elevations, vegetation type, conservation status, substrate type (microhabitat), and the traits of the specimens (color, growth form, fertility, etc.). All this information should be carefully recorded in the field on the paper bags and/or in a field notebook.

Upon returning to the laboratory, the paper bags should be opened, and the samples air-dried (checking them every day) as quickly as possible to avoid fungal growth. If necessary, very moist samples can be dried using an electric (or light-bulb) oven at low temperatures (40–60°C). Fresh material (maintained in plastic bags) should be stored at approximately 10°C until examined. Liverworts containing oil bodies should be studied within just a few days after collection, as those oil bodies disappear quite rapidly. The oil bodies should be measured, counted, and described before they vanish.

Dried plants can be stored in separate paper packets (with their respective herbarium specimen labels). Very tiny plants, sporophytes, or other fertile structures should be placed in mini-packets together with the main envelope. An A4 sheet of paper can be folded into a standard envelope (measuring about 11 × 15 cm) for preserving bryophyte specimens.

Dissecting and compound microscopes are useful for studying bryophytes in detail in the laboratory. The dry or moist plants can first be analyzed under a dissecting microscope to determine their growth form, leaf arrangements, reproductive structures, and all likely informative characters. Small pieces of plants (gametophyte and sporophyte, if present) may be separated using micro-forceps and very thin needles and placed in a drop of water on a glass slide. Still under the dissecting microscope, some leaves can be detached from the stem and thalloid plants can be sectioned (to be viewed under a coverslip). Many additional characteristics can be assessed under a compound microscope by observing prepared slides, such as cell shapes, the numbers of cells, leaf borders, costae, and teeth on the gametophyte leaves; sporophyte details such as the peristome and stomata should also be observed.

Local floras and specific revisions are very informative for taxonomic determinations. Contacts with bryophyte taxonomists specializing in specific groups will increase the reliability of specimen determinations. Taxonomic determinations are very important for researchers examining the chemical compounds produced by bryophyte species and must be very precise to ensure the quality of those studies.

7.3.2 PTERIDOPHYTES

A number of factors must be taken into consideration when collecting ferns and fern allies and depositing their vouchers in a herbarium. A practical guide concerning that subject is presented here; for more details and information refer to Fidalgo and Bononi (1989) and Peixoto and Maia (2013).

7.3.2.1   How Big is the Plant?

This may seem an obvious point, but if you are preparing an exsiccate, the plant must fit on an herbarium sheet. Therefore, the size of the plant matters! There are huge differences in fern shapes and sizes. Some film ferns may be just a few centimeters long, while some tree ferns may reach up to 3 m or more. How does one proceed in such cases?

When the plants are small, the best is to collect as many specimens as needed to fill the herbarium sheet. When the plant is larger than the sheet, you may fold the species into a "N" or "V" shape to better fit the sheet. However, be sure to leave the taxonomically important portions easily visible (such as sori). When the plants are much too big for a single sheet (such as tree ferns), the fronds may be cut into pieces that each fit on an herbarium sheet. In those cases, the base of the petiole, with the indument, a pair of basal pinnae, a pair of median pinnae, and the frond apex will compose the exsiccate. Additionally, you should note the intact frond size (length and width) and estimate the size of the rhizome. This information will be included as additional data on the exsiccate label.

7.3.2.2   What to Collect

The more information you provide the better. Therefore, the best is to collect the whole plant – with rhizome and fronds – providing more subsidies for accurate identifications. Since some groups of ferns can easily be identified even when sterile, whereas others are almost impossible, the best option is to collect fertile specimens. Among ferns they can be monomorphic (with identical sterile and fertile fronds), or dimorphic (when only a portion of the frond is fertile), or holodimorphic (with the fertile frond being distinct from the sterile frond). In the latter case, you should collect both sterile and fertile fronds.

7.3.2.3   How Should the Collected Material be Processed?

Once you have properly collected the plant(s) of interest, the herborization process starts by pressing the plants and then drying them. The recommendation is to proceed with the herborization as soon as possible to avoid plant dehydration and shriveling. The usual way to do that is to place the plants between sheets of newspaper. To avoid any confusion regarding the origins of the specimens numbering the newspaper sheets with the respective collection numbers of the plants is important. The next step is to intercalate the newspaper sheets containing the plants with cardboard, and then

Herbário BHCB   Universidade Federal de Minas Gerais
No.   **187872**

‖‖‖‖‖‖‖‖‖‖‖‖‖‖‖‖‖‖‖‖‖‖‖‖
BHCB187872

Family: **Gleicheniaceae**
Specie: *Dicranopteris rufinervis*   (Mart.) Ching
Det.: L.V.Lima   **Date:** 13 October 2016

Locality: Brasil, Minas Gerais, Catas Altas, Serra do Caraça, Caminho
                  para Capelinha.

            20°05'44" S, 43°29'03" W   **Elevation:** 1349m
Collector: L.V.Lima 199   **Date:** 13 October 2016

Notes: Rupiculous, along the road side

FIGURE 7.6   Example of a herbarium label properly filled out with the necessary information.

put them all in a press. The press is then tied tightly with rope and placed into a drying oven. The ideal temperature for drying plants is approximately 60°C; 2 days are usually enough to completely dry them (although some rhizomes may be thicker and require more time). To avoid plant carbonization, check the plants regularly to determine if they are sufficiently dry. After drying, the next step is preparing the exsicate. Different herbaria may have different techniques for fixing the plants to the final herbarium sheets, but the two main principles are: make sure the plant is well-attached to the sheet, and place a herbarium label on your sample.

7.3.2.4   What Information Goes on the Herbarium Label?

The format and layout of herbarium labels (Figure 7.6) may vary from one institution to the next, but the key data still need to be provided. Recording complete information about the plant origin on the label is very important. Those data consist of, location, collection date, name of the collector (followed by his/her collection number), and GPS coordinates (if available). Additional notes about the substrate (e.g. terricolous, rupicolous, or corticolous) should be included, as well as the frond and stipe dimensions (in the case of tree ferns). The herbarium label must contain the herbarium number (the ID number of the plant), the name of person who determined the plant's name, and the date of that determination.

7.3.2.5   How to Assign Names to Ferns and Fern Allies

There are many ways to identify a plant, but one of the most used methods is the consulting experts is strongly recommended, especially when working with potential medicinal plants. The taxonomy of ferns and fern allies can be very intricate, and species identifications may be inaccurate when performed by nonspecialists. Therefore the collection of good and complete specimens is very important, to make complete field notes, correctly fill out the herbarium label – and always consult an expert.

7.4   CHEMICAL COMPOUNDS IN BRYOPHYTES

Many different classes of chemical compounds, including terpenoids, steroids, alkaloids, flavonoids, acetogenins, and aromatic compounds have been described in different species of bryophytes throughout the world (Asakawa et al., 2013a). There have only been rare investigations in Brazil focusing on the biological activities of native bryophytes (Pinheiro et al., 1989). Herein, we discuss the potential medicinal uses of some taxa based on previous assays with specimens of the same (or related) taxa of Brazilian bryophytes.

Large varieties of chemical compounds are found in liverworts, mosses, and hornworts, although liverworts stand out in terms of the numbers of new and different compounds identified. Because of the essential oils stored in the oil bodies of liverworts, the chemical nature of that bryophyte group is very complex (Asakawa et al., 2013a). Mono- and sesquiterpenoids are rare in mosses and hornworts, but di- and triterpenoids have been isolated from certain mosses. Only ca. 5% of the total of known bryophytes worldwide have yet been studied chemically. Although liverworts are a less speciose group than mosses, new terpenoids and phenolic compounds with interesting biological activities are regularly isolated from them (Asakawa, 2007).

7.5 CHEMICAL COMPOUNDS IN MARCHANTIOPHYTA

Over 3,000 compounds have been found in Marchantiophyta, including more than 800 terpenoids (excluding triterpenoids and tetraterpenoids), and 300 aromatic compounds (not including flavonoids) (Asakawa et al., 2013a). Most of the compounds isolated from, or detected in, liverworts are lipophilic terpenoids (mono-, sesqui-, and diterpenoids) and aromatic compounds (Asakawa, 2007; Ludwiczuk and Asakawa, 2010, 2015). Approximately 80% of the sesqui- and diterpenoids found in liverworts, noteworthy, enantiomers of those found in higher plants (Asakawa, 2007), but pinguisane-type sesquiterpenoid compounds that have not been found in any other organisms have been detected in Marchantiophyta, especially in the Lejeuneaceae, Trichocoleaceae, Ptilidiaceae, Porellaceae, and Aneuraceae families(Ludwiczuk and Asakawa, 2008).

7.5.1 TERPENOIDS

Monoterpenoids, sesquiterpenoids, diterpenoids, triterpenoids, and steroids have been described from different species of bryophytes (Asakawa et al., 2013a). Terpenoids (Figures 7.7 and 7.8) are found in the three different groups of bryophytes, and liverworts contain the most diverse classes of diterpenoids and sesquiterpenoids (Andersen et al., 1977; Asakawa, 2001; Asakawa et al., 2013a). Some Brazilian species of liverworts appear as very promising sources of large diversities of terpenoids. Among this group in Brazil, the liverwort family Plagiochilaceae stands out with at least seven species containing identified terpenoids.

The essential oils of *Plagiochila bifaria* and *Plagiochila stricta* are rich in different monoterpenoids, including terpinolene (**1**) and β-phellandrene (**3**), and *allo*-ocimene (**4**), and *neo-allo*-ocimene (**5**) (Figueiredo et al., 2005). Extracts of *Plagiochila rutilans*, a species with a peppermint-like odor, contained a variety of monoterpenoids, including α-terpinene (**6**), terpinolene (**1**), limonene (**7**), *p*-cymene (**8**), β-phellandrene (**3**), *p*-cymen-8-ol (**9**), pulegone (**10**), 3,7-dimethyl-2,6-octadien-1,6-olide (**11**), menthone (**12**), isomenthone (**13**), sabinene (**15**), and β-pinene (**17**) (Rycroft and Cole, 2001).

Reboulia hemisphaerica (Aytoniaceae) is very rich in sesquiterpenoids (ca. 41 known compounds; Table 7.1), especially 1,(10) 8-aristoladiene (= Caespitene) (**18**) (Toyota et al., 1999). (–)-β-Barbatene (**19**), another sesquiterpenoid, is commonly found in leafy liverworts, but is also present in the thallose *R. hemisphaerica* and *Dumortiera hirsuta* (Bardón et al., 1999a; Warmers and, König, 1999). The liverwort *R. hemisphaerica* is a rich source of many sesquiterpenoids, such as a gymnomitrane, gymnomitr-3(15)-en-9-one (**20**), gymnomitrol (**21**), and (+)-gymnomitr-3(15)-en-4a-ol (**22**) (Ludwiczuk and Asakawa, 2008; Toyota et al., 1999).

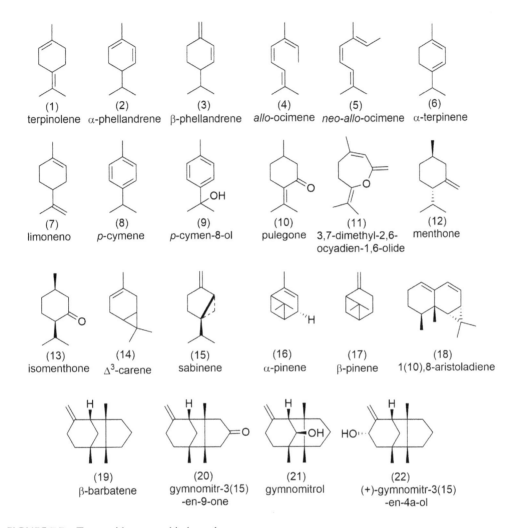

(1) terpinolene
(2) α-phellandrene
(3) β-phellandrene
(4) *allo*-ocimene
(5) *neo-allo*-ocimene
(6) α-terpinene

(7) limoneno
(8) *p*-cymene
(9) *p*-cymen-8-ol
(10) pulegone
(11) 3,7-dimethyl-2,6-ocyadien-1,6-olide
(12) menthone

(13) isomenthone
(14) Δ³-carene
(15) sabinene
(16) α-pinene
(17) β-pinene
(18) 1(10),8-aristoladiene

(19) β-barbatene
(20) gymnomitr-3(15)-en-9-one
(21) gymnomitrol
(22) (+)-gymnomitr-3(15)-en-4a-ol

FIGURE 7.7   Terpenoids reported in bryophytes.

Diterpenoids are also common in liverworts. *Odontoschisma denudatum* (Cephaloziaceae) produces the dolabellanes acetoxyodontoschismenol (**23**) and acetoxyodontoschismenetriol (**24**) as major components, along with other denudatenone diterpenoids (Asakawa et al., 2013a; Hashimoto et al., 1998a, 1998b). Additionally, the fusicoccane diterpenoid fusicorrugatol (**25**) has been isolated from *Plagiochila corrugata* (Tori et al., 1994).

Steroids such as stigmast-4-en-3-one (**26**) and sitost-4-en-3-one (**27**) have been isolated from the thallose liverwort *Ricciocarpos natans*, representing the first isolation of a steroid ketone from a bryophyte (Asakawa et al., 2013a; Yoshida et al., 1997). Stigmast-4-en-3-one and stigmast-4-en-3,6-dione (**28**) were also isolated from the leafy liverwort *Frullania brasiliensis* (Bardón et al., 2002).

Triterpenoids such as hopanoids including diploptene (= hop-22(29)-ene) (**30**), diplopterol (= hopan-22-ol) (**31**), and α-zeorin (= hopan-6α,22-diol) (**32**) are very common in liverworts of the orders Jungermanniales, Metzgeriales, and Marchantiales. The genera *Asterella*, *Reboulia*, and *Plagiochasma* (Aytoniaceae) stand out as rich sources of hopanoids. α-Zeorin, for example, has been found in *Reboulia hemisphaerica*, *Plagiochasma rupestre*, and *F. brasiliensis* (Wei et al., 1995; Bardón et al., 1999; Bardón et al., 2002).

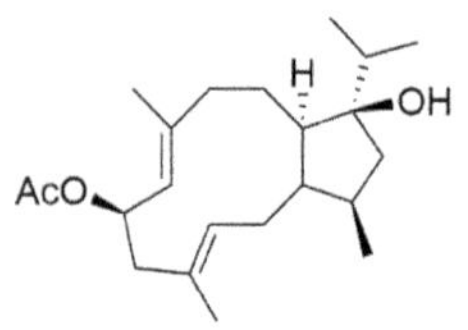

(23)
acetoxyodontoschismenol

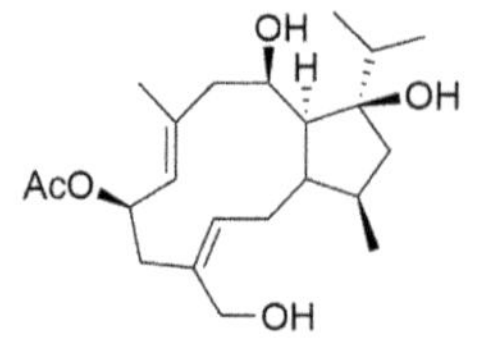

(24)
acetoxyodontoschismenetriol

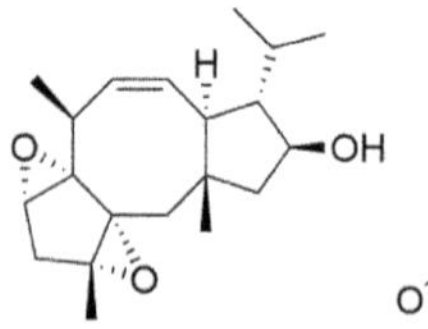

(25)
fusicorrugatol

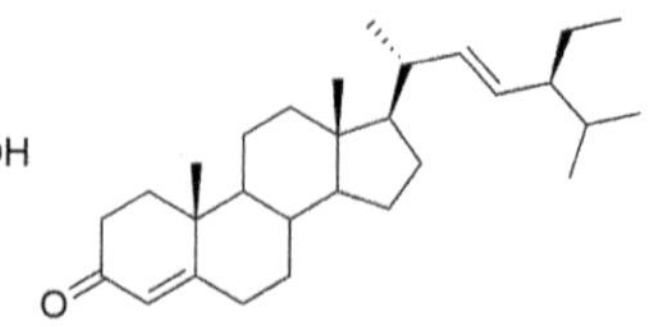

(26)
stigmast-4-en-3-one

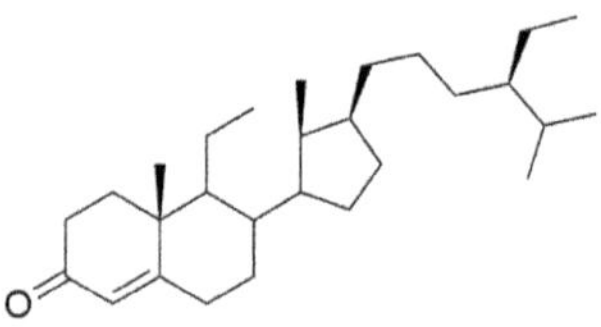

(27)
sitost-4-en-3-one

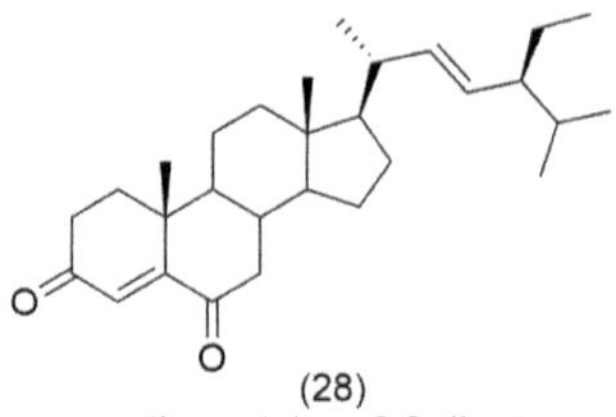

(28)
stigmast-4-en-3,6-dione

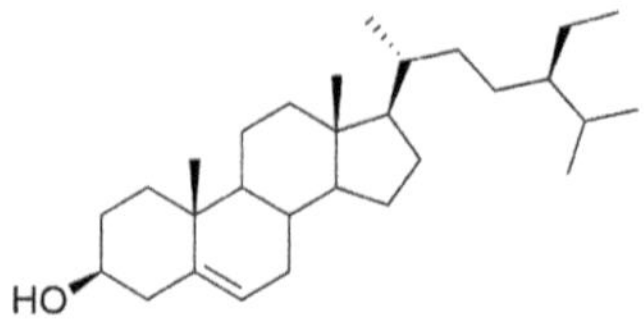

(29)
sitosterol

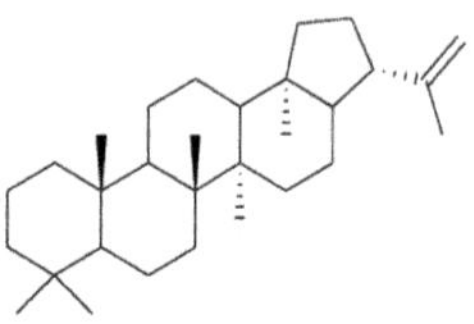

(30)
diploptene

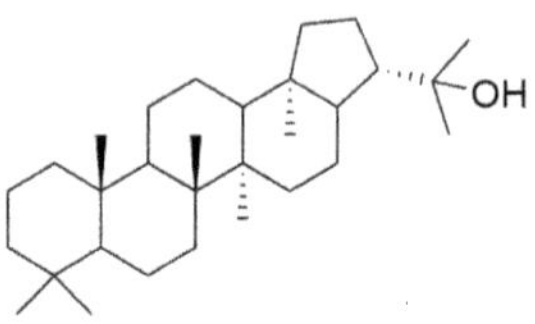

(31)
diplopterol

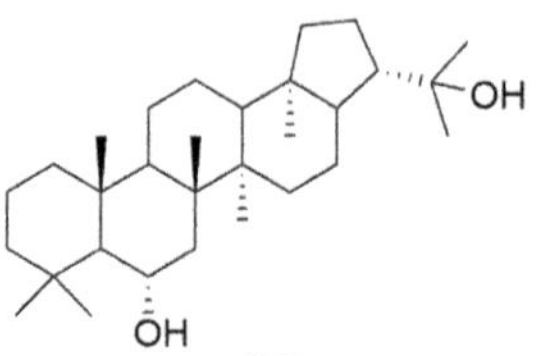

(32)
α-zeorin

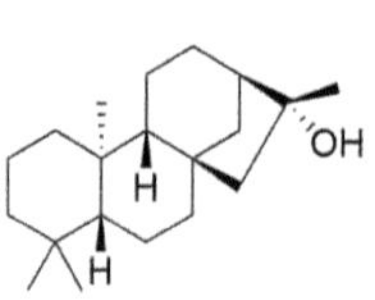

(33)
ent-16β–hydroxykaurane

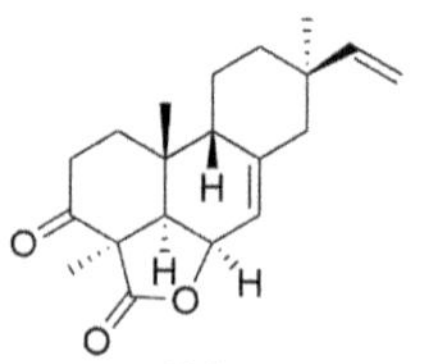

(34)
momilactones A

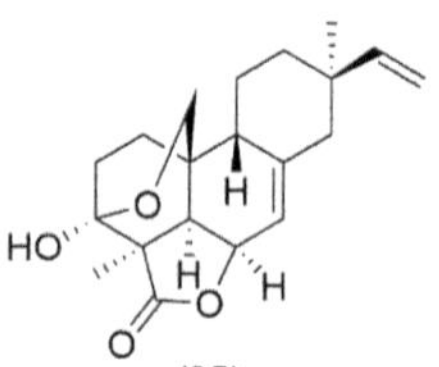

(35)
momilactones B

FIGURE 7.8   Terpenoids and steroids reported in bryophytes.

7.5.2 Aromatic Compounds

Marthantiophyta exhibits a wide diversity of bis-bibenzyls (Figure 7.9), mostly among Jungermannniales, Marthantiales, and Metzgeliares. Several *Marchantia* species are rich sources of bis-bibenzyls; dimeric bis-bibenzyls are significant components of *Riccardia* species; and *Radula* species are rich in bibenzyls and prenyl bibenzyls. *Corsinia coriandrina* is unique in producing nitrogen- and sulfur-containing compounds (Asakawa et al., 2013a). *Plagiochila diversifolia* produces three prebibenzyls: longispinone A (**36**), longispinone B (**37**), and longispinol (**38**) (Heinrichs et al., 2000). Prelunularin (**39**) has been recorded in the thallose liverworts *Marchantia polymorpha* and *R. natans* (Kunz and Becker, 1994). Lunularin (**40**) and lunularic acid (**41**) have been identified in different thallose liverworts, including *Lunularia cruciata, D. hirsuta, M. polymorpha,* and *R. natans* (Asakawa et al., 1996; Kunz and Becker, 1994; Lu et al., 2006).

TABLE 7.1

Examples of Brazilian Bryophyte Species with Chemical Compounds Recorded in the Literature (Asakawa et al., 2013a and References Therein).

Bryophyte Group/Family	Species	Number of Chemical Compounds			Brazilian Biomes
		Aromatic	Steroids and Terpenoids	Flavonoids	
Marchantiophyta					
Acrobolbaceae	*Lethocolea glossophylla*	5	–	–	AtF
Adelanthaceae	*Adelanthus decipiens*	10	–	1	AtF
Aneuraceae	*Aneura pinguis*	–	4	–	AtF, Pan
	Riccardia multifida	6	–	–	AtF
Aytoniaceae	*Asterella venosa*	–	7	–	AtF, Sav
	Corsinia coriandrina	10	4	–	AtF
	Isotachis aubertii	5	–	2	AtF
	Plagiochasma rupestre	4	5	–	AtF, Pam
	Reboulia hemisphaerica	5	42	–	AtF
Balantiopsidaceae	*Isotachis aubertii*	–	4	–	AtF
Cephaloziaceae	*Odontoschisma denudatum*	–	23	–	AmF, AtF, Sav
Corsiniaceae	*Corsinia coriandrina*	–	–	–	AtF
Dumortieraceae	*Dumortiera hirsuta*	9	45	5	AmF, Sav, AtF, Pan
Frullaniaceae	*Frullania arecae*	–	–	1	AtF, Sav
	Frullania brasiliensis	1	10	–	AtF, Sav
	Frullania serrata	–	7	–	AtF
Jungermanniaceae	*Anastrophyllum auritum*	–	18	–	AtF
Lejeuneaceae	*Bryopteris filicina*	1	20	–	AmF, Sav, AtF, Pan
	Cheilolejeunea trifaria	–	1	–	AmF, Sav, AtF, Pan
	Lejeunea flava	–	4	–	AtF, AmF, Caa, Sav, Pam, Pan
	Marchesinia brachiata	2	2	1	AmF, Sav, AtF
	Omphalanthus filiformis	1	5	–	AmF, AtF, Pan
Lepidoziaceae	*Bazzania nitida*	–	5	–	AtF
Lophocoleaceae	*Lophocolea bidentata*	–	3	–	AmF, AtF, Sav
Pelliaceae	*Noteroclada confluens*	–	6	–	AtF, Sav
Lunulariaceae	*Lunularia cruciata*	2	9	2	AtF
Marchantiaceae	*Marchantia chenopoda*	2	1	–	AmF, Sav, AtF, Pan
	Marchantia paleacea	9	6	4	AtF
	Marchantia polymorpha	29	18	14	AtF
Metzgeriaceae	*Metzgeria conjugata*	–	1	–	AtF
	Metzgeria furcata	–	6	–	AmF, Sav, AtF
Pallaviciniaceae	*Symphyogyna brasiliensis*	–	13	–	AmF, AtF, Sav
	Symphyogyna podophylla	–	6	–	AtF
	Plagiochila corrugata	–	1	–	AmF, Sav, AtF
	Plagiochila cristata	1	13	–	AmF, AtF
	Plagiochila diversifolia	10	7	4	AtF
	Plagiochila gymnocalycina	1	–	–	AtF
	Plagiochila rutilans	7	15	1	AmF, Sav, AtF
	Plagiochila stricta	9	45	2	AtF

(Continued)

TABLE 7.1 *(Continued)*

Examples of Brazilian Bryophyte Species with Chemical Compounds Recorded in the Literature (Asakawa et al., 2013a and References Therein).

| Bryophyte Group/Family | Species | Number of Chemical Compounds | | | Brazilian Biomes |
		Aromatic	Steroids and Terpenoids	Flavonoids	
Radulaceae	*Radula nudicaulis*	4	9	4	AtF
Ricciaceae	*Ricciocarpos natans*	3	5	8	AmF, AtF, Pan
Bryophyta					
Aulacomniaceae	*Aulacomnium palustre*	–	–	4	AtF
Hedwigiaceae	*Hedwigia ciliata*	–	–	1	AtF
Polytrichaceae	*Polytrichum commune*	5	1	2	AmF, Sav, AtF
Pottiaceae	*Eucladium verticillatum*	4	–	–	AtF, Sav

Brazilian Biomes: AtF, Atlantic Rainforest; AmF, Amazon Rainforest; Caa, Deciduous Caatinga; Sav, Cerrado Savanna; Pan, Pantanal Wetlands; Pam, Pampa Grasslands (Flora do Brasil, 2020)

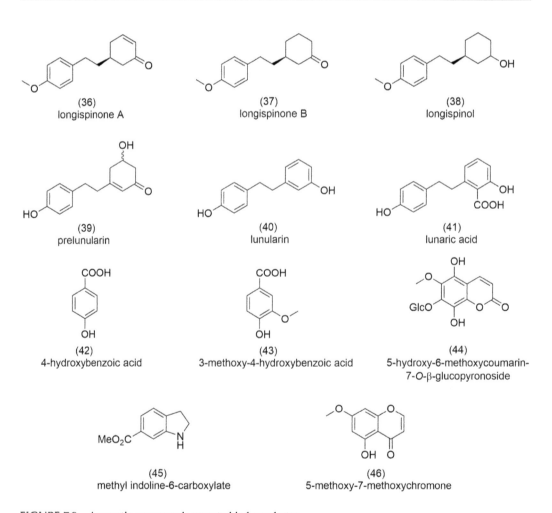

FIGURE 7.9 Aromatic compounds reported in bryophytes.

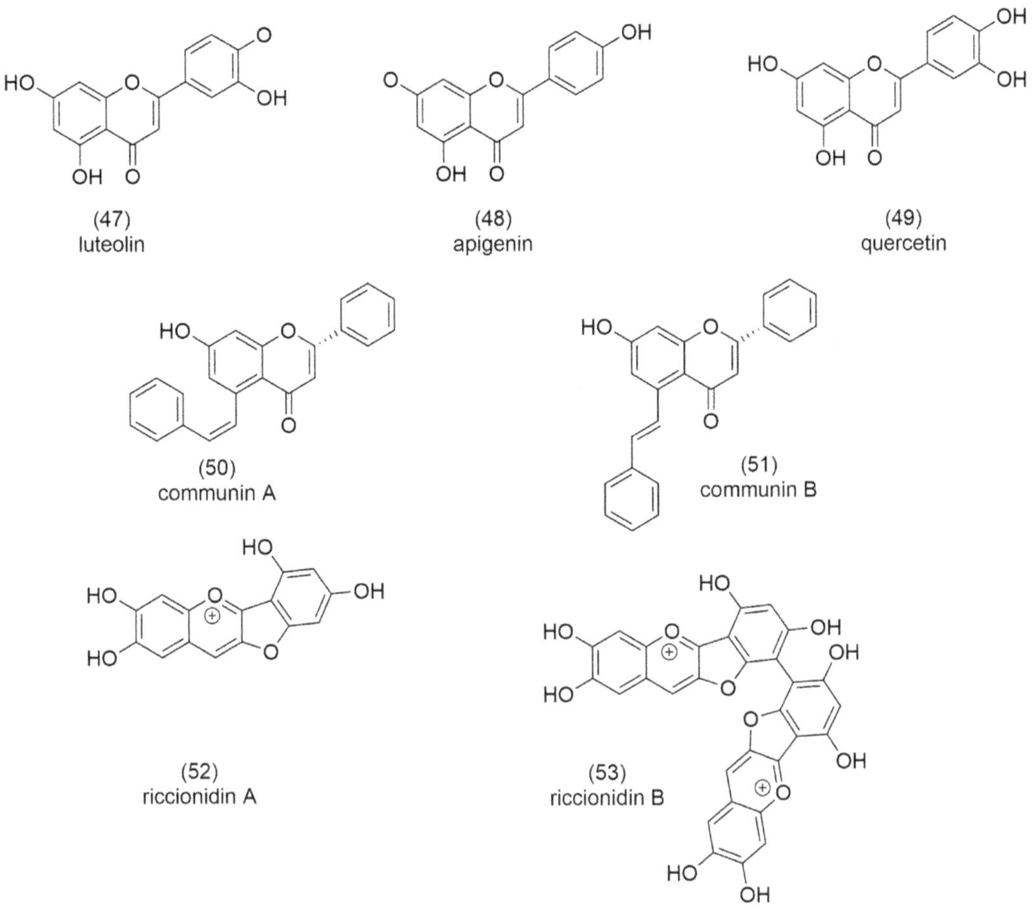

FIGURE 7.10 Flavonoids (47–51) and anthocyanins (52 and 53) reported in bryophytes.

7.5.3 FLAVONOIDS AND ANTHOCYANIDINS

All main orders of Marthantiophyta (Jungermanniales, Metzgeriales, and Marchantiales) contain flavonoids (Figure 7.10), although they are generally underrepresented in liverworts. Flavones are more common than flavanones, and luteolin (47) and apigenin (48) are the most abundant in the Marchantiophyta, including *M. polymorpha* (Adam and Becker, 1994). Luteolin-7-*O*-glucoside and quercetin (49) were found in *L. cruciata* (Jackovic et al., 2008). A new compound, an anthocyanidin named riccionidin A (52) and the dimer riccionidin B (53) were identified in the cell walls of *R. natans* grown under axenic conditions and these were also detected in the liverwort *M. polymorpha* (Kunz et al., 1994).

7.5.4 ACETOGENINS AND LIPIDS

Several liverworts, such as *P. rutilans*, produce the acetogenins 1-octen-3-ol (54) and/or 1-octen-3-yl acetate (55) (Asakawa et al., 2013a; Rycroft and Cole, 2001). In terms of the fatty acids found in liverworts, *M. polymorpha* cells grown under controlled conditions were found to contain linolenic (56), arachidonic (57), and eicosapentaenoic acids (58) (Saruwatari et al., 1999). Example of acetogenins and fatty acids reported in bryophytes is shown in Figure 7.11.

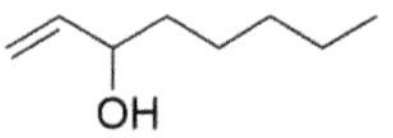

FIGURE 7.11   Acetogenins and fatty acids present in bryophytes.

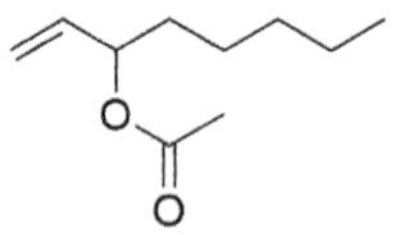

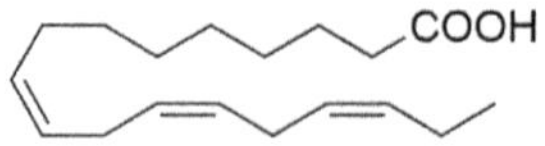

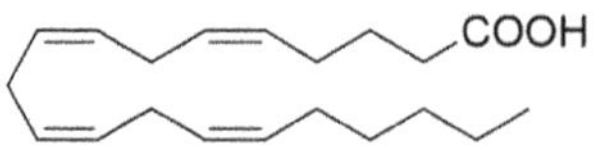

7.6   CHEMICAL COMPOUNDS IN BRYOPHYTA

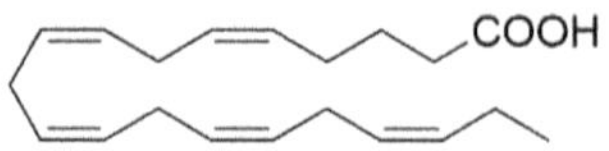

Although the many known species of mosses other than liverworts, only a small number of them have been chemically analyzed (Asakawa et al., 2013a). Only four monoterpene hydrocarbons (α-phellandrene (**2**), β-phellandrene (**3**), Δ³-carene (**14**), and α-pinene (**16**)) have been detected in *Sphagnum* species; and only four diterpenoids, ent-16β-hydroxykaurane (**33**), momilactones A (**34**) and B (**35**), and 18 monoterpenoids, five trinorsesquiterpenoids, 72 sesquiterpenoids, ten diterpenoids, and nine triterpenoids have been isolated from, or detected in, the entire Bryophyta group. Moss species such as *Polytrichum commune* may contain several compounds, including the steroid sitosterol (**29**), the aromatic compounds 4-hydroxybenzoic acid (**42**), 3-methoxy-4-hydroxybenzoic acid (**43**), 5-hydroxy-6-methoxycoumarin-7-*O*-β-glucopyranoside (**44**), methyl indoline-6-carboxylate (**45**), 5-hydroxy-7-methoxychromone (**46**), and the flavonoids communin A (**50**) and communin B (**51**) (Asakawa et al., 2013a).

7.7   CHEMICAL COMPOUNDS IN HORNWORTS

Although only few species of hornworts have been chemically analyzed, the chemical constituents of Anthocerotophyta are apparently very distinct from Marchantiophyta and Bryophyta. Several monoterpenoids, sesquiterpenes, sterols, aromatic compounds, and alkaloids have been detected in, or isolated from, certain hornworts. However, none of those species seem to occur in Brazilian ecosystems. For instance, *Anthoceros agrestis* contains glutamic acid amides, 4-hydroxybenzoic, protocatechuic, vanillic, isoferulic, and coumaric acids, and the new alkaloid anthocerodiazonin (**59**; Figure 7.12) (which contains a nine-membered ring system) (Asakawa et al., 2013a; Trennheuser et al., 1994).

It is important to stress that although the majority of the above cited species occur in Brazilian ecosystems (Table 7.1), many of the studies reported here were carried out using specimens from other localities around the world. Since the chemical compounds produced by plants can be influenced by environmental conditions, the sites where the plants were collected and the time of year when they were harvested will be important variables to be considered in biochemical studies. That observation reinforces the importance of collecting detailed field data and making precise specimen determinations. However, the fact that many plants are cultured *in vitro* before analyzing their extracts may mitigate field effects, and result in similar determinations of the classes of chemical compounds among the taxa studied.

Brazilian liverwort species such as *R. hemisphaerica* (Aytoniaceae), *D. hirsuta* (Dumortieraceae), *M. polymorpha* (Marchantiaceae), *P. bifaria,* and *P. stricta* (Plagiochilaceae)

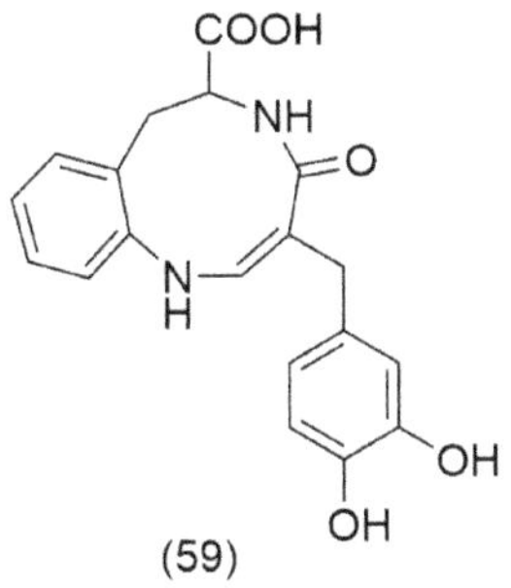

(59)
anthocerodiazonin

FIGURE 7.12   A new alkaloid present in the hornwort *Anthoceros agrestis*.

stand out in terms of the numbers of different chemicals they produce, including aromatic compounds, flavonoids, and terpenoids (Table 7.1). Those species are mostly present in the Atlantic
rain forest, although several also occur in Amazon rain forest, Cerrado (Brazilian savanna),
and Pantanal sites.

Since less than 1% of the total bryophyte diversity found in Brazil is represented in
Table 7.1, the urgency of studies examining bryophyte taxa chemical content throughout Brazil
is highlighted. Compared to other areas around the world, especially the temperate zones,
Brazil has a huge potential for harboring species containing new and interesting medicinal
compounds. Many of those bryophyte species should be found in the Atlantic rain forest –
the Brazilian biome that offers the most diverse combination of microclimatic conditions and
habitat heterogeneity.

7.8   CHEMICAL COMPOUNDS IN PTERIDOPHYTES

Despite the wide array of secondary metabolites present in ferns and fern allies, little information
is actually available concerning those phytochemicals and their potential pharmacological applications (Cao et al., 2017). Even with our current limited knowledge regarding the subject, a wide
range of alkaloids, flavonoids, polyphenols, terpenoids, and steroids has already been reported in
ferns and fern allies (Cao et al., 2017; Dong et al., 2012; Ho et al., 2011; Socolsky et al., 2007, 2012;
Xia et al., 2014). Additionally, differences in the structures of those compounds from those found in
angiosperms indicate a rich, unexplored, neglected, and potential field for pharmaceutical investigations (Cao et al., 2017).

7.8.1   TERPENOIDS

Terpenoids (Figures 7.13–7.16) are the largest chemical group found in ferns and fern allies. This
class includes triterpenoids, diterpenoids, and sesquiterpenoids (Ho et al., 2011). Triterpenoids such
as the filicenes are typical secondary metabolic constituents of those plants (Ho et al., 2011; Nakane
et al., 2002; Reddy et al., 2001). Many compounds have been isolated from club-moss species (Zhou
et al., 2003a, 2003b, 2004), especially the Serratenes group of naturally occurring pentacyclic triterpenoids with seven tertiary methyl groups (Ho et al., 2011); many of those same compounds have
been recorded in conifers (Tanaka et al., 2004; Wittayalai et al., 2012). Diterpenoids have mainly
been found in Pteridaceae, including ent-kaurane, ent-atisane, and ent-pimarane types (Alonso-
Amelot, 2002; Cao et al., 2017; Ho et al., 2011). Sesquiterpenoids are mainly represented in ferns
and fern allies by the indane and cadinane groups (Ho et al., 2011). Pterosines are a large group
of sesquiterpenes with an indane skeleton (Cao et al., 2017), many are characteristic constituents

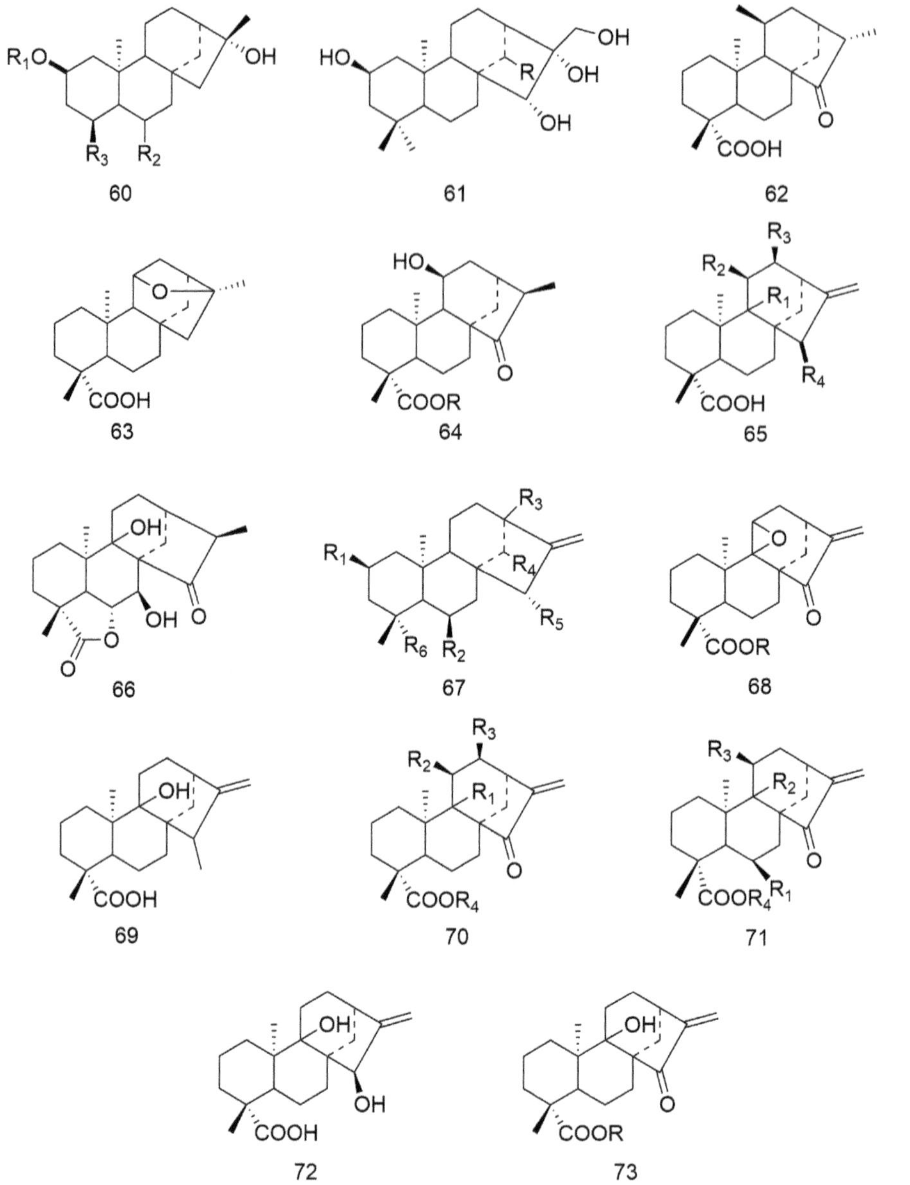

FIGURE 7.13 Basic structures of some serratene-type triterpenoids described in pteridophytes from the family Lycopodiaceae.

of Bracken ferns (including peterosine B) and are well-represented in that group (*Pteridium spp.*) (Hikino et al., 1970, 1971, 1972).

7.8.2 PHENOLICS

Phenolic compounds (Figure 7.17) are represented in ferns and fern allies mainly by chlorogenic, caffeic, ferulic, hydroxybenzoic, hydroxycinnamic, and vanillic acids. Extracts of *Pteris* (Pteridaceae) showed the presence of kaempferol 3-*O*-L-rhamnopyranoside-7-*O*-[-D-apio-

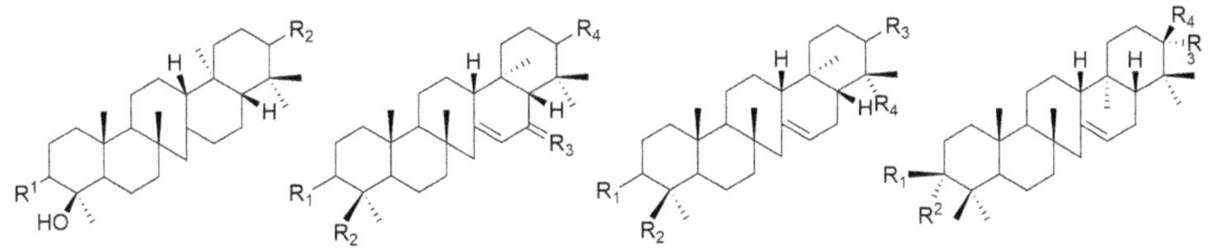

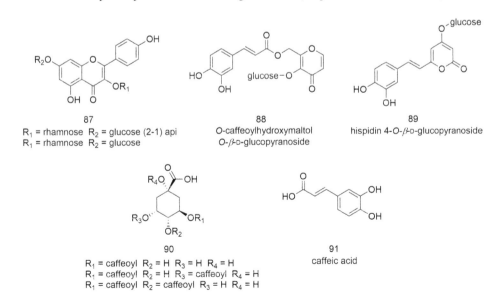

FIGURE 7.14 Basic structures of some *ent*-kaurane-type diterpenoids from the genus *Pteris*.

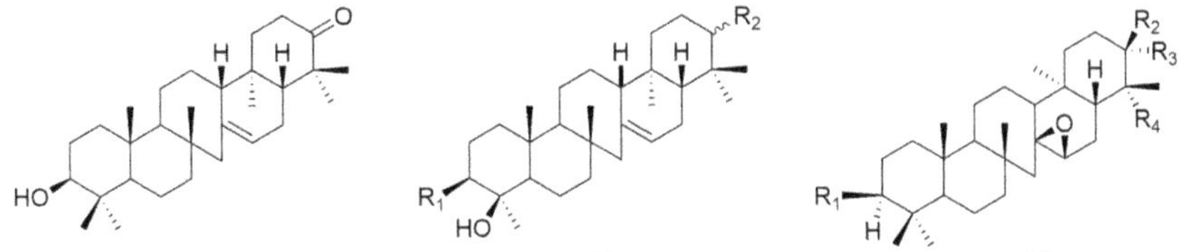

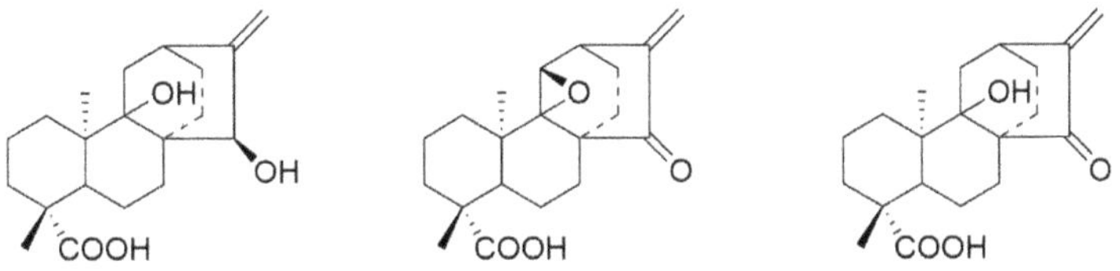

81 82 83

FIGURE 7.15 Basic structures of some atisane-type diterpenoids from the genus *Pteris*.

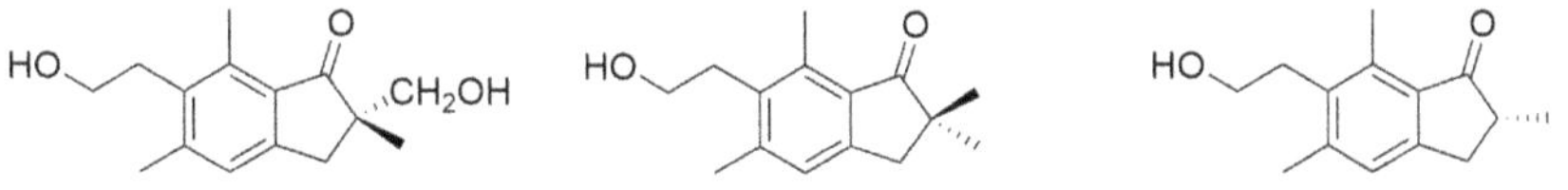

84
Pterosin A

85
Pterosin Z

86
Pterosin B

FIGURE 7.16 Examples of pterosins found in the genus *Pteris* (adapted from Cao et al., 2017).

87
R_1 = rhamnose R_2 = glucose (2-1) api
R_1 = rhamnose R_2 = glucose

88
O-caffeoylhydroxymaltol
O-*β*-ᴅ-glucopyranoside

89
hispidin 4-*O*-*β*-ᴅ-glucopyranoside

90
R_1 = caffeoyl R_2 = H R_3 = H R_4 = H
R_1 = caffeoyl R_2 = H R_3 = caffeoyl R_4 = H
R_1 = caffeoyl R_2 = caffeoyl R_3 = H R_4 = H

91
caffeic acid

FIGURE 7.17 Examples of phenolic compounds in *Pteris*.

furanosyl-(1-2)-*O*-D-glucopyranoside], 7-*O*-caffeoylhydroxymaltol 3-*O*-D-glucopyranoside, hispidin 4-*O*-D-glucopyranoside, kaempferol 3-*O*-L-rhamnopyranoside-7-*O*-D-glucopyranoside, caffeic acid, 5-caffeoylquinic acid, 3,5-dicaffeoylquinic acid, and 4,5-di-caffeoylquinic acid (Chen et al., 2007).

7.8.3 Flavonoids

Numerous flavonoid compounds (Figure 7.18) with medicinal properties have been identified in ferns and fern allies (Ho et al., 2011). Flavonoids can be divided into several classes, including anthocyanins, flavones, flavonols, flavanones, dihydroflavonols, chalcones, aurones, flavonons, flavan, proanthocyanidins, isoflavonoids, and bioflavonoids (Iwashina, 2000). Some Selaginella-derived flavonoids stand out in terms of their potential pharmacological uses, such as amentoflavone, hinokiflavone, heveaflavone, neocryptomerin, pulvinatabiflavone, and 7″-*O*-methylamentoflavone (Cheng et al., 2008; Zhang et al., 2012a, 2012b, 2012c, 2012d).

7.8.4 Alkaloids

Alkaloids (Figure 7.19) are well-represented in fern and fern allies, especially in club-mosses (Lycopodiaceae). The groups lycopodine, lycodine, and fawcettimine standout, especially huperzine, a lycodine with a quinolizidine skeleton (Ho et al., 2011). Other alkaloid compounds have been isolated from club-mosses, including clavolonine, flabelliformine, gnidioidine, lycocarinatine, lycodoline, miyoshianine, and phlegmariurine (Thorroad et al., 2014; Tong et al., 2003).

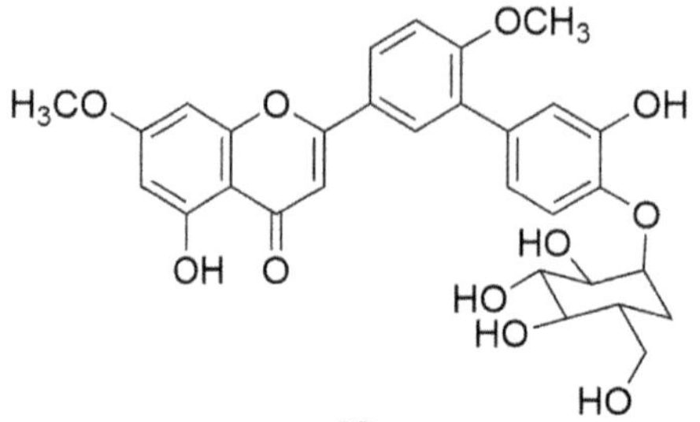

92
involvenflavone A

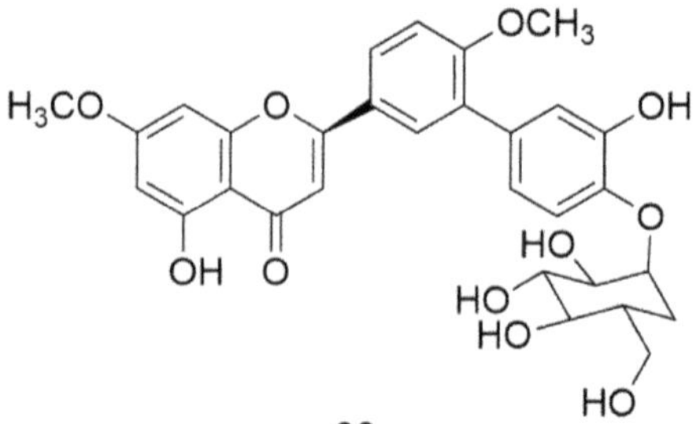

93
involvenflavone B

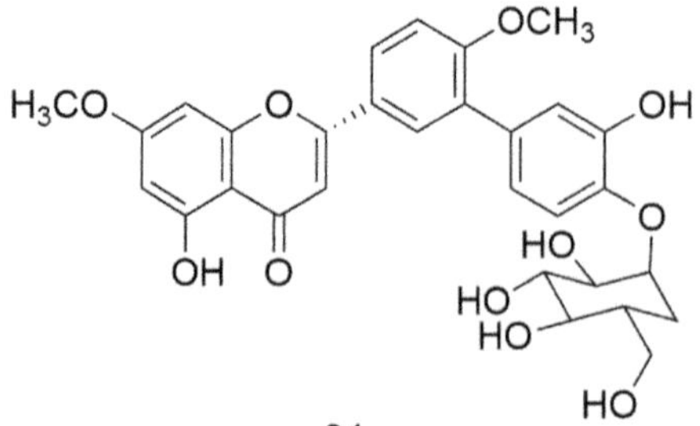

94
involvenflavone C

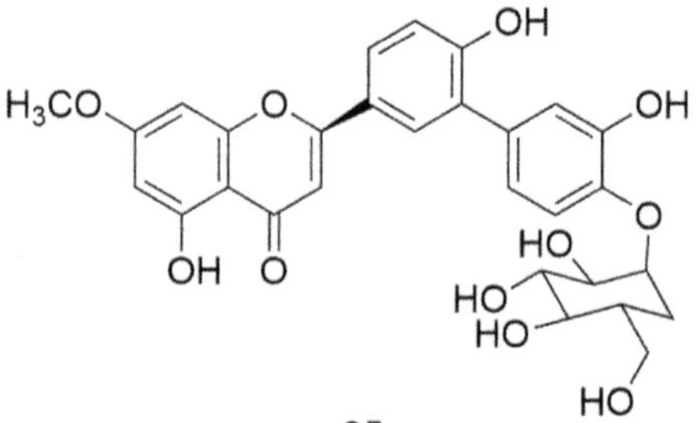

95
involvenflavone D

FIGURE 7.18 Examples of involvenflavones from *Selaginella* (adapted from Cao et al., 2017).

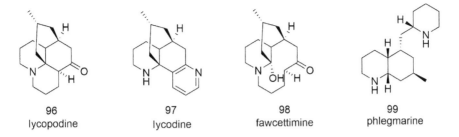

FIGURE 7.19   Examples of alkaloids from club-mosses (adapted from Cao et al., 2017).

7.9   BIOLOGICALLY ACTIVE COMPOUNDS AND THEIR POTENTIAL MEDICINAL USES

7.9.1   BRYOPHYTES

Chemical compounds isolated from bryophytes, especially from liverworts (Marchantiophyta), have unique characteristics. Liverworts produce a great variety of lipophilic terpenoids, aromatic compounds, and acetogenins that are responsible for fragrances, bitterness, pungency, and sweetness, as well as allergenic responses, contact dermatitis, cytotoxic effects, antimicrobial, antifungal, calmodulin inhibitory, cardiotonic, larvicidal, 5-lipoxygenase inhibitory, molluscicidal, muscle relaxant, neurotrophic, plant growth regulatory, superoxide release inhibitory, thromboxane synthase inhibitory, and vasopressin antagonist activities (Asakawa, 2007; Asakawa et al., 2013a).

When crushed, liverworts emit a very strong odor. Lipophilic terpenoids (such as monoterpenoids) and aromatic compounds held in oil bodies are responsible for the intense sweetwoody, turpentine, sweet-mossy, fungal-like, carrot-like, mushroom-like, or seaweed-like odors of liverworts. Almost all liverworts that smell of mushrooms contain 1-octen-3-ol and the corresponding acetate (Asakawa et al., 2013a). Allergenic contact dermatitis caused by some *Frullania* species is associated with sesquiterpenes with α-methylene-γ-lactone functionality (Asakawa et al., 2013a).

The essential oils of several liverworts exhibit anti-bacterial, fungal, and viral activities (Asakawa et al., 2013), including against *Escherichia coli*, *Staphylococcus aureus* (Lorimer and Perry, 1994), *Cladosporium cucumerinum* (Scher et al., 2004), *Candida albicans* (Wu et al., 2008, 2009, 2010), and he H1N1 and H5N1 influenza A virus (Iwai et al., 2011). A Brazilian study (Pinheiro et al., 1989) with ten liverworts and 15 moss species found that 40% of them produced chemical compounds that inhibited the growth of several bacterial strains. Extracts of *Calymperes lonchophyllum* moss and an unidentified *Bazzania* species significantly inhibited *E. coli* growth; and *Leucomium lignicola* extracts were active against *S. aureus*, *Proteus vulgaris*, *Klebsiella pneumoniae*, and *Edwardsiela tarda*. Dehydrocostus lactone (from the thallose liverwort *Targionia lorbeeriana*) likewise showed antifungal activity against *C. albicans*, and insecticidal activity against *Aedes aegypti* (Neves et al., 1999).

Several eudesmanolides, germacranolides, and guaianolides isolated from liverworts exhibit cytotoxic activity against KB nasopharyngeal and P-388 lymphocytic leukemia cells. Some *Marchantia* species produce marchantins A, B, D, perrottetin F, and paleatin B, which show DNA polymerase β inhibitory activity (IC_{50} range 14.4–97.5 μM), cytotoxicity against KB cells (IC_{50} range 3.7–20 μM), and anti-HIV-1 activity (IC_{50} range 5.3–23.7 mg/cm^3) (Asakawa et al., 2008, 2009).

Lunularic acid appears to have ABA-like activity in vascular plants, inhibiting the germination and growth of plants such as *Lepidium sativum* and *Lactuca sativa*. Yoshikawa et al. (2002) hypothesized that vascular plants altered their endogenous growth regulator from lunularic acid to abscisic acid during their evolution.

The *in vitro* cultivation of bryophytes appears to be the most appropriate route for large biomass productions and the isolation of compounds showing interesting biological activities (Sabovljevic et al., 2009).

7.9.2 Pteridophytes

The importance of ferns and fern allies to pharmacological and medical applications was noted by the Greek botanist Theophrastus (ca. 372–287 B.C.) and by the father of pharmacognosy Dioscorides (ca. 50 A.D.). Both the bracken fern (*Pteridium* spp.) and the male fern (*Dryopteris filix-mas*) were mentioned in Dioscorides' magna opus "De materia medica" (Banerjee and Sen, 1980).

In spite of the historical importance and wide use of plants in traditional medicines (Maciel et al., 2002), the numbers of medicinal ferns and lycophytes encountered in ethnobotanical studies are not consistent with the therapeutic potentials often attributed to them (Reinaldo et al., 2015).

The need for new drugs and reports of therapeutic effectiveness of compounds derived from ferns and ferns allies present open opportunities for pharmacological research (Cao et al., 2017). Many studies have demonstrated the pharmacological potentials of ferns and lycophytes due to the presence of metabolites with antioxidant, anti-inflammatory, analgesic, antimutagenic, immuno-modulatory, and neuromodulatory activities (Goldberg et al., 1975; Keller and Prance, 2015; Lee and Lin, 1988; Nonato et al., 2009; Tomšík 2014; Wu et al., 2005).

Investigation of the potential pharmaceutical uses of Brazilian ferns and fern allies is still incipient. Santos et al. (2010) estimated that only approximately 4.7% of the Brazilian pteridoflora has been examined in that light, even though many experiments with pteridophytes have shown their efficacy in treating human ailments.

Antioxidant activities have been reported in many genera that (also) occur in Brazil, such as *Hypolepis*, *Pteridium*, *Dryopteris*, *Polystichum*, *Dicranopteris*, *Lycopodium l.s.*, *Osmunda*, *Adiantum*, *Pteris*, *Lygodium*, *Selaginella*, and *Thelypteris l.s.* (Baskaran et al., 2018; Shin, 2010), due to the presence of substances such as 2,2-diphenyl-1-picrylhydrazyl activity and 2,2′-azino-bis-(3-ethyl benzthiazoline-6-sulphonie acid diammonium salt.

Anticancer activities have been reported in extracts of the genera *Selaginella* due to the actions of lycopodine and biflavones such as amentoflavone and isocryptomerin (Silva et al., 1995); other compounds with anticancer activities have been found in *Acrostichum* (Uddin et al., 1998) and the naturalized genera *Macrothelypteris* (Liu et al., 2012). Antihyperglycemic and analgesic properties were likewise observed in plants of the genera *Christella* and *Adiantum* (Paul et al., 2012; Sultana et al., 2014; Tanzin et al., 2013).

Anti-inflammatory properties were reported in *Selaginella* (Dhiman, 1998), *Ophioglossum*, *Lygodium* (Vasudeva, 1999), *Christella* (Gogoi, 2002), *Pteris* (Lee and Lin, 1988; Wu et al., 2005), *Dryopteris* (Otsuka et al., 1972), *Cyathea* (Benjamin and Manickam, 2007; Madhukiran and Ganga, 2011), *Blechnum* (Nonato et al., 2009), and others. Antimicrobial activities were likewise reported in *Nephrolepis* (Jimenez et al., 1979), *Lygodium* (Cambie and Ash, 1994), *Marattia* (de Boer et al., 2005), *Adiantum* (Reddy et al., 2001), *Equisetum* (Joksic et al., 2003; Radulovic et al., 2006), and others.

Therefore, that ferns and fern allies represent a huge unexplored group of plants with potential pharmaceutical and medical uses is undeniable, and even though very few studies have examined the Brazilian pteridoflora, a number of bioactive compounds have already been identified (Table 7.2). Additional studies have pointed out that extracts of Brazilian ferns and fern allies contain antioxidant, anticancer, antiviral, anti-inflammatory, and other activities (Table 7.3).

Table 7.4 summarizes some information about the Brazilian bryophytes and pteridophytes discussed in the chapter.

TABLE 7.2

Brazilian Species of Pteridophytes Containing Isolated Bioactive Compounds

Taxa	Family	Bioactive Compounds	Reference
Adiantopsis flexuosa	Pteridaceae	7-*O*-Glycosides of apigenin, aglycone, 7-*O*-glycosides of luteolin, 7-*O*-glycosides of chrysoeriol, 3-*O*-glycosides of kaempferol, 3-*O*-glycosides of kaempferol, 3-*O*-glycosides of quercetin	Salatino and Prado (1998)
*Adiantum capillus-veneris**	Pteridaceae	Adiantone, adiantoxide, astragalin, β-sitosterol, caffeic acids, caffeylgalactose, caffeylglucose, campesterol, carotenes, coumaric acids, coumarylglucoses, diplopterol, epoxyfilicane, fernadiene, fernene, filicanes, hopanone, hydroxyl-adiantone, hydroxyl-cinnamic acid, isoadiantone, isoquercetin, kaempferols, lutein, mutatoxanthin, naringin, neoxanthin, nicotiflorin, oleananes, populnin, procyanidin, prodelphinidin, quercetin, querciturone, quinic acid, rhodoxanthin, rutin, shikimic acid, violaxanthin, and zeaxanthin	Taylor (2003)
Cyathea phalerata	Cyatheaceae	Kaempferol-3-neohesperidoside, 4-*O*-β-D-glucopyranosyl caffeic acid, 4-*O*-β-D-glucopyranosyl *p*-coumaric acid, 3,4-spyroglucopyranosyl protocatechuic acid, sitosterol β-D-glucoside, β-sitosterol, kaempferol, and vitexin	Hort et al. (2008)
Doryopteris Concolor	Pteridaceae	7-*O*-Glycosides of apigenin, aglycone, 7-*O*-glycosides of luteolin, 7-*O*-glycosides of chrysoeriol, 3-*O*-glycosides of kaempferol, 3-*O*-glycosides of kaempferol, and 3-*O*-glycosides of quercetin	Salatino and Prado (1998)

(Continued)

TABLE 7.2 *(Continued)*

Brazilian Species of Pteridophytes Containing Isolated Bioactive Compounds

Taxa	Family	Bioactive Compounds	Reference
*Equisetum arvense**	Equisetaceae	Isoquercetin, quercetin 3-*O*-glucoside, quercetin 3-*O*-(6″-*O*-malonylglucoside), 5-*O*-caffeoyl mesotartaric acid, monocaffeoyl meso-tartaris acid, monocaffeoyl meso-tartaris acid, di-E-caffeoyl-meso-tartaric acid, hexahydrofarnesyl acetone, cis-geranyl scetone, thymol, and trans-phytol	Radulovic et al. (2006), Mimica et al. (2008), Milovanovic et al. (2007)
Lytoneuron ornithopus	Pteridaceae	7-*O*-Glycosides of apigenin, aglycone, 7-*O*-glycosides of luteolin, 7-*O*-glycosides of chrysoeriol, 3-*O*-glycosides of kaempferol, 3-*O*-glycosides of kaempferol, and 3-*O*-glycosides of quercetin	Salatino and Prado (1998)
Ormopteris cymbiformis	Pteridaceae	7-*O*-Glycosides of apigenin, aglycone, 7-*O*-glycosides of luteolin, 7-*O*-glycosides of chrysoeriol, 3-*O*-glycosides of kaempferol, 3-*O*-glycosides of kaempferol, and 3-*O*-glycosides of quercetin	Salatino and Prado (1998)
Ormopteris gleichenioides	Pteridaceae	7-*O*-Glycosides of apigenin, aglycone, 7-*O*-glycosides of luteolin, 7-*O*-glycosides of chrysoeriol, 3-*O*-glycosides of kaempferol, 3-*O*-glycosides of kaempferol, and 3-*O*-glycosides of quercetin	Salatino and Prado (1998)
Ormopteris pinnata	Pteridaceae	7-*O*-Glycosides of apigenin, aglycone, 7-*O*-glycosides of luteolin, 7-*O*-glycosides of chrysoeriol, 3-*O*-glycosides of kaempferol, 3–*O*-glycosides of kaempferol, and 3-*O*-glycosides of quercetin	Salatino and Prado (1998)
Ormopteris riedelii	Pteridaceae	7-*O*-Glycosides of apigenin, aglycone, 7-O-glycosides of luteolin, 7-*O*-glycosides of chrysoeriol, 3-*O*-glycosides of kaempferol, 3-*O*-glycosides of kaempferol, 3-*O*-glycosides of quercetin	Salatino and Prado (1998)
Pityrogramma calomelanos	Pteridaceae	2'6'-Dihydroxy-4,4'-dimethoxydihydrochalcone, kaempferol 7-methyl ether, apigenin 7-methyl ether	Star and Mabry (1971)

(Continued)

TABLE 7.2 *(Continued)*

Brazilian Species of Pteridophytes Containing Isolated Bioactive Compounds

Taxa	Family	Bioactive Compounds	Reference
Psilotum nudum	Psilotaceae	Quercetin, kaempferol, amentoflavone, hinokiflavone, vicenin-2 psilotin, 3'-hydroxypsilotin	Cambie and Ash (1994)
Pteridium aquilinum	Desdendticiaceae	*p*-Coumaric acid, *p*-hydroxybenzoic acid, caffeic acid, ferulic acid, vanillic acid, protocatechuic acid, kaempferol, quercetin, apigenin	Michael and Gillian (1984)
Pteris altissima	Pteridaceae	7-*O*-Glycosides of apigenin, aglycone, 7-*O*-glycosides of luteolin, 7-*O*-glycosides of chrysoeriol, 3-*O*-glycosides of kaempferol, 3-*O*-glycosides of kaempferol, 3-*O*-glycosides of quercetin	Salatino and Prado (1998)
Pteris angustata	Pteridaceae	7-*O*-Glycosides of apigenin, aglycone, 7-*O*-glycosides of luteolin, 7-*O*-glycosides of chrysoeriol, 3-*O*-glycosides of kaempferol, 3-*O*-glycosides of kaempferol, 3-*O*-glycosides of quercetin	Salatino and Prado (1998)
Pteris decurrens	Pteridaceae	7-*O*-Glycosides of apigenin, aglycone, 7-*O*-glycosides of luteolin, 7-*O*-glycosides of chrysoeriol, 3-*O*-glycosides of kaempferol, 3-*O*-glycosides of kaempferol, 3-*O*-glycosides of quercetin	Salatino and Prado (1998)
Pteris deflexa	Pteridaceae	7-*O*-Glycosides of apigenin, aglycone, 7-*O*-glycosides of luteolin, 7-*O*-glycosides of chrysoeriol, 3-*O*-glycosides of kaempferol, 3-*O*-glycosides of kaempferol, 3-*O*-glycosides of quercetin	Salatino and Prado (1998)
Pteris denticulata	Pteridaceae	7-*O*-Glycosides of apigenin, aglycone, 7-*O*-glycosides of luteolin, 7-*O*-glycosides of chrysoeriol, 3-*O*-glycosides of kaempferol, 3-*O*-glycosides of kaempferol, 3-*O*-glycosides of quercetin	Salatino and Prado (1998)
Pteris multifidi	Pteridaceae	Luteolin-7-*O*-glucoside, 16-hydroxy-kaurane-2-β-D-glucoside, luteolin, palmitic acid, apigenin4-*O*-α-L-rhamnoside, quercetin, hyperin, isoquercitrin, kaempferol, rutin, apigenin-7-*O*-β-D-glucoside, and pterosin	Hoang and Tran (2014), Lu et al. (1999), Liu and Qin (2002), Murakami and Machashi (1985), Shu et al. (2012), Wang et al. (2010)

(Continued)

TABLE 7.2 *(Continued)*
Brazilian Species of Pteridophytes Containing Isolated Bioactive Compounds

Taxa	Family	Bioactive Compounds	Reference
Pteris podophylla	Pteridaceae	7-*O*-Glycosides of apigenin, aglycone, 7-*O*-glycosides of luteolin, 7-*O*-glycosides of chrysoeriol, 3-*O*-glycosides of kaempferol, 3-*O*-glycosides of kaempferol, 3-*O*-glycosides of quercetin	Salatino and Prado (1998)
Pteris propinqua	Pteridaceae	7-*O*-Glycosides of apigenin, aglycone, 7-*O*-glycosides of luteolin, 7-*O*-glycosides of chrysoeriol, 3-*O*-glycosides of kaempferol, 3-*O*-glycosides of kaempferol, 3-*O*-glycosides of quercetin	Salatino and Prado (1998)
Pteris quadriaurita	Pteridaceae	7-*O*-Gycosides of apigenin, aglycone, 7-*O*-glycosides of luteolin, 7-*O*-glycosides of chrysoeriol, 3-*O*-glycosides of kaempferol, 3-*O*-glycosides of kaempferol, and 3-*O*-glycosides of quercetin	Salatino and Prado (1998)
Pteris splendens	Pteridaceae	7-*O*-Glycosides of apigenin, aglycone, 7-*O*-glycosides of luteolin, 7-*O*-glycosides of chrysoeriol, 3-*O*-glycosides of kaempferol, 3-*O*-glycosides of kaempferol, 3-*O*-glycosides of quercetin	Salatino and Prado (1998)
Pteris tripatita	Pteridaceae	α-Caryophyllene and octadecanoic acid	Baskaran and Jeyachandran (2010)
*Pteris vitata**	Pteridaceae	Rutin, kaempferol monoglycoside, kaempferol diglycoside, quercetin monoglycoside, quercetin diglycoside	Salatino and Prado (1998), Sigh et al. (2008)
Salvinia molesta	Salviniaceae	Salviniside I, montbretol, 5,6-dehydrosugiol, 7-methoxyrosmanol, montbretyl 12-methyl ether, 11-hydroxysugiol, sugiol, ferruginol, 7-hydroxyferruginol, 6,7-dehydroferruginol, 12-hydroxy simonellite, simonellite	Li et al. (2013)
Adiantum cuneatum	Pteridaceae	Filicene, filicenal	Bresciani et al. (2003)

* Cultivated plants.

TABLE 7.3
Brazilian Species Pteridophytes with Potential Pharmacological Utility

Taxon	Family	Activity	Reference
Adiantum radianum	Pteridaceae	Antioxidant	Lai and Lim (2011)
Nephrolepis biserrata	Nephrolepidaceae	Antioxidant	Lai and Lim (2011)
Microgramma vacciniifolia	Polypodiaceae	Antioxidant	Peres et al. (2009)
Phlebodium decumanum	Polypodiaceae	Anticancer	Chang et al. (2007)
*Selaginella willdenowii**	Selaginellaceae	Anticancer	Lee and Lin (1988)
Marsilea minuta	Marsileaceae	Anticancer	Sarker et al. (2011)
Christella dentata	Thelypteridaceae	Antiviral	Paul et al. (2012)
*Asplenium nidus**	Aspleniaceae	Antiviral	Chand et al. (2013), Singh (1999), Santhosh et al. (2014)
Blechnum occidentale	Blechnaceae	Anti-inflamatory	Nonato et al. (2009)
Adiantum caudatum	Pteridaceae	Antibiotic	Banerjee and Sen (1980), Lakshmi and Pullaiah (2006), Lakshmi et al. (2006), Singh et al. (2008a)
Selaginella pallescens	Selaginellaceae	Antibiotic	Haripriya et al. (2010), Rojas et al. (1999)
Lygodium venustum	Lygodiaceae	Antibiotic	Moraes-Braga et al. (2012)

* Cultivated plants.

TABLE 7.4
The Names of Plant Species Including Family and Common Names from Which Some Natural Products Originated and Are Presented in This Chapter

Scientific Name	Family	Common Names
Adelanthus decipiens (sin. *Marsupidium brevifolium*)	Icacinaceae	
Adiantopsis flexuosa	Pteridaceae	
Adiantum capillus-veneris (sin. *Adiantum remyanum*)	Pteridaceae	Common maidenhair fern, avenca
Adiantum caudatum (sin. *Adiantum lyratum*)	Pteridaceae	Common maidenhair fern, avenca
Adiantum cuneatum	Pteridaceae	Common maidenhair fern, avenca
Adiantum radianum	Pteridaceae	Common maidenhair fern, avenca
Anastrophyllum auritum (sin. *Anastrophyllum hintzeanum*)	Anastrophyllaceae	
Aneura pinguis	Aneuraceae	
Anthoceros agrestis	Anthocerotaceae	
Asplenium nidus (sin. *Neottopteris musaefolia*)	Aspleniaceae	
Asterella venosa (sin. *Fimbraria venosa*)	Solanaceae	
Aulacomnium palustre (sin. *Aulacomnium pygmaeum*)	Aulacomninaceae	
Bazzania nitida (sin. *Mastigobryum stephanii*)	Lepidoziaceae	
Blechnum occidentale (sin. *Blechnum rugosum*)	Blechnaceae	
Bryopteris filicina (sin. *Bryopteris brevis*)	Lejeuneaceae	
Cheilolejeunea trifaria (sin. *Euosmolejeunea robillardii*)	Lejeuneaceae	
Christella dentata (sin. *Cyclosorus dentatus var. violascens*)	Thelypteridaceae	
Corsinia coriandrina (sin. *Riccia coriandrina*)	Corsiniaceae	
Cyathea	Cyatheaceae	Tree fern, samambaiaçu
Cyathea phalerata (sin. *Trichipteris phalerata*)	Cyatheaceae	Tree fern, samambaiaçu

(Continued)

TABLE 7.4 *(Continued)*

The Names of Plant Species Including Family and Common Names from Which Some Natural Products Originated and Are Presented in This Chapter

Scientific Name	Family	Common Names
Dicranopteris	Gleicheniaceae	
Dicranopteris	Pteridaceae	
Doryopteris concolor (sin. *Cheilanthes concolor*)	Pteridaceae	
Dryopteris filix-mas (sin. *Aspidium veselskii*)	Pteridaceae	
Dumortiera hirsuta (sin. *Marchantia hirsuta*)	Dumortieraceae	
Elaphoglossum	Dryopteridaceae	Deer tongue fern
Equisetum	Equisetaceae	Horsetail, cavalinha
Equisetum arvense (sin. *Equisetum calderi*)	Equisetaceae	Horsetail, cavalinha
Equisetum giganteum (sin. *Equisetum bolivianum*)	Equisetaceae	Horsetail, cavalinha
Eucladium verticillatum (sin. *Eucladium angustifolium*)	Pottiaceae	
Frullania arecae (sin. *Frullania crispistipula*)	Jubulaceae	
Frullania brasiliensis	Jubulaceae	
Frullania serrata (sin. *Frullania lacerata*)	Jubulaceae	
Hedwigia ciliata (sin. *Hedwigia macowaniana*)	Burseraceae	
Hymenophyllum	Hymenophyllaceae	Film fern
Isotachis aubertii (sin. *Isotachis rutenbergii*)	Balantiopsaceae	
Lejeunea flava	Lejeuneaceae	
Lethocolea glossophylla	Acrobolbaceae	
Lophocolea bidentata (sin. *Lophocolea setacea*)	Lophocoleaceae	
Lunularia cruciata	Lunulariaceae	
Lygodium venustum (sin. *Lygodium commutatum*)	Lygodiaceae	
Macrothelypteris	Thelypteridaceae	
Macrothelypteris	Thelypteridaceae	
Marchantia chenopoda	Marchantiaceae	
Marchantia paleacea	Marchantiaceae	
Marchantia polymorpha	Marchantiaceae	
Marchesinia brachiata (sin. *Marchesinia aquatica*)	Lejeuneaceae	
Marsilea minuta (sin. *Marsilea perrieriana*)	Marsileaceae	Trevo de quatro folhas
Metzgeria conjugata	Metzgeriaceae	
Metzgeria furcata	Metzgeriaceae	
Microgramma vacciniifolia (sin. *Lepicystis vacciniifolia*)	*Microgramma*	
Nephrolepis biserrata (sin. *Nephrolepis mollis*)	Nephrolepidaceae	
Noteroclada confluens	Pelliaceae	
Odontoschisma denudatum (sin. *Jungermannia denudata*)	Cephaloziaceae	
Omphalanthus filiformis (sin. *Jungermannia filiformis*)	Lejeuneaceae	
Ormopteris gleichenioides (sin. *Pellaea gleichenioides*)	Pteridaceae	
Ormopteris cymbiformis (sin. *Pellaea cymbiformis*)	Pteridaceae	
Ormopteris pinnata (sin. *Cassebeera pinnata*)	Pteridaceae	
Ormopteris riedelii (sin. *Pellaea riedelii*)	Pteridaceae	
Pellaea mucronata (sin. *Lytoneuron ornithopus*)	Pteridaceae	
Phlebodium decumanum (sin. *Phlebodium multiseriale*)	Polypodiaceae	
Plagiochasma rupestre (sin. *Aytonia rupestris*)	Aytoniaceae	
Plagiochila bifaria (sin. *Jungermannia bifaria*)	Plagiochilaceae	
Plagiochila corrugata (sin. *Jungermannia corrugata*)	Plagiochilaceae	
Plagiochila cristata (sin. *Plagiochila secundifolia*)	Plagiochilaceae	
Plagiochila diversifolia	Plagiochilaceae	
Plagiochila gymnocalycina	Plagiochilaceae	

(Continued)

TABLE 7.4 *(Continued)*

The Names of Plant Species Including Family and Common Names from Which Some Natural Products Originated and Are Presented in This Chapter

Scientific Name	Family	Common Names
Plagiochila rutilans	Plagiochilaceae	
Plagiochila stricta	Plagiochilaceae	
Polytrichum commune (sin. *Polytrichum leonii*)	Dryopteridaceae	
Psilotum nudum (sin. *Psilotum domingense*)	Psilotaceae	
Pteridium aquilinum (sin. *Pteris lanuginosa*)	Dennstaedtiaceae	Samambaia-das-taperas
Pteris altissima (sin. *Litobrochia grandis*)	Pteridaceae	
Pteris angustata (sin. *Litobrochia angustata*)	Pteridaceae	
Pteris decurrens	Pteridaceae	
Pteris deflexa (sin. *Pteris gaudichaudii*)	Pteridaceae	
Pteris denticulata	Pteridaceae	
Pteris multifida (sin. *Pycnodoria multifida*)	Pteridaceae	
Pteris podophylla (sin. *Pteris inflexa*)	Pteridaceae	
Pteris propinqua (sin. *Pteris hostmanniana*)	Pteridaceae	
Pteris quadriaurita (sin. *Pteris prolifera*)	Pteridaceae	
Pteris splendens	Pteridaceae	
Pteris tripartita (sin. *Litobrochia tripartita*)	Pteridaceae	
Pteris vittata (sin. *Pteris vittata* f. *cristata*)	Pteridaceae	
Radula nudicaulis	Radulaceae	
Reboulia hemisphaerica	Aytoniaceae	
Riccardia multifida	Aneuraceae	
Ricciocarpos natans (sin. *Riccia natans*)	Ricciaceae	
Salvinia molesta (sin. *Salvinia adnata*)	Salviniaceae	
Selaginella pallescens (sin. *Selaginella cuspidata*)	Selaginellaceae	
Selaginella willdenowii (sin. *Lycopodium willdenowii*)	Selaginellaceae	
Symphyogyna brasiliensis (sin. *Symphyogyna lehmanniana*)	Pallaviciniaceae	
Symphyogyna podophylla (sin. *Jungermannia podophylla*)	Pallaviciniaceae	
Thelypteris	Thelypteridaceae	

7.10 FINAL REMARKS AND FUTURE PERSPECTIVES

That bryophytes and pteridophytes show great (but currently neglected) potential for pharmaceutical and medicinal uses is quite clear, especially Brazilian species. The taxa that have been sampled in phytochemical studies have usually been harvested near research centers, so that our knowledge of phytochemical diversity in different Brazilian biomes is still very limited. The native Brazilian flora is also under constant threat, with biomes such as the Atlantic forest having less than 8% of its original cover still intact.

The importance of floristic and ethnobotanical surveys cannot be over emphasized. Efforts must be applied to the mapping and cataloging of mosses and their allies – an essential step in their conservation and for a better understanding of their diversity and distribution patterns. Investigations must also focus on species used in traditional medicine, as well as other species within genera known to contain bioactive compounds.

Modern tools should also be tested and employed in bioprospecting bryophytes and pteridophytes, especially with the onset of the genomic era – which permits to investigate gene clusters that govern the biosynthesis of bioactive compounds (Zotchev et al., 2012) – to accelerate the processes of discovery and cataloging chemical compounds within the Brazilian flora.

ACKNOWLEDGMENTS

ASMS thanks Fundação de Amparo à Pesquisa do Estado de Minas Gerais (FAPEMIG, APQ-00395-14) and the Conselho Nacional de Desenvolvimento Científico e Tecnológico (CNPq, 459764/2014-4). This study was also sponsored by the Coordenação de Aperfeiçoamento de Pessoal de Nível Superior – Brasil (CAPES) – Finance Code 001 (88887.19244/2018-00).

REFERENCES

Adam, K. -P.; Becker, H. 1994. Phenanthrenes and other phenolics from in vitro cultures of *Marchantia polymorpha*. Phytochemistry, 35, 139–143.

Alonso-Amelot, M. E. 2002. The chemistry and toxicology of bioactive compounds in bracken fern (*Pteridium* sp.), with special reference to chemical ecology and carcinogenesis. Studies in Natural Products Chemistry, 26, 685–740.

Andersen, N. H.; Ohta, Y.; Liu, C. -B.; Michael Kramer, C.; Allison, K.; Huneck, S. 1977. Sesquiterpenes of thalloid liverworts of the genera *Conocephalum, Lunularia, Metzgeria* and *Riccardia*. Phytochemistry, 16, 1727–1729.

Asakawa, Y. 2001. Recent advances in phytochemistry of bryophytes-acetogenins, terpenoids and bis (bibenzyl) s from selected Japanese, Taiwanese, New Zealand, Argentinean and European liverworts. Phytochemistry, 56, 297–312.

Asakawa, Y. 2007. Biologically active compounds from bryophytes. Pure and Applied Chemistry, 79, 557–580.

Asakawa, Y.; Ludwiczuk, A. 2017. Chemical constituents of bryophytes: structures and biological activity. Journal of Natural Products, 81, 641–660.

Asakawa, Y.; Ludwiczuk, A.; Nagashima, F. 2013a. Chemical constituents of bryophytes, progress in the chemistry of organic natural products. In A. D. Kinghorn; O. H. Columbus; H. Falk; et al., eds., Progress in the Chemistry of Organic Natural Products, v. 95. Vienna: Springer.

Asakawa, Y.; Ludwiczuk, A.; Nagashima, F. 2013b. Phytochemical and biological studies of bryophytes. Phytochemistry, 91, 52–80.

Asakawa, Y.; Ludwiczuk, A.; Nagashima, F.; et al. 2009. Bryophytes: bio- and chemical diversity, bioactivity and chemosystematics. Heterocycles, 77, 99–150.

Asakawa, Y.; Toyota, M.; Nakaishi, E.; et al. 1996. Distribution of terpenoids and aromatic compounds in New Zealand liverworts. Journal of the Hattori Botanical Laboratory, 80, 271–295.

Banerjee, R. D.; Sen, S. P. 1980. Antibiotic activity of pteridophytes. Economic Botany, 34, 284–298.

Bardón, A.; Bovi, B. M.; Kamiya, N.; Toyota, M.; Asakawa, Y. 2002. Eremophilanolides and other constituents from the Argentine liverwort *Frullania brasiliensis*. Phytochemistry, 59, 205–213.

Bardón, A.; Kamiya, N.; Toyota, M.; Asakawa, Y. 1999a. A 7-nordumortenone and other dumortane derivatives from the Argentine liverwort *Dumortiera hirsuta*. Phytochemistry, 51, 281–287.

Bardón, A.; Kamiya, N.; Toyota, M.; Takaoka, S.; Asakawa, Y. 1999b. Sesquiterpenoids, hopanoids and bis(bibenzyls) from the Argentine liverwort *Plagiochasma rupestre*. Phytochemistry, 52, 1323–1329.

Baskaran, X.; Jeyachandran, R. 2010. Evaluation of antioxidant and phytochemical analysis of *Pteris tripartita* Sw. a critically endangered fern from South India. Journal of Fairy lake Botanical Garden, 9, 28–34.

Baskaran, X. -R.; Geo Vigila, A. -V.; Zhang, S. Z.; Feng, S. -X.; Liao, W. -B. 2018. A review of the use of pteridophytes for treating human ailments. Journal of Zhejiang University, 19, 1–35.

Benjamin A.; Manickam, V. S. 2007. Medicinal pteridophytes from the Western Ghats. Indian Journal of Traditional Knowledge, 6, 611–618.

Bresciani, L. F.; Priebe, J. P.; Yunes, R. A.; Dal Magro, J.; Delle Monache, F.; de Campos, F.; de Souza, M. M.; Cechinel-Filho, V. 2003. Pharmacological and phytochemical evaluation of *Adiantum cuneatum* growing in Brazil. Zeitschrift für Naturforschung, 58, 191–194.

Cambie, R. C.; Ash, J. 1994. Fijian Medicinal Plants. Melbourne: CSIRO.

Camus, J. M. 1990. Marattiaceae. In K. U. Kramer; P. S. Green, eds., Pteridophytes and Gymnosperms. The Families and Genera of Vascular Plants, vol. 1. Berlin and Heidelberg: Springer.

Cao, H.; Chai, T. T.; Wang, X.; Morais-Braga, M. F. B.; Yang, J. H.; Wong, F. C.; Yao, H.; et al. 2017. Phytochemicals from fern species: potential for medicine applications. Phytochemistry Reviews, 16, 379–440.

Chand-Basha S.; Sreenivasulu, M.; Pramod, N. 2013. Antidiabetic activity of *Actinopteris radiata* (Linn.). Asian Journal of Research in Pharmaceutical Sciences, 1, 1–6.

Chandra, S.; Chandra, D.; Barh, A.; Bhatt, P., Pandey, R. K.; Sharma, I. P. 2017. Bryophytes: hoard of remedies, an ethno-medicinal review. Journal of Traditional and Complementary Medicine, 7, 94–98.

Chang, H. C.; Huang, G. J.; Agrawal, D. C.; Kuo, C. L.; Wu, C. R.; Tsay, H. S. 2007a. Antioxidant activities and polyphenol contents of six folk medicinal ferns used as "Gusuibu". Botanical Studies, 48, 397–406.

Chen, Y. H.; Chang, F. R.; Lin, Y. J.; Wang, L.; Chen, J. F.; Wu, Y. C.; Wu, M. J. 2007. Identification of phenolic antioxidants from sword brake fern (*Pteris ensiformis* Burm.). Food Chemistry, 105, 48–56.

Cheng, X. L.; Ma, S. C.; Yu, J. D.; Yang, S. Y.; Xiao, X. Y.; Hu, J. Y.; Lu, Y.; Shaw, P. C.; But, P. P. H.; Lin, R. C. 2008. Selaginellin A and B, two novel natural pigments isolated from *Selaginella tamariscina*. Chemical Pharmaceutical Bulletin, 56, 982–984.

Chopra, R. N.; Kumra, P. K. 1988. Biology of Bryophytes. New Delhi: Wiley Eastern.

Christenhusz, M. J.; Byng, J. W. 2016. The number of known plants species in the world and its annual increase. Phytotaxa, 261, 201–217.

Crandall-Stotler, B.; Stotler, R. E.; Long, D. G. 2009. Morphology and classification of the Marchantiophyta. In B. Goffinet; A. J. Shaw, eds., Bryophyte Biology, pp. 1–54. Cambridge: Cambridge University Press.

Cristina Figueiredo, A.; Sim-Sim, M.; Costa, M. M.; Barroso, J. G.; Pedro, L. G.; Esquível, M. G.; Gutierres, F.; Lobo, C.; Fontinha, S. 2005. Comparison of the essential oil composition of four *Plagiochila* species: *P. bifaria, P. maderensis, P. retrorsa and P. stricta*. Flavour and Fragrance Journal, 20, 703–709.

de Boer, H. J.; Kool, A.; Broberg, A.; Mziray, W. R.; Hedberg, I.; Levenfors, J. 2005. Anti-fungal and anti-bacterial activity of some herbal remedies from Tanzania. Journal of Ethnopharmacology, 96, 461–469.

Dhiman A. K. 1998. Ethnomedicinal uses of some pteridophytic species in India. Indian Fern Journal, 15, 61–64.

Dong, L.; Yang, J.; He, J.; Luo, H. R.; Wu, X. D.; Deng, X.; Peng, L. Y.; Cheng, X.; Zhao, Q. S. 2012. Lycopalhine A, a novel sterically congested *Lycopodium* alkaloid with an unprecedented skeleton from *Palhinhaea cernua*. Chemical Communications, 48, 9038–9040.

Duraiswamy, H.; Nallaiyan, S.; Nelson, J.; Samy, P. R.; Johnson, M.; Varaprasadam, I. 2010. The effect of extracts of *Selaginella involvens* and *Selaginella inaequalifolia* leaves on poultry pathogens. Asian Pacific Journal of Tropical Medicine, 3, 678–681.

Fidalgo, O.; Bononi, V. L. R. 1989. Técnicas de coleta, preservação e herborização de material botânico. São Paulo: Instituto de Botânica.

Flora do Brasil. 2020. em construção. Jardim Botânico do Rio de Janeiro. http://floradobrasil.jbrj.gov.br (accessed on March, 2018).

Frahm, J. P. 2003. Manual of tropical bryology. Tropical Bryology, 23, 1–200.

Frahm, J. P. 2004. Recent developments of commercial products from bryophytes. The Bryologist, 107, 277–283.

Gerrienne, P.; Gonéz, P. 2011. Early evolution of life cycles in embryophytes: a focus on the fossil evidence of gametophyte/sporophyte size and morphological complexity. Journal of Systematics and Evolution, 49, 1–16.

Gifford, E. M.; Foster, A. S. 1987. Morphology and Evolution of Vascular Plants, 3rd ed. New York: WH Freeman.

Glime, J. M. 2017. Medical uses: biologically active substances. In J. M. Glime, ed., Bryophyte Ecology. Ebook sponsored by Michigan Technological University and the International Association of Bryologists. http://digitalcommons.mtu.edu/bryophyte-ecology (accessed on October 8, 2017).

Glime, J. M.; Wagner, D. H. 2013. Laboratory techniques: equipment. In J. M. Glime, ed., Bryophyte Ecology. Methods. Ebook sponsored by Michigan Technological University and the International Association of Bryologists. www.bryoecol.mtu.edu (accessed on September 7, 2013).

Goffinet, B.; Buck, W. R. 2012. The evolution of body form in bryophytes. Annual Plant Reviews, 45, 51–89.

Goffinet, B.; Buck, W. R.; Shaw, A. J. 2009. Morphology and classification of the Bryophyta. In B. Goffinet; A. J. Shaw, eds., Bryophyte Biology, pp. 55–138. Cambridge: Cambridge University Press.

Gogoi, R. 2002. Ethnobotanical studies of some ferns used by the Garo Tribals of Meghalaya. Advances in Plant Sciences, 15, 401–405.

Goldberg, D. J.; Begenisich, T. B.; Cooper, J. R. 1975. Effects of thiamine antagonists on nerve conduction. II. Voltage clamp experiments with antimetabolites. Developmental Neurobiology, 6, 453–462.

Gradstein, S. R.; Churchill, S. P.; Allen, N. S. 2001. Guide to the bryophytes of tropical America. Memoirs of The New York Botanical Garden, 86, 1–577.

Harris, E. S. J. 2008. Ethnobryology: traditional uses and folk classification of bryophytes. The Bryologist, 111, 169–217.

Harris, J. G.; Harris, M. W. 1994. Plant Identification Terminology: An Illustrate Glossary. Utah: Spring Lake Publishing.

Hashimoto, T.; Irita, H.; Yoshida, M.; Kikkawa, A.; Toyota, M.; Koyama, H., Motoike, Y.; Asakawa, Y. 1998a. Chemical constituents of the Japanese liverworts *Odontoschisma denudatum, Porella japonica, P. acutifolia* subsp. *tosana* and *Frullania hamatiloba*. Journal of the Hattori Botanical Laboratory, 84, 309–314.

Hashimoto, T.; Kikkawa, A.; Yoshida, M.; Tanaka, M.; Asakawa, Y. 1998b. Two novel skeletal diterpenoids, neodenudatenones A and B, from the liverwort *Odontoschisma denudatum*. Tetrahedron Letters, 39, 3791–3794.

Hauke R. L. 1990. Equisetaceae. pteridophytes and gymnosperms. In K. U. Kramer; P. S. Green, eds., The Families and Genera of Vascular Plants, vol. 1. Berlin: Springer Heidelberg.

Heinrichs, J.; Anton, H.; Gradstein, S. R.; Mues, R. 2000. Systematics of *Plagiochila* sect. *Glaucescens* Carl (Hepaticae) from tropical America: a morphological and chemotaxonomical approach. Plant Systematics and Evolution, 220, 115–138.

Hikino, H.; Takahashi, T.; Arihara, S.; Takemoto, T. 1970. Structure of pteroside b, glycoside of *Pteridium aquilinum* var. *latiusculum*. Chemical and Pharmaceutical Bulletin, 18, 1488–1491.

Hikino, H.; Takhashi, T.; Takemoto, T. 1971. Structure of pteroside Z and D, glycosides of *Pteridium aqullinum* var. *latiusculum*. Chemical and Pharmaceutical Bulletin, 19, 2424–2425.

Hikino, H.; Takhashi, T.; Takemoto, T. 1972. Structure of pteroside A and C, glycosides of *Pteridium aqullinum* var. *latiusculum*. Chemical and Pharmaceutical Bulletin, 20, 210–212.

Ho, R.; Teai, T.; Bianchini, J. P.; Lafont, R., Raharivelomanana, P. 2011. Ferns: from traditional uses to pharmaceutical development, chemical identification of active principles. In A. Kumar; H. Fernández; M. Revilla, eds., Working with Ferns, pp. 321–346. New York: Springer.

Hoang, L.; Tran, H. 2014. *In vitro* antioxidant and anti-cancer properties of active compounds from methanolic extract of *Pteris multifida* poir. leaves. European Journal of Medicinal Plants, 4, 292–302.

Hort, M. A.; Dalbo, S.; Brighente, I. M. C.; Pizzolatti, M. G.; Pedrosa, R. C.; Ribeiro-do-Valle, R. M. 2008. Antioxidant and hepatoprotective effects of *Cyathea phalerata* mart. (Cyatheaceae). Basic & Clinical Pharmacology & Toxicology, 103, 17–24.

Huneck, S. 1983. Chemistry and biochemistry of bryophytes. In R. M. Schuster, ed., New Manual of Bryology, pp. 1–116. Nichinan: Hattori Bot. Lab.

Iwai, Y.; Murakami, K.; Gomi, Y.; Hashimoto, T.; Asakawa, Y.; Okuno, Y.; Ishikawa, T.; Hatakeyama, D.; Echigo, N.; Kuzuhara, T. 2011. Anti-influenza activity of marchantins, macrocyclic bisbibenzyls contained in liverworts. PLoS One, 6, e19825.

Iwashina, T. 2000. The structure and distribution of the flavonoids in plants. Journal of Plant Research, 113, 287–299.

Jackovic, N. P. B.; Andrade, P.; Valentão, P.; Sabovljevic, M. 2008. HPLC-DAD of phenolics in bryophytes *Lunularia cruciata, Brachytheciastrum velutinum* and *Kindbergia praelonga*. Journal of the Serbian Chemical Society, 73, 1161–1167.

Jimenez, M. C. A.; Rojas, H. N. M.; Lopez, A. M. A. 1979. Biological evaluation of Cuban plants. IV. Revista Cubana de Medicina Tropical, 31, 29–35.

Joksic, G.; Stankovic, M.; Novak, A. 2003. Antibacterial medicinal plants *Equiseti herba* and *Ononidis radix* modulate micronucleus formation in human lymphocytes *in vitro*. Journal of Environmental Pathology, Toxicology and Oncology, 22, 41–48.

Keller, H. A.; Prance, G. T. 2015. The ethnobotany of ferns and lycophytes. Fern Gazette, 20, 1–13.

Kunz, S.; Becker, H. 1994 Bibenzyl derivatives from the liverwort *Ricciocarpos natans*. Phytochemistry, 36, 675–677.

Kunz, S.; Burkhardt, G.; Becker, H. 1994. Riccionidins A and B, anthocyanidins from the cell walls of the liverwort *Ricciocarpos natans*. Phytochemistry, 35, 233–235.

Lai, H. Y.; Lim Y. Y. 2011. Antioxidant properties of some Malaysian ferns. The 3rd International Conference on Chemical, Biological and Environmental Engineering IPCBEE, v. 20, Singapore: IACSIT Press.

Lakshmi, P. A.; Kalavathi, P., Pullaiah, T. 2006. Phytochemical and antimicrobial studies of *Adiantum latifolium*. Journal of Tropical Medicinal Plants, 7, 17–22.

Lakshmi, P. A.; Pullaiah, T. 2006. Phytochemicals and antimicrobial studies of *Adiantum incisum* on gram positive, gram negative bacteria and fungi. Journal of Tropical Medicinal Plants, 7, 275–278.

Lawrence, G. H. M. 1977. Taxonomia das plantas vasculares. Lisboa: Fundação Calouste.

Lee, H.; Lin, J. Y. 1988. Antimutagenic activity of extracts from anti-cancer drugs in Chinese medicine. Mutation Research, 204, 229–234.

Lellinger, D. B. 2002. A Modern Multilingual Glossary for Taxonomic Pteridology. Washington: Smithsonian Institution.

Li, S.; Wang, P.; Deng, G.; Yuan, W.; Su, Z. 2013. Cytotoxic compounds from invasive giant salvinia (*Salvinia molesta*) against human tumor cells. Bioorganic & Medicinal Chemistry Letters, 23, 6682–6687.

Ligrone, R.; Duckett, J. G.; Renzaglia, K. S. 2000. Conducting tissues and phyletic relationships of bryophytes. Philosophical Transactions of the Royal Society, 355, 795–813.

Liu, Q.; Qin, M. 2002. Studies on chemical constituents of rhizomes of *Pteris multifida* Poir. Chinese Traditional and Herbal Drugs, 33, 114.

Lorimer, S. D.; Perry, N. B. 1994. Antifungal hydroxyacetophenones from the New Zealand liverwort, *Plagiochila fasciculata*. Planta Medica, 60, 386–387.

Lu, H.; Xu, J.; Zhang, L. X. et al. 1999. Bioactive constituents from *Pteris multifida*. Planta Medica, 65, 586–587.

Lu, Z. -Q.; Fan, P. -H.; Ji, M.; Lou, H. X. 2006. Terpenoids and bisbibenzyls from Chinese liverworts *Conocephalum conicum* and *Dumortiera hirsuta*. Journal of Asian Natural Products Research, 8, 187–192.

Ludwiczuk, A.; Asakawa, Y. 2008. Chapter five: distribution of terpenoids and aromatic compounds in selected Southern Hemispheric liverworts. Fieldiana Botany, 47, 37–58.

Ludwiczuk, A.; Asakawa, Y. 2010. Chemosystematics of selected liverworts collected in Borneo. Bryophyte Diversity and Evolution, 31, 33–42.

Ludwiczuk, A.; Asakawa, Y. 2015. Chemotaxonomic value of essential oil components in liverwort species. a review. Flavour and Fragrance Journal, 30, 189–196.

Maciel, M. A. M.; Pinto, A. C.; Veiga, J. V.; Valdir, F.; Grynberg, N. F.; Echevarria, A. 2002. Plantas medicinais: a necessidade de estudos multidisciplinares. Química Nova, 25, 429–438.

Maciel-Silva, A. S.; Pôrto, K. C. 2014. Reproduction in Bryophytes. In K. G. Ramawat; J. -M. Mérillon; K. R. Shivanna, eds., Reproductive Biology of Plants, 57–84, Boca Raton: CRC Press.

Madhukiran, P.; Ganga, B. R. 2011. Anti-inflammatory activity of methanolic leaf extract of *Cyathea gigantea* (Wall. Ex Hook.). International Journal of Pharmaceutical Research and Development, 3, 64–68.

Magallón S.; Gómez-Acevedo, S.; Sánchez-Reyes, L. L.; Hernández-Hernández, T. 2015. A metacalibrated time-tree documents the early rise of flowering plant phylogenetic diversity. New Phytologist, 207, 437–453.

Michael, S. F.; Gillian, C. D. 1984. Anti-microbial activity of phenolic acids in *Pteridium aquilinium*. Americal Fern Journal, 74, 87–96.

Mickel, J. T.; Smith, A. 2004. The Pteridophytes of Mexico, vols. I & II. New York: The New York Botanical Garden Press.

Milovanovic, V.; Radulovic, N.; Todorovic, Z.; Stanković, M.; Stojanović, G. 2007. Antioxidant, antimicrobial and genotoxicity screening of hydro-alcoholic extracts of five Serbian Equisetum species. Plant Foods for Human Nutrition, 62, 113–119.

Mimica-Dukic, N.; Simin, N.; Cvejic, J.; Jovin, E.; Orcic, D.; Bozin, B. 2008. Phenolic compounds in field horsetail (*Equisetum arvense* L.) as natural antioxidants. Molecules, 13, 1455–1464.

Morais-Braga, M. F. B.; Souza, T. M.; Santos, K. K. A.; Andrade, J. C.; Guedes, G. M. M.; Tintino, S. R.; Sobral-Souza, C. E.; et al. 2012. Antimicrobial and modulatory activity of ethanol extract of the leaves from *Lygodium venustum* SW. American Fern Journal, 102, 154–160.

Murakami, T.; Machashi, N. T. 1985. Chemical and chemotaxonomical studies on filices. Journal of the Pharmacological Society of Japan 105, 640–648.

Nakane, T.; Maeda, Y.; Ebihara, H.; Arai, Y.; Masuda, K.; Takano, A.; Ageta, H.; Shiojima, K.; Cai, S. Q.; Abdel-Halim, O. B. 2002. Fern constituents: triterpenoids from *Adiantum capillus-veneris*. Chemical and Pharmaceutical Bulletin, 50, 1273–1275.

Neves, M.; Morais, R.; Gafner, S.; Stoeckli-Evans, H.; Hostettmann, K. 1999. New sesquiterpene lactones from the Portuguese liverwort *Targonia lorbeeriana*. Phytochemistry, 50, 967–972.

Nóbrega, G. A.; Prado, J. 2018. Equisetaceae in Flora do Brasil 2C em construção. Jardim Botânico do Rio de Janeiro. Disponível em: http://floradobrasil.jbrj.gov.br/reflora/floradobrasil/FB91154 (accessed on June 15, 2018).

Nonato, F. R.; Nonato, T. A.; Barros, A. M.; Oliveira, C. E.; Santos, R. R.; Soares, M. B.; Villarreal C. F. 2009. Antiinflammatory and antinociceptive activities of *Blechnum occidentale* L. extract. Journal Ethnopharmacology, 125, 102–107.

Øllgaard, B. 1990. Lycopodiaceae. In K. U. Kramer; P. S. Green, eds., Pteridophytes and Gymnosperms. The Families and Genera of Vascular Plants, vol. 1. Berlin: Springer.

Otsuka, H.; Tsuki, M.; Toyosato, T. 1972. Anti-inflammatory activity of crude drugs and plants. Takeda Kenkynsho Ho, 31, 238–246.

Paul, T.; Das, B.; Apte, K. G.; Banerjee, S.; Saxena, R. C. 2012. Evaluation of antihyperglycemic activity of adiantum philippense linn. A pteridophyte in alloxan induced diabetic rats. Journal of Diabetes & Metabolism, 3, 1–8.

Peixoto, A. L.; Maia, L. C. 2013. Manual de procedimentos para herbários. INCT-Herbário virtual para a flora e os fungos. Recife: Editora Universitária UFPE.

Peres, M. T. L. P.; Simionatto, E.; Hess, S. C.; Bonani, V. F. L.; Candido, A. C. S.; Castelli, C.; Poppi, N. R.; Honda, N. K.; Cardoso, C. A. L.; Faccenda, O. 2009. Chemical and biological studies of *Microgramma vacciniifolia* (Langsd. & Fisch.) Copel (Polypodiaceae). Química Nova, 32, 897–901.

Pigg, K. B. 1992. Evolution of Isoetalean lycopsids. Annals of the Missouri Botanic Garden, 79, 589–612.

Pinheiro, M. D. F. D. S.; Lisboa, R. C. L.; Brazão, R. D. V. 1989. Contribuição ao estudo de briófitas como fontes de antibióticos. Acta Amazonica, 19, 139–145.

PPG I. 2016. A community-derived classification for extant lycophytes and ferns. Journal of Systematic and Evolution. 54, 563–603.

Prado, J.; Sylvestre, L. S.; Labiak, P. H.; Windisch, P. G.; Salino, A.; Barros, I. C. L.; Hirai, R. Y.; et al. 2015. Diversity of ferns and lycophytes in Brazil. Rodriguésia, 66, 1073–1083.

Proctor, M. C.; Oliver, M. J.; Wood, A. J.; Alpert, P.; Stark, L. R.; Cleavitt, N. L.; Mishler, B. D. 2007. Desiccation-tolerance in bryophytes: a review. The Bryologist, 110, 595–621.

Pryer, K. M.; Schuettpelz, E.; Wolf, P. G.; Schneider, H.; Smith, A. R.; Cranfill, R. 2004. Phylogeny and evolution of ferns (monilophytes) with a focus on early-diverging lineages. American Journal of Botany, 91, 1582–1598.

Radulovic, N.; Stojanovic, G.; Palic, R. 2006. Composition and antimicrobial activity of *Equisetum arvense* L. essential oil. Phytotherapy Research, 20, 85–88.

Reddy, V. L.; Ravikanth, V.; Rao, T. P.; Diwan, P. V.; Venkateswarlu, Y. 2001. A new triterpenoid from the fern *Adiantum lunulatum* and evaluation of antibacterial activity. Phytochemistry, 56, 173–175.

Reinaldo, R. C. P.; Santiago, A. C. P.; Medeiros, P. M.; Albuquerque, U. P. 2015. Do ferns and lycophytes function as medicinal plants? A study of their low representation in traditional pharmacopoeias. Journal of ethnopharmacology, 175, 39–47.

Renzaglia, K. S.; Schuette, S.; Duff, R.; Ligrone, R.; Shaw, A. J.; Mishler, B. D.; Duckett, J. G. 2007. Bryophyte phylogeny: advancing the molecular and morphological frontiers. The Bryologist, 110, 179–213.

Renzaglia, K. S.; Villareal, J. C.; Duff, R. J. 2009. New insights into morphology, anatomy and systematics of hornworts. In B. Goffinet; A. J. Shaw, eds., Bryophyte Biology, pp. 139–171. Cambridge: Cambridge University Press.

Rojas, A.; Bah, M.; Rojas, J. I.; Serrano, V.; Pacheco, S. 1999. Spasmolytic activity of some plants used by the Otomi Indians of Queretaro (Mexico) for the treatment of gastrointestinal disorders. Phytomedicine, 6, 367–371.

Rycroft, D. S.; Cole, W. J. 2001. Hydroquinone derivatives and monoterpenoids from the neotropical liverwort *Plagiochila rutilans*. Phytochemistry, 57, 479–488.

Sabovljević; A.; Bijelović, A.; Grubišić, D. 2001. Bryophytes as a potential source of medicinal compounds. Pregledni članak – Review, 21, 17–29.

Sabovljevic, A.; Sabovljevic, M.; Jockovic, N. 2009. *In vitro* culture and secondary metabolite isolation in bryophytes. In S. M. Jain; P. Saxena, eds., Protocols for in Vitro Cultures and Secondary Metabolite Analysis of Aromatic and Medicinal Plants, pp. 117–128. New York: Humana Press.

Salatino, M. L. F.; Prado J. 1998. Flavonoid glycosides of Pteridaceae from Brazil. Biochemistry Systematic Ecology, 26, 761–769.

Santos, M. G.; Kelecom, A.; Paiva, S. R.; Moraes, M. G.; Rocha, L.; Garrett, R. 2010. Phytochemical studies in pteridophytes growing in Brazil: A review. American Journal of Plant Science and Biotechnology, 4, 113–125.

Sarker, M. A. Q.; Mondol, P. C.; Alam, M. J.; Parvez, M. S.; Alam, M. F. 2011. Comparative study on antitumor activity of three pteridophytes ethanol extracts. International Journal of Agricultural Technology, 7, 1661–1671.

Saruwatari, M.; Takio, S.; Ono, K. 1999. Low temperature-induced accumulation of eicosapentaenoic acid in *Marchantia polymorpha* cells. Phytochemistry, 52, 367–372.

Scher, J. M.; Speakman, J-B.; Zapp, J.; Becker, H. 2004. Bioactivity guided isolation of antifungal compounds from the liverwort *Bazzania trilobata* (L.) S.F. Gray. Phytochemistry, 65, 2583–2588.

Schofield, W. B. 1985. Introduction to Bryology. New York: Macmillan.

Sharpe, J. M.; Mehltreter, K.; Walker, L. R. 2010. Ecological importance of ferns. In K. Mehltreter; L. R. Walker; J. M. Sharpe, eds., Fern Ecology. Cambridge: Cambridge University Press.

Shin, S. L. 2010. Functional components and biological activities of pteridophytes as healthy biomaterials. PhD thesis, Chungbuk National University, Cheongju, Korea.

Shu, J. C.; Liu, J. Q., Zhong, Y. Q.; Pan, J.; Liu, L.; Zhang, R. 2012. Two new pterosin sesquiterpenes from *Pteris multifida* Poir. Phytochemistry Letters, 5, 276–279.

Silva, G. L.; Chai, H.; Gupta, M. P.; Farnsworth, N. R.; Cordell, G. A.; Pezzuto, J. M.; Beecher, C. W.; Kinghorn, A. D. 1995. Cytotoxic bioflavonoids from *Selaginella willdenowii*. Phytochemistry, 40, 129–134.

Singh, H. B. 1999. Potential medicinal pteridophytes of India and their chemical constituents. Journal Economic and Taxonomic Botany, 23, 63–78.

Singh, M.; Govindarajan, R.; Rawat, A. K. S.; Prem, K. 2008a. Antimicrobial flavonoid rutin from *Pteris vittata* L. against pathogenic gastrointestinal microflora. American Fern Journal, 98, 98–103.

Singh, M.; Singh, N.; Khare, P. B.; Rawat, A. K. 2008b. Antimicrobial activity of some important Adiantum species used traditionally in indigenous systems of medicine. Journal of Ethnopharmacology, 115, 327–329.

Socolsky, C.; Asakawa, Y.; Bardón, A. 2007. Diterpenoid glycosides from the bitter fern *Gleichenia quadripartita*. Journal of Natural Products, 70, 1837–1845

Star, A. E.; Mabry, T. J. 1971. Flavonoid frond exudates from two Jamaican ferns *Pityrogramma tartarea* and *Pityrogramma calmoelanos*. Phytochemistry, 10, 2817–2818.

Stearn, W. T. 1998. Botanical Latin. Devon: David & Charles Book.

Sultana, S.; Nandi, J. K.; Rahman, S.; Jahan, R.; Rahmatullah, M. 2014. Preliminary antihyperglycemic and analgesic activity studies with *Angiopteris evecta* leaves in Swiss *Albino mice*. World Journal of Pharmacy and Pharmaceutical Sciences, 3, 1–12.

Tanaka, R.; Minami, T.; Ishikawa, Y.; Tokuda, H.; Matsunaga, S. 2004. Cancer chemopreventive activity of serratane-type triterpenoids from *Picea jezoensis*. Chemical Biodiversity, 1, 878–885.

Tanzin, R.; Rahman, S.; Hossain, M. S. 2013. Medicinal potential of pteridophytes – an antihyperglycemic and antinociceptive activity evaluation of methanolic extract of whole plants of *Christella dentata*. Advances in Natural Applied Sciences, 7, 67–73.

Taylor, L. 2003. Herbal Secrets of the Rainforest, 2nd ed. California: Sage Press.

Thorroad, S.; Worawittayanont, P.; Khunnawutmanotham, N.; Chimnoi, N.; Jumruksa, A.; Ruchirawat, S.; Thasana, N. 2014. Three new *Lycopodium* alkaloids from *Huperzia carinata* and *Huperzia squarrosa*. Tetrahedron, 70, 8017–8022.

Tomšík, P. 2014. Ferns and lycopods – a potential treasury of anticancer agents but also a carcinogenic hazard. Phytotherapy Research, 28, 798–810.

Tong X. T.; Tan, C. H.; Ma, X. Q.; Wang, B. D.; Jiang, S. H.; Zhu, D. Y. 2003. Miyoshianines A and B, two new Lycopodium alkaloids from *Huperzia miyoshiana*. Planta Medica, 69, 576–579.

Tori, M.; Nakashima, K.; Takaoka, S.; Asakawa, Y. 1994. Fusicorrugatol from the Venezuelan liverwort *Plagiochila corrugata*. Chemical and Pharmaceutical Bulletin, 42, 2650–2652.

Toyota, M.; Konoshima, M.; Asakawa, Y. 1999. Terpenoid constituents of the liverwort *Reboulia hemispherica*. Phytochemistry, 52, 105–112.

Trennheuser, F.; Burkhardt, G.; Becker, H. 1994. Anthocerodiazonin, an alkaloid from *Anthoceros agrestis*. Phytochemistry, 37, 899–903.

Tryon, R. M.; Tryon, A. F. 1982. Ferns and Allied Plants, with Special Reference to Tropical America. New York: Springer Verlag.

Uddin, M. G.; Mirza, M. M.; Pasha, M. K. 1998. The medicinal uses of pteridophytes of Bangladesh. Bangladesh Journal of Plant Taxonomy, 5, 29–41.

Vanderpoorten, A.; Goffinet, B. 2009. Introduction to Bryophytes. Cambridge, England: Cambridge University Press.

Vanderpoorten, A.; Papp, B.; Gradstein, R. 2010. Sampling of bryophytes. In Eymann J.; DeGreef J.; Häuser C.; Monje J. C.; Samyn Y.; VandenSpiegel D., eds., Manual on Field Recording Techniques and Protocols for all Taxa Biodiversity Inventories and Monitoring, pp. 340–354. Available at http://www.abctaxa.be/volumes/volume-8-manual-atbi.

Vasudeva, S. M. 1999. Economic importance of pteridophytes. Indian Fern Journal, 16, 130–152.

Wang, H. B.; Wong, M. H.; Lan, C. Y.; Qin, Y.; Shu, W.; Qiu, R.; Ye, Z. 2010. Effect of arsenic on flavonoid contents in *Pteris* species. Biochemical Systematics and Ecology, 38, 529–537.

Warmers, U.; König, W. A. 1999. Gymnomitrane-type sesquiterpenes of the liverworts *Gymnomitron obtusum* and *Reboulia hemisphaerica*. Phytochemistry, 52, 1502–1505.

Webster, T. R. 1992. Developmental problems in *Seleginella* (Selaginellaceae) in an evolutionary context. Annals of the Missouri Botanical Garden, 79, 632–647.

Wei, H. -C.; Ma, S. -J.; Wu, C. -L. 1995. Sesquiterpenoids and cyclic bisbibenzyls from the liverwort *Reboulia hemisphaerica*. Phytochemistry, 39, 91–97

Wellman, C. H.; Osterloff, P. L., Mohiuddin, U. 2003. Fragments of the earliest land plants. Nature, 425, 282–285.

Wittayalai, S.; Sathalalai, S.; Thorroad, S.; Worawittayano, P.; Ruchirawat, S.; Thasana, N. 2012. Lycophlegmariols A–D: cytotoxic serratene triterpenoids from the club moss *Lycopodium phlegmaria* L. Phytochemistry, 76, 117–123.

Wu, M. J.; Weng, C. Y.; Wang, L.; Lian, T. W. 2005. Immunomodulatory mechanism of the aqueous extract of sword brake fern (*Pteris ensiformis* Burm.). Journal of Ethnopharmacology, 98, 73–81.

Wu, Q.; Yang, X. W.; Yang, S. H. 2007. Chemical constituents of *Cibotium barometz*. Natural Products Research and Development, 19, 240–243.

Wu, X. -Z.; Chang, W. -Q.; Cheng, A. -X.; et al. 2010. Plagiochin E, an antifungal active macrocyclic bis (bibenzyl), induced apoptosis in *Candida albicans* through a metacaspase-dependent apoptotic pathway. Biochimica et Biophysica Acta, 1800, 439–447.

Wu, X. -Z.; Cheng, A. -X.; Sun, L. -M.; Lou, H. -X. 2008. Effect of plagiochin e, an antifungal macrocyclic bis (bibenzyl), on cell wall chitin synthesis in *Candida albicans*. Acta Pharmacologica Sinica, 29, 1478–1485.

Wu, X. -Z.; Cheng, A. -X.; Sun, L. -M.; Lou, H. -X. 2009. Plagiochin E, an antifungal bis (bibenzyl), exerts its antifungal activity through mitochondrial dysfunction-induced reactive oxygen species accumulation in *Candida albicans*. Biochimica et Biophysica Acta, 1790, 770–777.

Xia, X.; Cao, J.; Zheng, Y.; Wang, Q.; Xiao, J. 2014. Flavonoid concentrations and bioactivity of flavonoid extracts from 19 species of ferns from China. Industrial Crops Products, 58, 91–98.

Yoshida, T.; Toyota, M.; Hashimoto, T. 1997. Chemical constituents of Japanese *Ricciocarpos natans*. Journal of Hattori Botanical Laboratory, 81, 259–262.

Yoshikawa, H.; Ichiki, Y.; Sakakibara, K.; Tamura, H.; Suiko, M. 2002. The biological and structural similarity between lunularic acid and abscisic acid. Bioscience, Biotechnology and Biochemistry, 66, 840–846.

Zhang, G. G.; Jing, Y.; Zhang, H. M.; Ma, E.; Guan, J. L.; Xue, F. -N.; Liu, H. -X.; Sun, X. -Y. 2012a. Isolation and cytotoxic activity of selaginellin derivatives and biflavonoids from *Selaginella tamariscina*. Planta Medica, 78, 390–392.

Zhang, L. B.; Wang, P. S.; Wang, X. Y. 2012b. *Selaginella longistrobilina* (Selaginellaceae), a new species from Guizhou, China, and *Selaginella prostrata*, a new combination and its lectotypification. Novon, 22, 260–263.

Zhang, M.; Cao, J.; Dai, X.; Chen, X.; Wang, Q. 2012c. Flavonoid contents and free radical scavenging activity of extracts from leaves, stems, rachis and roots of *Dryopteris erythrosora*. Iranian Journal of Pharmacological Research, 11, 991.

Zhang, X. Q.; Kim, J. H.; Lee, G. S.; Pyo, H. B.; Shin, E. Y.; Kim, E. G.; Zhang, Y. H. 2012d. In vitro antioxidant and in vivo anti-inflammatory activities of *Ophioglossum thermale*. The American journal of Chinese Medicine, 40, 279–293.

Zhou, H.; Jiang, S. H.; Tan, C. H.; Wang, B. D.; Zhu, D. Y. 2003a. New epoxyserratanes from *Huperzia serrata*. Planta Medica, 69, 91–94.

Zhou, H.; Tan, C. H.; Jiang, S. H.; Zhu, D. Y. 2003b. Serratene-type Triterpenoids from *Huperzia serrata*. Journal of Natural Products, 66, 1328–1332.

Zotchev, S. B.; Sekurova, O. N.; Katz, L. 2012. Genome-based bioprospecting of microbes for new therapeutics. Current Opinion in Biotechnology, 23, 941–947.

8 Chemical and Functional Properties of Amazonian Fruits

Elaine Pessoa, Josilene Lima Serra,
Hervé Rogez, and Sylvain Darnet
Centre for Valorization of Amazonian Bioactive Compounds
& Federal University of Pará, Belém, Pará, Brazil

CONTENTS

8.1   INTRODUCTION

As suggested by the last Food and Agriculture Organization (FAO) of the United Nations the projections for the next 20 years will be the key to food security and nutrition on a worldwide scale and a better use of biodiversity (FAO, 2017). The challenge is to develop more sustainable cultures in the context of global climate change for a growing urban population with a changing diet (Choudhury and Headey, 2017; FAO, 2018). Approximately 50% of all the calories consumed are from only three cereal crops, and the genetic erosion of crops and the loss of livestock, forest and aquatic sources are decreasing rapidly (FAO, 2018). The value of plant diversity is crucial. The list of edible plants provides approximately 30,000 species, but only 30 of these plants form the basis of human nutrition. The neglected and underutilized species should help to diversify nutritional sources (FAO, 2017). The new sources should supply food or calories, but a significant expectation is to increase the nutritional values in the diet, including a high content of vitamins, minerals and other micronutrients (FAO, 2018). The food diversification should include sustainability to preserve the biodiversity and the ecological foundations necessary to sustain life and rural livelihoods (FAO, 2018). One agricultural system that is environmentally friendly is agroforestry (FAO, 2018). Agroforestry has many advantages, such as the maintenance of water and soil quality, carbon sequestration, habitat provision for wild species and the facilitation of biological pest control and pollination (FAO, 2018). Another crucial aspect is to increase food security and to mitigate the over representation and use of annual plants in actual agricultural systems (Meldrum et al., 2018). Perennial crops should represent a good alternative for human nutritional diversification. Some neglected and underutilized types of plants with high nutritional value are trees bearing fruit from tropical regions in Africa, Asia, Central and South America (Meldrum et al., 2018).

The Amazonia rainforest, particularly the eastern region, is highly enriched in edible plants, such as fruits, and the legacy of 4500 years of polyculture agroforestry by the pre-Columbian population and biodiversity exploration (Maezumi et al., 2018). With the exception of two fruits, Brazilian nuts and açaí, the Amazon fruits are neglected and underutilized; however, over the last 30 years, scientific studies have highlighted the high nutritional value and medicinal properties of this legacy (Dutra et al., 2016; Neri-Numa et al., 2018; Oliveira et al., 2012). This chapter presents the functional studies from a list of valuable fruits of the Amazonian region. This list of valuable and underused fruit is very long, yet only few species have already been functionally explored. For example, *Theobroma grandiflorum* (cupuaçu), *Platonia insignis* (bacuri) and *Endopleura uchi* (uxi) are not included due to the lack of characterization and functional studies. *Theobroma cacao* is excluded from this chapter, although it originated in

FIGURE 8.1 Fruit of four Amazonian trees: (A) Brazil nut (*B. excelsa*), (B) genipap (*G. americana*), (C) camu-camu (*M. dubia*) and (D) yellow mombin (*S. mombin*).

the Amazon, because the functional studies focus on the byproduct, cocoa and chocolate, and the research was performed outside of the native South American region. In the first part of this chapter, we describe the functional research of the fruit of four trees, *Spondias mombin, Myrciaria dubia, Genipa americana* and the well-known Brazilian nut (*Bertholletia excelsa*) (Figure 8.1) (Table 8.1). In addition, palm trees are a primary element of the Amazonian landscape and an essential plant for the local population, and studies on four palm tree fruits are described, *Astrocaryum vulgare, Mauritia flexuosa, Bactris gasipaes* and the well-known açaí (*Euterpe oleracea*) (Figure 8.2) (Table 8.1) (Brokamp et al., 2011; Paniagua-Zambrana et al., 2015; Sosnowska and Balslev, 2009).

The phytochemistry of each fruit is described with the emphasis on bioactive compounds (Table 8.2), and an updated review shows the functional and medicinal properties using an

TABLE 8.1
Fruits from Amazonian Biome with Functional and Chemical Properties Studied in This Chapter

Scientific Name	Family	Common Names
Astrocaryum vulgare (*sin. Astrocaryum tucumoides*)	*Arecaceae*	Tucumã, awarra palm
Bactris gasipaes (sin. *Bactris dahlgreniana*)	*Arecaceae*	Pupunha, peach palm
Bertholletia excelsa (sin. *Bertholletia nobilis*)	*Lecythidaceae*	Castanha-do-Pará, Brazil nut
Euterpe oleracea (sin. *Euterpe cuatrecasana*)	*Arecaceae*	Açaí, Assai palm
Genipa americana (sin. *Genipa venosa*)	*Rubiaceae*	Jenipapo, genipap
Mauritia flexuosa (sin. *Mauritia minor*)	*Arecaceae*	Buriti, moriche palm
Myrciaria dubia (sin. *Myrciaria paraensis*)	*Myrtaceae*	Camu-camu
Spondias mombin (sin. *Myrobalanus lutea*)	*Anacardiaceae*	Cajá/Taperebá, yellow mombin

FIGURE 8.2 Fruit of four Amazonian palm trees: (A) açaí (*E. oleracea*), (B) tucumã (*A. vulgare*), (C) buriti (*M. flexuosa*) and (D) peach palm (*B. gasipaes*).

TABLE 8.2
Bioactive Compounds in Amazonian Fruit

Fruit	Main Bioactive Compound	Bioactive Compounds
Bertholletia excelsa	Selenium	**Nut (1-3)**
		Phenolic compounds: Gallic acid, gallocatechin protocatechuic acid, catechin vanillic acid,
		taxifolin, myricetin, ellagic acid, quercetin, tannins
		Fatty acids: (C12:0), (C14:0), (C16:0), (C16:1), (C18:0), (C18:1), (C18:2) and (C20:4)
		Sterols: Campesterol, stigmasterol, sitosterol
		Tocopherols: α-tocopherol, β-tocopherol, γ-tocopherol
		Minerals: Se, Mg and P
Genipa americana	Genipin (Blue pigment)	**Unripe fruit (4, 5)**
		Iridoids: Genipin, geniposide
		Ripe fruit (4, 6, 7)
		Iridoids: Gardoside, geniposidic acid, genipin-1-β-D-gentiobioside, caffeoyl geniposidic acid, p-coumaroyl geniposidic acid, feruloylgardoside, feruloylgenipin gentiobioside, genipacetal, genipamide, genipaol, genameside
		Phenolic compounds: Dicaffeoylquinic acids, 3,5-dicaffeoylquinic acid, 4,5-dicaffeoylquinic
		acid and 5-caffeoylquinic acid, quercetin, leucoanthocyanidins, catechins, flavanones
		Others: Anthraquinones, anthrone, coumarins, triterpenoids, steroids

(Continued)

TABLE 8.2 *(Continued)*
Bioactive Compounds in Amazonian Fruit

Fruit	Main Bioactive Compound	Bioactive Compounds
Myrciaria dubia	Vitamin C	**Leaves (8)** *Phenolic compounds:* Flavonoids, tannins **Whole fruit (9-12)** *Phenolic compounds:* Gallic acid, ferulic acid, *p*-coumaric, protocatechuic acid, catechin, Myricetin, Rutin, narigerin *Carotenoids:* β-carotene, all-*trans*-lutein, zeaxanthin. luteoxanthin, violaxanthin, neoxanthin **Pulp (11-15)** *Phenolic compounds:* Ellagitannin B, kaempferol, quercetin *Organic acids:* Ascorbic acid **Seed (16)** Betulinic acid **Seed coat (17)** Resveratrol
Spondias mombin	Carotenoids	**Leaves (18)** *Phenolic compounds:* Ellagic acid and chlorogenic acid *Sterol:* Sitosterol **Pulp (18-20)** *Carotenoids:* β-carotene, α-carotene, lutein, zeinoxanthin, β-criptoxanthin *Phenolic* compounds: Total phenolics, flavonoids
Astrocaryum vulgare	Carotenoids	**Pulp (21, 22, 42, 43)** *Fatty acids:* (C18:1), (C16:0), (C18:0), (C18:2) *Tocopherols:* α-tocopherol *Carotenoids:* β-carotene, α-carotene, γ-carotene, ζ-carotene, δ-carotene, lutein, phytoene *Phytosterols:* β-sitosterol, cycloartenol, arundoin **Kernel (40)** *Fatty acids:* (C12:0), (C14:0), (C18:1), (C18:2) *Phytosterols:* β-sitosterol, Δ5-avenasterol, campesterol
Bactris gasipaes	Carotenoids	**Mesocarp (23, 24, 27, 42-44)** *Fatty acids:* (C18:1), (C16:0) *Tocopherols:* α-tocopherol *Carotenoids:* β-carotene, α-carotene, δ-carotene, γ-carotene, lycopene, *Phytosterols:* β-sitosterol, fucosterol **Oil kernel (41)** *Fatty acids:* (C12:0), (C14:0) *Phytosterols:* β-sitosterol, Δ5-avenasterol, campesterol *Tocopherols:* α-tocopherol
Euterpe oleracea	Anthocyanins	**Pulp (45-55)** *Fatty acids:* (C18:1), (C16:0), (C18:2) *Carotenoids:* Lutein, β-carotene, α-carotene

(Continued)

TABLE 8.2 *(Continued)*

Bioactive Compounds in Amazonian Fruit

Fruit	Main Bioactive Compound	Bioactive Compounds
		Phytosterols: β-sitosterol
		Tocopherols: α-tocopherol
		Phenolic compounds: Cyanidin 3-*O*-glucoside, cyanidin 3-*O*-rutinoside, homoorientin, orientin, isovitexin, (–)-epicatechin, (+)-catechin, scoparin, taxifolin deoxyhexose, procyanidin, *p*-hydroxybenzoic acid, ferulic acid, ferulic acid, gallic acid, protocatechuic acid, ellagic acid, vanillic acid
		Seed (56-58)
		Phenolic compounds: Catechin, polymeric proanthocyanidins
Mauritia flexuosa	Carotenoids	**Pulp Oil (22, 26-39, 42-44)**
		Fatty acids: (C18:1), (C16:0), (C18:2)
		Tocopherols: β-tocopherol, α-tocopherol, γ-tocopherol
		Vitamins: A and E
		Carotenoids: β-carotene, α-carotene, lutein
		Phytosterols: β-sitosterol, stigmasterol, campesterol
		Pulp extract (32)
		Phenolic compounds: Protocatechuic, chlorogenic acid, caffeic acid, (+)-catechin, (–)-epicatechin, apigenin, luteolin, myricetin, kaempferol and quercetin
		Leaf extract (32)
		Phenolic compounds: Caffeic acid hexoside, naringenin, (–)-epicatechin, vitexin, scoparin, cyanidin-3*O*-rutinoside and cyanidin-3-*O*-glucoside
		Trunk extract (32)
		Phenolic compounds: Caffeic acid hexoside, naringenin, (–)-epicatechin and kaempferol

Source: (1) Jonh and Sahidi (2010); (2) Chunhieng et al. (2008); (3) Yang (2009); (4) Bentes and Mercadante (2014); (5) Náthia-Neves et al. (2017); (6) Ono et al. (2005); (7) Ono et al. (2007); (8) Alves et al. (2017); (9) Akter et al. (2011); (10) Zanatta and Mercadante (2007); (11) Rodrigues-Amaya et al. (2008); (12) Anhê et al. (2018); (13) Fujita et al. (2017); (14) Neri-Numa et al. (2018); (15) Genovese et al. (2008); (16) Yazawa et al. (2011); (17) Fidelis et al. (2018); (18) Cabral et al. (2016); (19) Zielinski et al. (2014); (20) Tiburski et al. (2011); (21) Bony et al. (2012); (22) Rodrigues et al. (2010); (23) Hempel et al. (2014); (24) Quesada et al. (2011); (25) Yuyama et al. (2003); (26) Santos et al. (2013); (27) Santos et al. (2013); (28) Silva et al. (2011); (29) Darnet et al. (2011); (30) Aquino et al. (2012); (31) Costa et al. (2010); (32) Koolen et al. (2013); (33) Cândido et al. (2015); (34) Aquino et al. (2015); (35) Manhães et al. (2015); (36) Lima et al. (2009); (37) Bataglion et al. (2014); (38) Speranza et al. (2016); (39) Medeiros et al. (2015); (40) Bereau et al. (2003); (41) Radice et al. (2014); (42) de Rosso and Mercadante (2007); (43) Santos et al. (2015); (44) Rojas-Garbanzo et al. (2011); (45) Schauss et al. (2006); (46) Mulabagal and Calderon (2012); (47) Gordon et al. (2012); (48) Dias et al. (2013); (49) Rogez et al. (2011); (50) Dias et al. (2012); (51) Carvalho et al. (2017); (52) Bichara and Rogez (2011); (53) Costa et al. (2010); (54) Ribeiro et al. (2018); (55) Romualdo et al. (2015); (56) de Bem et al. (2014); (57) de Moura et al. (2011); (58) de Oliveira et al. (2015).

in vitro model (Table 8.3) and an *in vivo* model with animals (Table 8.4) and humans (Table 8.5). Many beneficial effects have already been clearly demonstrated, including in humans, and the value of these fruits for the diversification of human nutrition. It is time to (re)discover Amazonian hidden treasures!

TABLE 8.3

In Vitro Studies Performed with Amazonian Fruit or Fraction

Scientific Name	Source	Observations	References
Bertholletia excelsa H.B.K.			
	Nuts	Allergic symptoms (vomiting, diarrhea and loss of consciousness) ↑ Level of specific IgE	Bartolomé et al. (1997)
	Nuts	Allergic symptoms after ingestion Brazil nut (anaphylactic shock and laryngeal edema) ↑ Level of specific IgE	Pastorello et al. (1998)
	Nuts	Incidence of specific IgE to Brazil nut in patients of different ages and sex	Pumphrey et al. (1999)
	Nuts	↑ Reception of cholesteryl esters by the HDL	Strunz et al. (2008)
	Nuts	↑ Plasma selenium level ↑ Selenium levels	Maranhão et al. (2011)
	Nuts	↓ Total cholesterol and LDL-c ↑ Plasma selenium levels after diet supplementation	Stockler-Pinto et al. (2012)
	Nuts	↑ Increase in HLD concentrations Improvement of the Castelli I and II indexes	Cominetti et al. (2012)
	Nuts	↑ Plasma selenium levels and HDL-c	Colpo et al. (2013)
	Nuts	↑ Plasma Se and GPx activity ↑ HDL-c levels ↓ Cytokines, 8-OHdG and 8-isoprostane plasma ↓ LDL-c levels	Stockler-Pinto et al. (2014)
	Partially defatted nut flour	↓ In serum total cholesterol and non-HDL-c levels	Carvalho et al. (2015)
	Nuts	↑ Plasma Se levels, rectal selenoprotein P (SePP) and β-catenin mRNA	Hu et al. (2016)
	Nuts	↑ GPX1 mRNA expression only in subjects with CC	Donadio et al. (2017)
Genipa americana L.			
	Fruit	None cytotoxicity or interference in cell differentiation ↑ Antitumor effect on choriocarcinoma-derived cells ↑ Effects on trophoblast metabolism through the MAPK pathway	Da Conceição et al. (2011)
	Leaf, fruit and peel	↑ Tyrosinase inhibitory activity	Souza et al. (2012)
	Peel, pulp and seed	↑ Antioxidant activity of extracts ↑ Acetylcholinesterase inhibition by thin layer chromatography ↑ Lipid peroxidation in membrane	Omena et al. (2012)
	Seed coat	↑ Antihypertensive activity	Fidelis et al. (2018)
	Fruit	↑ Stability at 12–20 °C and low pH (3.0–4.0) ↑Antioxidant capacity on digestion	Neri-Numa et al. (2018)
	Leaf	Cell death of epimastigote, trypomastigote and amastigote forms of *Trypanosoma cruzi*	Souza et al. (2018)
Myrciaria dubia (Kunth) McVaugh.			
	Pulp	↑ Levels of phenolics, ascorbic acid, proanthocyanidins, antioxidant and antimicrobial activity	Fujita et al. (2015)

(Continued)

TABLE 8.3 *(Continued)*
In Vitro Studies Performed with Amazonian Fruit or Fraction

Scientific Name	Source	Observations	References
	Pulp	↑ Properties antihyperglycemia and antimicrobial ↑ Cellular rejuvenation	Fujita et al. (2015)
	Peel and seeds	↑ Antimicrobial activities of the extracts (Acylphloroglucinol and rhodomyrtone and acylphloroglucinols)	Kaneshima et al. (2017)
	Seed coat	↑ Total phenolic content, antioxidant activity and lipid oxidation	Fidelis et al. (2018)
	Camu-camu extracts	↑ Antihypertensive activity and angiotensin-converting enzyme inhibitory activity	Azevedo et al. (2018)
Spondias mombin L.			
	Leaf	↓ Larval development on nematode eggs from sheep	Ademola et al. (2005)
	Leaf	Not activity against promastigotes of Leishmania chagasi ↑ Activity against amastigotes	Accioly et al. (2012)
	Leaf	↑ Inhibition of leukocyte migration on inflammation site	Cabral et al. (2016)
Astrocaryum vulgare Mart.			
	Unsaponifiable fraction of pulp oil	Anti-inflammatory properties with inhibition expression of COX-2, decreased NO, prostaglandin E_2, TNF-α, and IL-6 and -10 production	Bony et al. (2012)
Bactris gasipaes Kunth.			
	Mesocarp	Protective effects of lipid peroxidation on liver homogenates of rats with TBHP	Quesada et al. (2011)
	Peel, pulp and seeds oil	Antimicrobial against *Staphylococcus aureus*	Araújo et al. (2012); Araújo et al. (2013)
Euterpe oleracea Mart.			
	Freeze-dried açaí pulp powder	↑ Serum sulfhydryl groups and ↓ ROS formation in PMN cell	Kang et al. (2010)
	Concentrate açaí juice	Anti-lipidemic and anti-inflammatory effects on 3T3-L1 mouse adipocytes	Martino et al. (2016)
	Velutin isolated from açaí pulp	LPS-induced pro-inflammatory and IL-6 production in macrophages	Xie et al. (2012)
	Fruit pulp fractions	Attenuate inflammatory stress signaling on BV-2 mouse microglial cells	Poulose et al. (2012)
	Antochyanin-rich açaí extract	Antioxidant properties, antiproliferative properties on C-6 rat cell and no effect on the growth of MDA-468 human breast cancer cells	Hogan et al. (2010)
	Bark, seed and total açaí fruit	Cytotoxic effects in malignant cells lines	Silva et al. (2014) Barros et al. (2015)
	Extract of açaí seed	↓ Cell viability and necroptosis in the MCF-7 cell	Freitas et al. (2017)
	Açaí oil	Without genetic toxicity in rat cells	Marques et al. (2016)
	Polysaccharides isolated from açaí fruit	Induction of innate immune response	Holderness et al. (2011)

(Continued)

TABLE 8.3 *(Continued)*
In Vitro Studies Performed with Amazonian Fruit or Fraction

Scientific Name	Source	Observations	References
	Frozen fruit pulp	Antioxidant activity of H_2O_2 in the cerebral cortex, hippocampus and cerebellum of rats	Spada et al. (2009)
	Açaí extract	Antioxidant potential and protected human neuron-like cells (SH-SY5Y)	Torma et al. (2017)
	Açaí extract	Attenuates Mn-induced oxidative stress on rat astrocytes	da Silva Santos et al. (2014)
	Açaí stone extract	Vasodilator effect dependent on activation of NO-cGMP pathway in rat Mesenteric vascular	Rocha et al. (2007)
	Açaí extract	Protection in human vascular endothelial cells against glucose-induced oxidative stress and inflammation	Noratto et al. (2011)
	Açaí extract	Neuroprotective effects against β-amyloid exposure on CP12 rat cell	Wong et al. (2013)
	Açaí extract added into the sugar–yeast medium	↑ Transcript level of l(2)ef l(2)ef, GstD1 and MtnA ↓ Transcript level Pepck ↑ Lifespan of oxidative-stressed females caused by sod1 RNAi	Sun et al. (2010)
	Açaí extract added into yeast medium	Dampening stress-induced expression of the GSTD1 and eliminates paraquat induced circadian rhythm deficits	Vrailas-Mortimer et al. (2012)
	Açaí pulp	Suppressed IgE-mediated degranulation and transcription of the cytokine genes	Horiguchi et al. (2011)
	Açaí pulp	↓ *Bacteroides–Prevotella* spp. and the *Clostridium histolyticum* groups ↑ Short-chain fatty acids	Alqurashi et al. (2017)
	Aqueous açaí extract	↓ ROS; ↑ Polyglutamine protein aggregation; Prevention sulfhydryl level; Activation of gcs-1; ↓ Proteasome activity	Bonomo et al. (2014)
	Anthocyanin-rich extract	↓ ROS via DAF-16/FOXO translocation	Peixoto et al. (2016)
	Açaí berry extract	↓ Osteoclastogenesis and osteoclast activity; ↓ (IL)-1α, -6 and TNF-α; ↑ IL-3, -4, -13 and IFN-γ	Brito et al. (2016)
	Methanol/water açaí extract	Malvidin and cyanidin: Protects BALB/3T3 cells against UVA irradiation; Interferes with ROS generation; Keeps intracellular GSH; Lipid peroxidation close to normal cellular levels	Petruk et al. (2017)
	Açaí berry water extracts	↑ Migration of HS68 fibroblast cells; ↑ mRNA fibronectin expression.; ↑ mRNA fibronectin expression; ↓ mRNA MMP-1 expression	Kang et al. (2017)
Mauritia flexuosa L.f.			
	Buriti oil	Low cytotoxicity on HaCat and 3T3 cell lines humans and mice respectively	Zanatta et al. (2008)
	Buriti oil	Photoprotective potential in after sun formulations	Zanatta et al. (2010)
	Buriti peel oil extract	Antiplatelet/Antithrombotic activities	Fuentes et al. (2013)
	Peel, pulp and endocarp	Chemopreventive potentialities of fruits and their by-products; Peels with higher quantities of bioactive	Pereira-Freire et al. (2018)

(Continued)

TABLE 8.3 *(Continued)*

In Vitro Studies Performed with Amazonian Fruit or Fraction

Scientific Name	Source	Observations	References
	Leaves, trunk and green fruits extracts	Moderate antimicrobial activity from leaf extract against *Staphylococcus aureus* and *Pseudomonas aeruginosa*	Koolen et al. (2013)
	Extracts of stems, leaves and fruits	Stems and leaf extract inhibition the growth of *S. aureus* and antiproliferative activity on human tumor cell lines	Siqueira et al. (2014)

IgE: immunoglobulin E, HDL-c: high-density lipoprotein cholesterol, LDL-c: low-density lipoprotein cholesterol, GPx: glutathione peroxidase, mRNA: messenger ribonucleic acid, COX-2: cyclooxygenase-2, NO: nitric oxide, TNF-α: tumor necrosis factor-α, IL: interleukin, TBHP: tert-butyl hydroperoxide, PMN: neutrophil numbers, BALB/3T3: mouse embryonic fibroblast cell line, BV-2: microglial cells, LPS: lipopolysaccharide, MDA: malondialdehyde, embryonic fibroblast cell line, MCF-7: human breast adenocarcinoma cell line, SH-SY5Y: neuron-like cells, l(2)ef lethal (2): essential for life, MtnA: metallothionein A, Pepck: phosphoenolpyruvate carboxykinase, SOD: superoxide dismutase, RNAi: RNA de interferência, GstD1: glutathione S transferase D1, ROS: reactive oxygen species, GCS: glutamylcysteine synthetase, DAF-16: transcription factor required in lifespan extension in mutation of the insulin-like receptor daf-2, FOXO: Forkhead box protein O, HS68: human foreskin fibroblast, GSH: reduced glutathione, MMP: metalloproteinase.

TABLE 8.4

In Vivo Studies Performed with Amazonian Fruit or Fraction

Scientific Name	Source	Observation	References
Bertholletia excelsa H.B.K.			
	Stem	Trypanocidal activity and antioxidant activity	Campos et al. (2005)
Genipa americana L.			
	Leaf, Fruit and Peel	Tyrosinase inhibitory activity	Souza et al. (2012)
	Peel, pulp and seed	Lipid peroxidation membrane model in rats liver	Omena et al. (2012)
	Leaf	Neuroprotective effect in the brain morphology and oxidative markers mice behavioral models	Nonato et al. (2018)
Myrciaria dubia (Kunth) McVaugh.			
	Seed	↑ Suppression of paw edema formation	Yazada et al. (2011)
	Pulp	Assessment the antioxidant, genotoxic and antigenotoxic potential on blood cells of mice after acute, subacute and chronic treatments	da Silva et al. (2012)
	Pulp	↓ White adipose tissues, glucose, total cholesterol, triglycerides, LDL-c and insulin blood levels; ↑ HDL-c levels Not inflammatory markers and liver enzymes	Nascimento et al. (2013)
	Pulp	↑ Plasma antioxidant activity; ↓ triacylglycerol and total cholesterol	Gonçalves et al. (2014)
	Fruit	Improves the immune response and growth in Nile tilapia (*Oreochromis niloticus*)	Yunis-Aguinaga et al. 2016
	Pulp	Physiological parameters of tambaqui fed with proportion difference of camu-camu ↑ Cortisol, glucose, proteins and triglycerides	Aride et al. (2018)
	Fruit	↑ Prevention diet-induced obesity and ameliorate	Anhê et al. (2018)
	Fruit	↑ Colon protective effects against DMH damage ↓ DXR mutagenicity effect	Azevedo et al. (2018)

(Continued)

TABLE 8.4 *(Continued)*

In Vivo Studies Performed with Amazonian Fruit or Fraction

Scientific Name	Source	Observation	References
	Fruit	↓ Induced neurodegeneration in transgenic *Caenorhabditis elegans*	Azevêdo et al. (2015)
	Seed coat	↑ Inhibition of lipid oxidation induced by lipid peroxidation in rats	Fidelis et al. (2018)
Spondias mombin L.			
	Leaf	↑ Anthelmintic activity on rats infected with gastrointestinal nematode	Ademola et al. (2005)
	Leaf	↑ Anti-inflammatory activity in rats using intraplantar injection of carrageenan	Nworu et al. (2011)
	Leaf	Gastroprotective effects against indomethacin-induced gastric ulcer in rats	Sabiu et al. (2016)
	Leaf	Methyl methane sulfonate (MMS) induced genotoxicity in rats	Oyeyemi et al. (2015)
	Leaf	Anti-inflammatory activity on model carrageenan-induced peritonitis in mice	Cabral et al. (2016)
	Leaf	Detoxification of hepatic and macromolecular oxidants in acetaminophen-intoxicated rats	Saheed et al. (2017)
	Leaf	Gastric lesion models induced by absolute ethanol and indomethacin in rats	Brito et al. (2018)
Astrocaryum vulgare Mart.			
	Pulp oil	Anti-inflammatory properties in a mice model of endotoxic shock and a rat model of pulmonary inflammation ↓ Pro-inflammatory cytokines ↑ Anti-inflammatory cytokines	Bony et al. (2012)
	Unsaponifiable fraction of pulp oil	Anti-inflammatory properties on mice model of endotoxic shock ↑ TNF-α, IL-6 and IL-10 serum concentration	Bony et al. (2014)
	Tucuma oil	Hypoglycemic effect in alloxan-induced type 1 diabetic mice	Baldissera et al. (2017)
	Tucuma oil	Protective effect on memory, enzymatic activities (Na^+, K^+-ATPase) and AChE in the brain of alloxan-induced diabetic mice	Baldissera et al. (2017)
	Tucuma oil	Protective action against diabetes in the alloxan-induced diabetic mice Improves the immune system; Changes in the purinergic system	Baldissera et al. (2017)
	Tucuma oil	Effective protection against lipid oxidative damage in the liver tissue of diabetic mice Improved the enzymatic antioxidant defense system	Baldissera et al. (2018)
Bactris gasipaes Kunth.			
	Ration made with fruit	Bioavailability of vitamin A in rat liver	Yuyama et al. (1991)
	Diet with peach palm	↑ Concentration of vitamin A in the liver of the rats and bioavailable of zinc in the femurs	Yuyama and Cozzolino (1996)
	Pulp	↓ Body weight, total cholesterol in lactating rats	Carvalho et al. (2013)

(Continued)

TABLE 8.4 *(Continued)*

In Vivo Studies Performed with Amazonian Fruit or Fraction

Scientific Name	Source	Observation	References
Euterpe oleracea Mart.			
	ASE	Protective effect against emphysema in mice	de Moura et al. (2011)
	ASE	Potential reduction the inflammatory and oxidant actions of cigarette smoke in the mouse	Moura et al. (2012)
	Diet with açaí	Promotes healthy aging in SOD1-deficient female flies; ↓ 4-hydroxynonenal-protein adducts	Laslo et al. (2013)
	Açaí juice	Antioxidant and anti-inflammatory activities in ApoE deficient mice	Xie et al. (2011)
	Seed extract	Cardioprotective effects on rats subjected to myocardial infarction	Zapata-Sudo et al. (2014)
	Seed extract	Prevents vascular remodeling and endothelial dysfunction in spontaneously hypertensive rats	Cordeiro et al. (2015)
	ASE	Prevent hypertension in rats ↓ acetylcholine-induced vasodilation, ↓ MDA, carbonyl protein, SOD, CAT, GPx, SOD1, SOD2, eNOS and TIMP-1; Prevented vascular remodeling ↑ MMP-2	da Costa et al. (2012)
	Açaí stone	↓ Growth and survival of endometriotic lesions in female Sprague-Dawley rats	Machado et al. (2016)
	Açaí pulp	Protection against DXR-induced DNA damage in liver and kidney cells on rats	Ribeiro et al. (2010)
	Açaí pulp	Rat liver: ↑ mRNA levels for γ-GCS and GPx in hepatic tissue ↓ ROS by neutrophils	Guerra et al. (2011)
	ASE	Improved serum levels of urea, creatinine and Na excretion	de Bem et al. (2014)
	Açaí freeze-dried	Protective effect on chronic alcoholic hepatic injury in rats	Qu et al. (2014)
	Açaí extract	Attenuates glycerol-induced acute renal failure in rats; ↓ serum urea and serum; ↓ glutathione levels	Unis (2015)
	ASE	Protect mice from diet-induced obesity and fatty liver by regulating hepatic lipogenesis and cholesterol excretion	de Oliveira et al. (2015)
	Açaí extract	Attenuation of renal ischemia/reperfusion injury in a rat model; ↓ MDA and IFN-γ ↑ IL-10	El Morsy et al. (2015)
	Frozen açaí pulp	Attenuates hepatic steatosis via adiponectin-mediated effects on lipid metabolism in high-fat diet mice	Guerra et al. (2015)
	ASE	↓ Renal injury and prevented renal dysfunction in 2K1C rats	da Costa et al. (2017)
	ASE	Protects against renal injury in diabetic and spontaneously hypertensive rats	da Silva Cristino Cordeiro et al. (2018)
	Spray dried açaí powder	Inhibits the TCC development in male Swiss mice, due to its potential antioxidant action cells	Fragoso et al. (2012)
	Spray dried açaí powder	Protective effect on colon carcinogenesis on rat; ↓ number of invasive tumors; ↓ tumor Ki 67 cell proliferation and net growth index	Fragoso et al. (2013)
	Spray dried açaí powder	Protective effect on colon carcinogenesis in rat	Romualdo et al. (2015)

(Continued)

TABLE 8.4 *(Continued)*

In Vivo Studies Performed with Amazonian Fruit or Fraction

Scientific Name	Source	Observation	References
	Açaí oil	↓ Tumor in comparison to rat control group	Monge-Fuentes et al. (2017)
	ASE	Antiangiogenic and anti-inflammatory potential; Inhibition of DMBA carcinogenicity in breast cancer in female Wistar	Alessandra-Perini et al. (2018)
	ASE	↓ Plasma malondialdehyde levels, body weight, plasma triglyceride, total cholesterol, glucose levels and insulin resistance in mice fed a high-fat diet	de Oliveira et al. (2010)
	Açaí puree	Anti-atherosclerotic and anti-diabetic activity in hypercholesterolemic zebrafish	Kim et al. (2012)
	Açaí pulp	Hypocholesterolemic effect in rats; ↓ LDL, protein carbonylation and sulfhydryl groups; ↑ SOD ↑ paraoxonase activity	de Souza et al. (2010)
	Açaí pulp	Group HA of female fischer rats; ↓ serum total cholesterol, LDL and atherogenic index; ↑ HDL	de Souza et al. (2012)
	Açaí extract	Rabbits with diet-induced hypercholesterolemic: ↓ total cholesterol, non-HDL and TG; ↓ atherosclerotic plaque area in aorta; ↓ desmosterol/campesterol and desmosterol/β-sitosterol	Feio et al. (2012)
	Açaí fruit	Prevents electrophysiological deficits and oxidative stress induced by methyl-mercury in the rat retina	Brasil et al. (2017)
	Pulp extract	Restoration of stressor-induced calcium dysregulation and autophagy in rat; Recovery of depolarized brain cells from dopamine-induced Ca^{2+} influx	Poulose et al. (2014)
	Clarified açaí juice	Prevention MDA in the cerebral cortex in mice	Souza-Monteiro et al. (2015)
	ASE	Prevented chronic pain in a rat spinal nerve ligation model; Prevented thermal hyperalgesia and mechanical allodynia	Sudo et al. (2015)
	Açai frozen pulp	↑ Pro-inflammatory cytokines levels in different brain areas of Wistar rats	de Souza Machado et al. (2015)
	Freeze-dried açaí powder	↓ NOX2, NF-κB, ROS; ↑ Nrf2 hippocampus and frontal cortex of rats	Poulose et al. (2017)
	Lyophilized açaí fruit pulp	Improves cognition in aged rats; ↑ working memory in the Morris water maze; ↓ NO, TNF-α	Carey et al. (2017)
Mauritia flexuosa L.f.	Crude and refined buriti oil	Hypocholesterolemic effect in rats RB: ↓ total cholesterol, LDL, triglycerides and enzyme aspartate transaminase RB and CB: ↑ serum and hepatic retinol and tocopherol	Aquino et al. (2015)
	Cookies made with buriti oil	↑ β-carotene and monounsaturated fatty acids in the cookies ↑ serum and hepatic retinol levels in rats fed BOC and correlation serum retinol contents with hepatic retinol ↓ Total and LDL-c in rats fed BOC	Aquino et al. (2016)

(Continued)

TABLE 8.4 *(Continued)*

In Vivo Studies Performed with Amazonian Fruit or Fraction

Scientific Name	Source	Observation	References
	Crude and refined buriti oil	↑ Serum and liver retinol deposition among neonatal rats fed with CB and RB	Medeiros et al. (2015)
	Fruit pulp	Effect of MeHg on rat aversive memory acquisition and panic-like behavior	Leão et al. (2017)

HDL-c: high-density lipoprotein cholesterol, DMH: 1,2-dimethylhydrazine dihydrochloride, TNF-α: tumor necrosis factor-α, IL: interleukin, AChE: acetylcholinesterase, ASE: açaí stone extract, SOD: superoxide dismutase, ApoE: apolipoprotein E deficient (apoE–/–) mice, MDA: malondialdehyde, CAT: catalase, GPx: glutathione peroxidase, eNOS: endothelial constitutive nitric oxide synthase, TIMP-1: metallopeptidase inhibitor 1, MMP: metalloproteinase, DXR: doxorubicin, mRNA: messenger ribonucleic acid, γ-GCS: gamma-glutamylcysteine synthetase, ROS: reactive oxygen species, IFN-γ: interferon-gamma, TCC: transitional cell carcinoma, DMBA: 7,12-dimethylbenzanthracene, LDL-c: low-density lipoprotein cholesterol, Group HA: hypercholesterolemic diet supplemented with 2% açaí pulp, TG: triacylglycerol, NOX NADPH: (nicotinamide adenine dinucleotide phosphate)-oxidoreductase, NF-κB: nuclear factor κB, Nrf2: nuclear factor E2-related factor 2, NO: nitric oxide, RB: refined buriti oil, CB: crude buriti oil, BOC: cookies made with buriti oil, CB: crude buriti oil.

TABLE 8.5

Functional Characterization of Amazonian Fruit Performed in Human Studies

Scientific Name	Source	Observations	References
Bertholletia excelsa H.B.K			
	Nuts	Allergic symptoms (vomiting, diarrhea and loss of consciousness) ↑ Level of specific IgE	Bartolomé et al. (1997)
	Nuts	Allergic symptoms after ingestion Brazil nut (anaphylactic shock and laryngeal edema) ↑ Level of specific IgE	Pastorello et al. (1998)
	Nuts	Incidence of specific IgE to Brazil nut in patients of different ages and sex	Pumphrey et al. (1999)
	Nuts	↑ Reception of cholesteryl esters by the HDL	Strunz et al. (2008)
	Nuts	↑ Plasma selenium level ↑ Selenium levels	Maranhão et al. (2011)
	Nuts	↓ Total cholesterol and LDL-c ↑ Plasma selenium levels after diet supplementation	Stockler-Pinto et al. (2012)
	Nuts	↑ Increase in HLD concentrations Improvement of the Castelli I and II indexes	Cominetti et al. (2012)
	Nuts	↑ Plasma selenium levels and HDL-c	Colpo et al. (2013)
	Nuts	↑ Plasma Se and GPx activity ↑ HDL-c levels ↓ Cytokines, 8-OHdG and 8-isoprostane plasma ↓ LDL-c levels	Stockler-Pinto et al. (2014)
	Partially defatted nut flour	↓ In serum total cholesterol and non-HDL-c levels	Carvalho et al. (2015)
	Nuts	↑ Plasma Se levels, rectal selenoprotein P (SePP) and β-catenin mRNA	Hu et al. (2016)
	Nuts	↑ GPX1 mRNA expression only in subjects with CC	Donadio et al. (2017)

(Continued)

TABLE 8.5 *(Continued)*

Functional Characterization of Amazonian Fruit Performed in Human Studies

Scientific Name	Source	Observations	References
Myrciaria dubia (Kunth) McVaugh.			
	Fruit juice	Identification of seven biomarkers associated to genipapo consumption	Dickson et al. (2018)
Bactris gasipaes Kunth.			
	Carotenoids extracts	Promoted a low glycemic index	Quesada et al. (2011)
	Peach palm test meal	↑ β-carotene, γ-carotene, and lycopene and retinyl ester levels	Hempel et al. (2014)
Euterpe oleracea Mart.			
	Açaí pulp	↓ Glucose and insulin levels, total cholesterol and borderline significant reductions in LDL-c	Udani et al. (2011)
	Gel capsule of açaí	↓ Systolic blood pressure	Gale et al. (2014)
	Açaí pulp	↑ EGF and PAI-1 in overweight women; ↑ Body weight, BMI, % truncal fat and tríceps skinfold thickness in eutrophic women; ↓ Skinfold thickness and total body in overweight women	de Sousa Pereira et al. (2015)
	Açaí juice	↑ Total antioxidant capacity of plasma; attenuation of muscle damage caused by exercise and improvement of serum lipid profile	Sadowska-Krępa et al. (2015)
	Açai functional beverage	↑ Time for exhaustion in a shorter time with high intensity	Carvalho-Peixoto et al. (2015)
	Açaí pulp	↑ Catalase activity and total antioxidant capacity; ↓ Production of ROS	Barbosa et al. (2016)
	Açaí pulp	↓ROS, ox-LDL and malondialdehyde; ↑ Activity of antioxidative paraoxonase 1	Pala et al. (2018)
	Açaí-beverage	↓ Interferon gamma and urinary level of 8-isoprostane	Kim et al. (2018)
	Frozen açaí pulp	↓ Incremental açaí under the curve for total peroxide oxidative status	Alqurashi et al. (2016)

IgE: immunoglobulin E, HDL-c: high-density lipoprotein cholesterol, LDL: low-density lipoprotein cholesterol, GPx: glutathione peroxidase, mRNA: messenger ribonucleic acid, EGF: epidermal growth factor, PAI-1: plasminogen activator inhibitor-1, BMI: body mass index, ROS: reactive oxygen species, ox-LDL: oxidized low-density lipoprotein.

8.2 *BERTHOLLETIA EXCELSA*

8.2.1 Botanical Description

Bertholletia excelsa, most popularly known as the Brazil nut, belongs to the *Lecythidaceae* family. This species is a native plant from South America found in various Brazilian states, such as Maranhão, Mato Grosso, Pará, Rondônia and Amazonas. The fruit is approximately 2 kg in weight and has a woody shell, 8 to 12 mm thick that contains 8 to 24 seeds (nuts) (Figure 8.1A). Each seed is protected also by a woody, thick, indurate and rugose coat. An edible pale brownish-white kernel is found inside the seed (Figure 8.1A) (Lim, 2012).

8.2.2 Phytochemicals

Brazil nut is an excellent source of fats and rich in unsaturated fats, including monounsaturated fatty acids (MUFA) and polyunsaturated fatty acids (PUFA), phytosterols and tocopherols. However, the phenolic content is low (Table 8.2) (Chunhieng et al., 2008; Cicero et al., 2018; John and Shahidi, 2010). The cold-pressed Brazil nut oil contains approximately 20% of saturated fatty acids (SFA), 52% MUFA and 28% PUFA. The high unsaturation level is principally due to oleic acid (C18:1, n-9) and linoleic acid (C18:2, n-6), corresponding to 73% of the unsaturated fatty acids (UFA). The linoleic acid content is twofold higher than that of olive oil (Cicero et al., 2018).

The unsaturated lipid fraction of the Brazil nut is also rich in β-tocopherol. However, lower levels of α-tocopherol (11.3%) and γ-tocopherols (0.4%) are observed. The high β-tocopherol content, corresponding to 88% of the total tocopherols, is a peculiar characteristic of this oil and could be used as a marker of discrimination from other oils, such as soy oil and olive oil. The steroidal content is similar to olive oil, and β-sitosterol and squalene are the primary steroids found in the nut (Chunhieng et al., 2008; Cicero et al., 2018).

John et al. (2010) observed that the bound phenolic compound content is 86- and 19-fold higher in the brown skin than the kernel and whole nut, respectively. Additionally, the antioxidant activity is higher in the skin due to the phenolic compound content. The phenolic compounds found include gallic acid, gallocatechin, protocatechuic acid, catechin, vanillic acid, taxifolin, myricetin, ellagic acid and quercetin. Cicero et al. (2018) reported a lower content of phenolic compounds (<0.5 μg g^{-1}) in the cold-pressed Brazil nut oil, including p-coumaric acid, apigenin 7-O-glucoside, luteolin and p-hydroxybenzoic acid.

8.2.3 Mineral Content

The Brazil nut oil is rich in magnesium (1.4 g per 100 g), phosphorus (2.4 g per 100 g) and selenium (0.4 to 12.7 mg per 100 g). The high selenium content present in the nut protein fraction has an essential dietary antioxidant role. One Brazil nut of 5 g may contain up to 290 μg, which is six times the daily selenium requirement for an adult. However, selenium toxicity occurs with the consumption of more than 800 μg per day (Cardoso et al., 2017; Chunhieng et al., 2008; IOM, 2000; Stockler-Pinto et al., 2012).

The Se present in the Brazil nut is bound to proteins, known as selenoproteins, which are used as molecular biomarkers (Donadio et al., 2017). The Se of the protein fraction found in this nut is covalently linked to two amino acids resulting in selenomethionine and selenocysteine (Chunhieng et al., 2004). Jayasinghe and Caruso (2011) reported that the water-soluble protein fraction can be divided into two primary sub-groups, the first with high molecular weight Se-containing proteins and the second with low molecular weight Se-containing proteins with an abundance of methionine amino acids.

8.2.4 Biological and Pharmacological Activities

The Brazil nut is a good source of selenium, tocopherols and PUFA. *In vitro* (Table 8.3) and *in vivo* (Table 8.4) studies revealed their respective biological properties in support of beneficial effects on human health (Table 8.5).

In vitro assays revealed the antifungal activity of the Brazil nut oil. Five hundred microliters of the oil nut reduced the growth of *Aspergillus parasiticus* colonies by approximately 38% and inhibited aflatoxin production (Martins et al., 2014). Another study demonstrated a high trypanocidal activity against the trypomastigote form of the organic extracts (acetone and methanol) of *Bertholletia excelsa* stem barks at concentration of 500 μg. mL^{-1} (Campos et al., 2005).

The high concentrations of bioactive compounds and the low toxicity risks have increased the interests of various researches on the *in vivo* effects of the daily Brazil nut consumption in humans, especially on the lipid profile and reduction of cardiovascular risks. Stockler-Pinto et al. (2012)

showed that the consumption of one nut per day (approximately 5 g containing 290 µg of selenium) for 3 months increases the Se plasma levels in hemodialysis patients. After 12 months with the low supplementation levels, the Se plasma levels were significantly lower (Stockler-Pinto et al., 2012).

Cominetti et al. (2012) also showed that the daily consumption of one Brazil nut per day by obese patients for 8 weeks improved both the Se and lipid profiles, particularly high-density lipoprotein cholesterol (HDL-c). Similar results were found by Maranhão et al. (2011) during the daily supplementation of three to five Brazil nuts in the diet of obese adolescents for 16 weeks (Cominetti et al., 2012; Maranhão et al., 2011). Additionally, the daily consumption of defatted Brazil nut flour (13 g per day providing 227.5 µg of selenium per day) for 3 months showed the same effects on the lipoprotein profile of dyslipidemic and hypertensive patients, such as reductions in total cholesterol, VLDL and LDL. Therefore, the Brazilian nut is an alternative for healthy food market (Carvalho et al., 2015).

Supplementation with a higher amount of Brazil nut, 20 g(625 µg Se) and 50 g (1560 µg Se) per day in healthy people, demonstrated that after 9 hours of ingestion, the serum low-density lipoprotein cholesterol (LDL-c) decreased and HDL-c increased (Colpo et al., 2013). In a study by Strunz et al. (2008), the consumption of 11 Brazil nuts (865 µg Se) per day for 15 days increased the Se plasma level but did not alter the serum lipid profile. The consumption of high amounts of Brazil nut (more than two units) is not recommended due to the occurrence of Se toxicity in concentrations higher than 800 µg. Therefore, better health effects are observed with the daily supplementation of one Brazil nut for a period of 3 to 12 months (Stockler-Pinto et al., 2012).

Previous studies showed chemopreventive properties in colorectal cancer, as well as anti-inflammatory activity, suggesting that the beneficial health effects due to the consumption of the Brazil nut are associated with the nut's antioxidant potential and increased Se plasma levels (Hu et al., 2016; Stockler-Pinto et al., 2012). Stockler-Pinto et al. (2014) verified that the consumption of one Brazil nut per day for 3 months was adequate to reduce the inflammation, oxidative stress markers and the atherogenic risk in hemodialysis patients. This result suggests an increase in the antioxidant defenses of the patients (Stockler-Pinto et al., 2014). The supplementation of Brazil nuts (48 µg Se per day) in the diet of people belonging to a risk group for colorectal cancer helps to regulate colorectal cancer oncogenesis biomarkers, such as specific genes related to selenoproteins (SePP), WNT signaling (β-catenin), inflammation (NF-κB) and methylation (DNMT1), and consequently, should reduce cancer risk (Hu et al., 2016).

Furthermore, two primary Brazil nut proteins, Ber e 1 (2S albumin) and Ber e 2 (legumin), are considered allergens for some people who have a specific immunoglobulin E to Brazil nut (Geiselhart et al., 2018). Pastorello et al. (1998) verified the clinical symptoms of the allergens of Brazil nut in patients after the ingestion of two nuts associated with 2S albumin. Anaphylactic shock and laryngeal edema were the primary symptoms observed for symptomatic patients (Pastorello et al., 1998). Many studies also reported the allergenic effect through immunochemical methods, based on the IgE (Arshad et al., 2018; Bartolomé et al., 1997; Pumphrey et al., 1999).

The Ber e 1 allergen is a protein that possesses high stability during the pepsin digestion and is thermostable, showing heat denaturation at approximately 110°C (Van Boxtel et al., 2008). Therefore, the presence of Brazil nut is a required statement that must appear on food labels (FDA, 2011).

8.3   GENIPA AMERICANA

8.3.1   BOTANICAL DESCRIPTION

Genipa americana (Rubiaceae) is widely distributed in tropical regions and parts of subtropical regions, such as the lowlands of the Amazon Forest and regions of Central and South America. *Genipa americana* has different popular names, such as jagua, genipa, genipap and genipapo, depending on the region. The ripe fruits (Figure 8.1B) are edible and are primarily consumed as juice, liqueur and jelly. Fermentation of the fruit is used to produce the alcoholic beverage "cauí" (Oliveira et al., 2012; Ono et al., 2005; UNCTAD, 2005).

Unripe genipap fruits are widely used by indigenous tribes to extract the blue pigment, exposing the inside part of the fruit to the air. The genipap name originated from the Guarani language that means "fruit used to paint". The unripe fruits are lighter and shorter with a green color. The skin of the ripe fruits is yellow-reddish, and at the final stage of maturation, the blue pigment is absent (Bentes and Mercadante, 2014; Bentes et al., 2015).

8.3.2 PHYTOCHEMICALS

The genipap fruits are well known for their iridoid content, primarily genipin and geniposide (Figure 8.3) (Table 8.2). Both of these iridoids were found only in the unripe fruits. These compounds decrease more than 90% during ripening, explaining the absence of the formation of the blue pigment after the ripe fruits are opened. Geniposide is often used in Asian countries as a natural yellow dye. This compound represents more than 70% of the total iridoid content of the unripe fruit (Bentes and Mercadante, 2014). Genipin is a colorless iridoid from the monoterpene class, being an excellent source of blue pigment. Genipin reacts spontaneously with primary amines and proteins in the presence of oxygen, producing a water-soluble bluish-violet pigment. The endocarp and whole fruit present a higher genipin content than the other parts of fruit, such as the peel, mesocarp and seed (Náthia-Neves et al., 2017; Neri-Numa et al., 2017). In the ripe fruit, the genipin gentiobioside is the primary compound found in the endocarp (Bentes and Mercadante, 2014).

Moreover, other iridoids are present in genipap fruits, such as gardoside, geniposidic acid, genipin-1-β-D-gentiobioside, caffeoyl geniposidic acid, p-coumaroyl geniposidic acid, feruloylgardoside, feruloylgenipin gentiobioside, genipacetal, genipamide, genipaol and genamesides (Bentes and Mercadante, 2014; Ono et al., 2005; Ono et al., 2007).

Data on the phytochemical composition of genipap reveal that the peel and fruits contain leucoanthocyanidins, catechins, flavanones, anthraquinones, anthrone and coumarins, triterpenoids and steroids. The flavonoid quercetin predominates in the peel (48 µg g^{-1} DW) and seed (35 µg g^{-1} DW), while it is found in a lower amount in the pulp (9.8 µg g^{-1} DW) (Omena et al., 2012).

Bentes and Mercadante (2014) detected the presence of phenolic compounds only in the unripe fruit, including dicaffeoylquinic acid, 3,5-dicaffeoylquinic acid, 4,5-dicaffeoylquinic acid and 5-caffeoylquinic acid, while 5-caffeoylquinic acid is the primary phenolic compound found in this fruit.

The leaf extract of genipap is also a source of iridoids, and phytochemical studies confirmed the presence of flavonoids and tannins. Two iridoids were detected only in the leaves of genipap, 1-hydroxy-7-(hydroxymethyl)-1,4aH,5H,7aH-cyclopenta[c]pyran-4-carbaldehyde and iridoid

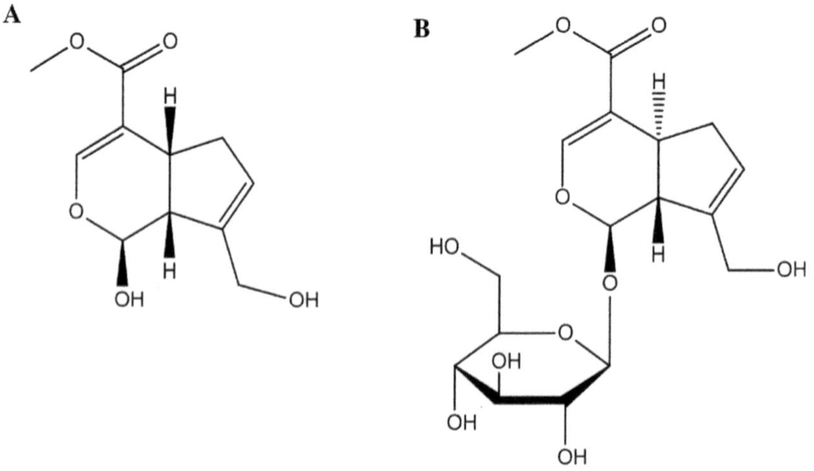

FIGURE 8.3   The chemical structure of genipin (A) and geniposide (B).

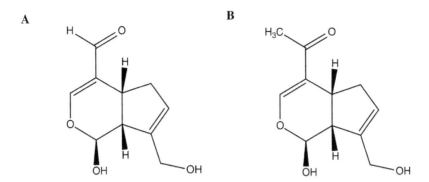

FIGURE 8.4 The chemical structure of iridoids 1-hydroxy-7-(hydroxymethyl)-1,4*aH*,5*H*,7*aH*-cyclopenta[*c*] pyran-4-carbaldehyde (A) and 7-(hydroximethyl)-1-methoxy-1*H*,4*aH*,5*H*,7*aH*-cyclopental[*c*]pyran-4-carbaldehyde (B).

7-(hydroxymethyl)-1-methoxy-1*H*,4a*H*,5*H*,7a*H*-cyclopenta[c]pyran-4-carbaldehyde (Figure 8.4) (Alves et al., 2017; Nogueira et al., 2014).

The pulp and seed are also excellent sources of phytosterols, such as campesterol, stigmasterol, β-sitosterol, sitostanol, Δ5-avenasterol, Δ7-stigmasterol and Δ7-avenasterol. The total phytosterol content in the pulp and seed oil of genipap (>200 mg. 100 g^{-1} FW) is higher than that of the Brazil nut (148 mg per 100 g FW) and açaí pulp (111 mg per 100 g FW) (Costa et al., 2010).

8.3.3 BIOLOGICAL AND PHARMACOLOGICAL ACTIVITIES

The primary compounds found in the genipap fruit are from the iridoid class. However, the bioactivity of other compounds is better described in the literature, such as non-phenolic compounds and steroids. The leaf extract is an excellent source of flavonoids, tannins and polysaccharides with antioxidant activity and bioactivity (Table 8.2).

The genipap iridoids are considered as a promising source of blue pigment for food applications, particularly acidic foods. The stability and antioxidant capacity were studied by *in vitro* simulated digestion of the unripe genipap endocarp extract. The extract is rich in iridoids, including genipin, genipin 1-β-gentiobioside, geniposide, gardenoside, 6′-*O*-*p*-coumaroyl geniposidic acid and 6′-*O*-feruloyl-geniposidic acid. The results revealed an increase in these compounds during the gastric phase (pH 2.0 at 37°C) and a decrease during the intestinal phase (pH 7.0 at 37°C). Genipin is not detected after the intestinal phase, demonstrating the instability of these compounds at neutral pH values. Interestingly, an increase of 17% to 18% was observed after *in vitro* digestion (Neri-numa et al., 2018).

Genipap pulp extract showed anti-acetylcholinesterase activity. However, this effect is not associated with the antioxidant activity and should be related to the non-phenolic compound. Acetylcholinesterase inhibitors are essential in the symptomatic treatment of Alzheimer's disease to elevate the levels of endogenous acetylcholine in the brain (Omena et al., 2012).

The treatment of BeWo cells with a fruit ethanolic extract dose of 100 μg. mL^{-1} inhibits and reactivates the mitogen-activated protein kinases (p38 MAPK). The consequence is an essential effect on trophoblast metabolism and consequently may have implications for placental development, such as the early termination of pregnancy, gestational abnormalities, or fetal growth defects. The steroids of the extracts should be considered to be a possible cause of the effects (da Conceição et al., 2011).

A study on the urinary metabolomics profile of volunteers who consumed 500 mL of genipap juice revealed seven compounds in the metabolic pathways of the iridoids and phenolic derivatives, including dihydroxyhydrocinnamic acid, hydroxyhydrocinnamic acid, genipic acid, 12-demethylated-8-hydroxygenipinic acid, 3(7)-dehydrogenipinic acid, genipic acid glucuronide, nonate, (1R,6R)-6-hydroxy-2-succinylcyclohexa-2,4-diene-1-carboxylate and 3,4-dihydroxyphenylacetate. These compounds were suggested to be biomarkers of genipap consumption in humans (Dickson et al., 2018).

The leaf aqueous extract in concentrations equal to or higher than 30 mg. mL^{-1} showed anthelmintic activity, and above 90% of the nematode's sheep were inhibited. The efficacy for the larval development inhibition is higher than that for the inhibition of egg eclosion. Phytochemical analysis of this extract using colorimetric methods identified flavonoids and tannins. However, *in vivo* studies are needed to assess the toxicity risks (Nogueira et al., 2014).

The antiparasitic action of the polysaccharide extracts (54% carbohydrates, with 21% being uronic acid) from genipap leaves on the epimastigote, trypomastigote and amastigote forms of *Trypanosoma cruzi* was also observed. The cell death of this protozoan may have relationship with reactive oxygen species (ROS) molecules that cause peroxidative damage to the trypanothione reductase and alter the redox balance (Souza et al., 2018).

Central inhibitory effect and anticonvulsant activity is another bioactivity of a heteropolysaccharide present in the polysaccharide extract of the genipap leaf. The extract confers brain protection against oxidative stress and impairment in the number of black hippocampal neurons (Nonato et al., 2018).

8.4 *MYRCIARIA DUBIA* (KUNTH) McVAUGH

8.4.1 Botanical Description

Myrciaria dubia (Myrtaceae)is a tree found in the margins of rivers, streams, lakes and swamps from the Amazon Region (Colombia, Venezuela, Peru and Brazil). The fruit is most popularly known as camu-camu, and when ripe, the berries present a color varying from red to purple (Figure 8.1C). The fruit is composed of 52.5% pulp, 21.2% skin and 26.3% seed. The berries are globoid and approximately 2.5 cm in diameter, with white pulp with high ascorbic acid content that prompts strong sour taste . Camu-camu is often consumed as juice, ice cream, cakes and liqueur (Azevedo et al., 2018; Genovese et al., 2008; Neri-numa et al., 2018).

8.4.2 Phytochemistry

Camu-camu has a unique phytonutrient profile, with abundance of vitamin C, phenolic compounds and carotenoids being considered a "superfruit", with the most abundant natural source of vitamin C in Brazil. Recently, Azevedo et al. (2018) determined the ascorbic acid content in the camu-camu, is approximately 1100 and 946 mg per 100 g in dry weight (DW), cultivated in Amazonas and Roraima states, respectively. Other studies have demonstrated higher contents of ascorbic acid, approximately 1882, 2585 and 6112 mg per 100 g of the fresh pulp in Belém, Amazonas and Roraima, respectively. Environmental factors and the difference in agroforestry systems should explain the variation in the vitamin C content. For example, camu-camu in the Amazonas state is cultivated in seasonally dry conditions on solid ground; in contrast, the camu-camu from the Roraima state is from the flooded environment near the Rio Branco river (Maeda et al., 2007; Rufino et al., 2010; Yuyama et al., 2002). Based on these data, only 10 g of fresh pulp is sufficient to provide the daily recommended amount of vitamin C for an adult (75 to 90 mg) (IOM, 2000). In comparison with other citrus fruits, the content of vitamin C in camu-camu nectar (340 mg per 100 g) is 3.6-fold higher than the content in orange juice (94.5 mg per 100 g) and 9.8-fold higher than that in lemon juice (34.5 mg per 100 g) (Maeda et al., 2007; TACO, 2011).

Others bioactive compounds present in camu-camu pulp include phenolic acids (gallic acid, ferulic acid, ellagic acid, ellagitannin B, *p*-coumaric and protocatechuic acid), flavonoids (myricetin, kaempferol, quercetin, rutin, naringenin, cyanidin-3-rutinoside, and cyanidin-3-*O*-glucoside) and carotenoids (β-carotene, lutein, β-cryptoxanthin, zeaxanthin, luteoxanthin, violaxanthin and neo-xanthin). The seed has two specific compounds, betulinic acid and trans-resveratrol, in the seed coat (Akter et al., 2011; Anhê et al., 2018; Fidelis et al., 2018; Genovese et al., 2008; Neri-numa et al., 2018; Rodrigues-Amaya et al., 2008; Zanatta and Mercadante, 2007).

The camu-camu fruit is an excellent natural source of antioxidants. *In vitro* studies have confirmed the antioxidant activity of ripe camu-camu fruit with average moisture content of 92%. Using the 2,2′-azino-*bis*(3-ethylbenzothiazoline-6-sulphonic acid (ABTS) and 2,2-diphenyl-1-picrylhydrazyl (DPPH) methods, the antioxidant capacity is 1418 and 1520 µmol trolox equivalents g^{-1} DW, respectively, and this value is lower in the unripe fruits. The average total polyphenol content found in camu-camu fruit extract is 6550 mg per 100 g DW. The major groups found are proanthocyanidins (1854 mg per 100 g DW), followed by ellagitannins, flavanols/flavonols and phenolic acids. Ellagic acid is found in significant proportions in the unripe fruit, but the antioxidant activity is inferior due to the lower anthocyanin and polyphenol concentrations (Anhê et al., 2018; Azevedo et al., 2018).

8.4.3 Biological and Pharmacological Activities

The bioactive compounds encountered in different parts of the camu-camu fruit, especially vitamin C, phenolic acids and flavonoids, are responsible for the functional properties and related to their high antioxidant activity. *In vitro* and *in vivo* studies using crude, aqueous or organic extracts obtained from the pulp, leaf, skin, seed and seed coat demonstrated the reduction of lipid peroxidation and positive effects on anti-inflammatory, anti-obesity, antiplasmodial and antileishmanial, antimutagenic and antihypertensive activities (Tables 8.3 to 8.5).

The aqueous extract of the seed and seed coat were shown as excellent anti-inflammatory agent during *in vitro* assays. This extract inhibited the lipid peroxidation induced by inflammatory mediators and suppressed the formation of carrageenan-induced paw edema in mice. Correlation analysis by principal component analysis reveals that the aqueous extract, with higher antioxidant activity and inhibition of lipid peroxidation, is composed of total phenolics, non-tannin phenolics, (−)-epicatechin, chlorogenic acid, 2,4-dihydroxybenzoic acid, 2,5-dihydroxybenzoic acid and gallic acid. The propanone extract of the seed coat also has excellent antihypertensive activity, possibly due its ability to chelate Cu^{2+}, and high levels of quercetin, quercetin-3-rutinoside (rutin), t-resveratrol, ellagic, caffeic, rosmarinic, ferulic and *p*-coumaric acids (Fidelis et al., 2018; Yazawa et al., 2011).

An antiobesity effect of camu-camu pulp has been observed in the treatment of diet-induced obese mice for 8 weeks. The rats were fed a high-fat high-sucrose diet and concomitantly started treatment with the ingestion of daily oral doses of camu-camu extract (200 mg kg^{-1}). Another group was treated with vitamin C (6.6 mg kg^{-1}) for comparison. A reduction of body weight, glucose, total cholesterol, triglycerides, LDL-c, insulin blood levels and an increase in the HDL-c levels was observed in rats treated with camu-camu extracts, but no alterations were observed in those treated with vitamin C. In addition, alterations in the gut microbiota were described during the treatment of obese mice with a high abundance of *Akkermansia muciniphila* and a reduction of *Lactobacillus*. The high incidence of *A. muciniphila* was positively correlated with lithocholic acid, deoxycholic acid and ursodeoxycholic acid in plasma bile acids (BAs), and the reduction of the *Lactobacillus* correlated with unconjugated BAs. A positive association between *Lactobacillus* and obesity was demonstrated by the researchers. The studies indicated that the consumption of camu-camu extract helps to prevent visceral and liver fat deposition as well as to generate alterations in the gut microbiota that contribute to the prevention of obesity (Anhê et al., 2018; Nascimento et al., 2013).

Whole fruit extracts of camu-camu demonstrated antimutagenic effects in mice, protecting the colon against the damage induced by 1,2-dimethylhydrazine dihydrochloride (DMH) doxorubicin chloridate (DXR) mutagenicity effects (Azevedo et al., 2018). In another study, the application of a fraction with low molecular weight extracted from fresh camu-camu was able to reduce the $A\beta_{1-42}$ peptide aggregation in the muscle tissues of *Caenorhabditis elegans*. In this Alzheimer's disease model, the predominance of polar basic bioactive compounds in the extract should explain the neuroprotective effect. A similar effect was observed in the MPP+-induced oxidative dopaminergic neurotoxicity model for Parkinson's disease in *C. elegans* (Azevêdo et al., 2015).

In humans, a study on the consumption of camu-camu juice in volunteers who smoked was conducted. One group ingested 70 mL of the juice daily (1050 mg of vitamin C), and another group ingested a vitamin C (1050 mg) tablet for 7 days for comparative effects. The researchers observed that camu-camu juice is an excellent antioxidant and anti-inflammatory agent (Inoue et al., 2008).

8.5   *SPONDIAS MOMBIN*

8.5.1   Botanical Description

The yellow mombin (*Spondias mombin*) is a species belonging to the Anacardiaceae family, which is distributed in tropical regions. The plant, popularly known as taperebá and cajá, is typical in the North and Northeast regions of Brazil. The fruit is a small ovoid drupe 2.5 to 4 cm in diameter with yellow skin (Figure 8.1D). The weight of the fruits varies from 13 to 33 g. A proportion of 70% to 82% of the fruit is edible, and the juicy pulp has a sour-sweet taste (Tiburski et al., 2011; Vasco et al., 2008).

8.5.2   Phytochemicals

The primary bioactive compound class of yellow mombin fruit is carotenoids (Table 8.2). The carotenoids are found in the fresh pulp and are responsible for the yellow color of the ripe fruits. The total chlorophyll content is low in the ripe fruit, and only some green spots are visible. Zielinski et al. (2014) observed a positive correlation between the yellow color of the fruit and the carotenoid content, particularly with β-carotene and lycopene. Hamano and Mercadante (2001) detected an average of 2.6 mg per 100 g FW of total carotenoids in the pulp. A lower value was detected in the juice (1.67 mg per 100 g FW). In most pulps evaluated by Silva et al. (2012), the total carotenoids were two times higher than in this study. The fruits used were harvested from six genotypes of yellow mombin trees, and five showed values of the total carotenoids ranging from 3.7 to 4.1 mg per 100 g FW (Hamano and Mercadante, 2001; Silva et al., 2012; Zielinski et al., 2014).

β-Cryptoxanthin is the principal carotenoid present in the fruit, pulp and juice, and represents more than 50% of the total vitamin A content. Other carotenoids found in the pulp are lutein, zeaxanthin, β-carotene and α-carotene. Of these compounds, only α-carotene, β-carotene and β-cryptoxanthin have provitamin A activity. The pulp should be considered to be a good source of provitamin A because 100 g provides 37.2% of the RDI for adults (Hamano and Mercadante, 2001; Tiburski et al., 2011).

Phenolic compounds are also present in the pulp with an average value of 249 mg gallic acid equivalents (GAE) per 100 g FW (Vasco et al., 2008). The flavonoid content (range 0.14 to 0.52 mg per 100 g FW) found by Silva et al. (2012) is relatively low in comparison to other fruits, and anthocyanins were not detected (Vasco et al., 2008; Zielinski et al., 2014).

The hydroethanolic extract of the yellow mombin leaf is a source of flavonoids, cinnamic derivatives, triterpenoids, steroids, mono- and sesquiterpenes, alkaloids, proanthocyanidins and leucoanthocyanidins. Chlorogenic acid, gallic acid, ellagic acid and isoquercetin are phenolic compounds more common in yellow mombin leaves and should be considered as excellent biomarkers for this genus (Brito et al., 2018; Cabral et al., 2016).

8.5.3   Biological and Pharmacological Activities

The biological properties of the yellow mombin leaf have been extensively studied in the literature. *In vitro* assays show a high antioxidant capacity, cytotoxicity activity against 3T3 fibroblast cells, anthelmintic activity reducing the larval development, leishmanicidal activity on promastigotes and antimicrobial activity (Accioly et al., 2012; Ademola et al., 2005; Cabral et al., 2016) (Table 8.3).

In vivo studies demonstrated that the biological effects are related to antioxidant activity. Brito et al. (2018) demonstrated that ethanolic extracts of yellow mombin leaf contribute to the chronic ulcer treatment mediated by the antioxidant activity. In particular, gallic acid and ellagic acid, isolated or associated, stimulate the gastric mucus production, while the presence of the sulfhydryl groups and nitric oxide consequently have antisecretory and anti-*Helicobacter pylori* activities (Brito et al., 2018).

Functional studies of yellow mombin leaf extract have highlighted the hepatoprotective effects in rats, limiting the damage caused by drugs, such as indomethacin and acetaminophen. The mechanism of action is associated with antioxidant systems of the extract that act as proton pump inhibitors (Sabiu et al., 2016; Saheed et al., 2017).

The oral supplementation of the leaf extract of yellow mombin can also alleviate inflammatory responses. Studies in mice demonstrated that ingestion of the extract in doses of 100 and 200 mg kg^{-1} *per os* reduced the tumor necrosis factor (TNF-α) levels and nitric oxide production between 2 and 4 hours, acting in the suppression of pro-inflammatory mediators (Nworu et al., 2011). Cabral et al. (2016) also suggested that the anti-inflammatory properties of yellow mombin extract in mice is due to the antioxidant properties and that chlorogenic acid and ellagic acid contribute to the pharmacological action. The absence of cytotoxicity in cell cultures of the extract was confirmed in both studies (Cabral et al., 2016).

Anxiolytic and antidepressant effects of yellow mombin leaf extract were observed in a zebrafish model (*Danio rerio*). Scototaxis and novel tank diving test (NTDT) tests revealed hypnotic and sedative effects by immersion (25 mg L^{-1}) and oral administration (25 mg kg^{-1}) of extract doses. The effects should be associated with the presence of isoquercitrin in the leaves of yellow mombin (Sampaio et al., 2018).

Oyeyemi et al. (2015) demonstrated that doses of 5000 mg kg^{-1} *per os* aqueous and hydro-ethanolic extracts of yellow mombin leaf did not induce acute toxicity in mice. In contrast, hydromethanolic extracts showed genotoxicity and antigenotoxicity action. The results of the genotoxicity tests in mice revealed genotoxic effects of hydromethanolic extracts with the potential to induce both somatic and germline genetic damage. The extracts also showed antigenotoxicity action and reduced genotoxicity induced by methyl methanesulfonate in bone marrow cells of the exposed mice. Therefore, studies suggested the therapeutic effects of yellow mombin leaf extract and that consumption for a long time should have toxic effects (Oyeyemi et al., 2015).

An *in vivo* study on yellow mombin juice consumption evaluated the effects of the incorporation of daily doses of 100 and 250 mg of the dry extract (pulp and skin) per kg of body weight reconstituted in 88.2% of water in the diet of rats on the cardiac remodeling process induced by exposure to tobacco smoke; these doses are equivalent to 329 and 610 g per day for a human of 60 kg, respectively. After 2 months, the results showed attenuation of this process with a reduction of the cardiac levels of lipid hydroperoxide, a reduction in glycolysis, and an increase in β-oxidation and oxidative phosphorylation (Lourenço et al., 2018).

8.6 *ASTROCARYUM VULGARE* MART. (ARECACEAE)

8.6.1 BOTANICAL ASPECTS AND OCCURRENCE

Astrocaryum vulgare, commonly known as tucumã, (Arecaceae) is primarily distributed in Brazil (Amazonas, Amapá, Goiás, Maranhão, Pará, Piauí, Tocantins), French Guiana, Guyana and Suriname (Henderson, 1995; Kahn, 2008). Tucumã is present in secondary vegetation, cerrado and forests, primarily on sandy soils (Kahn 2008; Lorenzi et al., 2010). Depending on the region, the fruit is designated tucumã, tucumã-do-Pará (Brazil); aouara (French Guiana); cumare (Colombia); awarra (Suriname) and cumare (Venezuela) (Kahn, 2008). Tucumã is a cespitose palm that forms a small cluster of unbranched stems reaching 4 to 10 m tall with spines in the trunks and stem parts (Oboh, 2009), with pistillate flower with calyx urn-shaped corolla as long as the calyx. Tucumã fruits are primarily used as sources of nutrition by populations where they occur due to the high carotenoid content found in the pulp oil (Kahn, 2008). In Amazonia, the fruits have been providing valuable nutrition to people for a long time (Smith, 2015).

8.6.2 Phytochemistry of Tucumã Pulp and Seeds

The tucumã palm fruit possesses a fleshy mesocarp (pulp) and a seed (Figure 8.2B), and both are rich in oil that can represent up to 58.6% in the pulp and 37.6% in the kernel (Bora, 2001; Oboh and Oderinde, 1989; Rodrigues et al., 2010). Pulp oil is very rich in carotenoids (Bony et al., 2012; Oboh, 2009) which imparts the characteristic orange-yellow color of this fruit.

The pulp contains 25 different fatty acids (Bora et al., 2001), and oleic acid (C18:1), a monosaturated fatty acid, can represent up to 65% of the total fatty acids, followed by SFA, including palmitic (C16:0) (25%) and stearic (C18:0) (3%) acids. Linoleic (C18:2) (2.6%) and linolenic (C:18:3) (0.2%) acids are the most abundant PUFA found (Bony et al., 2012; Bora et al., 2001; Rodrigues et al., 2010). The tucumã kernel oil has a high concentration of SFA in which lauric (C12:0) is prevalent (50% to 60%), followed by myristic acid (C14:0) (20% to 29%). The monosaturated fatty acids and PUFA present are oleic (C18:1) (13.6%) and linoleic (C18:2) (6%) acids, respectively (Bereau et al., 2003; Bora et al., 2001; Oboh and Oderinde, 1989). The composition indicates that the proteins are in a suitable concentration (8.44% of dry pulp) and kernels (2.06% on DW) (Bora et al., 2001).

The determination of the total carotenoid composition using High Performance Liquid Chromatography(HPLC) has revealed different isoforms of α, β, δ, γ and ζ-carotene, phytoene and phytofluene in tucumã pulp oil, reaching 1934 $\mu g\ g^{-1}$. β-Carotene is the most prevalent carotenoid in the pulp oil (60%), followed by phytoene (8.7%), and phytofluene isomers (4.33%) (Bony et al., 2012; Santos et al., 2015; Silva et al., 2011) (Table 8.2).

β-Sitosterol and campesterol are the essential phytosterols in the composition of the kernel oil (63% and 4%) and unsaponifiable extract pulp oil (32% and 8%). The presence of the Δ5-avenasterol (or fucosterol) and stigmasterol is 30% and 1.8% in the kernel oil, respectively. In an unsaponifiable extract of the pulp oil, arundoin and cycloartenol represent 16% and 11.4% of the total, respectively (Bereau et al., 2003; Bony et al., 2012). The presence of α-tocopherol, β-tocopherol, γ-tocopherol and vitamin C in pulp oil also has been shown (Bony et al., 2012; Dos Santos et al., 2015; Rodrigues et al., 2010) (Table 8.2).

8.6.3 Biological and Pharmacological Activities of Tucumã Pulp Oil

Tucumã pulp oil is an attractive therapeutic agent due to the high amount of carotene, phytosterol, and tocopherol and other bioactive compounds, all known for their antioxidant effects (Table 8.3). Bony et al. (2012) demonstrated the potential of tucumã pulp oil to act as an anti-inflammatory agent in an endotoxic shock and pulmonary inflammation model in rats (Table 8.4). In the endotoxic shock model, the cytokine production observed has a potent effect in serum, spleen and lung by decreasing interleukin-6 (IL-6) and TNF-α and increasing the IL-10 concentration. In the pulmonary inflammation model, a decrease in the inflammatory cell afflux in the lung was observed, particularly eosinophils and lymphocytes. Another study showed a similar effect of an ethanolic unsaponifiable fraction of tucumã pulp oil against the murine J774 macrophage cell line. The extract inhibits the *in vitro* production of both nitrite oxide and PGE2 by inhibiting iNOS expression and NO scavenging activity and inhibiting the activity and expression of the COX-2 enzyme, respectively. Additionally, this effect was confirmed in a model of endotoxin shock by modulating TNF-α, IL-6 and IL-10 serum concentration in mice (Bony et al., 2014).

The beneficial effects of tucumã oil have also been assessed using several parameters in alloxan-induced diabetic mice treated with this oil. The hypoglycemic effect improves insulin levels and antioxidant/oxidant status and protects against pancreatic damage (Baldissera et al., 2017). The extract promotes the capacity to maintain normal serum levels of ATP, ADP, AMP and adenosine, molecules that exhibit anti-inflammatory properties. The carotenoid compounds should be related to their beneficial effects (Baldissera et al., 2017) as adequate protection against neurotoxicity and consequent memory impairment by preventing lipid peroxidation, increasing the levels of the antioxidant enzymes catalase (CAT) and superoxide dismutase (SOD) and inhibiting Na^+, K^+-ATPase activity. All these mechanisms are involved in the memory deficits of mellitus diabetes (Baldissera et al., 2017).

8.7   *BACTRIS GASIPAES* KUNTH. (ARECACEAE)

8.7.1   Botanic Aspects and Occurrence

Bactris gasipaes, the peach palm or pupunha (Brazil), is the only Neotropical palm with domesticated populations in the Amazon region (Clement et al., 2010). Peach palm is endemic from the tropical forest with a natural distribution that extends from Panama to Bolivia (Leterme et al., 2005). Peach palm has a large variety of breeds and ecotypes, broadly distributed in areas, known as landraces, located in the humid Neotropics, particularly in Amazonia (Clement et al., 2010).

Peach palm is a monoecious, allogamous (cross-pollinated), spiny, multi-stemmed palm that may attain up to 20 m height with 15 to 25 pinnate fronds in the crown. The inflorescences appear among the axils of the senescent fronds. The fruit bunch contains 50 to 100 single-seeded fruits. The fruit is a drupe, which when ripe presents a fibrous red, orange or yellow epicarp, a moist starch/oily mesocarp and a single endocarp with a fibrous and oily white kernel (Figure 8.2D) (Clement et al., 2004; Valencia et al., 2015).

8.7.2   Phytochemistry of Peach Palm Fruit and Seed Oil

The peach palm oil has been analyzed in different populations and parts of the fruits, such as the mesocarp and seed. In the fruit mesocarp, the lipid content demonstrates considerable variation, ranging from 3.5% to 19%, while the kernel is more homogeneous , varying from 11.5% to 16.4%. The monounsaturated oleic acid (C18:1) is present in large amounts in peach palm mesocarp oil (42.8% to 60.8%), followed by saturated palmitic acid (24.1% to 42.3%) (Espinosa-Pardo et al., 2014; Santos et al., 2013; Yuyama et al., 2003). Kernel oil analysis highlighted the high content of the SFA (79.5%) with lauric (C12:0) and myristic (C14:0) acids as the prevalent constituents (Bereau et al., 2003; Radice et al., 2014) (Table 8.2).

Peach palm mesocarp is an essential source of carotenoids, varying from 198 to 357.4 μg. g^{-1} total carotenoid content (de Rosso and Mercadante, 2007; Santos et al., 2015). Although the carotenoid content is different among varieties (1.1 to 7.4 mg/100 g), β-carotene (31.2% to 42.2%) and γ-carotene (20.3% to 26.6%) are predominant (de Rosso and Mercadante 2007; Jatunov et al., 2010; Rojas-Garbanzo et al., 2011).

Santos, et al. (2013) analyzed the minor components in oils obtained from Amazonian peach palm fruit, and the predominant phytosterols are β-sitosterol and campesterol, present at 82.2% and 10.9%, respectively. The α-tocopherol was the unique tocopherol detected (117 μg g^{-1}). In the kernel oil, the phytosterols are primarily represented by sitosterol (73.3%) and fucosterol (22%) (Bereau et al., 2003), while the tocopherols include α-tocopherol (47 μg g^{-1}) and β-tocopherol (7 μg g^{-1}) (Radice et al., 2014).

8.7.3   Biological and Pharmacological Activities of Peach Palm Pulp Oil

The primary constituents of the peach palm have been investigated for their functional properties, since they are primarily used in ethnomedicine for their effects on 11 different diseases (Sosnowska and Balslev, 2009). Yuyama et al. (1991) showed that the vitamin A of peach palm was highly bioavailable in the plasma and liver in male Wistar pups. The *in vitro* protective effects of lipid peroxidation on the liver homogenates of rats induced to oxidative stress with tert-butyl hydroperoxide suggested that the carotenoids may contribute to an enhanced antioxidant defense (Quesada et al., 2011) (Table 8.3).

A supplementation study with red peach palm fruit on Wistar rats before and post-lactation showed an increase in HDL-c and a reduction of body weight, total cholesterol and triglycerides, indicating that the dietary intake of the red peach palm is healthy, especially during the lactating and post-lactating periods (Carvalho et al., 2013) (Table 8.4).

In humans, the ingestion of peach palm mesocarp promoted a lower glycemic index, similar to those presented by legumes, and is recommended in the prevention of several chronic diseases

(Quesada et al., 2011). The bioavailability of α-carotene, β-carotene, γ-carotene, lycopene and their isomers from peach palm fruits in humans was demonstrated by Hempel et al. (2014), who also correlated the significant increase in β-carotene, γ-carotene, and lycopene and retinyl ester levels with the conversion of the ingested provitamin A carotenoids to vitamin A (Table 8.5).

Different oil extracts were tested for their antimicrobial potential on the growth of *Staphylococcus aureus,* but only the bark oil inhibits *S. aureus* growth after 24 hours (Araújo et al., 2012). This result was confirmed in a similar study that evaluated the antimicrobial activity of the shell, pulp and seed oil of the peach palm on *Pseudomonas aeruginosa* and *S. aureus.* After 48 hours of oil treatment, the antimicrobial effect of the bark of pupunha was observed in *S. aureus* with a 10-mm halo of inhibition (Araújo et al., 2013).

8.8   *EUTERPE OLERACEA* MART. (ARECACEAE)

8.8.1   Botanic Aspects and Occurrence

The *Euterpe* genus has approximately 28 species distributed in Central and South America (Uhl and Dransfield, 1991). *E. oleracea* (açaí), a native palm of tropical South America, forms large spontaneous populations being notable for the commercialization of the species' fruits (Amsellem-Laufer, 2015). The plant's highest abundance occurs in the floodplains of the Brazilian Amazon Basin, particularly in the states of Pará and Amazonas (Kahn and de Granville, 1992).

E. oleracea is a multicaule palm that can reach up to 25 m in height (Henderson, 1995). At the end of each stem, there are spiral leaves pinned with pairs of leaflets. The inflorescence is of the bunch type constituted by a central axis (rachis) and lateral branches (rachilles), in which unisex sessile flowers are inserted, being a preferentially allogamous species performing cross-fertilization (Bovi and Castro, 1993; Jardim and Macambira, 1996). The fruits are drupes, grouped in clusters, in which the most popular variety is green during the initial maturation stages and when mature assumes their black coloration (Figure 8.2A). The mature fruits exhibit a diameter of up to 13.5 mm. The juicy exocarp is a thin layer, and the mesocarp is only 1 to 2 mm thick, while the seed represents 85% to 95% of the total fruit volume with a solid and homogeneous endosperm (Figure 8.2A) (Henderson, 1995; Pompeu et al., 2009; Rogez, 2000).

E. oleracea is the most valuable food resource species of wild fruit in Amazonia(Brokamp et al., 2011). In the population of the Amazon estuary, açaí fruits are the primary staple food base and can represent up to 42% of their diet in DW (Heinrich et al., 2011).

8.8.2   Phytochemistry of Açaí Fruits

During the past 10 years, the açaí fruit production has been scaled up from a minor local product to an international commodity (Brokamp et al., 2011), witha the pulp with oleaginous lipids representing from 40.7% to 60.4% of dry matter (DM) and has a high content of dietary fibers (20.9% to 21.8% of DM), proteins (6.7% to 10.5% of DM), vitamins and antioxidants (Bichara and Rogez, 2011).

The fatty acid composition of açaí revealed that MUFA compose 51.2% of all the fatty acids, with oleic acid predominating (49.7%). The SFA and PUFA are represented by palmitic acid (25.3%) and linoleic acid (13.5%), respectively (Rogez, 2000; Schauss et al., 2006). The predominance of UFA, particularly oleic acid, was also reported in other studies, including extracts of the pericarp, endocarp and fruit (Mantovani et al., 2003), enzymatic extraction process (Nascimento et al., 2008) and analysis of the oleic content profile among different samples (Luo et al., 2012) (Table 8.2).

The phenolic content in açaí has revealed approximately 30 compounds, including flavonoids and phenolic acids. The first studies demonstrated the presence of homoorientin, orientin, isovitexin, (–)-epicatechin, (+)-catechin, scoparin, taxifolin deoxyhexose, procyanidin, *p*-hydroxybenzoic acid, ferulic acid, ferulic acid, gallic acid, protocatechuic acid, ellagic acid and vanillic acid as non-anthocyanin polyphenols (Del Pozo-Insfran et al., 2004; Gallori, et al., 2004; Schauss et al., 2006).

Later studies have broadened the knowledge of the non-anthocyanin polyphenols in the açaí with the identification of additional compounds, including lignans, eriodictyol, (2S,3S)-dihydrokaempferol 3-O-β-D-glucoside and the isomer, (2R,3R)-dihydrokaempferol 3-O-β-D-glucoside, velutin, 5,4′-dihydroxy-7,3′,5′-trimethoxyflavone and hydroxymethylglutaryl-rhamnoside (Dias et al., 2013; Gordon et al., 2012; Kang et al., 2011; Lichtenthäler et al. 2005; Mulabagal and Calderon, 2012; Pacheco-Palencia et al., 2009).

Among the anthocyanins, the cyanidin 3-O-glucoside (C3G) and cyanidin 3-O-rutinoside (C3R) are prevalent reaching 1159 to 1609 μg g^{-1} of the total anthocyanins in the açaí fruit (Gallori et al., 2004; Rogez et al., 2011; Schauss et al., 2006). The non-anthocyanin phenolic content varies depending on the fruit origin, genotypes and commercial pulps. The C3G and C3R can reach 18,942 and 34,397 μg g^{-1}, respectively (Carvalho et al. 2017; Rogez et al., 2011).

During the fruit maturation, the profile of the anthocyanins C3G and C3R vary. In the beginning, the two anthocyanins are present in similar proportions, but in the latter stages, C3R is more abundant (Rogez et al., 2011). Dias et al. (2012) demonstrated that pelargonidin-3-glucoside (Pg3G), peonidin-3-O-glucoside (Pn3G) and peonidin 3-O-rutinoside (Pn3R) are minor anthocyanins in açaí fruits.

The tocopherol composition of açaí fruits has also been demonstrated, and α-tocopherol primarily predominates. Although the low content of 147.72 μg g^{-1} has been already reported, other studies highlighted higher and significant contents in açaí pulp (394.3 μg g^{-1} DM) or oil (1101.11 mg L^{-1}). Açaí should be considered an excellent source of vitamin E (Bichara and Rogez, 2011; Costa et al., 2010; Darnet et al., 2011; Ribeiro et al., 2018). The primary carotenoids are found in low concentrations and include lutein (1.5 to 7.17 μg g^{-1}), β-carotene (1.49 to 2.4 μg g^{-1}) and α-carotene (0.03 to 0.42 μg g^{-1}) in açaí pulp. The total phytosterol content is high (1110 μg g^{-1}), with β-sitosterol the most prevalent (940 μg g^{-1}) (Costa et al., 2010; Ribeiro et al., 2010; Romualdo et al., 2015) (Table 8.2)

8.8.3   Biological and Pharmacological Activities of Açaí Fruits

The high diversity of the phenolics in açaí has led to studies that have demonstrated their significant antioxidant properties (Table 8.3). The juice, usually called "pulp" by various authors, was characterized using different *in vitro* methodologies, such as measuring the antioxidant capacity against DPPH, hydroxyl, peroxyl and peroxynitrite radicals, anion superoxide, and the inhibition of liposomes (Carvalho et al., 2017; Chin et al., 2008; Del Pozo-Insfran et al., 2004; Gordon et al., 2012; Hassimotto et al., 2005; Kang et al., 2010; Lichtenthäler et al., 2005; Paz et al., 2015; Rufino et al., 2010; Torma et al., 2017). The healthy nutrient composition and antioxidant activity of the açaí have been supported by studies on their different physiological and beneficial effects (Tables 8.4 and 8.5).

8.8.3.1   Cardiovascular Effects

The endothelium-dependent vasodilator effects from the skin and seed extracts have been verified in the isolated mesenteric vascular bed of the rat (Table 8.4). The vasodilator effect was detected in both extracts, and açaí stone extract (ASE) was more potent, suggesting the possibility of using ASE to treat cardiovascular diseases (Rocha et al., 2007). In other experiments on hypertension, ASE also showed a significant antihypertensive effect (Cordeiro et al., 2015). In renovascular hypertensive rats, ASE oral administration prevents the increase in blood pressure and plasma renin levels and recovers the endothelial-dependent vasodilator effect of acetylcholine. SOD, CAT, and glutathione peroxidase (GPx) activities are also recovered, increasing the nitrite content and protein expression of endothelial constitutive nitric oxide synthase (eNOS) and decreasing the malondialdehyde (MDA) and carbonyl protein levels in the mesenteric vessels (da Costa et al., 2012). The mechanism of the antihypertensive effect of ASE is not entirely elucidated, but hemodynamic effects in normotensive healthy individuals treated with açaí demonstrated a significant reduction in systolic pressure (Gale et al., 2014). Similar results were previously reported (Udani et al., 2011).

Additionally, ASE has been associated with myocardial ischemia due to the induction of cardiac dysfunction and exercise intolerance in rats (Zapata-Sudo et al., 2014). In overweight men, the consumption of a flavonoid-rich açaí meal was associated with improvements in vascular function, which may lower the risk of a cardiovascular event (Alqurashi et al., 2016).

8.8.3.2 Renal Failure Effects

The beneficial effect of açaí skin has been observed in renal dysfunction experiments in rats with renal protective action by improving kidney function and decreasing the serum urea, creatinine and blood urea nitrogen. The effects were associated with açaí's antioxidant action, that significantly improved renal oxidative stress markers (Unis, 2015). In a study on renal ischemia and reperfusion injury (I\R), açaí skin acts to attenuate I\R induced renal damage, decreasing the blood urea nitrogen levels, serum creatinine and renal tissue content of kidney molecule-1 (KIM-1). The reduction of MDA, myeloperoxidase (MPO), IFN-γ, caspase-3, collagen IV and endothelin-1 were also observed (El Morsy et al., 2015).

Additionally, the oral administration of ASE during rat pregnancy prevents functional and structural changes, such as oxidative stress, increases nephron and glomerular number, glomerular volume, and the serum levels of renin, urea, creatinine and fractional excretion of sodium (de Bem et al., 2014). The protective effect of açaí fruit was also verified on chronic alcoholic hepatic injury in rats. A reduction was observed in serum alanine transferase (ALT) and aspartate aminotransferase (AST), MDA, triacylglycerol (TG), and serum TNF-α and IL-6 and an increase in reduced glutathione (GSH) and SOD (Qu et al., 2014). Based on the data in the literature, supplementation with açaí may reduce oxidative stress and inflammation and consequently chronic kidney disease (Martins et al., 2018).

8.8.3.3 Effects on Lipids and Diabetes Metabolism

In mice supplemented with a high-fat diet, the oral ASE treatment increases the body weight, plasma triglyceride, TC, glucose levels, oral glucose tolerance test and insulin resistance (HOMA index) (de Oliveira et al., 2010). The supplementation of rat hypercholesterolemic with açaí pulp caused a hypocholesterolemic effect by reducing the total and non-HDL-c (de Souza et al. 2010). Rats fed with açaí pulp also exhibited a significant decrease in serum TC, LDL-c and the atherogenic index. Açaí has a hypocholesterolemic effect in a rat model (de Souza et al., 2012).

Additionally, high-fat mice supplemented with açaí showed attenuated hepatic steatosis via adiponectin-mediated effects on lipid metabolism (Guerra et al., 2015). In humans, supplementation with açaí results in a favorable action on plasma HDL (Pala et al., 2018), a positive impact on the reduction of LDL oxidation with a tendency to increase plasma antioxidant capacity (Sampaio et al., 2006) and a beneficial overall role against atherosclerosis (Feio et al., 2012). In junior athletes, the juice blend has a hypocholesterolemic activity (Sadowska-Krępa et al., 2015).

8.8.3.4 Antitumor Effects

In an experimental cancer model in the rat, different supplementations with açaí showed evidence of both anticarcinogenic and chemopreventive effects. In the case of esophageal tumorigenesis, a progressive inhibition was observed with a reduction in the serum levels of the cytokines IL-5 and GRO/KC. The expression of rat IL-8, a direct homolog of that of humans, is lower, and the antioxidant capacity increases in serum (Stoner et al., 2010). Other studies reported the inhibition of urinary bladder carcinogenesis by açaí supplementation. The effects are the reduction of DNA damage, the expression of p63 and proliferation of cell nuclear antigen (PCNA), but the diet supplemented with açaí for 10 weeks did not significantly alter cytoplasmic and nuclear β-catenin (Fragoso et al., 2012). Açaí should be considered to have active anticancer activity. The açaí affects the human malignant MCF-7 cell line (Silva et al., 2014) and inhibits colon carcinogenesis induced by 1,2-dimethylhydrazine (DMH) in Wistar rats (Fragoso et al., 2013). In an experimental model of colon cancer induced by azoxymethane (AOM) with dextran sulfate sodium (DSS) in ICR mice, a downregulation of myeloperoxidase (MPO) and pro-inflammatory cytokines (TNF-α, IL-1β and IL-6) was described. The inhibition of COX-2, PCNA and Bcl-2 and an increase in Bad and cleaved caspase-3 expression was also observed (Choi et al., 2017).

Furthermore, açaí is a useful photosensitizer to reduce melanoma carcinogenesis by increasing the necrotic tissue per tumor area (Monge-Fuentes et al., 2017). Based on tumor diameter and weight, Nascimento et al. (2016) reported an anticarcinogenic effect of açaí in anorexia-cachexia syndrome induced by Walker-256 tumors due to antioxidant capacity.

In breast cancer induced chemically by 7,12-dimethylbenzanthracene (DMBA), experiments in female Wistar rats demonstrated the antiangiogenic and anti-inflammatory potential of açaí. A decrease in the number of inflammatory cells and positive macrophage cells were correlated with the inhibition of DMBA carcinogenicity (Alessandra-Perini et al., 2018). The effects of açaí in tumor cells have been suggested due to the anti-inflammatory, antiproliferative and proapoptotic properties (Alessandra-Perini et al., 2018).

8.8.3.5 Nontoxic Effects

Studies in experimental models reported the absence of toxicity. Parameters, such as animal body weight or food consumption, show no significant differences during açaí supplementation (Fragoso et al., 2012; Fragoso et al., 2013; Schauss et al., 2010). Additionally, açaí administration by gavage presented no genotoxic effect based on DNA damage evaluation induced by antitumor medication (Marques et al., 2016; Ribeiro et al., 2010; Schauss et al., 2010). A micronucleus test and a comet assay demonstrated the absence of genotoxic effects in mice supplemented with açaí. The results were based on parameters in the bone marrow and peripheral blood cells, polychromatic erythrocytes and liver and kidney cells. A protective role of açaí in human health was suggested by the reduction of DNA damage induced by doxorubicin (Ribeiro et al., 2010). Another study identified the same effects using a bacterial reverse mutation, a chromosomal aberration, a mammalian cell mutation assay and an *in vivo* micronucleus study (Schauss et al., 2010). No significant genotoxic effects in a comet assay and micronucleus test were verified in rat cells treated with the three doses of açaí. Similar results were reported (Marques et al., 2016).

8.8.3.6 Other Effects

Additional effects from the polysaccharide fraction of açaí were observed in the treatment of asthma and infectious disease (Holderness et al., 2011). A protective effect was demonstrated against emphysema by reducing oxidative and inflammatory reactions, and therefore, the inflammatory and oxidant actions of cigarette smoke in mice (de Moura et al., 2011; Moura et al., 2012). The anticonvulsant properties in mice (Souza-Monteiro et al., 2015) and a significant antinociceptive effect via a multifactorial mechanism of action suggest that açaí could be used to develop new analgesic drugs (Sudo et al., 2015).

8.9 *MAURITIA FLEXUOSA* L.F. (ARECACEAE)

8.9.1 Botanic Aspects and Occurrence

Mauritia flexuosa is a robust, solitary-stemmed palm tree that reaches 30 m in height, with palmate leaves that form a spherical crown, and the roots often bear a pneumatophore structure in flooded soils (Janick and Paull, 2008). This dioecious tropical palm bears many hanging inflorescences that are similar in both sexes. The fruit is spherical or ellipsoidal 4 to 5 cm in diameter and 5 to 7 cm long. Under brownish-red scales, the mesocarp is pulpy and varies from yellow, orange to reddish orange in color (Figure 8.2C). The spongy endocarp harbors the seed (Janick and Paull, 2008).

M. flexuosa is a palm tree native to South America, restricted to permanently or seasonally flooded soils, that often forms extensive, monodominant stands designated buritizais in Brazil (Koolen et al., 2013; Maria Pacheco Santos, 2005). The buriti is distributed in Brazil and occurs in the states of Pará, Amazonas, Maranhão, Piauí, Bahia, Ceará and Tocantins (Pereira Freire et al., 2016) and Colombia, Equator, Venezuela and Guyana (Delgado et al., 2007).

 For the population from the Amazon region, buriti pulp is a vital part of the diet. The fruit is used to produce juice, jam, compote, wine and ice cream (Manhães et al., 2015). Other tree palm parts are used, including trunks as bridges, fiber and seeds as handicraft and timber or leaves in construction (Brokamp et al., 2011).

8.9.2 PHYTOCHEMISTRY

Buriti fruit contains relatively high oil content (38.4% DM), similar to palm oil and other widespread oleaginous crop seeds, such as canola (40% to 45%) and sunflower (35% to 45%) (Darnet et al., 2011). Buriti oil is rich in monosaturated fatty acid (75.5% to 92.3%), in which oleic acid (C18:1) represents approximately 75%. The SFA (18.75% to 19.6%) include palmitic acid (C16:0), while the PUFA are approximately 2.1%, with linoleic acid (C18:2) predominating (Aquino et al., 2012; Darnet et al., 2011; Rodrigues et al., 2010). The comparison between the crude and refined buriti oil showed levels of PUFA and MUFA, although the refining process changes their profiles and reduces the content of carotenoids (Aquino et al., 2012; Medeiros et al., 2015) (Table 8.2).

The quantitative composition was determined among Amazonian samples using HPLC, and the results showed that the buriti has a high content of total carotenoids (513.87 to 1576 µg. g^{-1}). Twenty carotenoids were identified, with β-carotene prevalent (85.22% to 89.32%), followed by α-carotene (3.88% to 4.75%). The primary isomers are all-*trans*-β-carotene, 13-*cis*-β-carotene, 9-*cis*-β-carotene (72.45%, 11.52% and 3.61%, respectively) (de Rosso and Mercadante, 2007; Santos et al., 2015; Silva et al., 2011). The tocopherol content of the buriti fruit varies between 1129 and 1567 µg g^{-1}, with β-tocopherol representing approximately 50% to 67%, and a contribution of γ-tocopherol or α-tocopherol with approximately 78% and 70% of the total carotenoid content, respectively (Costa et al., 2010; Rodrigues et al., 2010; Santos et al., 2013; Silva et al., 2011). These results suggest that buriti can be considered an excellent source of vitamin E (Table 8.2).

β-Sitosterol, stigmasterol and campesterol are the primary phytosterols in the buriti fruit, reaching values of 84%, 20% and 9%, respectively, from the total content (1830 µg. g^{-1}). In the kernel, the phytosterol concentration is seven times smaller, with a campesterol content of 20% and β-sitosterol and stigmasterol approximately 24% (Costa et al., 2010; Dembitsky et al., 2011; Santos et al., 2013) (Table 8.2).

The phenolic compound characterization of the fruit pulp using UHPLC–ESI(–)-MS/MS revealed six phenolic acids, including *p*-coumaric acid, ferulic acid, caffeic acid, protocatechuic acid, chlorogenic acid and quinic acid. Seven flavonoids, such as (+)-catechin, (–)-epicatechin, apigenin, luteolin, myricetin, kaempferol and quercetin were identified. Protocatechuic and chlorogenic acids are the primary phenolic compounds (2175.93 and 11,154.15 µg g^{-1} DWP) (Bataglion et al., 2014). The analysis in the leaf extracts (LE), trunk extract (TE) and fruit extract (FE) showed that myricetin, (+)-catechin, chlorogenic acid, naringenin and rutin are present in all the extracts. Caffeic acid hexoside, naringenin, and (–)-epicatechin are present in both the LE and FE. Vitexin, scoparin, C3R and C3G are only detected in the FE and kaempferol in the TE (Koolen et al., 2013) (Table 8.2).

Buriti fruit is considered a good source of provitamin A with 7280 RE/100 g and contains relatively high values of total dietary fiber (22.8% FW) and protein contents (7.6% of DM) (de Rosso and Mercadante, 2007; Rodrigues et al., 2010).

8.9.3 BIOLOGICAL AND PHARMACOLOGICAL ACTIVITIES

The high content of β-carotene, α-tocopherol and oleic acid in the buriti mesocarp oil have been investigated for their effects in supplementation, protection against diseases and cosmetic formulation (Tables 8.3 and 8.4).

The antioxidant capacity of methanolic extracts has been demonstrated as higher in the leaf (iron reduction test) than in the fruit pulp (DPPH method) (Koolen et al., 2013). The same method showed that extracts from dried fruit exhibited higher antioxidant capacity than the fresh fruit extract (Gomes et al., 2016) (Table 8.3).

Young rats fed with crude buriti oil (CB) or refined buriti oil (RB) showed higher vitamin A content in serum and liver. The available vitamin A and E in young rats fed CB or RB is higher for all physiological parameters. The result suggests that buriti oil is an essential source of the antioxidant vitamins A and E and improves the lipid profile (Aquino et al., 2015). An increase in vitamin A was demonstrated after rats consumed cookies made with buriti oil. Consequently, the lipid profile and retinol content were improved, and blood glucose was not affected (Aquino et al., 2016). Pretreatment with enriched feed also prevents the neurocytotoxic and behavioral effects caused by MeHg. These results indicate the protective effect against cognitive deficits and the cytoplasmic membrane damage induced by lipid peroxidation in the rat hippocampal region (Leão et al., 2017).

Buriti oil emulsion is a potential vehicle as photo blocker by decreasing the cell damage caused by UVA and UVB radiation in X-rayed keratinocytes (Zanatta et al., 2010). The healing activity contributes to the formation and deposition of collagen fibers and provides cellular stimulation and proliferation (Batista et al., 2012). A protective effect is observed on platelet activation by increasing the antiplatelet and antithrombotic activities (Fuentes et al., 2013).

Buriti oil and extracts have been tested for antimicrobial activity .Batista et al. (2012) showed that buriti oil inhibited the bacterial growth of *Enterobacter aerogenes*, *Bacillus subtilis*, *Klebsiella pneumoniae* and *S. aureus*. *B. subtilis* has greater sensitivity to buriti oil (Batista et al., 2012). The results of the antimicrobial tests against *S. aureus*, *P. aeruginosa*, *Escherichia coli*, *Micrococcus luteus*, and *Bacillus cereus* revealed a moderate effect of the methanolic extracts on the inhibition of growth. The best results were obtained with the leaf extract against the pathogen *P. aeruginosa* with a minimum inhibitory concentration (MIC) of 50 µg. mL^{-1} (Koolen et al., 2013).

In another study, the antimicrobial activity against different fungi and bacteria was exhibited by triterpenes isolated from buriti roots. An MIC ranging from 50.8 to 203.5 µM was determined (Koolen et al., 2013). Stem ethanolic extracts from buriti demonstrated the growth inhibition of *S. aureus* (methicillin-susceptible *S. aureus* – MSSA; methicillin-resistant *S. aureus* – MRSA) (31.3 µg mL^{-1}), while the leaf extract showed activity against MRSA (62.5 µg mL^{-1}), demonstrating the antimicrobial potential of the extracts (Siqueira et al., 2014).

8.10 FINAL REMARKS

Amazon fruits and nuts are the most abundant sources of bioactive compounds with antioxidant action, such as phenolic compounds, carotenoids, tocopherols, vitamin C, UFA, terpenoids and steroids. Characteristic compounds, present in a higher amount, are a highlight for some fruits, such as vitamin C in camu-camu fruit, carotenoids in the peach palm and tucumã fruits, iridoids in genipap and selenium and UFA in Brazil nut. The synergistic effect of all these compounds showed clear evidence of the health benefits of the consumption of these fruits associated with their high antioxidant capacity.

ACKNOWLEDGMENTS

The authors thank the Federal University of Pará (PROPESP), Conselho Nacional de Desenvolvimento Científico e Tecnológico (CNPq) and Coordenação de Aperfeiçoamento de Pessoal de Nível Superior (CAPES) – Brasil (CAPES) – Finance Code 001 – for their financial support.

REFERENCES

Accioly, M. P.; Bevilaqua, C. M. L.; Rondon, F. C. M.; de Morais, S. M.; Machado, L. K. A.; Almeida, C. A.; de Andrade, H. F.; Cardoso, R. P. A. 2012. Leishmanicidal activity *in vitro* of *Musa paradisiaca* L. and *Spondias mombin* L. fractions. Veterinary Parasitology, 187, 79–84.

Ademola, I. O.; Fagbemi, B. O.; Idowu, S. O. 2005. Anthelmintic activity of extracts of *Spondias mombin* against gastrointestinal nematodes of sheep: studies *in vitro* and *in vivo*. Tropical Animal Health and Production, 37, 223–235.

Akter, M. S.; Oh, S.; Eun, J.-B.; Ahmed, M. 2011. Nutritional compositions and health promoting phytochemicals of camu-camu (*Myrciaria dubia*) fruit: a review. Food Research International, 44, 1728–1732.

Alessandra-Perini, J.; Perini, J. A.; Rodrigues-Baptista, K. C.; de Moura, R. S.; Junior, A. P.; dos Santos, T. A.; Souza, P. J. C.; Nasciutti, L. E.; Machado, D. E. 2018. *Euterpe oleracea* extract inhibits tumorigenesis effect of the chemical carcinogen DMBA in breast experimental cancer. BMC Complementary and Alternative Medicine, 18, 116.

Alessandra-Perini, J.; Rodrigues-Baptista, K. C.; Machado, D. E.; Nasciutti, L. E.; Perini, J. A. 2018. Anticancer potential, molecular mechanisms and toxicity of *Euterpe oleracea extract* (açaí): a systematic review. PLOS ONE, 13, e0200101.

Alqurashi, R. M.; Alarifi, S. N.; Walton, G. E.; Costabile, A. F.; Rowland, I. R.; Commane, D. M. 2017. *In vitro* approaches to assess the effects of açai (*Euterpe oleracea*) digestion on polyphenol availability and the subsequent impact on the faecal microbiota. Food Chemistry, 234, 190–198.

Alqurashi, R. M.; Galante, L. A.; Rowland, I. R.; Spencer, J. P. E.; Commane, D. M. 2016. Consumption of a flavonoid-rich açai meal is associated with acute improvements in vascular function and a reduction in total oxidative status in healthy overweight men. The American Journal of Clinical Nutrition, 104, 1227–1235.

Alves, J. S. F.; Medeiros, L. A. d.; Fernandes-Pedrosa, M. d. F.; Araújo, R. M.; Zucolotto, S. M. 2017. Iridoids from leaf extract of *Genipa americana*. Revista Brasileira de Farmacognosia, 27, 641–644.

Amsellem-Laufer, M. 2015. *Euterpe oleracea* Martius (Arecaceae): açaï. Phytothérapie, 13, 135–140.

Anhê, F. F.; Nachbar, R. T.; Varin, T. V.; Trottier, J.; Dudonné, S.; Le Barz, M.; Feutry, P.; et al. 2018. Treatment with camu camu (*Myrciaria dubia*) prevents obesity by altering the gut microbiota and increasing energy expenditure in diet-induced obese mice. Gut, 68, 453–464.

Aquino, J. S.; Pessoa, D. C. N. D.; Araújo, K. L. G. V.; Epaminondas, P. S.; Schuler, A. R.; Souza, A. G.; Stamford, T. L. M. 2012. Refining of buriti oil (*Mauritia flexuosa*) originated from the Brazilian cerrado: physicochemical, thermal-oxidative and nutritional implications. J Braz Chem Soc, 23, 212–219.

Aquino, J. S.; Soares, J. K.; Magnani, M.; Stamford, T. C.; Mascarenhas, R. J.; Tavares, R. L.; Stamford, T. L. 2015. Effects of dietary Brazilian palm oil (*Mauritia flexuosa* L.f.) on cholesterol profile and vitamin A and E status of rats. Molecules, 20, 9054–9070.

Aquino, J. S.; Vasconcelos, M. H.; Pessoa, D. C.; Soares, J. K.; Prado, J. P.; Mascarenhas, R. J.; Magnani, M.; Stamford, T. L. 2016. Intake of cookies made with buriti oil (*Mauritia flexuosa*) improves vitamin A status and lipid profiles in young rats. Food Function, 7, 4442–4450.

Araújo, M. L.; Costa Silva, C. F.; Medeiros Souza, R.; Melhoranca Filho, A. L. 2013. Antimicrobial activity of oils extracted from acai and pupunha on developing *Pseudomonas aeruginosa* and *Staphylococcus aureus*. Bioscience Journal, 29, 985–990.

Araújo, M. L.; Silva, C. F. C.; Melhorança Filho, A. L.; Souza, R. M.; Oliveira, W. S. 2012. Atividade antimicrobiana do óleo de duas espécies (*Bactris gasipaes* kunth. and *Bactris dahlgreniana*) de pupunha frente ao crescimento de *Staphylococcus aureus*. Ensaios e Ciência: Ciências Biológicas, Agrárias e da Saúde, 16, 21–27.

Aride, P. H. R.; Oliveira, A. M.; Batista, R. B.; Ferreira, M. S.; Pantoja-Lima, J.; Ladislau, D. S.; Castro. P. D. S.; Oliveira, A. T. 2018. Changes on physiological parameters of tambaqui (*Colossoma macropomum*) fed with diets supplemented with Amazonian fruit Camu camu (*Myrciaria dubia*), Brazilian Journal of Biology, 78, 360–367.

Arshad, S. H.; Malmberg, E.; Krapf, K.; Hide, D. W. 2018. Clinical and immunological characteristics of Brazil nut allergy. Clinical & Experimental Allergy, 21, 373–376.

Azevêdo, J. C. S.; Borges, K. C.; Genovese, M. I.; Correia, R. T. P.; Vattem, D. A. 2015. Neuroprotective effects of dried camu-camu (*Myrciaria dubia* HBK McVaugh) residue in *C. elegans*. Food Research International, 73, 135–141.

Azevedo, L.; de Araujo Ribeiro, P. F.; de Carvalho Oliveira, J. A.; Correia, M. G.; Ramos, F. M.; de Oliveira, E. B.; Barros, F.; Stringheta, P. C. 2018. Camu-camu (*Myrciaria dubia*) from commercial cultivation has higher levels of bioactive compounds than native cultivation (Amazon Forest) and presents antimutagenic effects in vivo. Journal of the Science of Food and Agriculture, 99, 624–631.

Baldissera, M. D.; Souza, C. F.; Doleski, P. H.; Grando, T. H.; Sagrillo, M. R.; da Silva, A. S.; Leal, D. B. R.; Monteiro, S. G. 2017. Treatment with tucumã oil (*Astrocaryum vulgare*) for diabetic mice prevents changes in seric enzymes of the purinergic system: improvement of immune system. Biomedicine & Pharmacotherapy, 94, 374–379.

Baldissera, M. D.; Souza, C. F.; Grando, T. H.; Cossetin, L. F.; Sagrillo, M. R.; Nascimento, K.; da Silva, A. S.; et al. 2017. Antihyperglycemic, antioxidant activities of tucumã oil (*Astrocaryum vulgare*) in alloxan-induced diabetic mice, and identification of fatty acid profile by gas chromatograph: new natural source to treat hyperglycemia. Chemico-Biological Interactions, 270, 51–58.

Baldissera, M. D.; Souza, C. F.; Grando, T. H.; Sagrillo, M. R.; Cossetin, L. F.; da Silva, A. S.; Stefani, L. M.; Monteiro, S.G. 2018. Tucumã oil (*Astrocaryum vulgare*) ameliorates hepatic antioxidant defense system in alloxan-induced diabetic mice. Journal of Food Biochemistry, 42, e12468.

Baldissera, M. D.; Souza, C. F.; Grando, T. H.; Sagrillo, M. R.; Cossetin, L. F.; da Silva, A. S.; Stefani, L. M.; Monteiro, S. G. 2018. Tucuma oil (*Astrocaryum vulgare*) ameliorates hepatic antioxidant defense system in alloxan-induced diabetic mice. Journal of Food Biochemistry, 42, e12468.

Baldissera, M. D.; Souza, C. F.; Grando, T. H.; Sagrillo, M. R.; da Silva, A. S.; Stefani, L. M.; Monteiro, S. G. 2017. The use of tucumã oil (*Astrocaryum vulgare*) in alloxan-induced diabetic mice: effects on behavior, oxidant/antioxidant status, and enzymes involved in brain neurotransmission. Molecular and Cellular Biochemistry, 436, 159–166.

Barbosa, P. O.; Pala, D.; Silva, C. T.; de Souza, M. O.; do Amaral, J. F.; Vieira, R. A. L.; Folly, G. A. d. F.; Volp, A. C. P.; de Freitas, R. N. 2016. Açai (*Euterpe oleracea* Mart.) pulp dietary intake improves cellular antioxidant enzymes and biomarkers of serum in healthy women. Nutrition, 32, 674–680.

Barros, L.; Calhelha, R. C.; Queiroz, M. J. R. P.; Santos-Buelga, C.; Santos, E. A.; Regis, W. C. B.; Ferreira, I. C. F. R. 2015. The powerful *in vitro* bioactivity of *Euterpe oleracea* Mart. seeds and related phenolic compounds. Industrial Crops and Products, 76, 318–322.

Bartolomé, B.; Mendez, J. D.; Armentia, A.; Vallverdu, A.; Palacios, R. 1997. Allergens from Brazil nut: immunochemical characterization. Allergologia et Immunopathologia, 25, 135–144.

Bataglion, G. A.; da Silva, F. M. A.; Eberlin, M. N.; Koolen, H. H. F. 2014. Simultaneous quantification of phenolic compounds in buriti fruit (*Mauritia flexuosa* L.f.) by ultra-high performance liquid chromatography coupled to tandem mass spectrometry. Food Research International, 66, 396–400.

Batista, J.; Olinda, R.; Medeiros, V.; Rodrigues, C.; Oliveira, A.; Paiva, E.; Freitas, C.; Medeiros, A. 2012. Antibacterial and healing activities of buriti oil *Mauritia flexuosa* L. Ciência Rural, 42, 136–141.

Bentes, A. d. S.; de Souza, H. A. L.; Amaya-Farfan, J.; Lopes, A. S.; de Faria, L. J. G. 2015. Influence of the composition of unripe genipap (*Genipa americana* L.) fruit on the formation of blue pigment. Journal of Food Science and Technology, 52, 3919–3924.

Bentes, A. d. S.; Mercadante, A. Z. 2014. Influence of the stage of ripeness on the composition of iridoids and phenolic compounds in genipap (*Genipa americana* l.). Journal of Agricultural and Food Chemistry, 62, 10800–10808.

Bereau, D.; Benjelloun-Mlayah, B.; Banoub, J.; Bravo, R. 2003. FA and unsaponifiable composition of five Amazonian palm kernel oils. Journal of the American Oil Chemists' Society, 80, 49–53.

Bichara, C. M. G.; Rogez, H. 2011. Postharvest Biology and Technology of Tropical and Subtropical Fruits. New Delhi: Woodhead Publishing.

Bonomo, L. d. F.; Silva, D. N.; Boasquivis, P. F.; Paiva, F. A.; Guerra, J. F. d. C.; Martins, T. A. F.; de Jesus Torres, Á. G.; et al. 2014. Açaí (*Euterpe oleracea* Mart.) modulates oxidative stress resistance in *Caenorhabditis elegans* by direct and indirect mechanisms. PLoS ONE, 9, e89933.

Bony, E.; Boudard, F.; Brat, P.; Dussossoy, E.; Portet, K.; Poucheret, P.; Giaimis, J.; Michel, A. 2012. Awara (*Astrocaryum vulgare* M.) pulp oil: chemical characterization, and anti-inflammatory properties in a mice model of endotoxic shock and a rat model of pulmonary inflammation. Fitoterapia, 83, 33–43.

Bony, E.; Boudard, F.; Dussossoy, E.; Portet, K.; Brat, P.; Giaimis, J.; Michel, A. 2012. Chemical composition and anti-inflammatory properties of the unsaponifiable fraction from awara (*Astrocaryum vulgare* M.) pulp oil in activated j774 macrophages and in a mice model of endotoxic shock. Plant Foods for Human Nutrition, 67, 384–392.

Bony, E.; Dussossoy, E.; Michel, A.; Brat, P.; Boudard, F.; Giaimis, J.; Barouh, N.; Piombo, G. 2014. Chemical composition and anti-inflammatory activities of an unsaponifiable fraction of pulp oil from awara (*Astrocaryum vulgare* M.). Acta Horticulturae, 1010, 43–48.

Bora, P. S.; Narain, N.; Rocha, R. V. M.; De Oliveira Monteiro, A. C.; De Azevedo Moreira, R. 2001. Characterisation of the oil and protein fractions of tucuma (*Astrocaryum vulgare* Mart.) fruit pulp and seed kernel. Ciencia y Tecnologia Alimentaria, 3, 111–116.

Bovi, M. L. A.; Castro, A. 1993. Assai. In J. W. Clay; C. R. Clement (Eds.) Income Generating Forests and Conservation in Amazonia, 58–67. Roma: FAO.

Brasil, A.; Rocha, F. A. F.; Gomes, B. D.; Oliveira, K. R. M.; de Carvalho, T. S.; Batista, E. J. O.; Borges, R.D.S.; Kremers. J.; Herculano, A. M. 2017. Diet enriched with the Amazon fruit acai (*Euterpe oleracea*) prevents electrophysiological deficits and oxidative stress induced by methyl-mercury in the rat retina. Nutritional Neuroscience, 20, 265–272.

Brito, C.; Stavroullakis, A. T.; Ferreira, A. C.; Li, K.; Oliveira, T.; Nogueira-Filho, G.; Prakki, A. 2016. Extract of acai-berry inhibits osteoclast differentiation and activity. Archives of Oral Biology, 68, 29–34.

Brito, S. A.; de Almeida, C. L. F.; de Santana, T. I.; da Silva Oliveira, A. R.; do Nascimento Figueiredo, J. C. B.; Souza, I. T.; de Almeida, L. L.; et al. 2018. Antiulcer activity and potential mechanism of action of the leaves of *Spondias mombin* L. Oxidative Medicine and Cellular Longevity, 2018, 1–20.

Brokamp, G.; Valderrama, N.; Mittelbach, M.; Grandez R, C. A.; Barfod, A. S.; Weigend, M. 2011 Trade in palm products in north-western South America. The Botanical Review, 77, 571–606.

Cabral, B.; Siqueira, E. M. S.; Bitencourt, M. A. O.; Lima, M. C. J. S.; Lima, A. K.; Ortmann, C. F.; Chaves, V. C.; et al. 2016. Phytochemical study and anti-inflammatory and antioxidant potential of *Spondias mombin* leaves. Revista Brasileira de Farmacognosia, 26, 304–311.

Campos, F. R.; Januário, A. H.; Rosas, L. V.; Nascimento, S. K. R.; Pereira, P. S.; França, S. C.; Cordeiro, M. S. C.; Toldo, M. P. A.; Albuquerque, S. 2005. Trypanocidal activity of extracts and fractions of *Bertholletia excelsa*. Fitoterapia, 76, 26–29.

Cândido, T. L. N.; Silva, M. R.; Agostini-Costa, T. S. 2015. Bioactive compounds and antioxidant capacity of buriti (*Mauritia flexuosa* L.f.) from the cerrado and Amazon biomes. Food Chemistry, 177, 313–319.

Cardoso, B. R.; Duarte, G. B. S.; Reis, B. Z.; Cozzolino, S. M. F. 2017. Brazil nuts: nutritional composition, health benefits and safety aspects. Food Research International, 100, 9–18.

Carey, A. N.; Miller, M. G.; Fisher, D. R.; Bielinski, D. F.; Gilman, C. K.; Poulose, S. M.; Shukitt-Hale, B. 2017. Dietary supplementation with the polyphenol-rich açaí pulps (*Euterpe oleracea Mart.* and *Euterpe precatoria Mart.*) improves cognition in aged rats and attenuates inflammatory signaling in BV-2 microglial cells. Nutritional Neuroscience, 20, 238–245.

Carvalho, A. V.; Ferreira Ferreira da Silveira, T.; Mattietto. R. A.; Padilha de Oliveira, M. D.; Godoy, H. T. 2017. Chemical composition and antioxidant capacity of acai (*Euterpe oleracea*) genotypes and commercial pulps. Journal of the Science of Food and Agriculture, 97, 1467–1474.

Carvalho, R. F.; Huguenin, G. V.; Luiz, R. R.; Moreira, A. S.; Oliveira, G. M.; Rosa, G. 2015. Intake of partially defatted Brazil nut flour reduces serum cholesterol in hypercholesterolemic patients – a randomized controlled trial. Nutrition Journal, 14, 59.

Carvalho, R. P.; Lemos, J. R.; de Aquino Sales, R. S.; Martins, M. G.; Nascimento, C. H.; Bayona, M.; Marcon, J. L.; Monteiro, J. B. 2013. The consumption of red pupunha (*Bactris gasipaes* kunth) increases HDL cholesterol and reduces weight gain of lactating and post-lactating wistar rats. The Journal of Aging Research & Clinical Pratice, 2, 257–260.

Carvalho-Peixoto, J.; Moura, M. R. L.; Cunha, F. A.; Lollo, P. C. B.; Monteiro, W. D.; Carvalho, L. M. J. d.; Farinatti, P. d. T. V. 2015. Consumption of açaí (*Euterpe oleracea* Mart.) functional beverage reduces muscle stress and improves effort tolerance in elite athletes: a randomized controlled intervention study. Applied Physiology, Nutrition, and Metabolism, 40, 725–733.

Chin, Y.-W.; Chai, H.-B.; Keller, W. J.; Kinghorn, A. D. 2008. Lignans and other constituents of the fruits of *Euterpe oleracea* (açai) with antioxidant and cytoprotective activities. Journal of Agricultural and Food Chemistry, 56, 7759–7764.

Choi, Y. J.; Choi, Y. J.; Kim, N.; Nam, R. H.; Lee, S.; Lee, H. S.; Lee, H.-N.; Surh, Y.-J.; Lee, D. H. 2017. Açaí berries inhibit colon tumorigenesis in azoxymethane/dextran sulfate sodium-treated mice. Gut and Liver, 11, 243–252.

Choudhury, S.; Headey, D. 2017. What drives diversification of national food supplies? A cross-country analysis. Global Food Security, 15, 85–93.

Chunhieng, T.; Hafidi, A.; Pioch, D.; Brochier, J.; Didier, M. 2008. Detailed study of Brazil nut (*Bertholletia excelsa*) oil micro-compounds: phospholipids, tocopherols and sterols. Journal of the Brazilian Chemical Society, 19, 1374–1380.

Chunhieng, T.; Petritis, K.; Elfakir, C.; Brochier, J.; Goli, T.; Montet, D. 2004. Study of selenium distribution in the protein fractions of the Brazil nut, *Bertholletia excelsa*. Journal of Agricultural and Food Chemistry, 52, 4318–4322.

Cicero, N.; Albergamo, A.; Salvo, A.; Bua, G. D.; Bartolomeo, G.; Mangano, V.; Rotondo, A.; Di Stefano, V.; Di Bella, G.; Dugo, G. 2018. Chemical characterization of a variety of cold-pressed gourmet oils available on the Brazilian market. Food Research International, 109, 517–525.

Clement, C.; De Cristo-Araújo, M.; Coppens D'Eeckenbrugge, G.; Alves Pereira, A.; Picanço-Rodrigues, D. 2010. Origin and domestication of native Amazonian crops. Diversity, 2, 72–106.

Clement, C. R.; Weber, J. C.; van Leeuwen, J.; Astorga Domian, C.; Cole, D. M.; Arévalo Lopez, L. A.; Argüello, H. 2004. Why extensive research and development did not promote use of peach palm fruit in Latin America. Agroforestry Systems, 61–62, 195–206.

Colpo, E.; Vilanova, C. D. d. A.; Brenner Reetz, L. G.; Medeiros Frescura Duarte, M. M.; Farias, I. L. G.; Irineu Muller, E.; Muller, A. L. H.; Moraes Flores, E. M.; Wagner, R.; da Rocha, J. B. T. 2013. A single consumption of high amounts of the Brazil nuts improves lipid profile of healthy volunteers. Journal of Nutrition and Metabolism, 2013, 1–7.

Cominetti, C.; de Bortoli, M. C.; Garrido, A. B.; Cozzolino, S. M. F. 2012. Brazilian nut consumption improves selenium status and glutathione peroxidase activity and reduces atherogenic risk in obese women. Nutrition Research, 32, 403–407.

Cordeiro, S. C. V.; Carvalho, L. C. R. M.; de Bem, G. F.; Costa, C. A.; Souza; Sousa, P. J. C.; de Souza, M. A. V.; Rocha, V. N.; Carvalho, J. J.; de Moura, R. S.; Resende, A. C. 2015. *Euterpe oleracea* Mart. extract prevents vascular remodeling and endothelial dysfunction in spontaneously hypertensive rats. International Journal of Applied Research in Natural Products, 8, 6–16.

Costa, P. A. d.; Ballus, C. A.; Teixeira-Filho, J.; Godoy, H. T. 2010. Phytosterols and tocopherols content of pulps and nuts of Brazilian fruits. Food Research International, 43, 1603–1606.

da Conceição, A. O.; Rossi, M. H.; de Oliveira, F. F.; Takser, L.; Lafond, J. 2011. *Genipa americana* (Rubiaceae) fruit extract affects mitogen-activated protein kinase cell pathways in human trophoblast–derived bewo cells: implications for placental development. Journal of Medicinal Food, 14, 483–494.

da Costa, C. A.; de Oliveira, P. R. B.; de Bem, G. F.; de Cavalho, L. C. R. M.; Ognibene, D. T.; da Silva, A. F. E.; dos Santos Valença, S.; et al. 2012. *Euterpe oleracea* Mart.-derived polyphenols prevent endothelial dysfunction and vascular structural changes in renovascular hypertensive rats: role of oxidative stress. Naunyn-Schmiedeberg's Archives of Pharmacology, 385, 1199–1209.

da Costa, C. A.; Ognibene, D. T.; Cordeiro, V. S. C.; de Bem, G. F.; Santos, I. B.; Soares, R. A.; de Melo Cunha, L. L.; Carvalho, L. C. R. M.; de Moura, R. S.; Resende, A. C. 2017. Effect of *Euterpe oleracea* Mart. seeds extract on chronic ischemic renal injury in renovascular hypertensive rats. Journal of Medicinal Food, 20, 1002–1010.

da Silva, F. C.; Arruda, A.; Ledel, A.; Dauth, C.; Romão, N. F.; Viana, R. N.; de Barros Falcão Ferraz, A.; Picada, J. N.; Pereira, P. 2012. Antigenotoxic effect of acute, subacute and chronic treatments with Amazonian camu–camu (*Myrciaria dubia*) juice on mice blood cells. Food and Chemical Toxicology, 50, 2275–2281.

da Silva Cristino Cordeiro, V.; de Bem, G. F.; da Costa, C. A.; Santos, I. B.; de Carvalho, L. C. R. M.; Ognibene, D. T.; da Rocha, A. P. M. et al. 2018. *Euterpe oleracea* Mart. seed extract protects against renal injury in diabetic and spontaneously hypertensive rats: role of inflammation and oxidative stress. European Journal of Nutricion, 57, 817–832.

da Silva Santos, V.; Bisen-Hersh, E.; Yu, Y.; Cabral, I. S. R.; Nardini, V.; Culbreth, M.; Teixeira da Rocha, J. B.; Barbosa, F.; Aschner, M. 2014. Anthocyanin-rich açaí (*Euterpe oleracea* Mart.) extract attenuates manganese-induced oxidative stress in rat primary astrocyte cultures. Journal of Toxicology and Environmental Health, Part A, 77, 390–404.

Darnet, S.; Serra, J. L.; Rodrigues, A. M. D.; da Silva, L. H. M. 2011. A high-performance liquid chromatography method to measure tocopherols in assai pulp (*Euterpe oleracea*). Food Research International, 44, 2107–2111.

Darnet, S. H.; Silva, L. H. M. d.; Rodrigues, A. M. d. C.; Lins, R. T. 2011. Nutritional composition, fatty acid and tocopherol contents of buriti (*Mauritia flexuosa*) and patawa (*Oenocarpus bataua*) fruit pulp from the Amazon region. Ciência e Tecnologia de Alimentos, 31, 488–491.

de Bem, G. F.; da Costa, C. A.; de Oliveira, P. R. B.; Cordeiro, V. S. C.; Santos, I. B.; de Carvalho, L. C. R. M.; Souza, M. A. V.; et al. 2014. Protective effect of *Euterpe oleracea* Mart (açaí) extract on programmed changes in the adult rat offspring caused by maternal protein restriction during pregnancy. Journal of Pharmacy and Pharmacology, 66, 1328–1338.

de Moura, R. S.; Pires, K. M. P.; Ferreira, T. S.; Lopes, A. A.; Nesi, R. T.; Resende, A. C.; Sousa, P. J. C.; da Silva, A. J. R.; Porto, L. C.; Valenca, S. S. 2011. Addition of açaí (*Euterpe oleracea*) to cigarettes has a protective effect against emphysema in mice. Food and Chemical Toxicology, 49, 855–863.

de Oliveira, P. R. B.; da Costa, C. A.; de Bem, G. F.; Cordeiro, V. S. C.; Santos, I. B.; de Carvalho, L. C. R. M.; da Conceição, E. P. S.; et al. 2015. *Euterpe oleracea* Mart.-derived polyphenols protect mice from diet-induced obesity and fatty liver by regulating hepatic lipogenesis and cholesterol excretion. PLOS ONE, 10, e0143721.

de Oliveira, P. R. B.; da Costa, C. A.; de Bem, G. F.; Marins de Cavalho, L. C. R.; de Souza, M. A. V.; de Lemos Neto, M.; da Cunha Sousa, P. J.; de Moura, R. S.; Resende, A. C. 2010. Effects of an extract obtained from fruits of *Euterpe oleracea* Mart. in the components of metabolic syndrome induced in c57bl/6j mice fed a high-fat diet. Journal of Cardiovascular Pharmacology, 56, 619–626.

de Rosso, V. V.; Mercadante, A. Z. 2007. Identification and quantification of carotenoids, by HPLC-PDA-MS/MS, from Amazonian fruits. Journal of Agricultural and Food Chemistry, 55, 5062–5072.

de Souza, M. O., Silva, M.; Silva, M. E.; de Oliveira, P. R.; Pedrosa, M. L. 2010. Diet supplementation with acai (*Euterpe oleracea* Mart.) pulp improves biomarkers of oxidative stress and the serum lipid profile in rats. Nutrition, 26, 804–810.

de Souza, M. O.; Souza e Silva, L.; de Brito Magalhães, C. L.; de Figueiredo, B. B.; Costa, D. C.; Silva, M. E.; Pedrosa, M. L. 2012. The hypocholesterolemic activity of açaí (*Euterpe oleracea* Mart.) is mediated by the enhanced expression of the ATP-binding cassette, subfamily G transporters 5 and 8 and low-density lipoprotein receptor genes in the rat. Nutrition Research, 32, 976–984.

de Souza Machado, F.; Marinho, J. P.; Abujamra, A. L.; Dani, C.; Quincozes-Santos, A.; Funchal, C. 2015. Carbon tetrachloride increases the pro-inflammatory cytokines levels in different brain areas of wistar rats: the protective effect of acai frozen pulp. Neurochemical Research, 40, 1976–1983.

de Sousa Pereira, I.; Moreira Cançado Mascarenhas Pontes, T. C.; Lima Vieira, R. A., de Freitas Folly, G. A.; Cacilda Silva, F.; Pereira de Oliveira, F. L.; Ferreira do Amaral, J. et al. 2015. The consumption of açaí pulp changes the concentrations of plasminogen activator inhibitor-1 and epidermal growth factor (EGF) in apparently healthy women. Nutrición Hospitalaria, 32, 931–945.

Del Pozo-Insfran, D.; Brenes, C. H.; Talcott, S. T. 2004. Phytochemical composition and pigment stability of açai (*Euterpe oleracea* Mart.). Journal of Agricultural and Food Chemistry, 52, 1539–1545.

Delgado, C.; Couturier, G.; Mejia, K. 2007. *Mauritia flexuosa* (Arecaceae: Calamoideae): an Amazonian palm with cultivation purposes in Peru. Fruits, 62, 157–169.

Dembitsky, V. M.; Poovarodom, S.; Leontowicz, H.; Leontowicz, M.; Vearasilp, S.; Trakhtenberg, S.; Gorinstein, S. 2011. The multiple nutrition properties of some exotic fruits: biological activity and active metabolites. Food Research International, 44, 1671–1701.

Dias, A. L. S.; Rozet, E.; Chataigné, G.; Oliveira, A. C.; Rabelo, C. A. S.; Hubert, P.; Rogez, H.; Quetin-Leclercq, J. 2012. A rapid validated UHPLC–PDA method for anthocyanins quantification from *Euterpe oleracea* fruits. Journal of Chromatography B, 907, 108–116.

Dias, A. L. S.; Rozet, E.; Larondelle, Y.; Hubert, P.; Rogez, H.; Quetin-Leclercq, J. 2013. Development and validation of an UHPLC-LTQ-Orbitrap MS method for non-anthocyanin flavonoids quantification in *Euterpe oleracea* juice. Analytical and Bioanalytical Chemistry, 405, 9235–9249.

Dickson, L.; Tenon, M.; Svilar, L.; Fança-Berthon, P.; Lugan, R.; Martin, J.-C.; Vaillant, F.; Rogez, H. 2018. Main human urinary metabolites after genipap (*Genipa americana* l.) juice intake. Nutrients, 10, 1155.

Donadio, J.; Rogero, M.; Cockell, S.; Hesketh, J.; Cozzolino, S. 2017. Influence of genetic variations in selenoprotein genes on the pattern of gene expression after supplementation with Brazil nuts. Nutrients, 9, 739.

Dos Santos, M. d. F.; Mamede, R. V.; Rufino, M. d. S.; de Brito, E. S.; Alves, R. E. 2015. Amazonian native palm fruits as sources of antioxidant bioactive compounds. Antioxidants (Basel), 4, 591–602.

Dutra, R. C.; Campos, M. M.; Santos, A. R. S.; Calixto, J. B. 2016. Medicinal plants in Brazil: pharmacological studies, drug discovery, challenges and perspectives. Pharmacological Research, 112, 4–29.

El Morsy, E. M.; Ahmed, M. A. E.; Ahmed, A. A. E. 2015. Attenuation of renal ischemia/reperfusion injury by açaí extract preconditioning in a rat model. Life Sciences, 123, 35–42.

Espinosa-Pardo, F. A.; Martinez, J.; Martinez-Correa, H. A. 2014. Extraction of bioactive compounds from peach palm pulp (*Bactris gasipaes*) using supercritical CO_2. The Journal of Supercritical Fluids, 93, 2–6.

Feio, C. A.; Izar, M. C.; Ihara, S. S.; Kasmas, S. H.; Martins, C. M.; Feio, M. N.; Maués, L. A.; et al. 2012. *Euterpe oleracea* (açai) modifies sterol metabolism and attenuates experimentally-induced atherosclerosis. Journal of Atherosclerosis and Thrombosis, 19, 237–245.

Fidelis, M.; Santos, J. S.; Escher, G. B.; Vieira do Carmo, M.; Azevedo, L.; Cristina da Silva, M.; Putnik, P.; Granato, D. 2018. In vitro antioxidant and antihypertensive compounds from camu-camu (*Myrciaria dubia* McVaugh, Myrtaceae) seed coat: a multivariate structure-activity study. Food and Chemical Toxicology, 120, 479–490.

Food and Agriculture Organization of the United Nations – FAO. 2017. The future of food and agriculture – Trends and challenges. http://www.fao.org/publications/fofa/en/.

Food and Agriculture Organization of the United Nations – FAO. 2018. Biodiversity for sustainable agriculture – Biodiversity for sustainable agriculture. http://www.fao.org/documents/card/en/c/85baf9c5-ea7f-4e25-812f-737755a8b320/.

Food and Drug Administration – FDA. 2011 Food allergen labelling and information requirements the under the EU food information for consumers regulation no. 1169/2011 1: Technical guidance. https://www.food.gov.uk/business-guidance/allergen-labelling-for-food-manufacturers.

Fragoso, M. F.; Prado, M. G.; Barbosa, L.; Rocha, N. S.; Barbisan, L. F. 2012. Inhibition of mouse urinary bladder carcinogenesis by açai fruit (*Euterpe oleraceae* Martius) intake. Plant Foods for Human Nutrition, 67, 235–241.

Fragoso, M. F.; Romualdo, G. R.; Ribeiro, D. A.; Barbisan, L. F. 2013. Açai (*Euterpe oleracea* Mart.) feeding attenuates dimethylhydrazine-induced rat colon carcinogenesis. Food and Chemical Toxicology, 58, 68–76.

Freitas, D. d. S.; Morgado-Díaz, J. A.; Gehren, A. S.; Vidal, F. C. B.; Fernandes, R. M. T.; Romão, W.; Tose, L. V.; et al. 2017. Cytotoxic analysis and chemical characterization of fractions of the hydroalcoholic extract of the *Euterpe oleracea* Mart. seed in the MCF-7 cell line. Journal of Pharmacy and Pharmacology, 69, 714–721.

Fuentes, E.; Rodriguez-Perez, W.; Guzman, L.; Alarcon, M.; Navarrete, S.; Forero-Doria, O.; Palomo, I. 2013. *Mauritia flexuosa* presents *in vitro* and *in vivo* antiplatelet and antithrombotic activities. Evidence-Based Complementary and Alternative Medicine, 2013, 653257.

Fujita, A.; Sarkar, D.; Wu, S.; Kennelly, E.; Shetty, K.; Genovese, M. I. 2015. Evaluation of phenolic-linked bioactives of camu-camu (*Myrciaria dubia* McVaugh) for antihyperglycemia, antihypertension, antimicrobial properties and cellular rejuvenation. Food Research International, 77, 194–203.

Fujita, A.; Souza, V. B.; Daza, L. D.; Fávaro-Trindade, C. S.; Granato, D.; Genovese, M. I. 2017. Effects of spray-drying parameters on *in vitro* functional properties of camu-camu (*Myrciaria dubia* McVaugh): a typical Amazonian fruit. Journal of Food Science, 82, 1083–1091.

Gale, A. M.; Kaur, R.; Baker, W. L. 2014. Hemodynamic and electrocardiographic effects of açaí berry in healthy volunteers: a randomized controlled trial. International Journal of Cardiology, 174, 421–423.

Gallori, S.; Bilia, A. R.; Bergonzi, M. C.; Barbosa, W. L. R.; Vincieri, F. F. 2004. Polyphenolic constituents of fruit pulp of *Euterpe oleracea* Mart. (açai palm). Chromatographia, 59, 739–743.

Geiselhart, S.; Hoffmann-Sommergruber, K.; Bublin, M. 2018. Tree nut allergens. Molecular Immunology, 100, 71–81.

Genovese, M. I.; Da Silva Pinto, M.; De Souza Schmidt Gonçalves, A. E.; Lajolo, F. M. 2008. Bioactive compounds and antioxidant capacity of exotic fruits and commercial frozen pulps from Brazil. Food Science and Technology International, 14, 207–214.

Gomes, S. M.; Ghica, M. E.; Rodrigues, I. A.; de Souza Gil, E.; Oliveira-Brett, A. M. 2016. Flavonoids electrochemical detection in fruit extracts and total antioxidant capacity evaluation. Talanta, 154, 284–291.

Gonçalves, A. E. S. S.; Lellis-santos, C.; Curi, R.; Lajolo, F. M.; Genovese, M. I. 2014. Frozen pulp extracts of camu-camu (*Myrciaria dubia* McVaugh) attenuate the hyperlipidemia and lipid peroxidation of Type 1 diabetic rats. Food Research International, 64, 1–8.

Gordon, A.; Cruz, A. P. G.; Cabral, L. M. C.; de Freitas, S. C.; Taxi, C. M. A. D.; Donangelo, C. M.; de Andrade Mattietto, R.; Friedrich, M.; da Matta, V. M.; Marx, F. 2012. Chemical characterization and evaluation of antioxidant properties of Açaí fruits (*Euterpe oleraceae* Mart.) during ripening. Food Chemistry, 133, 256–263.

Guerra, J. F. d. C.; Maciel, P. S.; de Abreu, I. C. M. E.; Pereira, R. R.; Silva, M.; Cardoso, L. d. M.; Pinheiro-Sant'Ana, H. M.; Lima, W. G. d.; Silva, M. E.; Pedrosa, M. L. 2015. Dietary açai attenuates hepatic steatosis via adiponectin-mediated effects on lipid metabolism in high-fat diet mice. Journal of Functional Foods, 14, 192–202.

Guerra, J. F. d. C.; Magalhães, C. L. d. B.; Costa, D. C.; Silva, M. E.; Pedrosa, M. L. 2011. Dietary açai modulates ROS production by neutrophils and gene expression of liver antioxidant enzymes in rats. Journal of Clinical Biochemistry and Nutrition, 49, 188–194.

Hamano, P. S.; Mercadante, A. Z. 2001. Composition of carotenoids from commercial products of caja (*Spondias lutea*). Journal of Food Composition and Analysis, 14, 335–343.

Hassimotto, N. M. A.; Genovese, M. I.; Lajolo, F. M. 2005. Antioxidant activity of dietary fruits, vegetables, and commercial frozen fruit pulps. Journal of Agricultural and Food Chemistry, 53, 2928–2935.

Heinrich, M.; Dhanji, T.; Casselman, I. 2011. Açai (*Euterpe oleracea* Mart.)—a phytochemical and pharmacological assessment of the species' health claims. Phytochemistry Letters, 4, 10–21.

Hempel, J.; Amrehn, E.; Quesada, S.; Esquivel, P.; Jimenez, V. M.; Heller, A.; Carle, R.; Schweiggert, R. M. 2014. Lipid-dissolved gamma-carotene, beta-carotene, and lycopene in globular chromoplasts of peach palm (*Bactris gasipaes* Kunth) fruits. Planta, 240, 1037–1050.

Henderson, A. 1995. The Palms of the Amazon. Oxford: Oxford University Press.

Hogan, S.; Chung, H.; Zhang, L.; Li, J.; Lee, Y.; Dai, Y.; Zhou, K. 2010. Antiproliferative and antioxidant properties of anthocyanin-rich extract from açai. Food Chemistry, 118, 208–214.

Holderness, J.; Schepetkin, I. A.; Freedman, B.; Kirpotina, L. N.; Quinn, M. T.; Hedges, J. F.; Jutila, M. A. 2011. Polysaccharides isolated from açai fruit induce innate immune responses. PLoS ONE, 6, e17301.

Horiguchi, T.; Ishiguro, N.; Chihara, K.; Ogi, K.; Nakashima, K.; Sada, K.; Hori-Tamura, N. 2011. Inhibitory effect of açaí (*Euterpe oleracea* Mart.) pulp on IgE-mediated mast cell activation. Journal of Agricultural and Food Chemistry, 59, 5595–5601.

Hu, Y.; McIntosh, G. H.; Le Leu, R. K.; Somashekar, R.; Meng, X. Q.; Gopalsamy, G.; Bambaca, L.; McKinnon, R. A.; Young, G. P. 2016. Supplementation with Brazil nuts and green tea extract regulates targeted biomarkers related to colorectal cancer risk in humans. British Journal of Nutrition, 116, 1901–1911.

Inoue, T.; Komoda, H.; Uchida, T.; Node, K. 2008. Tropical fruit camu-camu (*Myrciaria dubia*) has anti-oxidative and anti-inflammatory properties. Journal of Cardiology, 52, 127–132.

IOM. 2000. Dietary Reference Intakes for Vitamin C, Vitamin E, Selenium, and Carotenoids. Washington: National Academies Press (US).

Janick, J.; Paull, R. E. 2008. The Encyclopedia of Fruit and Nuts. Cambridge: CABI.

Jardim, M. A. G.; Macambira, M. L. J. 1996. Biologia floral do açaizeiro (*Euterpe oleracea* Martius). Boletim do Museu Paraense Emílio Goeldi, 12, 131–136.

Jatunov, S.; Quesada, S.; Díaz, C.; Murillo, E. 2010. Carotenoid composition and antioxidant activity of the raw and boiled fruit mesocarp of six varieties of *Bactris gasipaes*. Archivos Latinoamericanos de Nutrición, 60, 99–104.

Jayasinghe, S. B.; Caruso, J. A. 2011. Investigation of Se-containing proteins in *Bertholletia excelsa* H.B.K. (Brazil nuts) by ICPMS, MALDI-MS and LC-ESI-MS methods. International Journal of Mass Spectrometry, 307, 16–27.

John, J. A.; Shahidi, F. 2010. Phenolic compounds and antioxidant activity of Brazil nut (*Bertholletia excelsa*). Journal of Functional Foods, 2, 196–209.

Kahn, F. 2008. The genus *Astrocaryum* (Arecaceae). Revista Peruana de Biologia, 1(Suppl.), 31–48.

Kahn, F.; de Granville, J. J. 1992. Palms in forest ecosystems of Amazonia, ecological studies. Ecological Studies. Berlin Heidelberg: Springer Verlag.

Kaneshima, T.; Myoda, T.; Toeda, K.; Fujimori, T.; Nishizawa, M. 2017 Antimicrobial constituents of peel and seeds of camu-camu (*Myrciaria dubia*). Bioscience, Biotechnology, and Biochemistry, 81, 1461–1465.

Kang, J.; Li, Z.; Wu, T.; Jensen, G. S.; Schauss, A. G.; Wu, X. 2010. Anti-oxidant capacities of flavonoid compounds isolated from acai pulp (*Euterpe oleracea* Mart.). Food Chemistry, 122, 610–617.

Kang, J.; Xie, C.; Li, Z.; Nagarajan, S.; Schauss, A. G.; Wu, T.; Wu, X. 2011. Flavonoids from acai (*Euterpe oleracea* Mart.) pulp and their antioxidant and anti-inflammatory activities. Food Chemistry, 128, 152–157.

Kang, M. H.; Choi, S.; Kim, B.-H. 2017. Skin wound healing effects and action mechanism of acai berry water extracts. Toxicological Research, 33, 149–156.

Kim, H.; Simbo, S. Y.; Fang, C.; McAlister, L.; Roque, A.; Banerjee, N.; Talcott, S. T.; Zhao, H.; Kreider, R. B.; Mertens-Talcott, S. U. 2018. Acai (*Euterpe oleracea* Mart.) beverage consumption improves biomarkers for inflammation but not glucose- or lipid-metabolism in individuals with metabolic syndrome in a randomized, double-blinded, placebo-controlled clinical trial. Food & Function, 9, 3097–3103.

Kim, J. Y.; Hong, J. H.; Jung, H. K.; Jeong, Y. S.; Cho, K. H. 2012. Grape skin and loquat leaf extracts and acai puree have potent anti-atherosclerotic and anti-diabetic activity *in vitro* and *in vivo* in hypercholesterolemic zebrafish. International Journal of Molecular Medicine, 30, 606–614.

Koolen, H. H. F.; da Silva, F. M. A.; Gozzo, F. C.; de Souza, A. Q. L.; de Souza, A. D. L. 2013. Antioxidant, antimicrobial activities and characterization of phenolic compounds from buriti (*Mauritia flexuosa* L.f.) by UPLC–ESI-MS/MS. Food Research International, 51, 467–473.

Koolen, H. H. F.; Soares, E. R.; da Silva, F. M.; de Oliveira, A. A.; de Souza, A. Q.; de Medeiros, L. S.; Rodrigues-Filho, E.; et al. 2013. Mauritic acid: a new dammarane triterpene from the roots of *Mauritia flexuosa* L.f. (Arecaceae). Natural Product Research, 27, 2118–2125.

Laslo, M.; X. Sun, X.; Hsiao, C. T.; Wu, W.W.; Shen, R. F., Zou, S. 2013. A botanical containing freeze dried açai pulp promotes healthy aging and reduces oxidative damage in sod1 knockdown flies. Age, 35, 1117–1132.

Leão, L. K. R.; Herculano, A. M.; Maximino, C.; Brasil Costa, A.; Gouveia Jr, A.; Batista, E. O.; Rocha, F. F. et al. 2017. *Mauritia flexuosa* L. protects against deficits in memory acquisition and oxidative stress in rat hippocampus induced by methylmercury exposure. *Nutritional* Neuroscience 20, 297–304.

Leterme, P.; García, M.-F.; Londoño, A.-M.; Rojas, M.-G.; Buldgen, A.; Souffrant, W.-B. 2005. Chemical composition and nutritive value of peach palm (*Bactris gasipaes* Kunth) in rats. Journal of the Science of Food and Agriculture, 85, 1505–1512.

Lichtenthäler, R.; Rodrigues, R. B.; Maia, J. G. S.; Papagiannopoulos, M.; Fabricius, H.; Marx, F. 2005. Total oxidant scavenging capacities of *Euterpe oleracea* Mart. (Açaí) fruits. International Journal of Food Sciences and Nutrition, 56, 53–64.

Lim, T. K. 2012. Edible Medicinal and Non-medicinal Plants. Vol. 10. Dordrecht: Springer.

Lima, A. L. d. S.; Lima, K. d. S. C.; Coelho, M. J.; Silva, J. M.; Godoy, R. L. d. O.; Pacheco, S. 2009. Avaliação dos efeitos da radiação gama nos teores de carotenóides, ácido ascórbico e açúcares do futo buriti do brejo (*Mauritia flexuosa* L.). Acta Amazonica, 39, 649–654.

Lorenzi, H.; Noblick, L. R.; Kahn, F.; Ferreira, E. 2010. Flora Brasileira: Arecaceae (Palmeiras). São Paulo: Instituto Plantarum.

Lourenço, M. A. M.; Braz, M. G.; Aun, A. G.; Pereira, B. L. B.; Figueiredo, A. M.; da Silva, R. A. C.; Kazmarek, E. M.; et al. 2018. *Spondias mombin* supplementation attenuated cardiac remodelling process induced by tobacco smoke. Journal of Cellular and Molecular Medicine, 22, 3996–4004.

Luo, R.; Tran, K.; A. Levine, R.; M. Nickols, S.; M. Monroe, D.; U. O. Sabaa-Srur, A.; E. Smith, R. 2012. Distinguishing components in Brazilian acai (*Euterpe oleraceae* Mart.) and in products obtained in the USA by using NMR. The Natural Products Journal, 2, 86–94.

Machado, D. E.; Rodrigues-Baptista, K. C.; Alessandra-Perini, J.; Soares de Moura, R.; Santos T. A.; Pereira, K. G.; Marinho da Silva, Y. et al. 2016. *Euterpe oleracea* extract (açaí) is a promising novel pharmacological therapeutic treatment for experimental endometriosis. PLoS One, 11, e0166059.

Maeda, R. N.; Pantoja, L.; Yuyama, L. K. O.; Chaar, J. M. 2007. Estabilidade de ácido ascórbico e antocianinas em néctar de camu-camu (*Myrciaria dubia* (H. B. K.) McVaugh). Ciência e Tecnologia de Alimentos, 27, 313–316.

Maezumi, S. Y.; Alves, D.; Robinson, M.; de Souza, J. G.; Levis, C.; Barnett, R. L.; Almeida de Oliveira, E.; Urrego, D.; Schaan, D.; Iriarte, J. 2018. The legacy of 4,500 years of polyculture agroforestry in the eastern Amazon. Nature Plants, 4, 540–547.

Manhães, L.; Menezes, E.; Marques, A.; Sabaa Srur, A. 2015. Flavored buriti oil (*Mauritia flexuosa* Mart.) for culinary usage: innovation, production and nutrition value. Journal of Culinary Science & Technology, 13, 362–374.

Mantovani, I. S. B.; Fernandes, S. B. O.; Menezes, F. S. 2003. Constituintes apolares do fruto do açaí (*Euterpe oleracea* M. – Arecaceae). Revista Brasileira de Farmacognosia, 13, 41–42.

Maranhão, P. A.; Kraemer-Aguiar, L. G.; de Oliveira, C. L.; Kuschnir, M. C. C.; Vieira, Y. R.; Souza, M. G. C.; Koury, J. C.; Bouskela, E. 2011. Brazil nuts intake improves lipid profile, oxidative stress and microvascular function in obese adolescents: a randomized controlled trial. Nutrition & Metabolism, 8, 32.

Maria Pacheco Santos, L. 2005. Nutritional and ecological aspects of buriti or aguaje (*Mauritia flexuosa* Linnaeus filius): a carotene-rich palm fruit from Latin America. Ecology of Food and Nutrition, 44, 345–358.

Marques, E. S.; Froder, J. G.; Carvalho, J. C. T.; Rosa, P. C. P.; Perazzo, F. F.; Maistro, E. L. 2016. Evaluation of the genotoxicity of *Euterpe oleraceae* Mart. (Arecaceae) fruit oil (açaí), in mammalian cells in vivo. Food and Chemical Toxicology, 93, 13–19.

Martino, H. S. D.; Dias, M. M. d. S.; Noratto, G.; Talcott, S.; Mertens-Talcott, S. U. 2016. Anti-lipidaemic and anti-inflammatory effect of açai (*Euterpe oleracea* Martius) polyphenols on 3T3-L1 adipocytes. Journal of Functional Foods, 23, 432–443.

Martins, I. C. V. S.; Borges, N. A.; Stenvinkel, P.; Lindholm, B.; Rogez, H.; Pinheiro, M. C. N.; Nascimento, J. L. M.; Mafra, D. 2018. The value of the Brazilian açai fruit as a therapeutic nutritional strategy for chronic kidney disease patients. International Urology and Nephrology, 50, 2207–2220.

Martins, M.; Kluszczovski, A. M.; Scussel, V. M. 2014. *In vitro* activity of the Brazil nut (*Bertholletia excelsa* H.B.K.) oil in aflatoxigenic strains of *Aspergillus parasiticus*. European Food Research and Technology, 239, 687–693.

Medeiros, M. C.; Aquino, J. S.; Soares, J.; Figueiroa, E. B.; Mesquita, H. M.; Pessoa, D. C.; Stamford, T. M. 2015. Buriti oil (*Mauritia flexuosa* L.) negatively impacts somatic growth and reflex maturation and increases retinol deposition in young rats. International Journal of Developmental Neuroscience, 46, 7–13.

Meldrum, G.; Padulosi, S.; Lochetti, G.; Robitaille, R.; Diulgheroff, S. 2018. Issues and prospects for the sustainable use and conservation of cultivated vegetable diversity for more nutrition-sensitive agriculture. Agriculture, 8, 112.

Monge-Fuentes, V.; Muehlmann, L. A.; Longo, J. P. F.; Silva, J. R.; Fascineli, M. L.; de Souza, P.; Faria, F.; et al. 2017. Photodynamic therapy mediated by acai oil (*Euterpe oleracea* Martius) in nanoemulsion: a potential treatment for melanoma. Journal of Photochemistry and Photobiology B: Biology, 166, 301–310.

Moura, R. S.; Ferreira, T. S.; Lopes, A. A.; Pires, K. M.; Nesi, R. T.; Resende, A. C.; Souza, P. J.; et al. 2012. Effects of *Euterpe oleracea* Mart. (ACAI) extract in acute lung inflammation induced by cigarette smoke in the mouse. Phytomedicine, 19, 262–269.

Mulabagal, V.; Calderon, A. I. 2012. Liquid chromatography/mass spectrometry based fingerprinting analysis and mass profiling of *Euterpe oleracea* (acai) dietary supplement raw materials. Food Chemistry, 134, 1156–1164.

Nascimento, O. V.; Boleti, A. P. A.; Yuyama, L. K. O.; Lima, E. S. 2013. Effects of diet supplementation with camu-camu (*Myrciaria dubia* HBK McVaugh) fruit in a rat model of diet-induced obesity. Anais da Academia Brasileira de Ciências, 85, 355–363.

Nascimento, R. J. S. d.; Couri, S.; Antoniassi, R.; Freitas, S. P. 2008. Composição em ácidos graxos do óleo da polpa de açaí extraído com enzimas e com hexano. Revista Brasileira de Fruticultura, 30, 498–502.

Nascimento, V. H. N. d.; Lima, C. d. S.; Paixão, J. T. C.; Freitas, J. J. d. S.; Kietzer, K. S. 2016. Antioxidant effects of açaí seed (*Euterpe oleracea*) in anorexia-cachexia syndrome induced by Walker-256 tumor. Acta Cirurgica Brasileira, 31, 597–601.

Náthia-Neves, G.; Tarone, A. G.; Tosi, M. M.; Maróstica Júnior, M. R.; Meireles, M. A. A. 2017. Extraction of bioactive compounds from genipap (*Genipa americana* L.) by pressurized ethanol: iridoids, phenolic content and antioxidant activity. Food Research International, 102, 595–604.

Neri-Numa, I. A.; Angolini, C. F. F.; Bicas, J. L.; Ruiz, A. L. T. G.; Pastore, G. M. 2018. Iridoid blue-based pigments of *Genipa americana* l. (Rubiaceae) extract: influence of pH and temperature on color stability and antioxidant capacity during *in vitro* simulated digestion. Food Chemistry, 263, 300–306.

Neri-Numa, I. A.; Pessoa, M. G.; Paulino, B. N.; Pastore, G. M. 2017. Genipin: a natural blue pigment for food and health purposes. Trends in Food Science & Technology, 67, 271–279.

Neri-Numa, I. A.; Soriano Sancho, R. A.; Pereira, A. P. A.; Pastore, G. M. 2018. Small Brazilian wild fruits: nutrients, bioactive compounds, health-promotion properties and commercial interest. Food Research International, 103, 345–360.

Nogueira, F. A.; Nery, P. S.; Morais-Costa, F.; Oliveira, N. J. d. F.; Martins, E. R.; Duarte, E. R. 2014. Efficacy of aqueous extracts of *Genipa americana* L. (Rubiaceae) in inhibiting larval development and eclosion of gastrointestinal nematodes of sheep. Journal of Applied Animal Research, 42, 356–360.

Nonato, D. T. T.; Vasconcelos, S. M. M.; Mota, M. R. L.; de Barros Silva, P. G.; Cunha, A. P.; Ricardo, N. M. P. S.; Pereira, M. G.; Assreuy, A. M. S.; Chaves, E. M. C. 2018. The anticonvulsant effect of a polysaccharide-rich extract from *Genipa americana* leaves is mediated by GABA receptor. Biomedicine & Pharmacotherapy, 101, 181–187.

Noratto, G. D.; Angel-Morales, G.; Talcott, S. T.; Mertens-Talcott, S. U. 2011. Polyphenolics from açaí (*Euterpe oleracea* Mart.) and red muscadine grape (*Vitis rotundifolia*) protect human umbilical vascular endothelial cells (huvec) from glucose- and lipopolysaccharide (lps)-induced inflammation and target microrna-126. Journal of Agricultural and Food Chemistry, 59, 7999–8012.

Nworu, C. S.; Akah, P. A.; Okoye, F. B. C.; Toukam, D. K.; Udeh, J.; Esimone, C. O. 2011. The leaf extract of *Spondias mombin* L. displays an anti-inflammatory effect and suppresses inducible formation of tumor necrosis factor-α and nitric oxide (NO). Journal of Immunotoxicology, 8, 10–16.

Oboh, F. O.; Oderinde, R. A. 1989. Fatty acid and glyceride composition of *Astrocaryum vulgare* kernel fat. Journal of the Science of Food and Agriculture, 48, 29–36.

Oboh, F. O. J. 2009. The food potential of Tucum (*Astrocaryum vulgare*) fruit pulp. International Journal of Biomedical and Health Sciences, 5, 57–64.

Oliveira, V. B.; Yamada, L. T.; Fagg, C. W.; Brandão, M. G. L. 2012. Native foods from Brazilian biodiversity as a source of bioactive compounds. Food Research International, 48, 170–179.

Omena, C. M. B.; Valentim, I. B.; Guedes, G. d. S.; Rabelo, L. A.; Mano, C. M.; Bechara, E. J. H.; Sawaya, A. C. H. F.; et al. 2012. Antioxidant, anti-acetylcholinesterase and cytotoxic activities of ethanol extracts of peel, pulp and seeds of exotic Brazilian fruits. Food Research International, 49, 334–344.

Ono, M.; Ishimatsu, N.; Masuoka, C.; Yoshimitsu, H.; Tsuchihashi, R.; Okawa, M.; Kinjo, J.; Ikeda, T.; Nohara, T. 2007. Three new monoterpenoids from the fruit of *Genipa americana*. Chemical & Pharmaceutical Bulletin, 55, 632–634.

Ono, M.; Ueno, M.; Masuoka, C.; Ikeda, T.; Nohara, T. 2005. Iridoid glucosides from the fruit of *Genipa americana*. Chemical & Pharmaceutical Bulletin, 53, 1342–1344.

Oyeyemi, I. T.; Yekeen, O. M.; Odusina, P. O.; Ologun, T. M.; Ogbaide, O. M.; Olaleye, O. I.; Bakare, A. A. 2015. Genotoxicity and antigenotoxicity study of aqueous and hydro-methanol extracts of *Spondias mombin* L., *Nymphaea lotus* L. and *Luffa cylindrical* L. using animal bioassays. Interdisciplinary Toxicology, 8, 184–192.

Pacheco-Palencia, L. A.; Duncan, C. E.; Talcott, S. T. 2009. Phytochemical composition and thermal stability of two commercial açai species, *Euterpe oleracea* and *Euterpe precatoria*. Food Chemistry, 115, 1199–1205.

Pala, D.; Barbosa, P. O.; Silva, C. T.; de Souza, M. O.; Freitas, F. R.; Volp, A. C. P.; Maranhão, R. C.; Freitas, R. N. d. 2018. Açai (*Euterpe oleracea* Mart.) dietary intake affects plasma lipids, apolipoproteins, cholesteryl ester transfer to high-density lipoprotein and redox metabolism: a prospective study in women. Clinical Nutrition, 37, 618–623.

Paniagua-Zambrana, N.; Cámara-Leret, R.; Macía, M. J. 2015. Patterns of medicinal use of palms across northwestern South America. The Botanical Review, 81, 317–415.

Pastorello, E. A.; Farioli, L.; Pravettoni, V.; Ispano, M.; Conti, A.; Ansaloni, R.; Rotondo, F.; Incorvaia, C.; Bengtsson, A.; Rivolta, F. 1998. Sensitization to the major allergen of Brazil nut is correlated with the clinical expression of allergy. Journal of Allergy and Clinical Immunology, 102, 1021–1027.

Paz, M.; Gúllon, P.; Barroso, M. F.; Carvalho, A. P.; Domingues, V. F.; Gomes, A. M.; Becker, H.; Longhinotti, E.; Delerue-Matos, C. 2015. Brazilian fruit pulps as functional foods and additives: evaluation of bioactive compounds. Food Chemistry, 172, 462–468.

Peixoto, H.; Roxo, M.; Krstin, S.; Röhrig, T.; Richling, E.; Wink, M. 2016. An anthocyanin-rich extract of acai (*euterpe precatoria* Mart.) increases stress resistance and retards aging-related markers in *Caenorhabditis elegans*. Journal of Agricultural and Food Chemistry, 64, 1283–1290.

Pereira Freire, J. A.; Barros, K. B. N. T.; Lima, L. K. F.; Martins, J. M.; Araújo, Y. d. C.; da Silva Oliveira, G. L.; de Souza Aquino, J.; Ferreira, P. M. P. 2016. Phytochemistry profile, nutritional properties and pharmacological activities of *Mauritia flexuosa*. Journal of Food Science, 81, R2611–R2622.

Pereira-Freire, J. A.; Oliveira, G. L. d. S.; Lima, L. K. F.; Ramos, C. L. S.; Arcanjo-Medeiros, S. R.; Lima, A. C. S. d.; Teixeira, S. A.; et al. 2018. *In vitro* and *ex vivo* chemopreventive action of *Mauritia flexuosa* products. Evidence-Based Complementary and Alternative Medicine, 2018, 1–12.

Petruk, G.; Illiano, A.; Del Giudice, R.; Raiola, A.; Amoresano, A.; Rigano, M. M.; Piccoli, R.; Monti, D. M. 2017. Malvidin and cyanidin derivatives from açai fruit (*Euterpe oleracea* Mart.) counteract UV-A-induced oxidative stress in immortalized fibroblasts. Journal of Photochemistry and Photobiology B: Biology, 172, 42–51.

Pompeu, D. R.; Silva, E. M.; Rogez, H. 2009. Optimisation of the solvent extraction of phenolic antioxidants from fruits of *Euterpe oleracea* using response surface methodology. Bioresource Technology, 100, 6076–6082.

Poulose, S. M., Bielinski, D. F.; Carey, A.; Schauss, A. G.; Shukitt-Hale, B. 2017. Modulation of oxidative stress, inflammation, autophagy and expression of Nrf2 in hippocampus and frontal cortex of rats fed with acai-enriched diets. Nutritional Neuroscience, 20, 305–315.

Poulose, S. M.; Fisher, D. R.; Bielinski, D. F.; Gomes, S. M.; Rimando, A. M.; Schauss, A. G.; Shukitt-Hale, B. 2014. Restoration of stressor-induced calcium dysregulation and autophagy inhibition by polyphenol-rich açaí (*Euterpe* spp.) fruit pulp extracts in rodent brain cells *in vitro*. Nutrition, 30, 853–862.

Poulose, S. M.; Fisher, D. R.; Larson, J.; Bielinski, D. F.; Rimando, A. M.; Carey, A. N.; Schauss, A. G.; Shukitt-Hale, B. 2012. Anthocyanin-rich açai (*Euterpe oleracea* Mart.) fruit pulp fractions attenuate inflammatory stress signaling in mouse brain bv-2 microglial cells. Journal of Agricultural and Food Chemistry, 60, 1084–1093.

Pumphrey, R. S.; Wilson, P. B.; Faragher, E. B.; Edwards, S. R. 1999. Specific immunoglobulin E to peanut, hazelnut and Brazil nut in 731 patients: similar patterns found at all ages. Clinical and Experimental Allergy, 29, 1256–1259.

Qu, S. S.; Zhang, J. J.; Li, Y. X.; Zheng, Y.; Zhu, Y. L.; Wang, L. Y. 2014. Protective effect of açaí berries on chronic alcoholic hepatic injury in rats and their effect on inflammatory cytokines. Zhonggue Zhong Yao Za Zhi, 39, 4869–4872.

Quesada, S.; Azofeifa, G.; Jatunov, S.; Jiménez, G.; Navarro, L.; Gómez, G. 2011. Carotenoids composition, antioxidant activity and glycemic index of two varieties of *Bactris gasipaes*. Emirates Journal of Food and Agriculture, 23, 482–489.

Radice, M.; Viafara, D.; Neill, D.; Asanza, M.; Sacchetti, G.; Guerrini, A.; Maietti, S. 2014. Chemical characterization and antioxidant activity of Amazonian (Ecuador) *Caryodendron orinocense* karst. and *Bactris gasipaes* kunth seed oils. Journal of Oleo Science, 63, 1243–1250.

Ribeiro, J. C.; Antunes, L. M. G.; Aissa, A. F.; Darin, J. D. a. C.; De Rosso, V. V.; Mercadante, A. Z.; Bianchi, M. d. L. P. 2010. Evaluation of the genotoxic and antigenotoxic effects after acute and subacute treatments with açai pulp (*Euterpe oleracea* Mart.) on mice using the erythrocytes micronucleus test and the comet assay. Mutation Research/Genetic Toxicology and Environmental Mutagenesis, 695, 22–28.

Ribeiro, P. R. E.; Santos, R. C.; Chagas, E. A.; de Melo Filho, A. A.; Montero, I. F.; Chagas, P. C.; Abreu, H. D. F.; de Melo, A. C. G. R. 2018. α-Tocopherol in Amazon fruits. Chemical Engineering Transactions, 64, 229–234.

Rocha, A. P. M.; Carvalho, L. C. R. M.; Sousa, M. A. V.; Madeira, S. V. F.; Sousa, P. J. C.; Tano, T.; Schini-Kerth, V. B.; Resende, A. C.; Soares de Moura, R. 2007. Endothelium-dependent vasodilator effect of *Euterpe oleracea* Mart. (Açaí) extracts in mesenteric vascular bed of the rat. Vascular Pharmacology, 46, 97–104.

Rodrigues, A. M. d. C.; Darnet, S.; Silva, L. H. M. d. 2010. Fatty acid profiles and tocopherol contents of buriti (*Mauritia flexuosa*), patawa (*Oenocarpus bataua)*, tucuma (*Astrocaryum vulgare*), mari (*Poraqueiba paraensis*) and inaja (*Maximiliana maripa*) fruits. Journal of the Brazilian Chemical Society, 21, 2000–2004.

Rodrigues-Amaya, D. B.; Kimura, M.; Amaya-Farfan, J. 2008. Fontes Brasileiras de carotenóides: Tabela brasileira de composição de carotenóides em alimentos. Brasilia: MMA/SBF.

Rogez, H. 2000. Açaí: Preparo, Composição e Melhoramento da Conservação. Belém, Pará: Universidade Federal do Pará – EDUPA.

Rogez, H.; Pompeu, D. R.; Akwie, S. N. T.; Larondelle, Y. 2011. Sigmoidal kinetics of anthocyanin accumulation during fruit ripening: a comparison between açai fruits (*Euterpe oleracea*) and other anthocyanin-rich fruits. Journal of Food Composition and Analysis, 24, 796–800.

Rojas-Garbanzo, C.; Pérez, A. M.; Bustos-Carmona, J.; Vaillant, F. 2011. Identification and quantification of carotenoids by HPLC-DAD during the process of peach palm (*Bactris gasipaes* H.B.K.) flour. Food Research International, 44, 2377–2384.

Romualdo, G. R.; Fragoso, M. F.; Borguini, R. G.; de Araújo Santiago, M. C. P.; Fernandes, A. A. H.; Barbisan, L. F. 2015. Protective effects of spray-dried açaí (*Euterpe oleracea* Mart.) fruit pulp against initiation step of colon carcinogenesis. Food Research International, 77, 432–440.

Rufino, M. d. S. M.; Alves, R. E.; de Brito, E. S.; Pérez-Jiménez, J.; Saura-Calixto, F.; Mancini-Filho, J. 2010. Bioactive compounds and antioxidant capacities of 18 non-traditional tropical fruits from Brazil. Food Chemistry, 121, 996–1002.

Sabiu, S.; Garuba, T.; Sunmonu, T. O.; Sulyman, A. O.; Ismail, N. O. 2016. Indomethacin-induced gastric ulceration in rats: ameliorative roles of *Spondias mombin* and *Ficus exasperata*. Pharmaceutical Biology, 54, 180–186.

Sadowska-Krępa, E.; Kłapcińska, B.; Podgórski, T.; Szade, B.; Tyl, K.; Hadzik, A. 2015. Effects of supplementation with acai (*Euterpe oleracea* Mart.) berry-based juice blend on the blood antioxidant defence capacity and lipid profile in junior hurdlers. A pilot study. Biology Sport, 32, 161–168.

Saheed, S.; Taofik, S. O.; Oladipo, A. E.; Tom, A. A. O. 2017. *Spondias mombin* L. (Anacardiaceae) enhances detoxification of hepatic and macromolecular oxidants in acetaminophen-intoxicated rats. Pakistan Journal of Pharmaceutical Sciences, 30, 2109–2117.

Sampaio, P. B.; Rogez, H.; Souza, J. N. S.; Rees, J. F.; Larondelle, Y. 2006. Antioxidant properties of açai (*Euterpe oleracea*) in human plasma. In H. Rogez; R. S. Pena (Eds.) (in press) Olhares cruzados sobre açai. Belém: EDUFPA.

Sampaio, T. I.; de Melo, N. C.; de Freitas Paiva, B. T.; da Silva Aleluia, G. A.; da Silva Neto, F. L. P.; da Silva, H. R.; Keita, H.; et al. 2018. Leaves of *Spondias mombin* L. a traditional anxiolytic and antidepressant: pharmacological evaluation on zebrafish (*Danio rerio*). Journal of Ethnopharmacology, 224, 563–578.

Santos, M.; Alves, R.; Roca, M. 2015. Carotenoid composition in oils obtained from palm fruits from the Brazilian Amazon. Grasas y Aceites, 66, e086.

Santos, M. F. G.; Alves, R. E.; Ruíz-Méndez, M. V. 2013. Minor components in oils obtained from Amazonian palm fruits. Grasas y Aceites, 64, 531–536.

Schauss, A. G. 2010. Açaí (*Euterpe oleracea* Mart.): a macro and nutrient rich palm fruit from the Amazon rain forest with demonstrated bioactivities *in vitro* and *in vivo*. In R. R. Watson and V. R. Preedy (Eds.) Bioactive Foods in Promoting Health: Fruits and Vegetables. 479–490. Oxford: Academic Press.

Schauss, A. G.; Clewell, A.; Balogh, L.; Szakonyi, I. P.; Financsek, I.; Horváth, J.; Thuroczy, J.; Béres, E.; Vértesi, A.; Hirka, G. 2010. Safety evaluation of an açai-fortified fruit and berry functional juice beverage (MonaVie Active®). Toxicology, 278, 46–54.

Schauss, A. G.; Wu, X.; Prior, R. L.; Ou, B.; Huang, D.; Owens, J.; Agarwal, A.; Jensen, G. S.; Hart, A. N.; Shanbrom, E. 2006. Antioxidant capacity and other bioactivities of the freeze-dried Amazonian palm berry, *Euterpe oleraceae* Mart. (acai). Journal of Agricultural and Food Chemistry, 54, 8604–8610.

Silva, D. F.; Vidal, F. C. B.; Santos, D.; Costa, M. C. P.; Morgado-Díaz, J. A.; do Desterro Soares Brandão Nascimen, M.; de Moura, R. S. 2014. Cytotoxic effects of *Euterpe oleracea* Mart. in malignant cell lines. BMC Complementary and Alternative Medicine, 14.

Silva, F. V. G. d.; Silva, S. d. M.; Silva, G. C. d.; Mendonça, R. M. N.; Alves, R. E.; Dantas, A. L. 2012. Bioactive compounds and antioxidant activity in fruits of clone and ungrafted genotypes of *Yellow mombin* tree. Food Science and Technology, 32, 685–691.

Silva, S. M.; Rocco, S. A.; Sampaio, K. A.; Taham, T.; da Silva, L. H. M.; Ceriani, R.; Meirelles, A. J. A. 2011. Validation of a method for simultaneous quantification of total carotenes and tocols in vegetable oils by HPLC. Food Chemistry, 129, 1874–1881.

Siqueira, E.; Andrade, A.; de Souza-Fagundes, E.; Ramos, J.; Kohlhoff, M.; Nunes, Y.; Cota, B. 2014. *In vitro* antibacterial action on methicillin-susceptible (MSSA) and methicillin resistant (MRSA) *Staphylococcus aureus* and antitumor potential of *Mauritia flexuosa* L.f. Journal of Medicinal Plants Research, 8, 1408–1417.

Smith, N. 2015. Palms and people in the Amazon. Geobotany Studies, 1st ed. New York: Springer International Publishing.

Sosnowska, J.; Balslev, H. 2009. American palm ethnomedicine: a meta-analysis. Journal of Ethnobiology and Ethnomedicine, 5, 43.

Souza, P. M.; Elias, S. T.; Simeoni, L. A.; de Paula, J. E.; Gomes, S. M.; Guerra, E. N. S.; Fonseca, Y. M.; Silva, E. C.; Silveira, D.; Magalhães, P. O. 2012. Plants from Brazilian cerrado with potent tyrosinase inhibitory activity. PLoS ONE, 7, e48589.

Souza, R. O. d. S.; Sousa, P. L.; Menezes, R. R. P. P. B. d.; Sampaio, T. L.; Tessarolo, L. D.; Silva, F. C. O.; Pereira, M. G.; Martins, A. M. C. 2018. Trypanocidal activity of polysaccharide extract from *Genipa americana* leaves. Journal of Ethnopharmacology, 210, 311–317.

Souza-Monteiro, J. R.; Hamoy, M.; Santana-Coelho, D.; Arrifano, G. P. F.; Paraense, R. S. O.; Costa-Malaquias, A.; Mendonça, J. R.; et al. 2015. Anticonvulsant properties of *Euterpe oleracea* in mice. Neurochemistry International, 90, 20–27.

Spada, P. D. S.; Dani, C.; Bortolini, G. V.; Funchal, C.; Henriques, J. A. P.; Salvador, M. 2009. Frozen fruit pulp of *Euterpe oleraceae* Mart. (acai) prevents hydrogen peroxide-induced damage in the cerebral cortex, cerebellum, and hippocampus of rats. Journal of Medicinal Food, 12, 1084–1088.

Speranza, P.; De Oliveira Falcão, A.; Alves Macedo, J.; Da Silva, L. H. M.; Da C. Rodrigues, A. M.; Alves Macedo, G. 2016. Amazonian buriti oil: chemical characterization and antioxidant potential. Grasas y Aceites, 67, e135.

Stockler-Pinto, M. B.; Lobo, J.; Moraes, C.; Leal, V.O.; Farage, N. E.; Rocha, A. V.; Boaventura, G. T.; Cozzolino, S. M. F.; Malm, O.; Mafra, D. 2012. Effect of Brazil nut supplementation on plasma levels of selenium in hemodialysis patients: 12 months of follow-up. Journal of Renal Nutrition, 22, 434–439.

Stockler-Pinto, M. B.; Mafra, D.; Moraes, C.; Lobo, J.; Boaventura, G. T.; Farage, N. E.; Silva, W. S.; Cozzolino, S. F.; Malm, O. 2014. Brazil nut (*Bertholletia excelsa*, H.B.K.) improves oxidative stress and inflammation biomarkers in hemodialysis patients. Biological Trace Element Research, 158, 105–112.

Stoner, G. D.; Wang, L.-S.; Seguin, C.; Rocha, C.; Stoner, K.; Chiu, S.; Kinghorn, A. D. 2010. Multiple berry types prevent n-nitrosomethylbenzylamine-induced esophageal cancer in rats. Pharmaceutical Research, 27, 1138–1145.

Strunz, C. C.; Oliveira, T. V.; Vinagre, J. C. M.; Lima, A.; Cozzolino, S.; Maranhão, R. C. 2008. Brazil nut ingestion increased plasma selenium but had minimal effects on lipids, apolipoproteins, and high-density lipoprotein function in human subjects. Nutrition Research, 28, 151–155.

Sudo, R. T.; Neto, M. L.; Monteiro, C. E. S.; Amaral, R. V.; Resende, Â. C.; Souza, P. J. C.; Zapata-Sudo, G.; Moura, R. S. 2015. Antinociceptive effects of hydroalcoholic extract from *Euterpe oleracea* Mart. (Açaí) in a rodent model of acute and neuropathic pain. BMC Complementary and Alternative Medicine, 15, 208.

Sun, X.; Seeberger, J.; Alberico, T.; Wang, C.; Wheeler, C. T.; Schauss, A. G.; Zou, S. 2010. Açai palm fruit (*Euterpe oleracea* Mart.) pulp improves survival of flies on a high fat diet. Experimental Gerontology, 45, 243–251.

TACO – Núcleo de Estudos e Pesquisas em Alimentação – NEPA/UNICAMP. 2011. Tabela brasileira de composição de alimentos. www.cfn.org.br/wp-content/uploads/2017/03/taco_4_edicao_ampliada_e_revisada.pdf.

Tiburski, J. H.; Rosenthal, A.; Deliza, R.; de Oliveira Godoy, R. L.; Pacheco, S. 2011. Nutritional properties of yellow mombin (*Spondias mombin* L.) pulp. Food Research International, 44, 2326–2331.

Torma, P. d. C. M. R.; Brasil, A. V. S.; Carvalho, A. V.; Jablonski, A.; Rabelo, T. K.; Moreira, J. C. F.; Gelain, D. P.; Flôres, S. H.; Augusti, P. R.; Rios, A. d. O. 2017. Hydroethanolic extracts from different genotypes of açaí (*Euterpe oleracea*) presented antioxidant potential and protected human neuron-like cells (SH-SY5Y). Food Chemistry, 222, 94–104.

Udani, J. K.; Singh, B. B.; Singh, V. J.; Barrett, M. L. 2011. Effects of Açai (*Euterpe oleracea* Mart.) berry preparation on metabolic parameters in a healthy overweight population: a pilot study. Nutrition Journal, 10, 45.

Uhl, C.; Dransfield, J. 1991. Genera *Palmarum*: A Classification of Palms Based on the Work of Harold and Moore. Kansas: Allen Press.

Unis, A. 2015. Açaí berry extract attenuates glycerol-induced acute renal failure in rats. Renal Failure, 37, 310–317.

United Nations Conference On Trade And Development – UNCTAD. 2005. Market brief in the European Union for selected natural ingredients derived from native species: *Genipa americana* (Jagua, huito). www.biotrade.org/.../biotradebrief-genipaamericana.pdf.

Valencia, G. A.; Moraes, I. C. F.; Lourenço, R. V.; Bittante, A. M. Q. B.; Sobral, P. J. d. A. 2015. Physicochemical, morphological, and functional properties of flour and starch from peach palm (*Bactris gasipaes* K.) fruit. Starch – Stärke, 67, 163–173.

van Boxtel, E. L.; Koppelman, S. J.; van den Broek, L. A. M.; Gruppen, H. 2008. Heat denaturation of Brazil nut allergen Ber e 1 in relation to food processing. Food Chemistry, 110, 904–908.

Vasco, C.; Ruales, J.; Kamal-Eldin, A. 2008. Total phenolic compounds and antioxidant capacities of major fruits from Ecuador. Food Chemistry, 111, 816–823.

Vrailas-Mortimer, A.; Gomez, R.; Dowse, H.; Sanyal, S. 2012. A survey of the protective effects of some commercially available antioxidant supplements in genetically and chemically induced models of oxidative stress in *Drosophila melanogaster*. Experimental Gerontology, 47, 712–722.

Wong, D. Y. S.; Musgrave, I. F.; Harvey, B. S.; Smid, S. D. 2013. Açaí (*Euterpe oleraceae* Mart.) berry extract exerts neuroprotective effects against β-amyloid exposure *in vitro*. Neuroscience Letters, 556, 221–226.

Xie, C.; Kang, J.; Burris, R.; Ferguson, M. E.; Schauss, A. G.; Nagarajan, S.; Wu, X. 2011. Açaí juice attenuates atherosclerosis in ApoE deficient mice through antioxidant and anti-inflammatory activities. Atherosclerosis, 216, 327–333.

Xie, C.; Kang, J.; Li, Z.; Schauss, A. G.; Badger, T. M.; Nagarajan, S.; Wu, T.; Wu, X. 2012. The açaí flavonoid velutin is a potent anti-inflammatory agent: blockade of LPS-mediated TNF-α and IL-6 production through inhibiting NF-κB activation and MAPK pathway. The Journal of Nutritional Biochemistry, 23, 1184–1191.

Yang, J. 2009. Brazil nuts and associated health benefits: a review. LWT – Food Science and Technology, 42, 1573–1580.

Yazawa, K.; Suga, K.; Honma, A.; Shirosaki, M.; Koyama, T. 2011. Anti-inflammatory effects of seeds of the tropical fruit camu-camu (*Myrciaria dubia*). Journal of Nutritional Science and Vitaminology, 57, 104–107.

Yunis-Aguinaga, J.; Fernandes, D. C.; Eto, S. F.; Claudiano, G. S.; Marcusso, P. F.; Marinho-Neto, F. A.; Fernandes, J. B. K.; de Moraes, F. R.; de Moraes, J. R. E. 2016. Dietary camu camu, *Myrciaria dubia*, enhances immunological response in *Nile tilapia*. Fish & Shellfish Immunology, 58, 284–291.

Yuyama, K.; Aguiar, J. P. L.; Yuyama, L. K. O. 2002. Camu-camu: um fruto fantástico como fonte de vitamina C1. Acta Amazonica, 32, 169–174.

Yuyama, L. K.; Aguiar, J. P.; Yuyama, K.; Clement, C. R.; Macedo, S. H.; Favaro, D. I.; Afonso, C.; et al. 2003. Chemical composition of the fruit mesocarp of three peach palm (*Bactris gasipaes*) populations grown in central Amazonia, Brazil. International Journal of Food Sciences and Nutrition, 54, 49–56.

Yuyama, L. K. O.; Cozzolino, S. M. F. 1996. Effect of supplementation with peach palm as source of vitamin A: study with rats. Revista de Saúde Pública, 30, 61–66.

Yuyama, L. K. O.; Favaro, R. M. D.; Yuyama, K.; Vannucchi, H. 1991. Bioavailability of vitamin a from peach palm (*Bactris gasipaes* H.B.K.) and from mango (*Mangifera indica* L.) in rats. Nutrition Research, 11, 1167–1175.

Zanatta, C.; Mercadante, A. 2007. Carotenoid composition from the Brazilian tropical fruit camu–camu (*Myrciaria dubia*). Food Chemistry, 101, 1526–1532.

Zanatta, C. F.; Mitjans, M.; Urgatondo, V.; Rocha-Filho, P. A.; Vinardell, M. P. 2010. Photoprotective potential of emulsions formulated with buriti oil (*Mauritia flexuosa*) against UV irradiation on keratinocytes and fibroblasts cell lines. Food and Chemical Toxicology, 48, 70–75.

Zanatta, C. F.; Ugartondo, V.; Mitjans, M.; Rocha-Filho, P. A.; Vinardell, M. P. 2008. Low cytotoxicity of creams and lotions formulated with buriti oil (*Mauritia flexuosa*) assessed by the neutral red release test. Food and Chemical Toxicology, 46, 2776–2781.

Zapata-Sudo, G.; da Silva, J. S.; Pereira, S. L.; Souza, P. J. C.; de Moura, R. S.; Sudo, R. T. 2014. Oral treatment with *Euterpe oleracea* Mart. (açaí) extract improves cardiac dysfunction and exercise intolerance in rats subjected to myocardial infarction. BMC Complementary and Alternative Medicine, 14, 227.

Zielinski, A. A. F.; Ávila, S.; Ito, V.; Nogueira, A.; Wosiacki, G.; Haminiuk, C. W. I. 2014. The association between chromaticity, phenolics, carotenoids, and *in vitro* antioxidant activity of frozen fruit pulp in Brazil: an application of chemometrics. Journal of Food Science, 79, C510–C516.

9 Plant Species from the Atlantic Forest Biome and Their Bioactive Constituents

*Rebeca Previate Medina, Carolina Rabal Biasetto,
Lidiane Gaspareto Felippe, Lilian Cherubin Correia,
Marília Valli, Afif Felix Monteiro, Alberto José Cavalheiro,
Ângela Regina Araújo, Ian Castro-Gamboa, Maysa Furlan,
Vanderlan da Silva Bolzani, and Dulce Helena Siqueira Silva*
Univ. Estadual Paulista, Núcleo de Bioensaios, Biossíntese e Ecofisiologia
de Produtos Naturais – NuBBE, Araraquara, Brazil

CONTENTS

9.1 INTRODUCTION

Throughout history, humans have benefited from Nature's resources and relied on natural products (NPs) for the treatment of a wide variety of diseases. Plant-derived preparations or mixtures constituted the primary source of biologically active NPs and served as the basis for the foundation of medicinal systems. Even nowadays, traditional medicinal practices and ethnomedicinal knowledge play crucial roles in health care and are important for the discovery of lead compounds and drug development programs (Cragg and Newman, 2013; Newman and Cragg, 2016).

9.1.1 THE ATLANTIC FOREST BIOME

Atlantic Forest is a biogeographic region that covers mostly Brazil and parts of Paraguay, Argentina and Uruguay. The Brazilian Atlantic Forest was originally found along the Atlantic Ocean coast from Rio Grande do Norte through Rio Grande do Sul states (Morellato and Haddad, 2000). This biome is considered an important global hotspot and a priority for biodiversity conservation, that has been severely degraded through the past five centuries, and hosts, nevertheless, *ca.* 15,782 plant species, distributed in 2,257 genera and 348 families, which accounts for 5% of the world's flora and 2% of the global endemic vascular plants (Myers et al., 2000; Stehmann et al., 2009). The uniqueness and large proportion of endemic plant and animal species found in the Atlantic Forest biome has been partly due to the isolation from the Amazon Rainforest by drier biomes as the Cerrado or the even drier Caatinga (Joly et al., 2014).

Atlantic Forest is divided into two major regions, which are covered by typical vegetation types: the Atlantic Rain Forest and the Atlantic Semi-deciduous Forest. The Atlantic Rain Forest occupies a narrow strip of land along the coast covering mostly low to medium elevations (0–1,000 m) from southern to northeastern Brazil and experiences a warm and wet climate without a dry season. On the other hand, the Atlantic Semi-deciduous Forest occupies the inlands across a plateau (usually higher than 600 m of elevation) from the country's center to the southeast and experiences a seasonal climate with a relatively severe dry season (generally from April to September). Additional smaller ecosystems within Atlantic Forest include mangrove forests, high-altitude grasslands (*campos rupestres*) and coastal dunes (*restingas*) with associated transitional vegetation types, which contributes to the biome uniqueness (Morellato and Haddad, 2000).

Originally, the extension of Atlantic Forest covered around 150 million hectares. However, about 88% of the biomes' original distribution area has been lost as a result of extensive human occupation (Ribeiro et al., 2009). At the same time, the extraction of natural resources from Atlantic Forest dates back to the colonization era when the Brazilian tree *pau-brasil* (*Caesalpinia echinata*, Fabaceae) was extensively collected for extraction of red dye from the heartwood, composed of brazilin (**1**) and brazileine (**2**) (Figure 9.1) pigments. Unfortunately, this resource was not exploited in a sustainable way, and nowadays, the species is still classified as endangered. Subsequent deforestation mainly

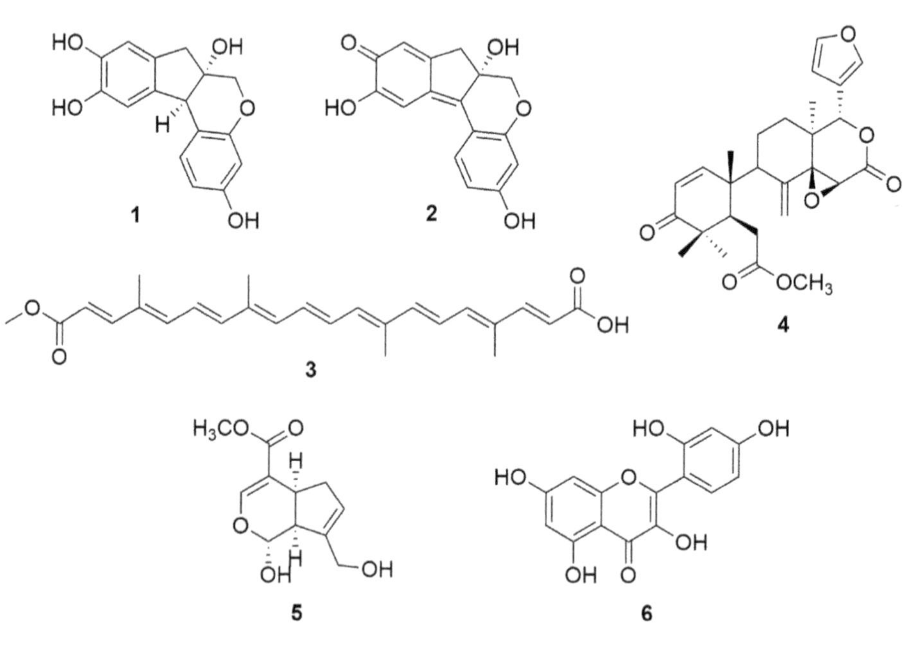

FIGURE 9.1   Natural pigments and related compounds from Brazilian plants.

due to sugarcane, coffee and soybean plantations across 16th to 20th centuries have left less than 12% of the original vegetation. Geographic elements with little use to agriculture as the ridges "Serra do Mar" and "Serra da Mantiqueira" in São Paulo, Rio de Janeiro and Minas Gerais states, have played a key role in the vegetation conservation in these areas.

Although only small fragments of the Atlantic Forest remain, the Brazilian government and non-governmental organizations (NGOs) have been taking actions for conservation and restoration to mitigate this situation (Silva et al., 2010).

Bioprospecting studies have been conducted at NuBBE[1] laboratories in the last 20 years aiming at novel pharmacologically active plant-derived compounds from this beautiful biome. Important results from Biota-FAPESP Program-funded research have largely contributed with reasonable and science-based inputs to the establishment of efficient conservational policies, and recent data have suggested that, over the past few decades, this ecosystem has been experiencing a positive balance of forest change (Costa et al., 2017; Metzger and Casatti, 2006; Ribeiro et al., 2009).

The huge biodiversity of Atlantic Forest is attested by a recent research study (Flora do Brasil, 2019), which describes 25 medicinal plant species (Table 9.1) naturally occurring in this biome out of 71 plants still used as folk medicines in Brazil. These plants are indicated by the RENISUS, a national list of medicinal plants to be adopted by the public healthcare system in the country (National List of Medicinal Plants of Interest to the Unique System Health (SUS) – Brasil, 2009).

The 25 species from the Atlantic Forest are listed as follows: *Anacardium occidentale* ("cajueiro"), *Ananas comosus* ("abacaxi"), *Apuleia ferrea* ("pau-ferro"), *Arrabidaea chica* ("crajiru"), *Baccharis trimera* ("carqueja"), *Bauhinia* spp. ("pata-de-vaca"), *Bidens pilosa* ("picão-preto"), *Casearia sylvestris* ("guaçatonga"), *Copaifera* spp. ("Copaíba"), *Cordia* spp. ("erva-baleeira"), *Costus* spp. ("cana-do-brejo"), *Erythrina mulungu* (mulungu"), *Eugenia uniflora* ("pitanga"), *Jatropha gossypiifolia* ("pinhão-roxo"), *Lippia sidoides* ("alecrim-pimenta"), *Maytenus* spp. ("espinheira-santa"), *Mikania* spp. ("guaco"), *Passiflora* spp. ("maracujá"), *Phyllanthus* spp. ("quebra-pedra"), *Portulaca pilosa* ("amor-crescido"), *Schinus terebinthifolius* ("aroeira-vermelha"), *Solanum paniculatum* ("jurubeba"), *Solidago microglossa* ("arnica-brasileira"), *Tabebuia avellanedae* ("ipê-roxo") and *Vernonia* spp. ("assa-peixe").

Many of these species have been the subject of multidisciplinary studies by research groups throughout Brazil and abroad, which confirmed their remarkable chemical diversity and potential medicinal uses. Notably, Atlantic Forest holds a fantastic biodiversity with many still-unknown species, and thus, with great potential for chemical prospecting of active biomolecules to inspire lead compounds for drug development and phytomedicines.

Aiming at the discovery of novel pharmacologically active plant-derived compounds from Atlantic Forest, bioprospection studies have been conducted at NuBBE laboratories in the last 20 years. The most prominent achievements on the chemical diversity and biological investigations from the plants collected in the Atlantic Forest obtained up to 2008 have been previously described (Silva et al., 2010), whereas results from the past 10 years are reviewed in this chapter. All the plants cited in this chapter are described as follows.

9.1.2 NATURAL PRODUCTS IN BRAZIL: HISTORICAL BENCHMARKS

Since the initial period right after the discovery of Brazil by Portuguese navigators in 1500, the exploration of mineral and vegetal wealthy products from Brazilian lands was set and focused initially on pau-brasil (*Caesalpinia echinata*), a plant species from Fabaceae known as "ibirapitanga" to the native Indians, meaning red wood tree. The species trunk wood yields a bright red pigment used since that period for tinting fabrics and as writing ink. The pigment red

---

[1] NuBBE, Nucleus for Bioassays, Biosynthesis and Ecophysiology of Natural Products located at the Institute of Chemistry, Sao Paulo State University (UNESP) in Araraquara, SP, Brazil.

TABLE 9.1

Summary of the Names of All Plant Species, Their Respective Family and Common Names Discussed in This Chapter

Scientific Name	Family	Common Names
Anacardium occidentale (sin. *Anacardium microcarpum*)	Anacardiaceae	Cajueiro
Ananas comosus (sin. *Ananas sativa*)	Bromeliaceae	Abacaxi
Aniba canelilla (*sin. Aniba eliiptica*)	Lauraceae	casca-preciosa
Apuleia ferrea	Fabaceae	pau-ferro
Baccharis trimera (sin. *Baccharis genistelloides var. trimera*)	Asteraceae	Carqueja
Banisteriopsis caapi (sin. *Banisteria caapi*)	Malpighiaceae	caapi or yage
Bauhinia spp.	Fabaceae	pata-de-vaca
Bidens pilosa (sin. *Bidens abadiae*)	Asteraceae	picão-preto
Bixa orellana (sin. *Bixa platycarpa*)	Bixaceae	Urucum
Caesalpinia echinata (sin. *Paubrasilia echinata*)	Fabaceae	pau-brasil
Carapa guianensis (sin. *Carapa llanocarti*)	Meliaceae	Andiroba
Carapichea ipecacuanha (sin. *Psychotria ipecacuanha*)	Rubiaceae	ipecacuanha, cagosanga, poaia or ipeca
Casearia sylvestris (sin. *Casearia subsessiliflora*)	Salicaceae	Guaçatonga
Cinnamomum triplinerve (sin. *Cinnamomum australe*)	Lauraceae	Canela
Copaifera langsdorfii (sin. *Copaiba langsdorfii*)	Fabaceae	copaíba, copaibeira or pau-de-óleo
Copaifera spp.	Fabaceae	Copaiba
Cordia curassavica (sin. *Cordia verbenacea*)	Boraginaceae	erva-baleeira
Cordia spp.	Boraginaceae	erva-baleeira
Costus spp.	Costaceae	cana-do-brejo
Croton heliotropiifolius (sin. *Croton salviifolius*)	Euphorbiaceae	Velame
Cryptocarya mandioccana (sin. *Oreodaphne polyantha*)	Lauraceae	noz-moscada-do-Brasil
Cryptocarya. moschata	Lauraceae	Canela batalha
Dipterix odorata	Fabaceae	Cumaru
Erythrina verna (sin. *Erythrina mulungu*)	Fabaceae	Mulungu
Esenbeckia leiocarpa	Rutaceae	guarantã or guarataiá-vermelha
Eugenia uniflora	Myrtaceae	Pitanga
Fridericia chica (sin. *Arrabidaea chica*)	Bignoniaceae	Crajiru
Geissospermum leave (sin. *Geissospermum martianum*)	Apocynaceae	pau-pereira, quinarana or pau-forquilha
Genipa Americana (sin. *Genipa venosa*)	Rubiaceae	Jenipapo
Himatanthus lancifolius (sin. *Plumeria lancifolia*)	Apocynaceae	quina-mole or quina-branca,
Hymenaea courbaril (sin. *Hymenaea multiflora*)	Fabaceae	Its resins are known as jatobá or jutaí
Jatropha curcas (sin. *Curcas lobata*)	Euphorbiaceae	mandubiguaçu or purgueira or pinhão-manso
Jatropha gossypiifolia (sin. *Jatropha jacquinii*)	Euphorbiaceae	pinhão-roxo
Jatropha ribifolia (sin. *Adenoropium ribifolium*)	Euphorbiaceae	Pinhão manso
Lippia sidoides	Verbenaceae	alecrim-pimenta
Maclura tinctoria (sin. *Chlorophora tinctoria*)	Moraceae	tatajuba, taiúva, espinheiro bravo
Maquira sclerophylla (sin. *Olmedioperebea sclerophylla*)	Moraceae	pau-tanino, rapé-de-índio or pau-de-índio
Maytenus ilicifolia (sin. *Maytenus ilicifolia* var. *boliviana*)	Celastraceae	espinheira santa,
Maytenus spp.	Celastraceae	espinheira-santa
Mikania spp.	Astereaceae	Guaco
Mimosa tenuiflora (sin. *Mimosa hostilis*)	Fabaceae	Vinho-de-Jurema
Moringa oleifera (sin. *Hyperanthera moringa*)	Moringaceae	drumstick tree
Ouratea multiflora	Ochnaceae	
Passiflora spp.	Passifloraceae	Maracuja

(Continued)

TABLE 9.1 *(Continued)*
Summary of the Names of All Plant Species, Their Respective Family and Common Names Discussed in This Chapter

Scientific Name	Family	Common Names
Peperomia obtusifolia (sin. *Peperomia petenensis*)	Piperaceae	Pepper face
Phyllanthus spp.	Phyllanthaceae	quebra-pedra
Pilocarpus microphyllus	Rutaceae	jaborandi, from the Tupi language Ya-bor-andi
Piper crassinervium (sin. *Piper novae-helvetiae*)	Piperaceae	Pariparoba, jaguarandi or jaguarandy
Piper tuberculatum (sin. *Artanthe decurrens*)	Piperaceae	pimenta-longa
Piptadenia peregrina (sin. *Anadenanthera peregrina*)	Fabaceae	Paricá, yopo, jopo or cohoba
Plectranthus barbatus (sin. *Plectranthus pseudobarbatus*)	Lamiaceae	Brazilian-boldo or false-boldo.
Porcelia macrocarpa (sin. *Porcelia goyazensis*)	Annonaceae	pindaíba, pindaíba-do-mato or banana-de-macaco
Portulaca pilosa (sin. *Portulaca mundula*)	Portulacaceae	amor-crescido
Psychotria viridis (sin. *Uragoga viridis*)	Rubiaceae	chacruna or chacrona
Pterogyne nitens	Fabaceae	tipá, viraró, cocal or amendoinzeiro
Schinus terebinthifolia (sin. *Schinus mellisii*)	Anacardiaceae	aroeira-vermelha
Solanum grandiflorum	Solanaceae	Lobeira
Solanum paniculatum (sin. *Solanum chloroleucum*)	Solanaceae	Jurubeba
Solidago chilensis (sin. *Solidago microglossa*)	Asteraceae	arnica-brasileira
Spondias tuberosa	Anacardiaceae	Umbu
Stemodia foliosa	*Plantaginaceae*	Meladinha
Strychnos castelnaeana (sin. *Strychnos castelnaei*)	Loganiaceae	
Strychnos guianensis (sin. *Toxicaria americana*)	Loganiaceae	
Swartzia langsdorffii (sin. *Swartzia brasiliensis*)	Fabaceae	banana-de-papagaio, jacarandá-banana, or jacarandá-de-sangue
Tabebuia avellanedae	Bignoniaceae	ipê-roxo
Tetrapterys mucronata (sin. *Tetrapterys silvatica*)	Malpighiaceae	
Vernonia spp.	Asteraceae	assa-peixe
Virola calophylla (sin. *Virola lepidota*)	Myristicaceae	Epená
Virola calophylloidea	Myristicaceae	
Virola theiodora (sin. *Virola rufula*)	Myristicaceae	yakee, epena and nyakwana

color is mainly associated to the presence of brazilin (**1**) from the crude extract, which gives brazilein (**2**) (Figure 9.1), that is the compound's oxidation derivative formed during the extraction procedure (Morsingh and Robinson, 1970).

Important historical benchmarks from previous centuries contributed to the scenario that favored the great navigations period and the discovery of America in 1492 and Brazil in 1500. Among examples is included the Crusades, which expanded the limits of known lands to the East and brought vast information of the abundance of valuable goods as gold, pigments and spices, especially from India and China since the 8th century. Marco Polo's excursions and his reports on the precious products he found during the years he lived in China gave additional impulse to the Europe interest on the exploration and trade of eastern products, which brought wealth and prosperity to thousands of European traders for a long period. Nevertheless, the taking of Constantinople by the Turkish in 1453 blocked the path to the East and brought collapse to a relevant part of Europe's economy at the time. Such events gave the necessary impulse to navigators, especially from Portugal, Spain and Italy, to find a new way to India, China and further eastern regions which had been providing Europe with goods to support its economic activities and relative political stability (Polo, 1958/1982).

Several fleets departed from Portugal, Spain and Italy in the end of the 15th century as part of a huge effort for the discovery of a new maritime route to the East Indies. The arrival of Columbus in Central America in 1492 was initially thought to take place in the extreme orient, as the navigators made their early incursions to the inlands to find cinnamon-smelling bushes and rhubarb, a precious Chinese medicinal plant used as a cathartic. Such findings mistakenly corroborated the navigators' expectations to have discovered the so much desired new path to the East, which was accomplished years later by Vasco da Gama, who finally rode across the Cape of Storms, later renamed as Cape of Good Hope in South Africa (Butler and Moffett, 1995).

The motivation to gain access to the valuable eastern spices, pigments and other valuable goods fairly explains the avidity on the exploration of pau-brasil as soon as Portugal realized that the discovery of Brazil and profiteering from the colony represented an additional important source of commercial products and wealth to the reign. In addition to pau-brasil tincture, Portuguese colonizers explored pigments from *Bixa orellana* (Bixaceae), known as "urucum", which means red in Tupi language, and *Genipa americana* (Rubiaceae), known as "jenipapo", in addition to "andiroba" oil, used for several purposes including preparation of soaps, protection of furniture against insects attack, and as lighting fuel. Each specimen of *B. orellana* may bear thousands of sea-urchin-like fruits plenty of seeds rich in bixin (**3**) (Figure 9.1). This norcarotenoid is still used as a colorant for food and as a sunscreen component and was exported to Europe in huge amounts during the 16th and 17th centuries. The extraction of bixin was often carried out using andiroba oil, obtained from *Carapa guianensis* (Meliaceae) fruits, which are rich in limonoids as andirobin (**4**) (Figure 9.1), a tetranortriterpene derived from euphane triterpenes. Jenipapo ripe fruits contain genipin (**5**) (Figure 9.1), an iridoid from the fruit sap used in body paintings by native Indians. Although genipin is colorless, when reacting with skin proteins produces a black color extensively employed for tattoos associated to native Indians rituals and religious ceremonies (Barber et al., 1961; Djerassi et al., 1961; Ollis et al., 1970).

Portuguese colonizers rapidly realized the potential of Brazilian flora as a source of novel interesting products as well as the valuable knowledge on plant species provided by native Indians, especially those associated to plant pigments and poisons. *Chlorophora tinctoria* (Moraceae) was known in the Indian language as "tatajuba", which means firewood or fire-colored wood. The flavonol morin (**6**) was shown to be responsible for the color (Figure 9.1) and became soon became a major commodity exported to Europe to be used as a fabric pigment (Pinto, 1995).

Resins, balms and spices were also the object of great interest and value to the Portuguese and Spanish colonizers. Cinnamon was one of the most valuable spices and the discovery of cinnamon-smelling trees by G. Pizzaro during expeditions in the Amazon basin, led him to think he would finally break the spices monopoly by Eastern traders. The newly discovered species in fact was not *Cinnamomum australe*, the original source of cinnamon found in China and India, but *Aniba canelilla* (Lauraceae). Interesting comparative phytochemical studies on such species disclosed the marked differences in their chemical constitution. Cinnamic aldehyde has been shown as the major component of the true cinnamon bark extract, whereas the studies carried out by Gottlieb and Magalhaes (1959) disclosed the presence of nitrophenylethane (**7**) as the major constituent of *A. canelilla* along with eugenol (**8**) and methyl-eugenol (**9**) (Figure 9.2). Nitrophenylethane is responsible for the cinnamon smell of *A. canelilla* and the first odoriferous nitro-derivative so far described in the literature. The numerous usages of resins, essences and balms as pain relievers and in religious ceremonies contributed to the enormous interest in the discovery of novel sources, and Brazilian plant species played a major role in this effort. The importance of rosin in ancient civilizations as in Greece, Macedonia and Egypt continued through the years to the navy industry in England who were compelled to explore North America forests in the search for new sources. Rosin or crude turpentine has been obtained as a resinous gum rich in abietic acid (**10**) (Figure 9.2) among other terpenoids mainly from *Pinus* trees and other conifers. Copal from Brazil has been considered as a rosin equivalent, especially in the Amazon region, where this resin has been used as incense and in boat construction as a sealant or to improve soldering quality. Such resins were

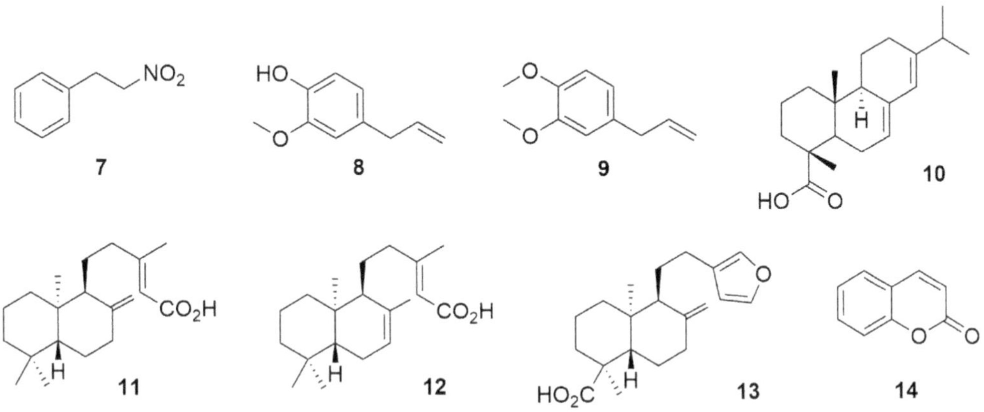

FIGURE 9.2   Flavoring and resin-derived compounds from Brazilian plants.

known as "jatobá" or "jutaí" and were obtained mainly from *Hymenea courbaril*, which presents copalic acid (**11**) (Figure 9.2) as amajor constituent. This compound has also been isolated from *Copaifera langsdorffii*, considered the main source of copaiba oil, a balm that has been extensively used to treat inflammation and as a wound-healing agent. The chemical studies on Brazilian *Copaifera* specieswere prompted by continuous and current use as folk medicine through the years which disclosed the presence of further diterpene acids such as cativic (**12**) and danelic acids (**13**) (Figure 9.2) in their resin and gave important contribution on the detection of adulterants in copaiba oils commercialized in Brazil. *Dipterix odorata*, a native tree to the Amazon basin that may reach 50 m in height, is known as cumaru and is widespread in adjacent regions to the Amazon. The fruits are known as fava tonka and were exported to Europe for their coumarin content (**14**) (Figure 9.2), and have been used as a tobacco odorant and is still widely used in the food industry as a flavoring agent (Duke and duCellier, 1993; Gottlieb and Magalhaes, 1959; Gottlieb and Mors, 1978; Nakano and Djerassi, 1961; Veiga et al., 1997).

Native Indians developed peculiar and efficient strategies for animal chasing, which included the use of venom, known as "curare" on the arrow tip. Several types of curare were used by different tribes, which were innocuous by oral administration, but paralyzed the animal within seconds when injected into the blood circulation, as described by Gottlieb and Mors (1978). "Curare" or "urari" was obtained from the "urariuva" tree in the region of Orinoco river. Indians from the "Ticunas" tribe used curare from the plant *Strychnos castelnaeana*, whose extract was taken to Europe in 1745 for scientific investigation of the species chemical constituents. *Strychnos guianensis*, formerly described as *Toxicaria americana*, and *Strychnos toxifera* were also used as sources of curare and these plants yielded D-tubocurarine (**15**) and toxiferine (**16**) (Figure 9.3) as their active compounds, respectively (Cannali and Vieira, 1967; Repke and Torres, 2006). Hallucinogenic beverages and snuffs also played an important role in cultural and religious practices of South American Indians. "Paricá" was also known as "yopo", "jopo" or "cohoba" and was prepared from roasted and grounded seeds of *Piptadenia peregrina* (synonym of *Anadenanthera peregrina*, Fabaceae), which produced bufotenine (**17**) (Figure 9.3) as active ingredient. Whereas the analog dimethyl-tryptamine (**18**) (Figure 9.3) was shown to be a major active constituent of both snuffs prepared from *Olmedioperebea sclerophylla* and "Vinho-de-Jurema" (*Mimosa hostilis*, Fabaceae), a widespread beverage used mainly by Pankararu, Xucuru and Kariri-xocó Indian tribes among others (Ott, 2002; Pachter, et al., 1959).

Additional sources of hallucinogenic snuffs have been associated with the Myristicaceae family, especially from the genus *Virola*. Although myristicaceous seeds, especially from *Virola* and *Iryanthera* species have been extensively investigated by South American researchers, their chemical profiles are mostly associated with lignans, neolignans and other shikimic acid-derived

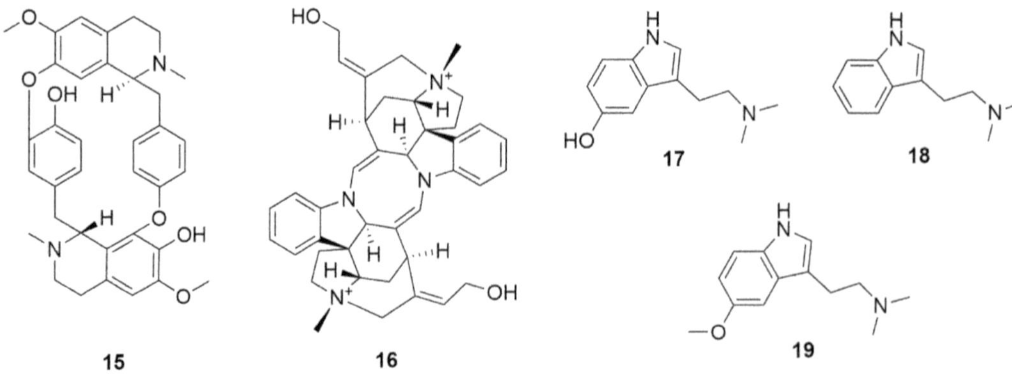

FIGURE 9.3   Poisonous and hallucinogenic compounds from Apocynaceae, Fabaceae and Myristicaceae Brazilian plant species.

biosynthetic pathways. These natural products failed to support the hallucinogenic properties of *Virola* species used in snuff preparation. Indian tribes of the northwest Amazon Basin as "Puinaves" and "Waiká" were reported to use the blood-red bark resin of *Virola calophylla*, *V. calophylloidea* and *V. theiodora* in the preparation of snuffs known as "yakee", "epena" and "nyakwana", where tryptamines are present in high concentrations, with 5-methoxy-*N,N*-dimethyltryptamine (**19**) (Figure 9.3) as the major constituent. Hallucinogenic preparations still in current use in religious rituals in Brazil as "ayhuasca" include plants containing tryptamines as *Psychotria viridis* and *Banisteria caapi* (Barker et al., 2012; Schultes, 1969).

Important scientific expeditions took place during the 18th century aiming the discovery and description of Brazilian geography, flora and fauna. The work "Historia Naturalis Brasiliae" resulted from observations of the European researchers Georg Marcgrave, Johannes de Laet and the physician Willem Piso, whose remarkable contribution to the knowledge on medicinal plants from South America represents our first natural history compendium. Subsequent expeditions played important roles in the gathering and systematization of accumulated information, as those carried out by Johann B. Spix and the botanist Carl Friederich von Martius, who suggested the invitation to the German pharmacist Theodore Peckolt to join the effort in the study of Brazilian flora (Freedberg, 1999). Peckolt came to Brazil in 1847 and was a pioneer in the systematic study of medicinal plants aimed mostly to the preparation and commercialization of remedies, which were often carried out in the laboratories of ancient pharmacies or "boticas". Remarkable results from this period include the study of *Plumeria lancifolia* (Apocynaceae), a medicinal plant used by the Guarani Indians to treat malaria. *P. lancifolia*, known as "quina-mole" or "quina-branca", was also used as a folk medicine to treat inflammation, gastric problems and women's' reproductive organs diseases. The plant's chemical study led to the isolation of plumeride (**20**, formerly named as ago-niadin) (Figure 9.4) by Peckolt, but the structural elucidation only occurred 88 years later and is considered the first isolated iridoid from a natural source (Santos et al., 1998; Halpen and Schmid, 1958). Additional important scientific work was carried out by the pharmacist Ezequiel Correia dos Santos, who obtained pereirin in 1838 from *Geissospermum leavis* (Apocynaceae) barks, a medicinal plant known as "pau-pereira", "quinarana" or "pau-forquilha", and used to treat fever and malaria. Pereirin has been considered the first alkaloid isolated in Brazil and several studies on pau-pereira that revealed to be a mixture of indole alkaloids with geissospermine (**21**) (Figure 9.4) as the major constituent. Recent studies have demonstrated antiviral properties and potential against AIDS and herpes infections (Pinto et al., 2002).

By the end of the 19th and beginning of the 20th centuries, phytochemical investigations on Brazilian medicinal plants at the School of Medicine in Rio de Janeiro and School of Pharmacy in Sao Paulo were initiated and gave important contributions as the studies on *Solanum grandiflorum*,

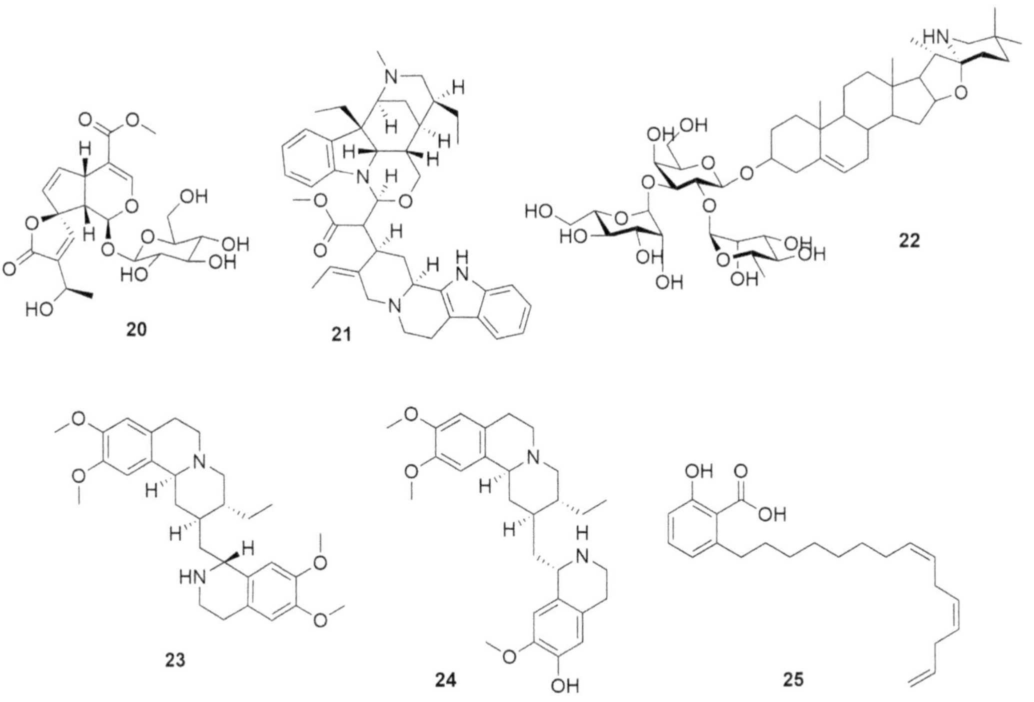

FIGURE 9.4 Bioactive compounds from Brazilian medicinal plants.

known as "lobeira", by Domingos José Freire Junior, and coffee (*Coffea arabica*) by Pedro Batista de Andrade. *S. grandiflorum* bears fruits rich in vitamins which constitute the basis of "guará" wolf diet, hence the popular name is wolf fruit ("fruta do lobo"). The plant's pulp has been used to control diabetes and afforded the steroidal alkaloid grandiflorin, later renamed as solasonine (**22**) (Figure 9.4) (Mors, 1997; Motidome et al., 1970).

Additional studies on Brazilian plants were carried out mainly due to their medicinal folk uses and have revealed several interesting natural products and confirmed their marked chemo-diversity. Ipecacuanha (*Psychotria ipecacuanha*, Rubiaceae) also known as "cagosanga", "poaia", or "ipeca", is a plant widespread throughout North and Northeastern regions in Brazil. Syrup prepared from the roots presented strong emetic properties and drew attention of Europeans for use as a purgative and antidote for poisoning. The high alkaloids content such as emetin (**23**) and cephaeline (**24**) (Figure 9.4) was associated to the emetic activity. Further studies on *P. ipecacuanha* described the antiprotozoal potential of emetin and led to the development of dehydroemetin to treat amebiasis with less nausea side effects than emetin (Cushny, 1918; Gupta and Siminovitch, 1977).

Cashew fruits (*A. occidentale*) have long been used by native Indians as an ingredient of fermented beverages. Their nut peel is strongly allergenic due to the high content of lipophilic acetyl-salicylic acid derivatives in its oil, similar to urushiol. Such compounds are known as anacardic acids (**25**) (Figure 9.4) and represent a mixture of organic acids with saturated or unsaturated C15 to C17 side chains. Their strong bactericidal activity against *Streptococcus mutans* and other gram-positive bacteria, including methicillin-resistant *Staphylococcus aureus* (MRSA) strains, demonstrated the potential to treat dental cavity and tuberculosis (Mathias, 1975; WHO, 2016).

Pilocarpus microphyllus (Rutaceae) is a shrub widespread in North and Northeast regions in Brazil, known as "jaborandi", from the Tupi language *Ya-bor-andi*, which means slobber-inducing plant. The extracts have been used as a folk medicine to treat bronchitis and rheumatism and induce

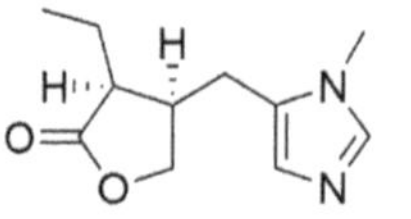

FIGURE 9.5   Bioactive compounds from Brazilian medicinal plants used in pharmaceutical products.

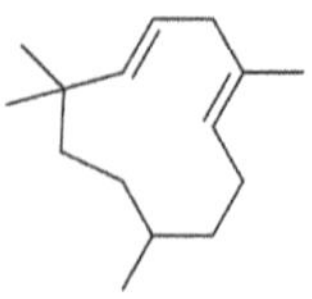

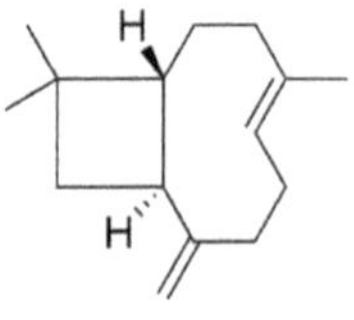

intense salivation and sweat due to the alkaloid pilocarpine (**26**) (Figure 9.5), a muscarinic receptor agonist found in the leaves. Pilocarpine has been used to treat glaucoma for over 100 years and is on the World Health Organization's (WHO's) List of Essential Medicines. *P. microphyllus* extractivism is associated to plantations in Maranhao, to provide leaves for extraction and commercial production of pilocarpine, especially by Merck Company, which has dominated this active principle market for a long period (WHO, 2016).

Although Brazilian plant diversity is enormous and provides a plethora of bioactive natural products with equally huge chemo-diversity, this biodiversityis still under-explored considering the development of phytotherapeutic products. Traditional knowledge on medicinal plants has played an important role in the selection of promising plant species for detailed chemical-pharmacological-toxicological aspects, essential for phytoceuticals development. Nevertheless, very few plant species have gone through systematic investigations to afford effective and safe commercial products. Among these, *Cordia verbenaceae*, popularly known as "erva-baleeira", constitutes an emblematic example, considering the plant's beneficial properties to treat inflammation have long been described in "De Medicina Brasiliensi" by Willem Piso in the 18th century. Additionally, fishermen and coastal populations have also used "erva-baleeira" extract or macerate for wound healing and treating arthritis, rheumatism and muscle pain. Integrated multidisciplinary studies disclosed the sesquiterpenes α-humulene (**27**) and *trans*-cariophyllene (**28**) (Figure 9.5) as the active constituents of *C. verbenaceae* essential oil, associated with the described anti-inflammatory properties. Further investigations on the chemistry, pharmacology, toxicology, pharmacokinetics and additional related issues completed the requirements for registering the new phytotherapy product containing the essential oil of *C. verbenaceae*. Joint efforts of researchers at public universities in Sao Paulo State, public research funding agencies and Aché Pharmaceutical Company have thus proven successful and resulted in a topical anti-inflammatory cream to treat myofascial pain, repetitive effort lesion, arthrosis and other painful inflammatory conditions (Basile et al., 1989; Fernandes et al., 2007).

Such findings suggest a key role played by plant extracts and their constituents in the expansion of knowledge on the chemistry and pharmacology of Brazilian biodiversity since the early years in Brazil colonization through the last five centuries and stimulates further bioprospection efforts by means of integrative and collaborative research. This scenario contributes to the discovery of novel bioactive natural products through a rational approach, which might lead to value-added bioproducts from Brazilian biodiversity along with conservational actions toward protection of biomes.

9.2   BIOPROSPECTING PLANTS FROM ATLANTIC FOREST

As part of the NuBBE research program devoted to bioprospecting bioactive metabolites of plant species from the Brazilian Atlantic Forest, several plants have been collected, identified, extracted, biologically screened and chemically analyzed focusing on bioactive compounds. In this chapter we describe compounds with several activities such as antioxidant, anti-inflammatory, chemopreventive, antifungal, antiprotozoal, acetylcholinesterase inhibitor, antiparasitic, cytotoxic (*in vitro* toxicity to malignant cells), antitumor (*in vivo*) and anticancer.

9.2.1 Natural Products Active on Redox Processes, Inflammation, Chemoprevention and Related Processes

Redox reactions are involved in important physiological processes associated to function and defense of organisms and include the generation of reactive oxygen species (ROS), such as free radical superoxide anion ($O_2^{\bullet-}$), hydrogen peroxide (H_2O_2) and hypochlorous acid (HOCl), in addition to reactive nitrogen species (RNS) as peroxynitrite ($ONOO^-$) and nitric oxide ($NO^{\bullet}$). Normal functioning of living systems requires a precise and delicate balance between pro- and antioxidant agents, which is often disturbed by biotic and abiotic conditions such as irradiation, atmospheric and food pollutants or products from metabolism. These conditions may enhance ROS and RNS generation, and ultimately lead to oxidative stress. Such species are known to participate in degenerative processes related to aging and in the etiology or progress of several diseases as myocardial and cerebral ischemia, arteriosclerosis, diabetes, rheumatoid arthritis, inflammation and cancer. Therefore, the search for natural antioxidants may represent a strategy for the discovery of new drugs, since they could decrease the oxidative stress caused by ROS/RNS excessive production (Vellosa et al., 2011).

Several assays were developed to screen natural products as antioxidant agents exploring their abilities to scavenge free radicals. The free radical scavenging assays using DPPH$^{\bullet}$ (2,2-diphenyl-1-picrylhydrazyl-hydrate) and ABTS$^{\bullet+}$ [2,2′-azinobis (3-ethylenebenzothiazoline-6-sulfonic acid)] radicals have been often used to evaluate the anti-radical potential of plant extracts and isolated compounds. The DPPH$^{\bullet}$ assay is based on the transfer of a hydrogen radical to DPPH$^{\bullet}$, which reacts as a probe, and the antiradical activity is measured through the decrease of absorbance at 517 nm. In the ABTS$^{\bullet+}$ assay, the reaction mechanism is based only on electron transfer and the decrease of absorbance is measured at 734 nm at several sample concentrations (Zeraik et al., 2016a,b). Additional scavenging assays include the use of H_2O_2, HOCl, taurine chloramine (TauCl), $O_2^{\bullet-}$ and $NO^{\bullet}$. H_2O_2 is a lesive agent due to the compound's easy decomposition to $HO^{\bullet}$, whereas HOCl is a strong oxidant, which could react with important constituents of biological systems such as nucleotides and amino acids. The reaction between biological amines and HOCl generates chloramines, oxidant agents whose reactivity depends on the substance's lipophilicity. Superoxide anion ($O_2^{\bullet-}$) is a free radical generated during cellular respiration or pathological processes as inflammatory diseases, and may react with $NO^{\bullet}$ to generate peroxinitrite ($ONOO^-$), which triggers oxidation of low-density lipoprotein (LDL), a key process in atherosclerosis etiology (Vellosa et al., 2015).

In addition to the assays described above, which involve scavenging of radical species in the evaluation of antioxidant properties, the use of electrochemical methods has been demonstrated as a useful alternative, which could integrate the evaluation of antioxidant properties by cyclic voltammetry and the identification and quantification of active compounds by high-performance liquid chromatography coupled with electrochemical detection (HPLC-ED). In this case, the electrochemical response is directly related to the structure of the antioxidant and the potential required for the product's oxidation, and establishes the chemical profile based on redox properties of the analyzed sample (Santos et al. 2010b).

Cellular models have also been used for the evaluation of antioxidant activity, for example, the investigation of oxidative damage in biological membranes using erythrocytes, since they present high polyunsaturated fatty acid contents in their membranes, in addition to high cellular oxygen and hemoglobin concentrations. Erythrocyte lipid peroxidation is associated with a variety of pathological events; thus, this evaluation may bring useful information for diagnosis and treatments for such conditions (Vellosa et al., 2011).

The antioxidant properties of extracts and their partially purified fractions, in addition to isolated compounds from plants collected in the Atlantic Forest biome, have been evaluated by using the assays described above, and represent an important alternative toward a rational exploration of Brazilian biodiversity by adding value to natural products.

Pterogyne nitens (Fabaceae) is a plant popularly known in Brazil as "tipá", "viraró", "cocal", "amendoinzeiro" and "amendoim-bravo", which can be found in Cerrado, Caatinga and Atlantic

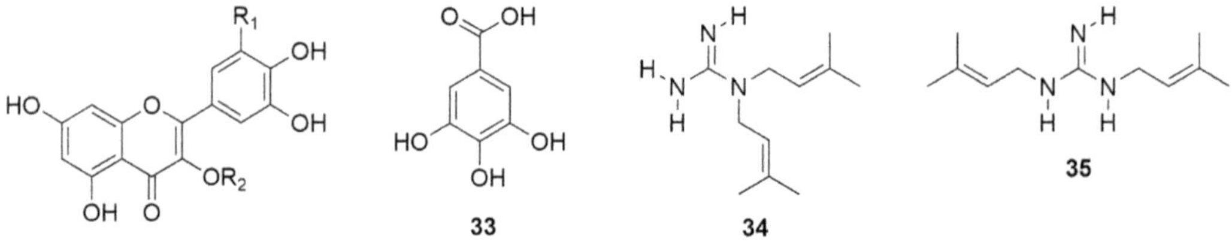

29 R_1 = OH; R_2 = H
30 R_1 = OH; R_2 = Rhamnose
31 R_1 = H; R_2 = Rhamnose
32 R_1 = H; R_2 = Sop

FIGURE 9.6 Polyphenols and guanidine alkaloids isolated from stem barks and flowers of *Pterogyne nitens*.

Forest biomes. Twenty extracts and fractions obtained from roots, branches, unripe fruits and stem bark of *P. nitens*, collected in the Botanic Garden of São Paulo (Atlantic Forest biome, São Paulo, Brazil) were screened for free radical scavenging activity using ABTS·+ and DPPH· radicals colorimetric assays and a TLC bleaching test nebulized with β-carotene solution. The ethyl acetate fraction from stem bark presented the strongest activity and was selected for further investigation. Subsequent chromatographic fractionation of this sample yielded three flavonols, myricetin (**29**), mirycetrin (**30**) and quercitrin (**31**) (Figure 9.6), which showed potent antiradical activity against DPPH· and ABTS·+, besides inhibiting β-carotene bleaching, which explains the strong antioxidant activity observed for the original fraction (Regasini et al., 2008a).

Flowers of *P. nitens* afforded nine phenol derivatives and two guanidine alkaloids which were assessed for the ability to scavenge free radicals and to inhibit myeloperoxidase (MPO), an abundant heme-enzyme in polymorphonuclear cells (PMNs), considered a key macromolecule in redox processes and in the nonspecific immune response to several agents. MPO triggers the conversion of H_2O_2 and chloride anion to water and HOCl. The excess of MPO activity and subsequent overproduction of HOCl leads to oxidative stress to PMNs, which is associated to several inflammatory processes, including rheumatoid arthritis and cystic fibrosis. Therefore, MPO inhibitors might be considered as prototypes for the development of anti-inflammatory agents. Among the tested compounds, quercetin-3-*O*-sophoroside (**32**) and gallic acid (**33**) (Figure 9.6) displayed strong MPO inhibition and antiradicalar activity against DPPH· and ABTS·+. On the other hand, the antioxidant activity of guanidine alkaloids, pterogynine (**34**) and pterogynidine (**35**) (Figure 9.6) was also evaluated, but they did not scavenge DPPH· or ABTS·+ radicals efficiently, which corroborates the importance of structural features as phenol hydroxy groups for a strong antioxidant activity (Regasini et al., 2008b).

The mutagenic and antimutagenic potential of ethyl acetate (EtOAc), n- butanol (BuOH) and hydroalcoholic (HA) fractions of the ethanol extract from *P. nitens* leaves were evaluated using *Tradescantia pallida* micronuclei assay. This is a simple and reliable assay, where *T. pallida* cuttings were treated with EtOAc, BuOH and HA fractions independently, and tetrads from the inflorescences were examined for micronuclei after exposure to test samples. Fractions BuOH and HA demonstrated mutagenic effects. Since the EtOAc fraction showed mutagenicity only at the higher concentration tested (0.460 mg mL⁻¹), the sample's antimutagenic potential was investigated and detected at lower concentrations, 0.115 and 0.230 mg mL⁻¹. BuOH and HA samples were fractionated by column chromatography resulting in the reisolation of guanidine alkaloids **34** and **35**, previously obtained from *P. nitens* flowers (Ferreira et al., 2009).

The ethanol extract from *P. nitens* leaves exhibited strong antioxidant potential against ABTS·+, DPPH· and HOCl, as indicated by low IC_{50} values. In the presence of 2′-azobis(2-amidinopropane) hydrochloride (AAPH) radical, a hemolysis-stimulating agent in erythrocytes (red blood cells), the ethanol extract and the flavonol glucoside rutin, often used as

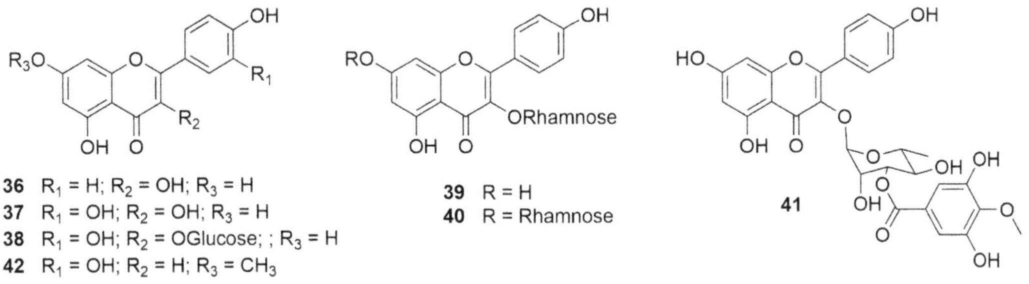

36 R_1 = H; R_2 = OH; R_3 = H
37 R_1 = OH; R_2 = OH; R_3 = H
38 R_1 = OH; R_2 = OGlucose; ; R_3 = H
42 R_1 = OH; R_2 = H; R_3 = CH$_3$

39 R = H
40 R = Rhamnose

41

FIGURE 9.7   Flavonoids isolated from leaves and fruits of *Pterogyne nitens*.

positive control, exhibited anti-hemolytic activity only at low concentrations. However, in the absence of AAPH radical, the tested sample triggered hemolysis over erythrocytes (Pasquini-Netto et al., 2012).

Additional flavonoids isolated from leaves of *P. nitens* as kaempferol (36), quercetin (37) and isoquercitrin (38) (Figure 9.7) were also assessed as ROS scavenging agents using H_2O_2, HOCl, TauCl, $O_2^{\cdot-}$ and $NO^{\cdot}$ assays. The evaluated samples exhibited moderate potential on the inhibition of $NO^{\cdot}$, although none of them were able to interact with H_2O_2, which would be evidence of the complexity of such interactions and corroborates the influence of structural features other than phenolic hydroxy groups for effective activity toward different radical species. Quercetin (37), one of the strongest antioxidant flavonoids, was also the most efficient agent against HOCl, TauCl and $O_2^{\cdot-}$ (Vellosa et al., 2011). Afzelin (39), kaempferitrin (40) and pterogynoside (41) (Figure 9.7), isolated from fruits of *P. nitens*, were also evaluated and showed moderate scavenging abilities toward $O_2^{\cdot-}$, HOCl and TauCl (Vellosa et al., 2015).

In addition, compounds 36–41 were evaluated in red blood cells as well, and displayed hemolytic effects, and they inhibited hemolysis in the presence of AAPH. However, these substances intensified the hemolytic activity when tested in a mixture with HOCl. Such data could be explained by a possible higher affinity between flavonoids and AAPH, reducing the hemolytic effects of both agents and producing less damaging products (Vellosa et al., 2011, 2015).

Flavonoids 39–41 inhibited TauCl, produced by the stimulation of neutrophils using phorbol 12-myristate13-acetate (PMA), although these flavonoids have also promoted the death of neutrophils in the presence or absence of PMA. Neutrophils are important ROS-generating systems, acting via oxidative burst, and constitute important components of tissue injury during inflammatory response. Therefore, despite their well-known scavenging action toward free radicals and oxidants, these compounds could be harmful to living organisms at the tested concentrations, through their action over erythrocytes and neutrophils (Vellosa et al., 2015).

The flavonol isoquercitrin (38) and flavone pedalitin (42) (Figure 9.7), also isolated from *P. nitens*, exhibited strong antioxidant activity by inhibiting β-carotene bleaching in a TLC assay. In addition, a fast, low-cost and convenient cyclic voltammetry screening, and the combination of HPLC with an electrochemical detector (HPLC-ED), confirmed the antioxidant activity of these flavonoids. Structural features as *ortho*-dihydroxy groups, α-, β-unsaturated carbonyl moiety (ring C) and β-hydroxyketone (rings A and B) are known to play a key role in antioxidant properties, since they enhance the radical stabilization after the first oxidation steps. In this regard, the use of HPLC-ED, already established as a qualitative and quantitative technique to detect antioxidant small molecules, is considered as a useful assay for the determination of antioxidants in complex matrixes without previous sample preparation or pre-concentration, since this technique offers high sensitivity and easy operation (Okumura et al., 2012).

Plectranthus barbatus Andrews (Lamiaceae) is a popular medicinal plant used to treat gastrointestinal and hepatic disorders and is known as "Brazilian-boldo" or "false-boldo". The aqueous extract from *P. barbatus* leaves, collected in Cajobi (São Paulo, Brazil), presented a significant

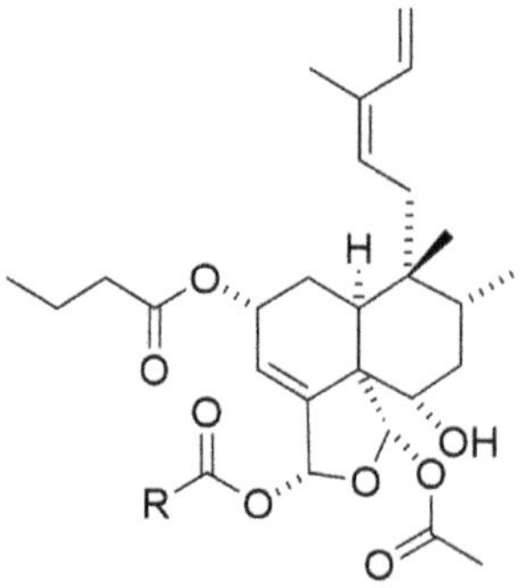

43 - R = CH₃
44 - R = C₃H₇

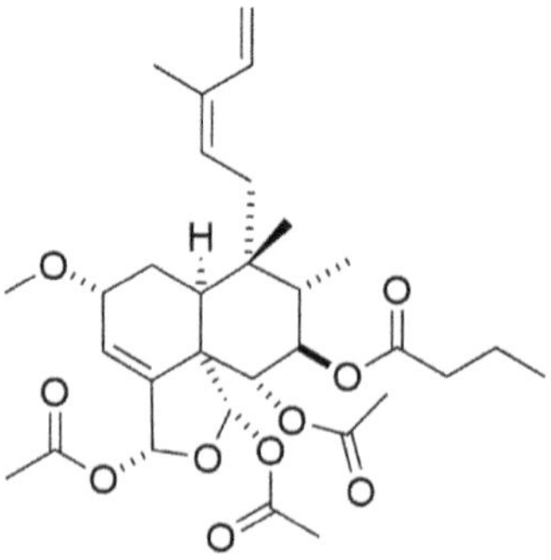

45

FIGURE 9.8   Clerodane diterpenes isolated from *Casearia sylvestris*.

free radical scavenging activity toward DPPH· and ·OH, besides iron chelating mediated activity, preventing the formation of the oxidant Fe^{2+}-bathophenanthroline disulfonic acid (BPS) complex. Moreover, this extract protected mitochondria against Fe^{2+}-citrate-mediated membrane lipid peroxidation, since cell swelling and malondialdehyde production was avoided with the activity persisting even after simulation of the product's passage through the digestive tract (Maioli et al., 2010).

C. *sylvestris* Swartz (Salicaceae) is a tree widely distributed in a variety of ecosystems from the Cerrado to the tropical Atlantic Forest and the equatorial Amazon forest known as "guaçatonga" and widely used in folk medicine to treat ulcer, inflammation and tumors. Their leaves are rich in clerodane diterpenes as casearins and caseargrewiins, which have shown cytotoxic and antifungal activities in several studies (Oberlies et al., 2002; Santos et al., 2010a).

Reinvestigation of C. *sylvestris* afforded caseargrewiin F (**43**) (Figure 9.8) from the leaves' ethanol extract (ELCS), collected at Parque Estadual Carlos Botelho (São Miguel Arcanjo, São Paulo, Brazil). The evaluation of chemopreventive properties suggested a protective effect of ELCS in micronucleus (MN) test and comet assay in mice. Cyclophosphamide (CP) was used in both tests to damage DNA and compound **43** showed a protective effect only in the comet assay. The MN test reveals more drastic lesions in chromosome level (mutagenicity), while the comet assay detects genomic lesions that are susceptible to DNA repair (genotoxicity). The tests were also performed without the injection of CP, indicating that ELCS and compound **43** triggered DNA damage only at high concentrations (Oliveira et al., 2009).

Further investigation on the mutagenic properties of C. *sylvestris* was carried out using *Tradescantia* micronucleus assay, the MN test in mouse bone marrow cells, and the comet assay. The same extract (ELCS) and casearin X (**44**) (Figure 9.8), another clerodane diterpene isolated from this extract, were evaluated as protective agents against the harmful effects of airborne pollutants from sugarcane burning. The mutagenic agent in this case was total suspended particulate (TSP) from air, collected near Araraquara (São Paulo, Brazil) during the sugarcane-burning season. ELCS exhibited antimutagenic activity in the *Tradescantia* micronucleus assay and was able to reduce DNA damage caused by TSP in the MN test and in the comet assay, while compound **44** reduced only DNA damage assessed by the comet assay. Such data suggested that C. *sylvestris* extract and the isolated diterpenes might act by different mechanisms to protect DNA against damage, including repairable and non-repairable damages (Prieto et al., 2012).

Casearin B (**45**) (Figure 9.8), also isolated from leaves of C. *sylvestris*, showed genotoxicity in HepG2 cells (comet assay) at concentrations higher than 0.30 μM when incubated with the formamidopyrimidine-DNA glycosylase (FPG) enzyme.Whereas, DNA damage caused by H_2O_2 in HepG2 cells in both pre- and posttreatment experiments was reduced. Compound **45** was not mutagenic to S. *typhimurium* strains TA98 and TA102, used in the Ames test, and exhibited strong inhibitory

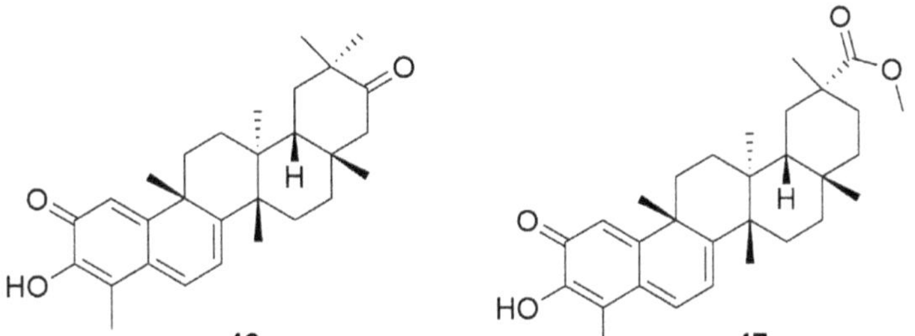

FIGURE 9.9   Quinonemethide triterpenes isolated from *Maytenus ilicifolia*.

activity against aflatoxin B1 in TA 98, in addition to moderate inhibitory activity against mytomicin C in TA 102 (antimutagenicity assays). Casearin B also displayed antioxidant activity, since the compound was able to reduce ROS generated by H_2O_2 in the 2′,7′-dichlorodihydrofluorescein diacetate (DCFDA) assay. However, compound **45** was less effective in the inhibition of DCFH oxidation than the positive control quercetin. In this test, DCFDA, which is a redox inactive compound, is converted to DCFH (active ROS) by an esterase inside the cell (HepG2 cell). H_2O_2 and other ROS oxidize intracellular DCFH (nonfluorescent) to DCF (fluorescent), which is then measured in the fluorescence assay (Prieto et al., 2013).

Maytenus ilicifolia (Celastraceae), known as "espinheira santa", is spread in tropical and subtropical parts of the Atlantic Forest and is widely used in traditional medicine as anti-inflammatory, analgesic and antiulcerogenic (Costa et al., 2008; Jorge et al., 2004). Their roots are known to accumulate quinonemethide triterpenes, which constitute chemo-taxonomical markers of this genus and present several biological activities. HPLC-DAD analyses of *M. ilicifolia* extracts obtained from root barks of adult plants (E2) and roots of seedlings (E4) indicated the presence of quinonemethide triterpenes and phenolic compounds as main chemical constituents. The DPPH• assay indicated that rutin, a major flavonoid from the extracts under investigation, exhibited higher antioxidant activity than the quinonemethide triterpenes maytenin (**46**) and pristimerin (**47**) (Figure 9.9), isolated from the root barks of *M. ilicifolia*. Such results are probably related to different structural features, since quinonemethide triterpenes present an α,β-unsaturated carbonyl moiety with extended conjugation through ring B, whereas rutin bears a catechol moiety in ring B in addition to an α,β-unsaturated carbonyl moiety conjugated to ring B. Their antioxidant properties were also monitored by voltammetry screening, which corroborated the results obtained by the DPPH• scavenging assay. In addition, cyclic voltammograms associated to HPLC-ED analyses suggested a synergistic interaction between quinonemethide triterpenes and flavonols, as indicated by the mixture of rutin with compound **47**, or rutin with *M. ilicifolia* extracts, which demonstrated, in both cases, enhanced antioxidant activity than the individual samples (Santos et al. 2010b).

Spondias tuberosa (Anacardiaceae) is a native plant from Northeast of Brazil spread all over arid and semi-arid regions and is especially useful for accumulating water in the species' tuberous roots, which provides continuous fruit loads even in drought seasons. The fruits are popularly known as "umbu" and may be consumed fresh, as juice, ice cream, sweet, jam or as the traditional "umbu-zada" (fruit pulp boiled with milk and sugar). Fractionation of umbu pulp methanol extract yielded two novel phenolic glucosides, 3,4-dihydroxyphenylethanol-5-β-D-glucose (**48**) and 5-hydroxyl,4-methoxy-3-*O*-β-D-glucose benzoic acid (**49**) (Figure 9.10), along with five known compounds, **33** (Figure 9.6) and **50**–**53** (Figure 9.10). The isolated substances exhibited strong antioxidant properties, which were evaluated by DPPH•, ABTS•+ and ORAC assays, while compound **48** showed the highest capacity to prevent the oxidative effects of radicals generated by AAPH on fluorescein in the ORAC assay. The dichloromethane extract exhibited chemopreventive activity, evidenced by strong quinone reductase (QR) enzyme induction in Hepa1c1c7 cells when compared to the positive control 4′-bromoflavone. The fractionation of the extract provided the isolation of an anacardic acid

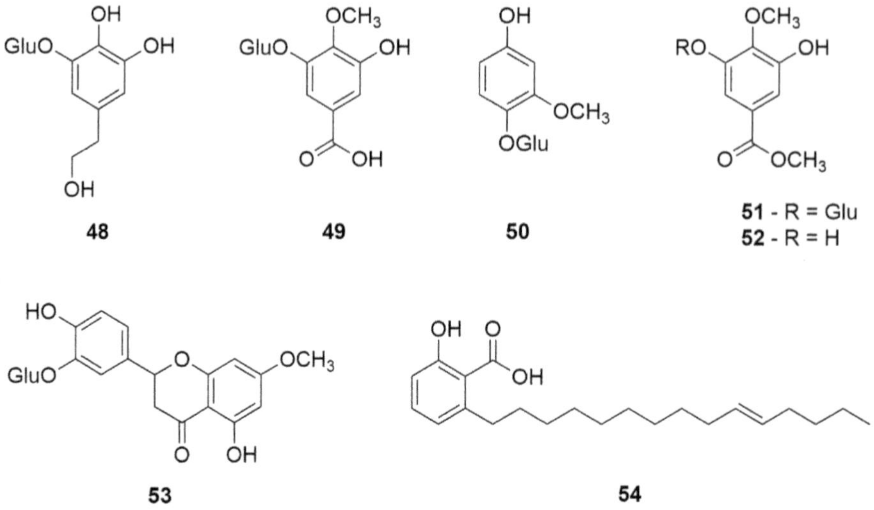

FIGURE 9.10 Phenolic compounds isolated from *Spondias tuberosa* ("umbu").

derivative (**54**) (Figure 9.10), which, however, was not effective in QR induction. The elevation of phase II enzymes such as QR could be correlated with protection against chemical-induced carcinogenesis in animal models in the stages of initiation and promotion (Zeraik et al., 2016a,b).

9.2.2 CYTOTOXIC COMPOUNDS

Cancer is a disease characterized by abnormal cell proliferation that can invade nearby tissues and spread to other parts of the body through the blood and lymph system (NCI, 2018). According to the WHO (2018), cancer is the second leading cause of death globally and accounted for 8.8 million deaths in 2015. Therefore, because of numerous facts, such as cancer severity, mortality, economic issues, lack of more selective and less toxic drugs and evolving resistance to currently available therapeutic agents, novel effective and accessible molecules are urgently required for the treatment of this disease. New strategies are also necessary for cancer prevention, ranging from a healthier lifestyle to prevention by chemical means, as the promotion of chemopreventive agent's consumption, either from vegetables such as broccoli, garlic and berries or nutraceuticals.

Concerning cancer treatment, plants have directly afforded natural drugs, or precursors to semi-synthetic derivatives, in addition to prototypes which inspired purely synthetic therapeutic agents, currently in clinical use, such as *vinca* alkaloids (i.e. vinblastine and vincristine), etoposide and teniposide (semi-synthetic derivatives of the NP *epi*-podophyllotoxin). Probably, one of the most recognized and noteworthy example is paclitaxel (Taxol®), an anticancer drug isolated from the leaves of several *Taxus* species along with the precursor compounds(Cragg and Newman, 2013).

The studies carried out at NuBBE have initially addressed crude extracts, which have been submitted to preliminary assays for the detection of cytotoxicity against tumor cell lines and allowed the selection of promising samples for further chemical and biological investigation to afford partially purified fractions and isolated compounds. Their evaluation against tumor cell lines, in addition to complementary assays in the case of promising samples, led to the discovery of active compounds belonging to several natural products classes.

In some studies, such as that of *Ouratea multiflora* (Ochnaceae), the resulting compounds did not exhibit remarkable cytotoxic activity. In other cases, potent bioactive compounds have been discovered as those exemplified by *M. ilicifolia* (Celastraceae) quinonemethides.

The chromatographic separation of chemical constituents from the ethanol extract of *O. multiflora* leaves, a medicinal plant used to treat inflammatory diseases such as rheumatism and

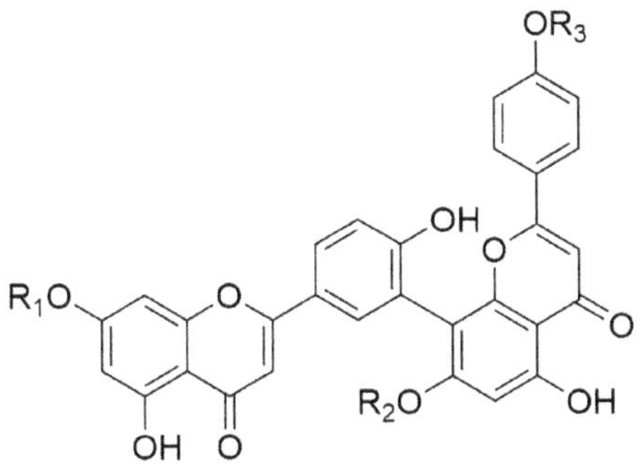

	R$_1$	R$_2$	R$_3$
55	CH$_3$	CH$_3$	CH$_3$
56	H	CH$_3$	CH$_3$
57	H	H	CH$_3$
58	H	H	H

FIGURE 9.11   Structures of flavone dimers **55–58**.

arthritic disorders (Carbonari et al., 2006), led to the isolation of four flavonoid dimers, namely, heveaflavone (**55**), amentoflavone-7″,4‴-dimethyl ether (**56**), podocarpusflavone-A (**57**) and amentoflavone (**58**) (Figure 9.11). The biflavonoids were evaluated for cytotoxicity against mouse lymphoma (L5178) and melanoma (KB) cancer cell lines. However, none of these metabolites was active in this assay (Carbonezi et al., 2007).

Pristimerin (47) (Figure 9.9), a quinonemethide triterpene exhibiting cytotoxic activity against various cancer cell lines (Deeb et al., 2014), was isolated from the ethanol extract of *M. ilicifolia* root barks with cytotoxic potential evaluated through MTT (3-(4,5-dimethyl-2-thiazolyl)-2,5-diphenyl-2*H*-tetrazolium bromide) assay (Mosmann, 1983) in five additional human tumor cell lines: HL-60 (promyelocytic leukemia), k-562 (chronic myelocytic leukemia), SF-295 (glioblastoma), HCT-8 (colon cancer) and MDA/MB-435 (melanoma). The selectivity of pristimerin (**47**) was also evaluated toward a normal proliferating cell line, by performing the Alamar Blue assay with human peripheral blood mononuclear cells (PBMC), after 72 hours of drug exposure. The mechanism of the action of pristimerin in leukemia cell (HL-60) cytotoxicity was also investigated. For this purpose, the following experiments were performed: the cell viability was determined by the trypan blue dye exclusion test, inhibition of DNA synthesis was assessed determining the amount of BrdU (5-bromo-2′-deoxyuridine) incorporated into DNA (Pera et al., 1977), inhibitory effects of pristimerin on human topoisomerase I were measured using a Topo I Drug Kit (TopoGEN, Inc), acridine orange/ethidium bromide (AO/EB) staining assay (McGahon et al., 1995) was performed in order to evaluate the cell death pattern induced by increasing concentrations of pristimerin, HL-60 cell membrane integrity was evaluated by exclusion of propidium iodide and then cell fluorescence was measured by flow cytometry; internucleosomal DNA (lysed) was also analyzed by flow cytometry. Pristimerin (47) displayed cytotoxic activity toward the five tumor cell lines tested, with IC$_{50}$ values ranging from 0.55 μM to 3.2 μM in MDA/MB-435 and k-562, respectively. The IC$_{50}$ values over PBMC from pristimerin and doxorubicin were 0.88 μM and 1.66 μM, respectively. Subsequent experiments conducted on HL-60 cells aimed to elucidate the mechanism of action of pristimerin in this cell line. The trypan blue test revealed that pristimerin reduced the number of viable cells and increased the number of nonviable cells in a concentration-dependent way, presenting morphological alterations consistent with apoptosis. However, pristimerin was shown not to be selective to cancer cells when compared with a normal cell line, since PBMC was inhibited at an IC$_{50}$ of 0.88 μM. The assessment of DNA synthesis inhibition by BrdU incorporation in HL-60 cells was 70% (0.4 μM) and 83% (0.8 μM), whereas pristimerin was not able to inhibit topoisomerase I. AO/EB staining showed that all tested concentrations of pristimerin were able to reduce the number of viable cells, with the occurrence of necrosis and apoptosis in a dependent concentration way, which represents results quite consistent with trypan blue exclusion insights. Furthermore, the analysis of membrane integrity and internucleosomal DNA fragmentation through flow cytometry in the presence of pristimerin suggested that treated cells underwent apoptosis. Therefore, these results highlight the importance of pristimerin as a representative compound of an emerging class of cytotoxic metabolites against several cancer cell lines, displaying antiproliferative effect by inhibiting DNA synthesis and triggering cell death likely by apoptosis (Costa et al., 2008).

A second quinonemethide triterpene, named maytenin (**46**) (Figure 9.9) was also isolated from *M. ilicifolia* (Santos et al., 2010b). As well as **47**, maytenin was assessed for cytotoxic effects by MTT assay toward human keratinocytes (NOK cells of the oral mucosa) while they were additionally assayed for antifungal activity, as antifungal agents require a broad spectrum of action and no toxicity over the cells. Both compounds exhibited cell viability higher than 80%, suggesting that these metabolites were not cytotoxic in the assay (Gullo et al., 2012).

Styrylpyrones isolated from *Cryptocarya spp.* (Lauraceae) and their derivatives have demonstrated antiproliferative activity in a range of human cell lines, therefore, representing antitumor potential even though the mechanism of action involved remains unknown. *Cryptocaria* species may be commonly found in both Brazilian phytogeographic regions Cerrado and Atlantic Forest. *C. mandiocanna* is a tree similar to *C. moschata*, and the latter species is popularly entitled the "Brazilian nutmeg" ("noz-moscada-do-Brasil") for their highly aromatic seeds. In this context, the styrylpyrone cryptomoschatone D2 (**59**) (Figure 9.12) was isolated from the methylene chloride extract from leaves of *Cryptocarya mandiocanna*, collected in Atlantic Forest, by chromatographic methods and subsequently assessed for cytotoxic activity in HPV-infected HeLa (HPV18) and SiHa (HPV16), as well as uninfected (C33A) human cervical carcinoma cell lines and in lung fibroblast MRC-5 cell line. The cells were submitted to a treatment with different concentrations of compound 59 (15 μM, 30 μM, 60 μM or 90 μM) for 6, 24 hours, and for 6 hours followed by a post-treatment recovery period of 24, 48, or 72 hours. High dose- and time-dependent cytotoxicity was observed for all cell lines. Furthermore, unlike the infected cells (HeLa and SiHa), C33A cells were unable to recover their proliferative ability proportionally to the posttreatment recovery time (Giocondo et al., 2009).

The Piperaceae family comprises groups of species, which produce amides, phenylpropanoids, lignans, benzoic acids and chromenes, some of the main classes of natural products found in *Piper* and *Peperomia* that represent the largest genera in Piperaceae.

Piperlongumine (**60**), also known as piplartine (5,6-dihydro-1-[1-*oxo*-3-(3,4,5-trimethoxyphenyl)-2-propenyl]-2(1*H*) pyridinone), is an amide alkaloid (Figure 9.13) widespread in *Piper* species. This metabolite displays strong cytotoxic activity over tumor cell lines (e.g. HL-60, k562, Jurkat and Molt-4), in addition to other bioactivities. Further investigation on piperlongumine cytotoxic properties was carried out after its re-isolation from the root extract of *Piper tuberculatum*. Piperlongumine was evaluated for genotoxic (mutagenic) effects and induction of apoptosis in V79 cell line (derived from Chinese hamster lung fibroblasts), as well as for mutagenic and recombinant potential in *Saccharomyces cerevisiae*. The extract was found to induce dose-dependent cytotoxicity in *S. cerevisiae* cultures, in addition to exhibiting weak mutagenic effect on cells during the

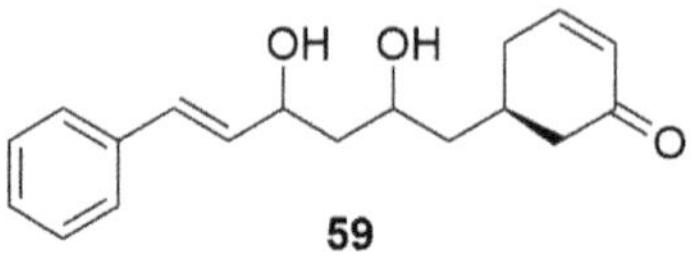

59

FIGURE 9.12   Styrylpyrone cryptomoschatone D2 isolated from *Cryptocarya mandiocanna*.

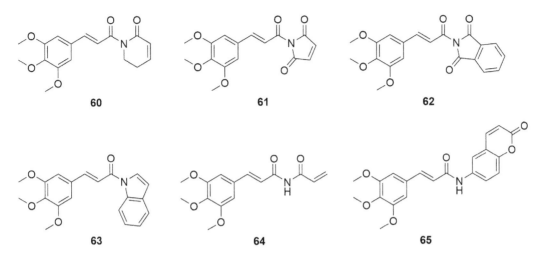

FIGURE 9.13 Structures of piperlongumine (60) isolated from *P. tuberculatum* and the synthetic analogs (61–65).

exponential growth phase in a buffer solution, although an increase on the frequency of mutations during growth in the medium was observed. The neutral and alkaline comet assays revealed that piperlongumine induced G2/M cell cycle arrest, probably due to triggering DNA double strain breaks and repair. Furthermore, treatment with this metabolite induced dose-dependent apoptosis, which was detected by a decrease in mitochondrial membrane potential in contrast to an increase in internucleosomal DNA fragmentation. Finally, cells surviving piperlongumine-induced DNA damage could accumulate mutations as this compound proved to be mutagenic and recombinogenic in *S. cerevisiae* (Bezerra et al., 2008).

The mutagenic and antimutagenic effects of piperlongumine were also evaluated in *Salmonella typhimurium* strains TA97a, TA98, TA100 and TA102 by the Ames test (Morandim-Giannetti et al., 2011). The results indicated piperlongumine efficacy for protecting genetic material against damage caused by mutagenic agents, and therefore suggested promising antitumor uses for this compound due to the close relationship between mutagenesis and carcinogenesis (Morandim-Giannetti et al., 2011).

Since 2011, several studies have focused on piperlongumine due to the compound's selective antitumor properties, comprising about 80 articles published worldwide, including some aimed to encourage and guide clinical trials. However, no studies addressing piperlongumine metabolism in human organism were known, until de Lima Moreira et al. (2016) investigated the compound's *in vitro* oxidation by Cytochrome P450 (CYP450) enzyme, in addition to the enzymatic kinetic profile catalyzed by CYP enzymes and the prediction of *in vivo* pharmacokinetic parameters. The structures of four piperlongumine metabolic products were also proposed by employing liquid chromatography coupled to high-resolution mass spectrometry (HR-LC-ESI-MS) analyses in both negative and positive ion modes and were confirmed on the basis of their fragmentation patterns in LC-IT-MS spectra. Subsequent isolation through LC-SPE-NMR (liquid chromatography-solid phase extraction-nuclear magnetic resonance) allowed ¹H NMR assignments for the metabolites' characterization, which corroborated their structures. Phenotypic studies and possible piperlongumine-drug interactions were reported as well. Altogether, these results propitiate a useful guide to further clinical studies aimed at a rational exploration of piperlongumine potential in the development of antitumor drugs.

Metastasis is the process in which cancer cells spread from the original tumor to other parts of the body through the blood or lymph system and initiates the formation of a new tumor. The process of cancer cell migrations to distant tissues or organs is a crucial step in metastasis (new tumor) formation. Most of the cancer drugs currently available target inhibition of cell proliferation and killing cancer cells, instead of cell migration. Based on piperlongumine's (**60**) previously reported activities (Bezerra et al., 2008, Morandim-Giannetti et al., 2011, de Lima Moreira et al., 2016), and on a cell-based screening, it's the potential to inhibit breast cancer cell line (MDA-MB-231) migration by the Boyden chamber assay and for cytotoxic activity against normal (MCF10A) and cancer (MDA-MB-231 and DU-145) cell lines (Valli et al., 2017) was evaluated. Furthermore, a series of five analogs (**61–65**) (Figure 9.13) was designed using the concepts of molecular simplification and hybridization, synthesized and evaluated in cell migration and cytotoxicity assays. Piperlongumine (**60**) inhibited the migration of MDA-MB-231 cells with EC_{50} of 3.0 ± 1.0 μM in the Boyden chamber assay, which is comparable to the activity of colchicine, used as positive control. Boyden chamber consists of a two-compartmental system separated by a plastic membrane and allows the quantitative determination of a compound effect on cell migration. Piperlongumine analog **64**, which was designed by molecular simplification, was the most active from the series ($EC_{50} = 9.0±1.0$ μM) and showed selective cytotoxicity toward normal breast cell line MCF10A, with a selectivity index (SI) of 4.4. Finally, piperlongumine did not show interaction with microtubules in the tubulin polymerization assay, which indicates that the activity must have a different mechanism of action (Valli et al., 2017).

9.2.3 NATURAL PRODUCTS ACTIVE ON THE CENTRAL NERVOUS SYSTEM

Natural products have been used to cause effects on the central nervous system (CNS) since ancient times. Populations all over the world have been making use of plants for rituals and medicinal purposes and such knowledge was passed through generations, although much information may have been lost. Ethnopharmacological research on CNS-related disorders has also been based on traditional knowledge and is very useful for guiding natural products research on such complex conditions.

Some plant families have been greatly reported to treat CNS diseases, such as Fabaceae, Asteraceae, Rubiaceae and Solanaceae, and many plants were studied having ethnopharmacology as a basis (Cooper, 1987; Ghedini et al., 2002; Mendes and Carlini, 2007; Seidler, 2001; Valli et al., 2016). CNS disorders are neurological illnesses impairing the function or structure of the brain, spinal cord and retina. They include Alzheimer's disease, Parkinson's disease, depression, anxiety, epilepsy, multiple sclerosis (MS) and schizophrenia (Ballios et al., 2011; DiNunzio and Williams III, 2008). The growing number of cases of these diseases in the past few years has been often associated to the ageing of world population (Varma et al., 2016). The use of pharmaceutical drugs is still one of the most widespread strategies for treating CNS diseases. The search for natural products for the development of novel therapeutic agents aimed at the treatment of CNS diseases has been one of the focuses of our research group, and herein we present a few interesting results.

In a screening of native Brazilian plants, the ethanol crude extract of *Esenbeckia leiocarpa* stems showed acetylcholinesterase inhibition and was submitted to bioassay-guided fractionation to isolate the biologically active compounds. Acetylcholinesterase inhibition has been considered a major target in the discovery of therapeutic agents to treat patients with Alzheimer's disease as it increases the availability of acetylcholine for synaptic transmission in the brain. The bioactivity-guided fractionation of the ethanol extract from stems afforded six alkaloids: leiokinine A (**66**), leptomerine (**67**), kokusaginine (**68**), skimmianine (**69**), maculine (**70**) and flindersiamine (**71**) (Figure 9.14). All isolated compounds displayed *in vitro* acetylcholinesterase inhibition activity in the Ellman TLC assay (Ellman et al., 1961). Leptomerine (**67**) showed the highest activity ($IC_{50} = 2.5$ μM), when compared to the reference compound galanthamine ($IC_{50} = 1.7$ μM) (Cardoso-Lopes et al., 2010).

Tetrapterys mucronata Cav. (Malpighiaceae) is one of the plants used in the preparation of ayahuasca in some regions of Brazil. Ayahuasca is a psychotropic plant decoction with a long cultural history of uses (Carlini, 2003), especially by devotees of some religions.

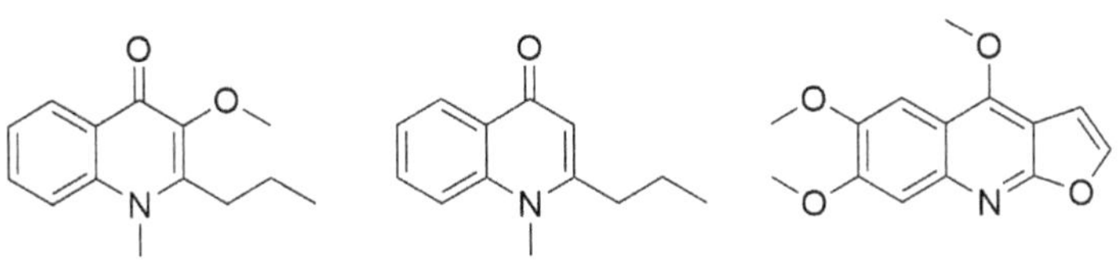

66 **67** **68**

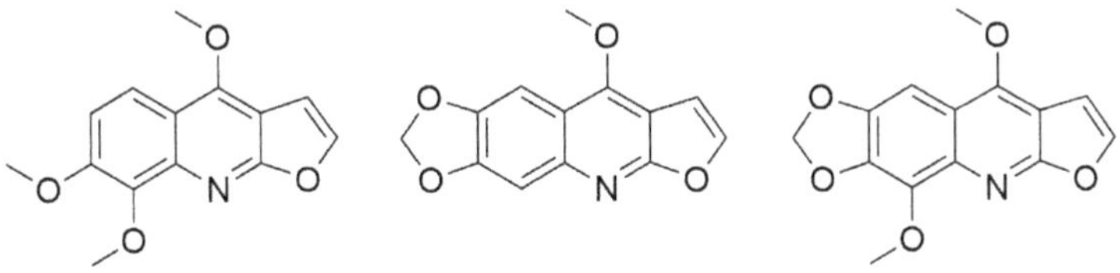

69 **70** **71**

FIGURE 9.14 Acetylcholinesterase inhibitory alkaloids isolated from *Esenbeckia leiocarpa*.

The chemical composition of *T. mucronata* was determined with the constituents evaluated as acetylcholinesterase inhibitors. The ethanol extract of *T. mucronata* barks exhibited *in vitro* acetylcholinesterase inhibition in a TLC bioautography assay (Atta-ur-Rahman et al., 2005; Di Giovanni et al., 2008; Ellman et al., 1961). The active constituents were identified and among the twenty-two isolated compounds, the tryptamine alkaloids, **17** (Figure 9.3) and **72–75** (Figure 9.15), inhibited acetylcholinesterase with IC_{50} values below 15 μM and were compared to the positive controls galanthamine (IC_{50} = 2.4 μM) and tacrine (IC_{50} = 0.09 μM) (Queiroz et al., 2014a).

Toxic and hallucinogenic properties were evaluated in a study performed with a water decoction, which mimics the ayahuasca preparation, to determine the decoction chemical profile and content of the main tryptamine alkaloids in *T. mucronata* stem barks. The extraction of stem barks with ethanol afforded bufotenine (**17**) (Figure 9.3), 5-methoxy-*N*-methyltryptamine (**72**), 5-methoxy-bufotenine (**73**) and 2-methyl-6- methoxy-1,2,3,4-tetrahydro-β-carboline (**76**) (Figure 9.15). A comparison with the water decoction revealed slightly lower levels of these constituents. These four alkaloids have been described for their toxic and hallucinogenic properties, especially bufotenine and 5-methoxy-bufotenine. Although some previous studies have indicated that the risk of intoxication by consuming ayahuasca is minimal, lethal cases have been reported. Therefore, this study was important to indicate that consumption of *T. mucronata* as an ingredient in ayahuasca preparations may present a risk to consumers (Queiroz et al., 2015).

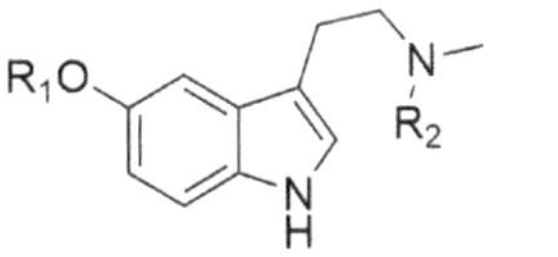

72 $R_1 = CH_3$ $R_2 = H$
73 $R_1 = CH_3$ $R_2 = CH_3$

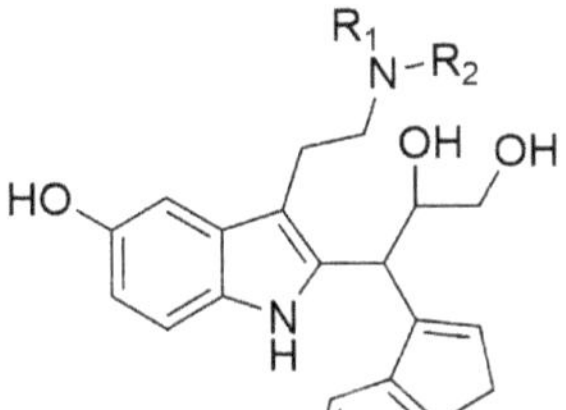

74 $R_1 = CH_3$ $R_2 = CH_3$
75 $R_1 = H$ $R_1 = H$

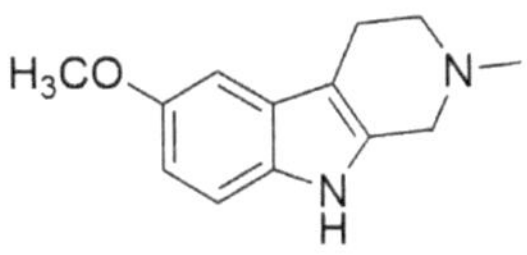

76

FIGURE 9.15 Tryptamine alkaloids from the stem bark of *T. mucronata*.

9.2.4 Antifungal Compounds

The increase in fungal infections represents a serious concern considering the insufficient amounts of antifungal drugs in addition to problems with toxicity and increased resistance to fungi. Natural compounds from the Brazilian biodiversity and their derivatives may represent an alternative in the search for new and effective antifungal therapeutic agents (Funari et al., 2012a; Newman et al., 2000).

Phytochemical studies on Brazilian plants developed in our research group in recent years, especially those from Atlantic Forest, have disclosed the presence of several antifungal secondary metabolites, mainly triterpenes, diterpenes, saponins, flavonoids, alkaloids and polyketides.

The pharmacological activities already reported for *M. ilicifolia* extracts and pure compounds instigated the study of this species against human pathogenic fungi, especially opportunistic species of *Aspergillus, Candida, Cryptococcus, Histoplasma, Fusarium, Trichophyton* and *Paracoccidioides* genera, which have been mainly associated with infection in immunocompromised patients (Shoham and Levitz, 2005). In this context, maytenin (**46**) and pristimerin (**47**) (Figure 9.9), previously isolated from the root barks of adult *M. ilicifolia* plants (Santos et al., 2010b), were evaluated using a qualitative analysis of a fungal viability and microdilution method (M27-S3 – CLSI – Clinical and Laboratory Standards Institute (2008)) with modifications to calculate their minimum inhibitory concentration (MIC) and minimum fungicide concentration (MFC). Maytenin showed potent antifungal activity against both yeasts and filamentous fungi with MICs ranging from 0.12 mg L^{-1} to 62.5 mg L^{-1}. Compounds **46** and **47** showed same MIC values against *Histoplasma capsulatum* and *Paracoccidioides brasiliensis* (0.48 mg L^{-1} and <0.12 mg L^{-1}, respectively), when compared to a positive control itraconazole, which exhibited MIC 0.25–2.00 mg L^{-1} for *H. capsulatum* and MIC lower than 0.0039 mg L^{-1} for *P. brasiliensis,* and amphotericin B, which exhibited MIC values of 0.06–0.25 and 0.015–0.25 mg L^{-1} for *H. capsulatum* and *P. brasiliensis*, respectively. Maytenin showed the best results for all tested fungal strains, while pristimerin displayed strong activity against *Candida krusei* (MIC 7.81 mg L^{-1}) and *Cryptococcus neoformans* (MIC 0.97 mg L^{-1}) and moderate activity against filamentous fungi with MICs ranging from 0.12 to 250 mg L^{-1}. In addition, maytenin displayed a selectivity index (SI) above 1.0 for all fungal strains tested, wherein the higher the SI, the greater the safety of the tested compound. Pristimerin exhibited high SI against *C. neoformans* and *H. capsulatum* fungal strains, but maytenin displayed the best structural features associated with a selective effect against the tested human pathogenic fungal strains (Gullo et al., 2012).

Croton heliotropiifolius Kunth (Euphorbiaceae) is popularly known as "velame" with the leaves and barks used in folk medicine as pills or infusions to treat gastrointestinal problems and for weight loss (Govaerts et al., 2000). The bio-guided fractionation of ethanol extract from *C. heliotropiifolius* stem barks, using chromatographic techniques to detect and isolate compounds with antifungal activity against *Candida albicans*, led to the isolation of nine compounds (**77–85**) (Figure 9.16). Velamone (**84**) and the compound's analog, velamolone acetate (**83**), showed weak antifungal activity, whereas spruceanol (**85**) demonstrated a relevant minimal inhibitory quantity (MIQ), when tested against the mutant strain DSY2621 and the wild strain. Spruceanol was the most active compound against *C. albicans* that was first reported on antifungal potential (Queiroz et al., 2014b).

The antifungal activity of the *Swartzia langsdorffii* (Fabaceae) extract against *C. albicans, C. krusei, C. parapsilosis* and *C. neoformans* strains was evaluated using a microdilution assay (CLSI, 2008) leading to the isolation of bioactive constituents. *S. langsdorffii* is popularly known as "banana-de-papagaio", "jacarandá-banana" or "jacarandá-de-sangue", and the bio-guided fractionation of an ethanol extract of the leaves using chromatographic techniques afforded pentacyclic triterpenes and saponins (Figure 9.17) with antifungal activity. The isolated substances were also evaluated by bioautography against phytopathogenic fungal strains, and the saponins oleanolic acid 3-sophoroside (**86**) and 3-*O*-β-D-(6′-methyl)-glucopyranosyl-28-*O*-β-D-glucopyranosyl-oleanate

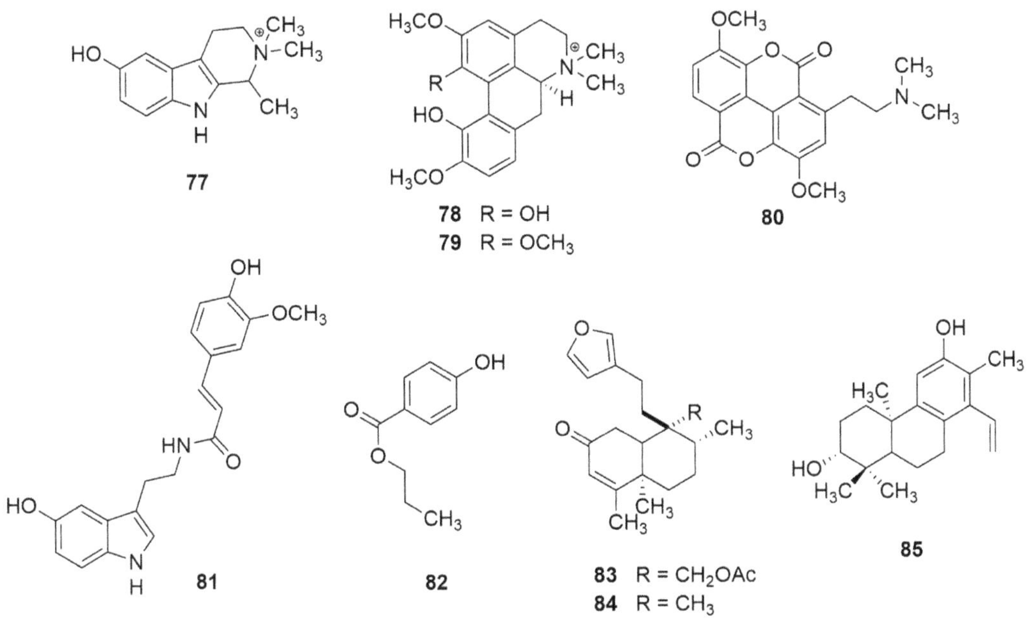

FIGURE 9.16   Compounds (**77–85**) isolated from stem bark of *Croton heliotropiifolius*.

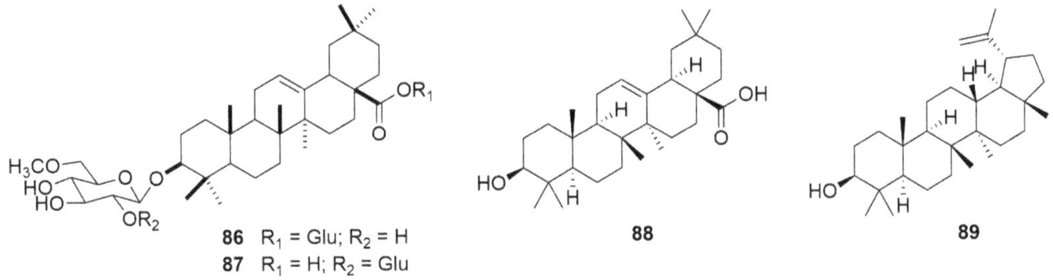

FIGURE 9.17   Saponins and triterpenes isolated from *Swartzia langsdorffii*.

(**87**) showed moderate activity against the phytopathogens *Cladosporium cladosporioides* and *C. sphaerospermum* (MIC 100.0 µg mL^{-1}), whereas oleanolic acid (**88**) exhibited weak activity (MIC 200.0 µg mL^{-1}) and lupeol (**89**) was not active against the pathogenic strains used in this study (Marqui et al., 2008).

The extracts and fractions obtained from *P. nitens* were evaluated by microdilution method and exhibited activity against the tested strains *C. albicans*, *C. krusei*, *C. parapsilosis* and *C. neoformans*. The n-butanol fractions from branches (MIC 15.6 µg mL^{-1}) and roots (MIC 31.2 µg mL^{-1}) exhibited the most potent activities against *C. krusei*. Such samples were selected for chemical investigation, which led to the isolation of four guanidine alkaloids (Figure 9.18), *N*-1,*N*-2,*N*-3-triisopentenylguanidine (**90**), and nitensidines A-C (**91-93**), with moderate antifungal activity against *C. krusei* (MIC 5 µg mL^{-1}) and *C. parapsilosis* (MIC 31.2 µg mL^{-1}). Additionally, all extracts, fractions and isolated compounds were evaluated against the four fungal strains and showed MFC values greater than 1,000 µg mL^{-1}, indicating their weak fungistatic behavior (Regasini et al., 2010).

Additional work on the antifungal properties of *P. nitens* flavonoids was carried out. The synergistic effects of pedalitin (**42**) (Figure 9.7) (Regasini et al., 2008b) and amphotericin B were evaluated against *C. neoformans* by *in vitro* and *in vivo* tests using the alternative animal model *Galleria mellonella*, addressing three parameters: survival curve, fungal burden and histological analysis

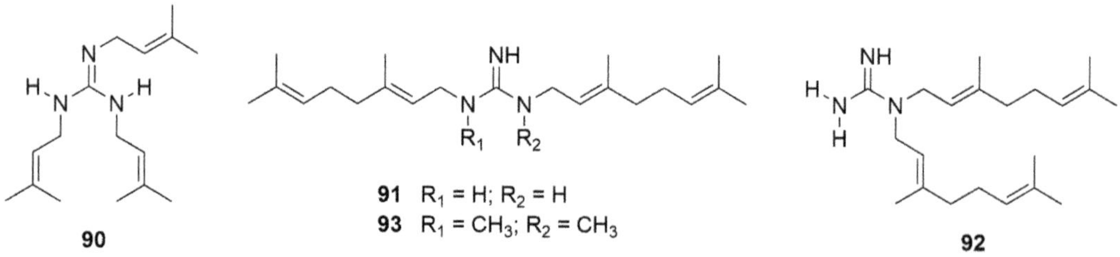

FIGURE 9.18   Prenylated guanidine alkaloids isolated from *Pterogyne nitens*.

(Sangalli-Leite et al., 2016). In the *in vitro* assay amphotericin B (AmB) and **42** were tested alone and showed MIC of 0.125 mg L^{-1} and 3.9 mg L^{-1}, respectively. This assay was performed by a micro-dilution method described by CLSI (2008) with modifications (Scorzoni et al., 2007). The combined treatment with AmB and **42** was performed by the checkerboard broth microdilution method. The fractional inhibitory concentration index (FICI) was calculated using the equation: $\Sigma FIC = FICA + FICB$, where FIC is the ratio of MIC of the drug in combination with MIC alone (Odds, 2003; White et al., 1996). The combination was considered synergistic at FICI ≤ 0.5, indifferent at FICI >1 and ≤ 4 and antagonistic at FICI >4.0. The same formula was used to calculate the fractional fungi-cidal concentration index (FFCI), using MFC values instead of MIC (Odds, 2003). The combina-tions tested decreased the MIC value by fourfold compared with AmB and **42** (0.03 mg L^{-1} and 1 mg L^{-1}, respectively). The synergistic effect was considered promising by the results of FICI and FFCI. In the synergistic treatment, all the combinations of AmB doses (1, 2 and 4 mg kg^{-1}) and pedalitin (6.25, 12.5, 25 and 40 mg kg^{-1}) were able to increase the survival of infected larvae ($P < 0.05$). Treatment with 0.3 mg kg^{-1} of AmB and 10 mg kg^{-1} of **42**, alone, led to survival of larvae up to the sixth day of the experiment by 18.7% and 0%, respectively. However, for the combined compounds, >56% of larvae were alive at the end of the experiment. The combination of AmB at 0.3 mg kg^{-1} + **42** at 10 mg kg^{-1}after 4 days resulted in almost 100% reduction of the fungal burden. Synergism efficacy of the treatment was also observed by histopathology of untreated and treated larvae used in the experiment. Histopathology data showed a reduction in the number of yeasts after 14 days of treatment with AmB, **42** or combination therapy.

The results for all trials were promising and the best time-kill results were obtained after 8 hours of exposure to the tested substances. After contact with AmB and **42** either alone or in combination, the yeast death rate was 100%. Alternative animal models such as *G. mellonella* were used to perform the *in vivo* antifungal assay. Before infecting *G. mellonella* larvae, the toxicity test of compounds and solvents was performed and compound **42** showed a toxic effect on larvae at doses ≥ 50 mg kg^{-1}. Experiments using either alone or combined compounds showed similar activities on larvae survival. To compare the results of synergism efficacy using the alternative animal model *G. mellonella*, the treatment of **42** + AmB in murine model was also performed and showed an increase in mice sur-vival. As observed in *G. mellonella*, combined treatment with AmB and **42** significantly increased the survival of mice, which suggested that this treatment was as efficient as AmB monotherapy at higher doses, suggesting the combination of antifungal compounds as an interesting alternative that can increase the efficiency of fungicidal treatment (Sangalli-Leite et al., 2016).

9.2.5   NATURAL PRODUCTS ACTIVE ON NEGLECTED DISEASES' PARASITES

In the last century, humanity has suffered from climate change all over the world with deterioration of biodiversity and deforestation of biomes that harbor disease vectors as major contributors that have significantly altered epidemic processes. This is no different in the Atlantic Forest, which is the natural habitat of numerous vectors and transmitters of tropical diseases, especially those con-sidered as neglected ones, such as leishmaniasis and Chagas' disease (Wood et al., 2014). However, nature itself can give us support to find alternatives to problems caused by these vectors.

A growing number of natural products derived from plants collected in the remaining areas of the Brazilian Atlantic Forest are known for their bioactivities against the causative agents of neglected diseases.

Many examples can be cited, with some related to plant species of *Peperomia* and *Piper* genera from the Piperaceae family. *Peperomia obtusifolia* is a well-known ornamental plant distributed from Mexico to South America. Despite the plant's predominant ornamental usage, some communities in Central America use the leaves', stems' and fruits' extracts to treat insect and snake bites and as a skin cleanser (Batista et al., 2017). Previous phytochemical investigation on *P. obtusifolia* aerial parts showed the presence of prenylated chromans, lignans, amides, flavonoids and other phenolic derivatives (Batista et al., 2011; Mota et al., 2009; Tanaka et al., 1998). Crude extracts and fractions of *P. obtusifolia* leaves and stems showed potent trypanocidal activity and their chemical investigation afforded seven compounds (**94–100**), including chromanes, furofuran lignans and flavone C-diglycosides (Figure 9.19) (Mota et al., 2009). This study revealed that the most active compounds were the chromanes, peperobtusin A (**94**) and the carboxy derivative 3,4-dihydro-5-hydroxy-2,7-dimethyl-8-(2″-methyl-2″-butenyl)-2-(4′-methyl1′,3′-pentadienyl)-2*H*-1-benzopyran-6-carboxylic acid (**95**), with IC_{50} values of 3.1 μM (almost three times more active than the positive control benznidazole, IC_{50} 10.4 μM) and 27.0 μM, respectively. The potent trypanocidal activity observed for these compounds seems to be related to a benzopyran nucleus substituted with isoprenyl moieties. The assay was performed measuring the proliferation of Y strain epimastigotes growing in axenic culture and the number of remaining viable protozoa was established by counting the parasites in a Neubauer chamber. Cytotoxicity assays using peritoneal murine macrophages indicated that the chromanes were not toxic at the level of the IC_{50} for trypanocidal activity (Mota et al., 2009), evidencing a selective index compatible with their use as prototypes for the development of therapeutic agents for Chagas disease.

Piper crassinervium was shown to accumulate antifungal, trypanocidal and antioxidant C-geranylated metabolites derived from both benzoic acid and *p*-hydroquinone (López et al., 2010). The chemical study was reported by Lopes et al. (2008) which resulted in the identification of two prenylated benzoic acid derivatives (**101** and **102**), one prenylated hydroquinone (**103**) and two

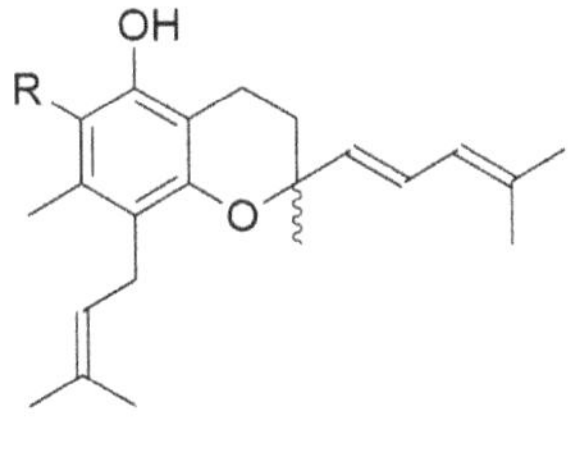

94 R = H
95 R = COOH

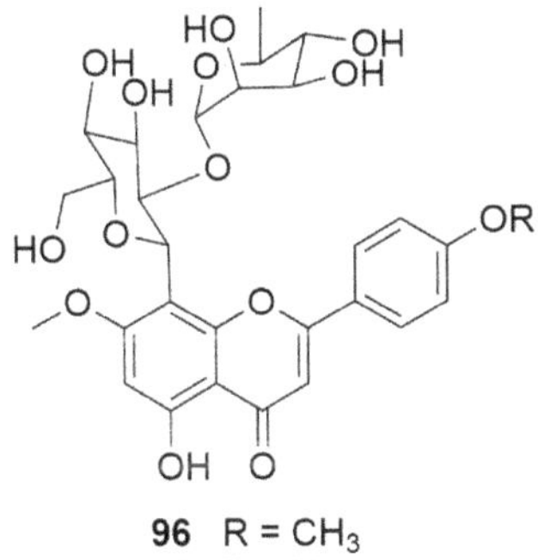

96 R = CH₃
97 R = H

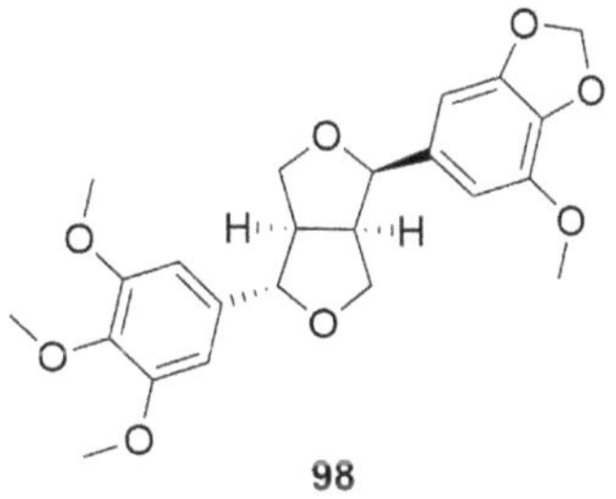

98

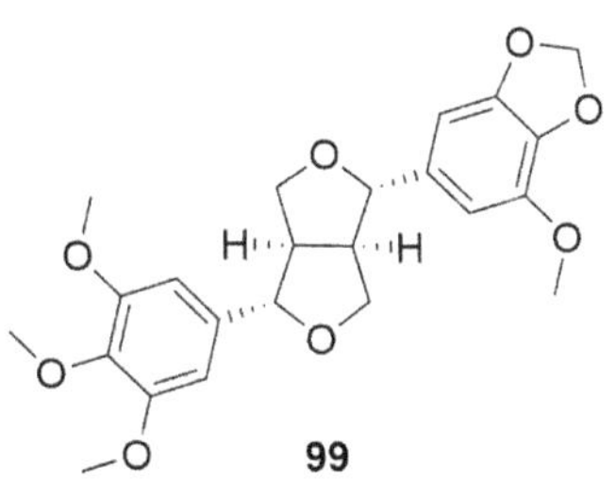

99

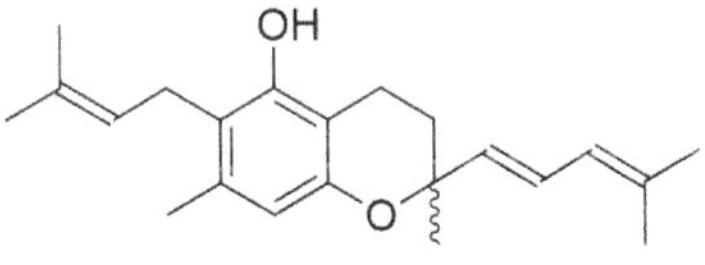

100

FIGURE 9.19   Chromanes, flavonoids and lignans isolated from *Peperomia obtusifolia*.

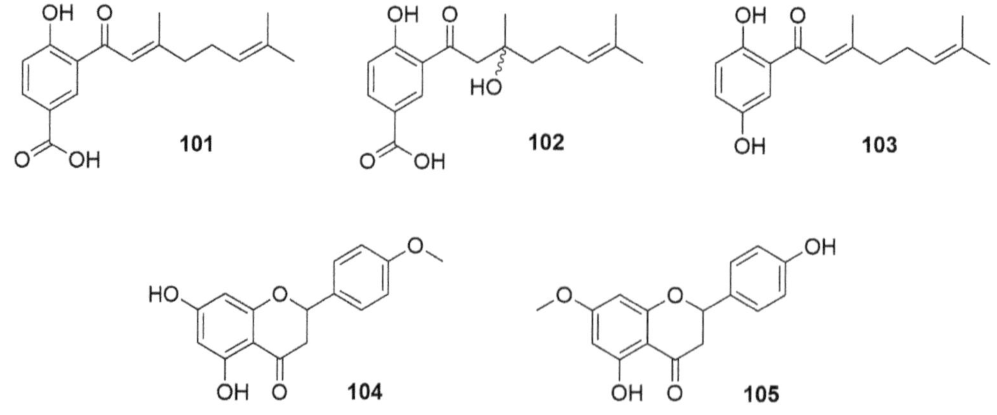

FIGURE 9.20   Phenolic compounds isolated from *Piper crassinervium*.

flavanones (**104** and **105**) (Figure 9.20). *In vitro* trypanocidal assays showed that the most active compound was the prenylated hydroquinone (**103**) with an IC_{50} value of 6.10 µg mL^{-1}, comparable to the positive control benznidazole (IC_{50} 1.60 µg mL^{-1}). The presence of lipophilic geranyl moiety oxygenated in the benzyl position was regarded as a key structural moiety essential to trypanocidal activity (Lopes et al., 2008). These results were consistent with previous data about the importance of isoprenyl moieties for the trypanocidal activity (Batista et al., 2008; Mota et al., 2009).

Other families from Atlantic Forest flora have also been the subject of bioprospecting studies. *Porcelia macrocarpa* belongs to the Annonaceae family and is widespread in the southeastern region from Brazil (Santos et al., 2015). Nonpolar extracts of the species' seeds displayed *in vitro* activity against *Trypanosoma cruzi* trypomastigotes. Thus, the crude bioactive extract was subjected to chromatographic fractionation procedures to afford an acetylene fatty acid, 12,14-octadecadiynoic acid/macrocarpic acid (**106**), and two acetylene di/triacylglycerol derivatives, α,α'-dimacrocarpoyl-β-oleylglycerol (**107**) and α-macrocarpoyl-α'-oleylglycerol (**108**) (Figure 9.21), which had their trypanocidal potential evaluated using a MTT colorimetric assay (Muelas-Serrano et al., 2000).

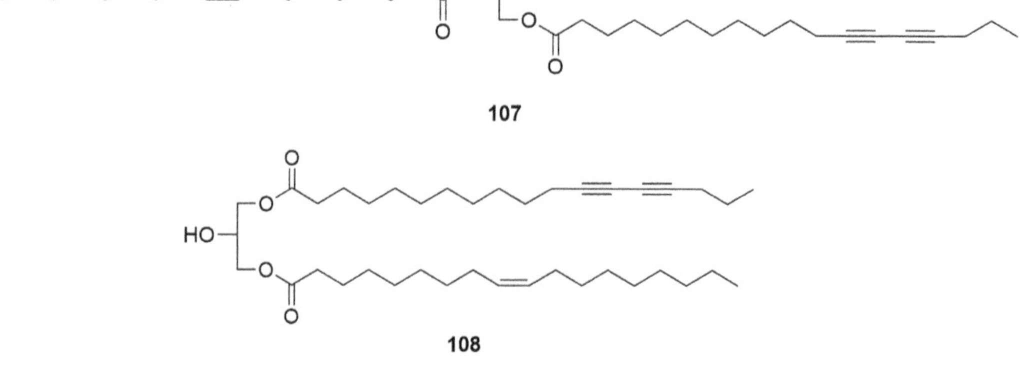

FIGURE 9.21   Acetylenic compounds **106–108** isolated of *Porcelia macrocarpa*.

Compound **106** displayed *in vitro* activity against *T. cruzi* trypomastigotes, while compounds **107** and **108** were inactive. The importance of unsaturation for the observed bioactivity was verified as compound **106** was hydrogenated, and the resulting product was reevaluated, and shown to be inactive (Santos et al., 2015).

The antiprotozoal potential of *Moringa oleifera* (Moringaceae) has also been investigated. *M. oleifera*, also known as the "drumstick tree", is recognized as a multipurpose and affordable source of phytochemicals, with potential applications in medicines and functional food preparations, water purification, in addition to biodiesel production (Saini et al., 2016). The flower's chemical studies led to the detection of flavonoids in the ethanol extract, and a trypsin inhibitor (MoFTI) was concentrated in a fraction from the ethanol extract. The flavonoids were evaluated against *T. cruzi* and for cytotoxicity to mammalian cells. Promising results were obtained both for the extract enriched with flavonoids and MoFTI, which triggered lysis of *T. cruzi* trypomastigotes with $LC_{50/24\,h}$ of 54.2 and 41.2 µg mL^{-1}, respectively. High selectivity indices for *T. cruzi* cells were found for the extract and MoFTI evidencing this compound is a trypanocidal principle of the flower extract from *M. oleifera* (Pontual et al., 2018).

Santos et al. (2012) reported the isolation of four sesquiterpene pyridine alkaloids, ilicifoliunines A (**109**) and B (**110**), aquifoliunine E-I (**111**) and mayteine (**112**) (Figure 9.22) from the root bark of *M. ilicifolia* (Celastraceae). An antileishmanial assay using promastigote forms of *Leishmania amazonensis* and *L. chagasi* in addition to an antitrypanosomal assay employing *T. cruzi* epimastigotes were performed using a MTT colorimetric method (Muelas-Serrano et al., 2000) with these isolated alkaloids. Among the tested compounds, alkaloid **111** presented activity against *L. chagasi* and *T. cruzi*, with IC_{50} values of 1.4 µM and 41.9 µM, respectively. Such data indicate the antiprotozoal high potential of **111**, as compared to the positive controls pentamidine (IC_{50} 5.1 µM) and benznidazole (IC_{50} 42.7 µM), drugs that are currently employed for the treatment of leishmaniasis and trypanosomiasis, respectively. Alkaloid **109** displayed potent antitrypanosomal activity, with an IC_{50} value of 27.7 µM. However, this compound was inactive against both *Leishmania* species. Compounds **110** and **112** did not exhibit activity against the protozoan species tested at 100 µM. Such results evidence the bioactivity dependence on the benzoyl substituent concomitant to the α-carbonyl 1,2-dimethyl-ethyl moiety, present in compounds **109** and **111**, but not in **110** and **112**. Interestingly, cytotoxic activity tests in mammalian normal cells, using murine peritoneal macrophages, demonstrated that the two active alkaloids **109** and **111** were more selective than the standard drug (Santos et al., 2012), which gives additional support to their use as prototypes for antiparasitic drugs.

The antiparasitic capacity of quinonemethide triterpenes maytenin (**46**) and pristimerin (**47**) (Figure 9.9), isolated from root barks of *M. ilicifolia*, were evaluated as described above

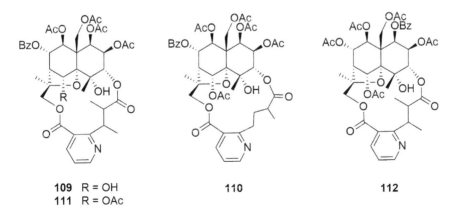

109 R = OH
111 R = OAc

110

112

FIGURE 9.22   Sesquiterpene-pyridine alkaloids isolated from *Maytenus ilicifolia*.

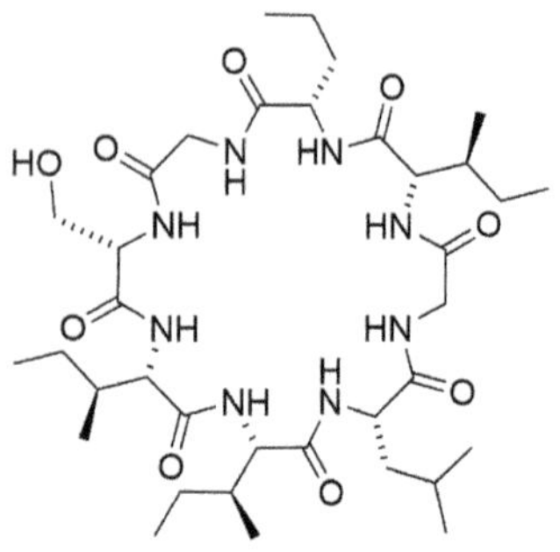

FIGURE 9.23 Cyclic peptide ribifolin (**113**) and the amino acid sequence of linear ribifolin (**114**) from *Jatropha ribifolia*.

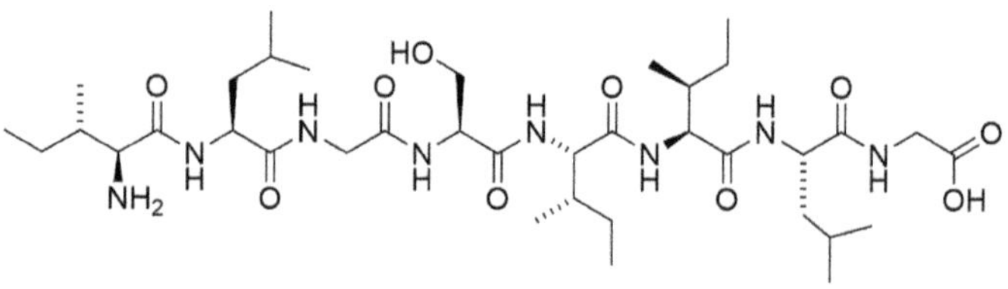

(Santos et al., 2012). These compounds showed potent *in vitro* activity against *L. amazonensis* and *L. chagasi* promastigotes as well as *T.cruzi* epimastigotes, with IC values in the nano-gram range. The IC_{50} values obtained for compounds **46** and **47** were 0.09 nM and 0.05 nM for *L. amazonensis* promastigotes and 0.46 nM and 0.41 nM for *L. chagasi* promastigotes, respectively. The IC_{50} values for *T. cruzi* epimastigotes were 0.25 nM and 0.30 nM, respectively. These two quinonemethide triterpenes showed stronger activity when compared to the positive controls pentamidine for *L. amazonensis* (IC_{50} 6.75 nM) and *L. chagasi* (IC_{50} 4.0 nM), and benznidazole for *T. cruzi* (IC_{50} 31.20 µM). The selectivity index (SI), a relevant characteristic for defining hit compounds, was calculated for compounds **46** and **47** by dividing their cytotoxic activity against murine macrophages (LC_{50}) by their leishmanicidal or trypanocidal activities. SIs for *L. amazonensis* and *L. chagasi* were 243.65 and 46.61 for **46** and 193.63 and 23.85 for **47**, whereas for *T. cruzi* epimastigotes the SIs were 85.00 for **46** and 332.37 for **47**. Such results indicated that both compounds present good selectivity for trypanosomatid extracellular forms and might represent attractive prototypes for antiprotozoal drugs development (Santos et al., 2013).

Further investigation of Atlantic Forest's plants with antiplasmodial activities concentrated on the species *Jatropha ribifolia* (Euphorbiaceae) and its potential against malaria. Although malaria is not formally considered a neglected disease, it is closely related and many challenges are still faced when treating and controlling this disease. The latex of *J. ribifolia* is widely used throughout north-eastern Brazil as a traditional herbal medicine for the antivenom activity, furthermore the seeds are sold in markets in the region for oil production as a purgative for veterinary use (Agra et al., 1996; Devappa et al., 2011). An orbitide, identified as ribifolin (**113**) and a linear ribifolin analog (**114**) determined in chemical studies (Figure 9.23) was prepared to be tested against malaria using the same assay. The synthetic linear (**114**) and cyclic (**113**) peptides were evaluated toward *Plasmodium falciparum*, a protozoan that causes the most severe form of the malaria disease (Sabandar et al., 2013), with chloroquine being used as positive control (IC_{50} 0.3µM) for these experiments. The cyclic peptide **113** was moderately effective against the parasite, with an IC_{50} of 42 µM, whereas the linear analog **114** showed weak activity with an IC_{50} of 519 µM, which provided evidence for the importance of cyclization to improve biological activity in this case. Their cytotoxic activity was also measured, but none of the tested compounds exhibited any cytotoxicity against human normal cells, which gives additional support to the use of compound **113** as a model for antiparasitic drugs (Pinto et al., 2015).

9.2.6 Miscellaneous Bioactive Compounds

Compounds isolated from plants' crude extracts or partially purified fractions can provide an alternative approach to new therapies, since they may present high chemical diversity and milder or

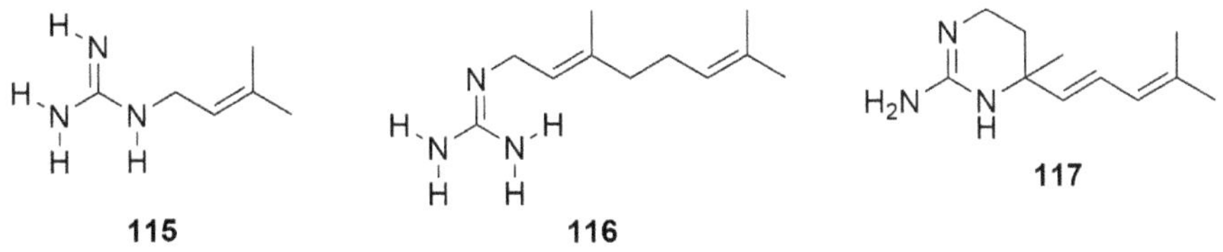

FIGURE 9.24 Galegine (**115**) and additional prenylated guanidine alkaloids (**116** and **117**) isolated from *Pterogyne nitens*.

inexistent side effects compared with conventional treatments (Jardim et al., 2015). The following studies revealed that Brazilian flora represents a vast, largely untapped resource of potential antibiotic, protease inhibitors and antiviral compounds.

The flavonol glucoside kaempferitrin (**40**) (Figure 9.7) (Regasini et al., 2008c) and the guanidine alkaloid galegine (**115**) (Figure 9.24) (Regasini et al., 2009), isolated from *P. nitens*, have shown hypoglycemic effect in vivo, which might be related to a possible antidiabetic effect. *Diabetes mellitus* is a metabolic disease in which the body is affected by hyperglycemia triggering changes in carbohydrate, lipid and protein metabolism pathways. In this context, the effect of treatment with *P. nitens* on diabetic rats regarding glycemic levels and physiological parameters was evaluated. Unfortunately, the plant crude extract did not change serum glucose levels or other physiological parameters, water or food intake and body weight, and did not improve the diabetic condition (Souza et al., 2009). Further studies evaluated hepatobiliary toxicity and biochemical markers levels in urine, during treatment with *P. nitens* extract on diabetic rats. Nevertheless, the treatment also had no therapeutic effect on the diabetic condition (Souza et al., 2010). The effect could be further evaluated with different doses of the extract, route of administration or severity of the induced diabetes to confirm the results obtained previously.

The phytochemical studies on leaves, flowers, fruits, bark and roots of *P. nitens* led to the isolation of pterogynidine (**35**) (Figure 9.6) and additional unusual prenylated guanidine alkaloids **90–93** (Figure 9.18) and **115–117** (Figure 9.24), which had their antibacterial activity evaluated against six clinically relevant multi-drug-resistant bacteria strains. Antibiotics were initially developed for therapeutic use in the 40s and were responsible for a remarkable increase in life expectancy. Nevertheless, their indiscriminate use has triggered the development of resistant bacterial strains to the available antibiotics. Methicillin-resistant *S. aureus* (MRSA) is a resistant bacteria strain to many commonly used antibiotics and is a major public health problem. The development of new classes of antibiotics is thus of urgent and great need. Prenylated guanidine alkaloids isolated from *P. nitens* showed strong activity against *S. aureus* strains, comparable to that of the positive control norfloxacin, except for alchorneine (**90**) (Figure 9.18), which was weakly active. The most promising compounds were galegine (**115**) and pterogynidine (**35**; Figure 9.6), which exhibited MIC of 31.4 and 20.5 µM respectively, for all tested strains. The side chain length and substitution pattern represented important chemical features for the antibacterial activity. Both compounds **35** and **115** exhibited bactericidal or bacteriostatic effect as evaluated in the minimum bactericidal concentration (MBC) assay. Their MBC values were the same as observed in the MIC assay, and the results evidenced that both compounds killed the bacteria, rather than just inhibiting their growth, which indicates a bactericidal activity. Such results indicated that guanidine alkaloids may be considered promising molecular models and could be used for further medicinal chemistry studies in the development of antibacterial therapeutic agents (Coqueiro et al., 2014).

Additional antibacterial compounds were isolated from *Stemodia foliosa* Benth. (*Plantaginaceae*), popularly known as "meladinha"' and used in Brazilian folk medicine as bioinsecticide and to treat respiratory infections. The ethanol extract from the aerial parts of *S. foliosa*, afforded three labdane diterpenoids, 6α-acetoxymanoyl oxide (**118**), 6α-malonyloxymanoyl oxide (**119**) and

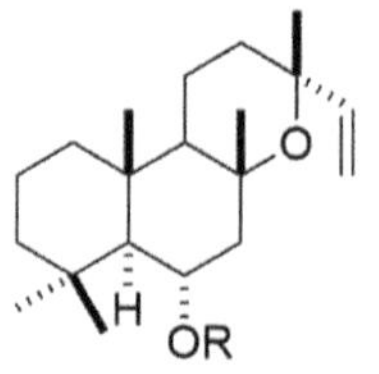

118 R = COCH$_3$
119 R = COCH$_2$CO$_2$H
120 R = COCH$_2$CO$_2$C$_4$H$_9$

FIGURE 9.25   Labdane diterpenoids isolated from the aerial parts of *Stemodia foliosa*.

6α-malonyloxy-*n*-butylestermanoyl oxide (**120**) (Figure 9.25), along with the triterpenes betulinic acid and lupeol, and the steroids stigmasterol and sitosterol. The isolated diterpenes were evaluated using a disc diffusion assay against Gram-positive bacteria *S. aureus*, *Bacillus cereus*, *B. subtilis*, *B. anthracis*, *Micrococcus luteus*, *Mycobacterium smegmatis* and *M. phlei*. Compound **119** exhibited moderate activity against these strains (MIC 7–20 µg mL^{-1}) which might be associated to the reported plant use in traditional medicine to treat respiratory infections (Silva et al., 2008).

Inappropriate treatment of hepatitis C virus (HCV) infection, a serious health problem, can cause liver cirrhosis with risk of hepatocellular carcinoma. The high costs of the available therapy as well as the potential for development of resistance evidence the need for alternative treatments and stimulated the search of new and efficient antiviral compounds. *M. ilicifolia* afforded flavonoids and sesquiterpene-pyridine alkaloids as major chemical constituents. Compounds **46** and **47** (Figure 9.9), **111** (Figure 9.22) and **121** (Figure 9.26) have been isolated from the root bark of *M. ilicifolia* and were subjected to studies on HCV genome replication, using the luciferase assay and Huh 7.5 cells, based on HCV sub-genomic replicons (SGR) of genotypes 2a (JFH-1), 1b and 3a. The sesquiterpene pyridine alkaloid aquifoliunine E-I (**111**) dramatically inhibited HCV SGR replication, and the western blotting assay confirmed these results, as the alkaloid also inhibited HCV protein expression. In addition, compound **111** presented activity against a daclatasvir resistance mutant subgenomic replicon and reduced the production of infectious JFH-1 virus (Jardim et al., 2015).

Jatropha species have shown a great potential as source of small cyclic peptides, which have attracted much attention due to their chemo-diversity and variety of important biological activities (Baraguey et al., 2001; Mongkolvisut et al., 2006). Chemical studies on the latex of *Jatropha curcas* led to isolation, structural elucidation and conformational studies of the cyclic peptide jatrophidin I (**122**) (Figure 9.27). The compound's biological evaluation showed strong inhibitory activity in a fluorometric protease inhibition assay using pepsin as a molecular model for aspartic protease inhibition (Gold et al., 2007), with IC$_{50}$ value of 0.88 µM, when compared to standard pepstatin A (IC$_{50}$ 0.40 µM). However, the cyclic peptide did not inhibit the serine protease subtilisin, evidencing that the observed inhibitory activity was specific for aspartic proteases (Altei et al., 2014).

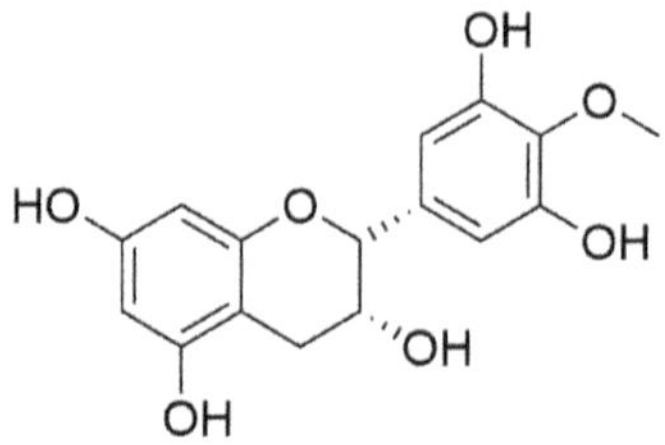

121

FIGURE 9.26   Catechin isolated from *Maytenus ilicifolia*.

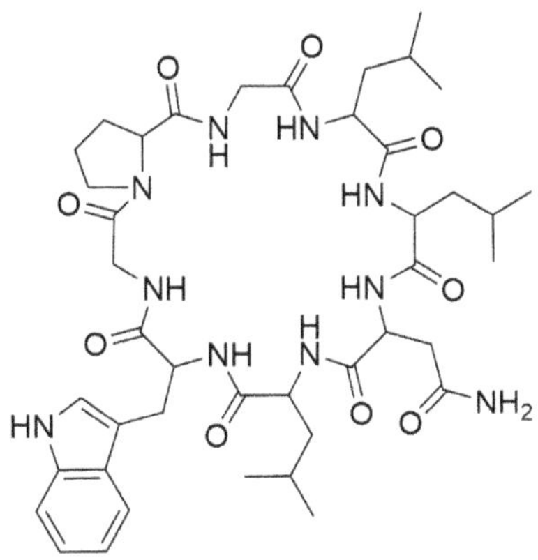

122

FIGURE 9.27   Cyclic peptide jatrophidin I (**122**) isolated from *Jatropha curcas.*

Proteases fulfill multiple roles in health and disease, and considerable interest has been expressed in the design and development of synthetic inhibitors of disease-related proteases. Virus-encoded proteases have been shown to be involved in the replication of many virus types. Considering proteases importance, they have become important drug targets (Kang et al., 2017). *Jatropha* species may thus represent an interesting alternative for bioprospecting studies aimed at the discovery of protease inhibitory cyclic peptides.

9.3   SOME COMMENTS ON RECENT ADVANCES IN NATURAL PRODUCTS RESEARCH

With the continued interest and reemergence of natural products in drug discovery, the development of novel techniques for extraction, as well as for data acquisition and processing during purification and identification procedures, is fundamental to the finding of promising lead compounds requiring sophisticated and efficient chemical assessment of natural sources.

Traditionally, crude extracts from plants, microorganisms or animals were obtained via classical extraction procedures, screened in bioassays, and the main compounds often passed through various isolation steps to purity for further spectroscopic analysis and identification. Considering that plant extracts are complex mixtures, this strategy leads unfortunately to the re-isolation and identification of many known secondary metabolites, representing an expensive and time-consuming process (Harvey et al., 2015). However, modern techniques, database mining and bioassay guided fractionation are being used to continue the natural products research in the Atlantic Forest and elsewhere.

In order to arrive at the structures presented herein, dereplication (Hanka et al., 1978) processes are used. Furthermore, for analyses and isolation the dereplication process relies on the following hyphenated techniques: GC-MS, LC-UV, LC-MS, LC-MS/MS and LC-NMR. Data are acquired and compared to reference compounds in commercial or in-house databases and include the Dictionary of Natural Products (Buckingham, 1997), NuBBE$_{DB}$ (Pilon et al., 2017; Valli et al., 2013), SuperNatural (Dunkel, 2006), PubChem (Bolton et al., 2008), ChEBI (Hastings et al., 2016), NAPROC (López-Pérez et al., 2007), CAS (Chemical Abstracts Service), ChemBank, SciFinder and ChemSpider (Hubert et al., 2017).

Strategies in bioinformatics, chemometrics and statistical tools are being refined to speed up analyses and improve the reliability of dereplication results (Funari et al., 2013). Among the modern dereplication tools, the Global Natural Products Social Molecular Networking (GNPS, http://gnps.ucsd.edu) is a powerful platform for large-scale data analysis in mass spectrometry (MS). GNPS is an open-access knowledge base, allowing the scientific community to share raw, processed or identified MS/MS data. The platform contains more than 93 million MS/MS spectra available from natural products and pharmacologically active substances. Nowadays, GNPS is considered the largest collection of public MS/MS data and the spectral libraries are growing through user's contributions (Wang et al., 2016).

Natural products dereplication is challenging due to the high chemo-diversity of metabolites, especially for metabolomics studies (Schwab, 2003). Metabolomics consist of comprehensive qualitative and quantitative analyses of the metabolome, which represents all metabolites present in an organism, at a specific time and under specific conditions (Yuliana et al., 2011). To achieve optimum results, one of the most exciting techniques for separation of complex samples is multidimensional chromatography, either gas or liquid, where multiple chromatographic separations are performed on a given mixture (Carr and Stoll, 2015; Li et al., 2015). Unlike two-dimensional gas chromatography (2D-GC), the utilization of 2D-LC remains relatively uncommon, nevertheless the technology's potential as a promising tool for metabolomics, shall probably increase use in near future(Funari et al., 2012a, 2012b; Wolfender et al., 2015).

The inherently low sensitivity of NMR techniques has been overcome by advances in hardware, software and pulse programs, and nowadays, high-quality spectra are possible to obtain on the microgram scale (Zani and Carroll, 2017). For this reason, NMR spectroscopy has been considered an excellent detection technique for hyphenation with liquid chromatography for separation and identification of compounds in natural products samples. (Cieśla and Moaddel, 2016; Kesting et al., 2011).

The advances in instrumentation also allow the coupling of complex instruments, and nowadays, hyphenation of LC-DAD-MS-SPE-NMR as a feasible option. For dereplication, the combination of complementary tools as DAD, MS and NMR contributes to a significant reduction of time, smaller amounts of sample required and an increase in the identification accuracy of natural products samples (Marshall and Powers, 2017).

Additional important aspects of natural products research include extraction and fractionation techniques. Extraction is an essential process to convert the matrix into a sample suitable for analytical studies. This procedure is the first step in the chemical investigation of natural products, and the success of subsequent steps of any bioassay, chromatographic or spectrometric experiment still depends on an efficient, reproducible and reliable extraction (Belwal et al., 2018); and in light of green chemistry, *use of alternative solvents to ensure a safe and high-quality extract are considered* (Chemat et al., 2012). For example, modern extraction techniques for natural products samples include supercritical fluid extraction (SFE), microwave-assisted extraction (MAE), ultrasound-assisted extraction (UAE), accelerated solvent extraction (ASE™), pulsed electric field (PEF)-assisted extraction, enzyme-assisted extraction (EAE), instant controlled pressure drop (DIC) (Allaf et al., 2013; Belwal et al., 2018; Funari et al., 2014; Selvamuthukumaran and Shi, 2017).

A recent strategy to couple the sample extraction with chromatographic systems is the online extraction (OLE), developed at the NuBBE laboratories. OLE was invented and patented by Ferreira et al. and consists of inserting dry or fresh plant material inside a chamber in a security guard holder. The precolumn is then connected to a liquid chromatographic system with a six-port valve. As the chromatographic analysis starts, the valve is switched so that the mobile phase flows through the precolumn containing the plant material prior to entering into the chromatographic column; therefore, the mobile phase is used both for extraction and separation. The new technique OLE-LC was compared with conventional sample preparations of medicinal plants and similar peak capacity and number of peaks were found for both methods, indicating that OLE-LC is a feasible technique encompassing sample pretreatment integrated with chromatographic analysis (Ferreira et al., 2016). Although this procedure is not automated yet, some papers have been published using OLE-LC with promising results (Russo et al., 2018; Tong et al., 2018a, 2018b).

9.4   CONCLUDING REMARKS

In conclusion, in this chapter, many examples of natural products from the Atlantic Forest are presented. Not enough emphasize can express how critical and important this research work is to understanding the biome, seeking answers to the challenges of biodiversity and the ever present potential for finding promising new drugs.

Modern techniques and equipment in addition to optimized strategies to investigate chemical composition, biological and toxicological properties of natural products, among many complementary aspects of bioprospecting and research have proven essential for unravelling the fascinating chemo-diversity and achieving the interesting results obtained so far. Phytochemical studies on Brazilian plants developed in our research group from Atlantic Forest, have disclosed the presence of many diverse secondary metabolites including triterpenes, diterpenes, saponins, flavonoids, alkaloids and polyketides presented herein and exhibiting a multitude of biological and pharmacological activity and more remains to be discovered.

Such scenario and goals require not only state-of-the-art equipment and methodologies but well-trained scientists who may greatly benefit from intensive collaborative work to strengthen research networks as well as to pave the way to continue consistent and systematic phytochemical studies for a deeper and broader knowledge on natural resources. The essential role of secondary metabolites in the species survival in addition to the maintenance of environmental equilibrium reinforces the need of urgent and thorough commitment to preservation policies, especially of biodiversity hotspots, as is the case of Atlantic Forest. They also evidence the utmost importance of sustainable procedures toward the exploration of this rich biome so that value-added products may outcome alongside biodiversity protection initiatives.

REFERENCES

Agra, M. F.; Locatelli, E.; Rocha, E. A.; Baracho, G. S.; Formiga, S. C. 1996. Medicinal plants of Cariris Velhos, Paraíba, part II: subclass magnoliidae, caryophyllidae, dilleniidae and rosidae. Brazilian Journal of Pharmacognosy, 77, 97–102.

Allaf, T.; Tomao, V.; Ruiz, K.; Chemat, F. 2013. Instant controlled pressure drop technology and ultrasound assisted extraction for sequential extraction of essential oil and antioxidants. Ultrasonics Sonochemistry, 20, 239–246.

Altei, W. F.; Picchi, D. G.; Abissi, B. M.; Giesel, G. M.; Flausino Jr., O.; Reboud-Ravaux, M.; Verli, H.; et al. 2014. Jatrophidin I, a cyclic peptide from Brazilian *Jatropha curcas* L.: isolation, characterization, conformational studies and biological activity. Phytochemistry, 107, 91–99.

Atta-ur-Rahman, Choudhary, M. I.; Thomsen, W. J. 2005. Bioassay Techniques for Drug Development. Amsterdam: Taylor & Francis e-Library.

Ballios, B. G.; Baumann, M. D.; Cooke, M. J.; Shoichet, M. S. 2011. Central nervous system. In A. Atala; R. Lanza; J. A. Thomson; R. Nerem, eds., Principles of Regenerative MedicinePhiladelphia: Elsevier.

Baraguey, C.; Blond, A.; Cavelier, F.; Pousset, J. L.; Bodo, B.; Auvin-Guette, C. 2001. Isolation, structure and synthesis of mahafacyclin B, a cyclic heptapeptide from the latex of *Jatropha mahafalensis*. Journal of the Chemical Society, 1, 2098–2103.

Barber, M. S.; Hardisson, A.; Jackman, L. M.; Weedon, B. C. L. 1961. Studies in nuclear magnetic resonance. Part IV. Stereochemistry of the bixins. Journal of the Chemical Society, 1625–1630.

Barker, S. A.; Mcilhenny, E. H.; Strassman, R. 2012. A critical review of reports of endogenous psychedelic N, N-dimethyltryptamines in humans: 1955–2010. Drug Testing and Analysis, 4, 617–635.

Basile, A. C.; Sertié, J. A. A.; Oshiro, T.; Caly, K. D. V.; Painizza, S. 1989. Topical anti-inflammatory activity and toxicity of *Cordia verbenacea*. Fitoterapia, 60, 260–263.

Batista, A. N. L.; Santos-Pinto, J. R. A.; Batista, J. M.; Souza-Moreira, T. M.; Santoni, M. M.; Zanelli, C. F.; Kato, M. J.; López, S. N.; Palma, M. S.; Furlan, M. 2017. The combined use of proteomics and transcriptomics reveals a complex secondary metabolite network in *Peperomia obtusifolia*. Journal of Natural Products, 80, 1275–1286.

Batista, J. M.; Batista, A. N. L.; Rinaldo, D.; Vilegas, W.; Ambrósio, D. L.; Cicarelli, R. M.; Bolzani, V. S.; et al. 2011. Absolute configuration and selective trypanocidal activity of gaudichaudianic acid enantiomer. Journal of Natural Products, 74, 1154–1160.

Batista, J. M.; Lopes, A. A.; Ambrósio, D. L.; Regasini, L. O.; Kato, M. J.; Bolzani, V. S.; Cicarelli, R. M. B.; Furlan, M. 2008. Natural chromenes and chromenes derivatives as potential anti-trypanosomal agents. Biological & Pharmaceutical Bulletin, 31, 538–540.

Belwal, T.; Ezzat, S. M.; Rastrelli, L.; Bhatt, I. D.; Daglia, M.; Baldi, A.; Devkota, H. P.; et al. 2018. A critical analysis of extraction techniques used for botanicals: trends, priorities, industrial uses and optimization strategies. Trends in Analytical Chemistry, 100, 82–102.

Bezerra, D. P.; Moura, D. J.; Rosa, R. M.; Vasconcellos, M. C.; Silva, A. C.; Moraes, M. O.; Silveira, E. R.; et al. 2008. Evaluation of the genotoxicity of piplartine, an alkamide of *Piper tuberculatum*, in yeast and mammalian V79 cells. Mutationm Research/Genetic Toxicology and Environmental Mutagenesis, 652, 164–174.

Bolton, E. E.; Wang, Y.; Thiessen, P. A.; Bryant, S. H. 2008. PubChem: integrated platform of small molecules and biological activities. In R. A. Wheeler; D. C. Spellmeyer, eds., Annual Reports in Computational Chemistry, vol. 4. Amsterdam: Elsevier.

Buckingham, J. 1997. Dictionary of Natural Products, Supplement 4. Boca Raton: CRC Press.

Butler, A. R.; Moffett, J. 1995. Pass the rhubarb. Chemistry in Britain, 31, 462–465.

Cannali, J.; Vieira, J. 1967. Efeito de alguns curares naturais e da d-Tubocurarina retardando o tempo de coagulação e o tempo de protrombina do sangue humano. Memórias do Instituto Oswaldo Cruz, 65, 167–173.

Carbonari, K. A.; Ferreira, E. A.; Rebello, J. M.; Felipe, K. B.; Rossi, M. H.; Felício, J. D.; Filho, D. W.; Yunes, R. A.; Pedrosa, R. C. 2006. Free-radical scavenging by *Ouratea parviflora* in experimentally-induced liver injuries. Redox Report, 11, 124–130.

Carbonezi, C. A.; Hamerski, L.; Gunatilaka, A. A. L.; Cavalheiro, A.; Castro-Gamboa, I.; Silva, D. H. S.; Furlan, M.; Young, M. C. M.; Lopes, M. N.; Bolzani, V. S. 2007. Bioactive flavone dimers from *Ouratea multiflora* (Ochnaceae). Revista Brasileira de Farmacognosia, 17, 319–324.

Cardoso-Lopes, E. M.; Maier, J. A.; Silva, M. R.; Regasini, L. O.; Simote, S. Y.; Lopes, N. P.; Pirani, J. R.; Bolzani,V. S.; Young, M. C. 2010. Alkaloids from stems of *Esenbeckia leiocarpa* Engl. (Rutaceae) as potential treatment for Alzheimer disease. Molecules, 15, 9205–9213.

Carlini, E. A. 2003. Plants and the central nervous system. Pharmacology Biochemistry and Behavior, 75, 501–512.

Carr, P. W.; Stoll, D. R. 2015. Two Dimensional Liquid Chromatography: Principles, Practical Implementation and Applications. Waldbronn: Agilent Technologies.

Chemat, F.; Vian, M. A.; Cravotto, G. 2012. Green extraction of natural products: concept and principles. International Journal of Molecular Sciences, 13, 8615–8627.

Cieśla, L.; Moaddel, R. 2016. Comparison of analytical techniques for the identification of bioactive compounds from natural products. Natural Product Reports, 33, 1131–1145.

Clinical and Laboratory Standards Institute (CLSI). 2008. Reference Method for Broth Dilution Antifungal Susceptibility Testing of Yeasts; Approved Standard, 3rd ed. Wayne: CLSI. Document M27-A3.

Cooper, J. M.; 1987. Estimulantes e narcóticos. In D. Ribeiro (org.), ed., Handbook of South American Indians. Petrópolis: FINEP.

Coqueiro, A.; Regasini, L. O.; Stapleton, P.; Bolzani, V. S.; Gibbons, S. 2014. *In Vitro* antibacterial activity of prenylated guanidine alkaloids from *Pterogyne nitens* and synthetic analogues. Journal of Natural Products, 77, 1972–1975.

Costa, P. M.; Ferreira, P. M. P.; Bolzani, V. S.; Furlan, M.; Freitas, V. A. F. M. S.; Corsino, J.; Moraes, M. O.; Costa-Lotufo, L. V.; Montenegro, R. C.; Pessoa, C. 2008. Antiproliferative activity of pristimerin isolated from *Maytenus ilicifolia* (Celastraceae) in human HL-60 cells. Toxicology In Vitro, 22, 854–863.

Costa, R. L.; Prevedello, J. A.; Souza, B. G.; Cabral, D. C. 2017. Forest transitions in tropical landscapes: a test in the Atlantic Forest biodiversity hotspot. Applied Geography, 82, 93–100.

Cragg, G. M.; Newman, D. J. 2013. Natural products: a continuing source of novel drug leads. Biochimica et Biophysica Acta (BBA) – General Subjects, 1830, 3670–3695.

Cushny, A. R. 1918. A Textbook of Pharmacology and Therapeutics, or the Action of Drugs in Health and Disease. Philadelphia and New York: Lea and Febiger.

de Lima Moreira, F.; Habenschus, M. D.; Barth, T.; Marques, L. M. M.; Pilon, A. C.; Bolzani, V. S.; Vessecchi, R.; Lopes, N. P.; Oliveira, A. R. M. 2016. Metabolic profile and safety of piperlongumine. Scientific Reports, 6, 33646

Deeb, D.; Gao, X.; Liu, Y. B.; Pindolia, K.; Gautam. S. C. 2014. Pristimerin, a quinonemethide triterpenoid, induces apoptosis in pancreatic cancer cells through the inhibition of pro-survival Akt/NF-κB/mTOR signaling proteins and anti-apoptotic Bcl-2. International Journal of Oncology, 44, 1707–1715.

Devappa, R. K.; Makkar, H. P. S.; Becker, K. 2011. Jatropha diterpenes: a review. Journal of the American Oil Chemists' Society, 88, 301–322.

Di Giovanni, S.; Borloz, A.; Urbain, A.; Marston, A.; Hostettmann, K.; Carrupt, P. A.; Reist, M. 2008. *In vitro* screening assays to identify natural or synthetic acetylcholinesterase inhibitors: thin layer chromatography versus microplate methods. European Journal of Pharmaceutical Sciences, 33, 109–119.

DiNunzio J. C.; Williams III, R. O. 2008. CNS disorders-current treatment options and the prospects for advanced therapies. Drug Development and Industrial Pharmacy, 34, 1141–1167.

Djerassi, C.; Nakano, T.; James, A. N.; Zalkow, L. H.; Eisenbraun, E. J.; Shoolery, J. N. 1961. Terpenoids. XLVII. The structure of Genipin. Journal of Organic Chemistry, 26, 1192–1206.

Duke, J. A.; duCellier, J. L. 1993. CRC Handbook of Alternative Cash Crops. Boca Raton: CRC Press.

Dunkel, M. 2006. SuperNatural: a searchable database of available natural compounds. Nucleic Acids Research, 34, D678–D683.

Ellman, G. L.; Courtney, K. D.; Andres, V.; Featherstone, R. M. 1961. A new and rapid colorimetric determination of acetylcholinesterase activity. Biochemical Pharmacology, 7, 88–91.

Fernandes, E. S.; Passos, G. F.; Medeiros, R.; Cunha, F. M.; Ferreira, J.; Campos, M. M.; Pianowski, L. F.; Calixto, J. B. 2007. Anti-inflammatory effects of compounds alpha-humulene and *trans*-caryophyllene isolated from the essential oil of *Cordia verbenacea*. European Journal of Pharmacology, 569, 228–236.

Ferreira, F. G.; Regasini, L. O.; Oliveira, A. M.; Campos, J. A. D. B.; Silva, D. H. S.; Cavalheiro, A. J.; dos Santos, R. A.; Bassi, C. L.; Bolzani, V. S.; Soares, C. P. 2009. Avaliação de mutagenicidade e antimutagenicidade de diferentes frações de *Pterogyne nitens* (Leguminosae), utilizando ensaio de micronúcleo em *Tradescantia pallida*. Brazilian Journal of Pharmacognosy, 19, 61–67.

Ferreira, V. G.; Leme, G. M.; Cavalheiro, A. J.; Funari, C. S. 2016. Online extraction coupled to liquid chromatography analysis (OLELC): eliminating traditional sample preparation steps in the investigation of solid complex matrices. Analytical Chemistry, 88, 8421–8427.

Flora do Brasil. 2019. Jardim Botânico do Rio de Janeiro. Available at http://floradobrasil.jbrj.gov.br. Accessed on 16 August 2019.

Flora do Brasil. 2020. under construction. Algas, Fungos e Plantas. Jardim Botânico do Rio de Janeiro. http://floradobrasil.jbrj.gov.br. Accessed on Aug 15, 2019.

Freedberg, D. 1999. Ciência, Comércio e Arte em o Brasil dos Holandeses. In P. Herkenhoff, org. Rio de Janeiro: GMT Editores Ltda.

Funari, C. S.; Carneiro, R. L.; Cavalheiro, A. J.; Hilder. E. F. 2014. A tradeoff between separation, detection and sustainability in liquid chromatographic fingerprinting. Journal of Chromatography A, 1354, 34–42.

Funari, C. S.; Castro-Gamboa, I.; Cavalheiro, A. J.; Bolzani V. S. 2013. Metabolômica, uma abordagem otimizada para exploração da biodiversidade brasileira: estado da arte, perspectivas e desafios. Química Nova, 36, 1605–1609.

Funari, C. S.; Eugster, P. J.; Martel, S.; Carrupt, P. A.; Wolfender, J. L.; Silva, D. H. S. 2012b. High resolution ultra high pressure liquid chromatography-time-of-flight mass spectrometry dereplication strategy for the metabolite profiling of Brazilian Lippia species. Journal of Chromatography A, 1259, 167–178.

Funari, C. S.; Gullo, F. P.; Napolitano, A.; Carneiro, R. L.; Mendes-Giannini, M. J.; Fusco-Almeida, A. M.; Piacente, S.; Pizza, C.; Silva, D. H. 2012a. Chemical and antifungal investigations of six *Lippia* species (Verbenaceae) from Brazil. Food Chemistry, 135, 2086–2094.

Ghedini, P. C.; Dorigoni, P. A.; Almeida, C. E.; Ethur, A. B. M.; Lopes, A. M. V.; Zachia, R. A. 2002. Survey of data on medicinal plants of popular use in the county of São João do Polesine, Rio Grande do Sul State, Brazil. Revista Brasileira de Plantas Medicinais, 5, 46–55.

Giocondo, M. P.; Bassi, C. L.; Telascrea, M.; Cavalheiro, A. J.; Bolzani, V. S.; Silva, D. H. S.; Agustoni, D.; Mello, E. R.; Soares, C. P. 2009. Cryptomoschatone D2 from *Cryptocarya mandioccana*: cytotoxicity against human cervical carcinoma cell lines. Revista de Ciências Farmacêuticas Básica e Aplicada, 30, 315–322.

Gold, N. D.; Deville, K.; Jackson, R. M. 2007. New opportunities for protease ligand-binding site comparisons using sitesbase. Biochemical Society Transaction, 35, 561–565.

Gottlieb, O. R.; Magalhaes, M. 1959. Occurrence of 1-nitro-2-phenylethane in *Ocotea pretiosa* and *Aniba canelilla*. Journal of Organic Chemistry, 24, 2070–2071.

Gottlieb, O. R.; Mors, W. B. 1978. *Fitoquímica* amazônica: uma *apreciação em perspectiva*. Interciência, 3(4), 252–263.

Govaerts, R.; Frodin, D. G.; Radcliffe-Smithv, A. 2000. World Check-List and Bibliography of Euphorbiaceae. Kew: Royal Botanic Gardens.

Gullo, F. P.; Sardi, J. C. O.; Santos, V. A. F. F. M.; Sangalli-Leite, F.; Pitangui, N. S.; Rossi, S. A.; Paula e Silva, A. C. A.; et al. 2012. Antifungal activity of maytenin and pristimerin. Evidence-Based Complementary and Alternative Medicine, 340787, 10.1155/2012/340787.

Gupta, R. S.; Siminovitch, L. 1977. The molecular basis of emetine resistance in Chinese hamster ovary cells: alteration in the 40S ribosomal subunit. Cell, 10, 61–66.

Halpen, O.; Schmid, H. 1958. Zur Kenntnis des Plumierids. 2. Mitteilung. Helvetica Chimica Acta, 41, 1109–1154.

Hanka, L. J.; Kuentzel, S. L.; Martin, D. G.; Wiley, P. F.; Neil, G. L. 1978. Detection and assay of antitumor antibiotics. In S. K. Carter; H. Umezawa; J. Douros; Y. Sakurai, eds., Antitumor Antibiotics. Recent Results in Cancer Research. Berlin and Heidelberg: Springer Berlin Heidelberg.

Harvey, A. L.; Edrada-Ebel, R.; Quinn, R. J. 2015. The re-emergence of natural products for drug discovery in the genomics era. Nature Reviews Drug Discovery, 14, 111–129.

Hastings, J.; Owen, G.; Dekker, A.; Ennis, M.; Kale, N.; Muthukrishnan, V.; Turner, S.; Swainston, N.; Mendes, P.; Steinbeck, C. 2016. ChEBI in 2016: improved services and an expanding collection of metabolites. Nucleic Acids Research, 44, D1214–1219.

Hubert, J.; Nuzillard, J. M.; Renault, J. H. 2017. Dereplication strategies in natural product research: how many tools and methodologies behind the same concept? Phytochemistry Reviews, 16, 55–95.

Jardim, A. C. G.; Igloi, Z.; Shimizu, J. F.; Santos, V. A.; Felippe, L. G.; Mazzeu, B. F.; Amako, Y.; Furlan, M.; Harris, M.; Rahal, P. 2015. Natural compounds isolated from Brazilian plants are potent inhibitors of hepatitis C virus replication *in vitro*. Antiviral Research, 115, 39–47.

Joly, C. A.; Metzger, J. P.; Tabarelli, M. 2014. Experiences from the Brazilian Atlantic Forest: ecological findings and conservation initiatives. New Phytologist, 204, 459–473.

Jorge, R. M.; Leite, J. P.; Oliveira, A. B.; Tagliati, C. A. 2004. Evaluation of antinociceptive, anti-inflammatory and antiulcerogenic activities of *Maytenus ilicifolia*. Journal of Ethnopharmacology, 94, 93–100.

Kang, C.; Keller, T. H.; Luo, D. 2017. Zika virus protease: an antiviral drug target. Trends in Microbiology, 25, 797–808.

Kesting, J. R.; Johansen, K. T.; Jaroszewski, J. W. 2011. Hyphenated NMR techniques. In A. J. Dingley; S. M. Pascal, eds., Biomolecular NMR Spectroscopy, vol. 3, pp. 413–434. Amsterdam: IOS Press.

Li, D.; Jakob, C.; Schmitz, O. 2015. Practical considerations in comprehensive two-dimensional liquid chromatography systems (LCxLC) with reversed-phases in both dimensions. Analytical and Bioanalytical Chemistry, 407, 153–167.

Lopes, A. A.; Lopez, S. N.; Regasini, L. O.; Junior, J. M.; Ambrósio, D. L.; Kato, M. J.; Bolzani, V. S.; Cicarelli, R. M.; Furlan, M. 2008. *In vitro* activity of compounds isolated from *Piper crassinervium* against *Trypanosoma cruzi*. Natural Product Research, 2, 1040–1046.

López, S. N.; Lopes, A. A.; Batista Junior, J. M.; Flausino Jr., O.; Bolzani, V. S.; Kato, M. J.; Furlan, M. 2010. Geranylation of benzoic acid derivatives by enzimatic extracts from *Piper crassinervium* (Piperaceae). Bioresource Technology, 101, 4251–4260.

López-Pérez, J. L.; Therón, R.; del Olmo, E.; Díaz, D. 2007. NAPROC-13: a database for the dereplication of natural product mixtures in bioassay-guided protocols. Bioinformatics, 23, 3256–3257.

Maioli, M. A.; Alves, L. C.; Campanini, A. L.; Lima, M. C.; Dorta, D. J.; Groppo, M.; Cavalheiro, A. J.; Curtie, C.; Mingatto, F. E. 2010. Iron chelating-mediated antioxidant activity of *Plectranthus parbatus* extract on mitochondria. Food Chemistry, 122, 203–208.

Marqui, S. R.; Lemos, R. B.; Santos, L. A.; Castro-Gamboa, I.; Cavalheiro, A. J.; Bolzani, V.S.; Young, M. C. M.; Torres, L. M. B. 2008. Saponinas antifúngicas de *Swartzia langsdorffii*. Química Nova, 31, 828–831.

Marshall, D. D.; Powers, R. 2017. Beyond the paradigm: combining mass spectrometry and nuclear magnetic resonance for metabolomics. Progress in Nuclear Magnetic Resonance Spectroscopy, 100, 1–16.

Mathias, S. 1975. Cem Anos de Química no Brasil. In Coleção da Revista de História, pp. 29–34. São Paulo: Universidade de São Paulo.

McGahon, A. J.; Martin, S. M.; Bissonnette, R. P.; Mahboubi, A.; Shi, Y.; Mogil, R. J.; Nishioka, W. K.; Green, D. R. 1995. The end of the (cell)line: methods for the study of apoptosis in vitro. Methods in Cell Biology, 46, 153–185.

Mendes, F. R.; Carlini, E. L. A. 2007. Brazilian plants as possible adaptogens: an ethnopharmacological surveys of books edited in Brazil. Journal of Ethnopharmacology, 109, 493–500.

Metzger, J. P.; Casatti, L. 2006. From diagnosis to conservation: the state of the art of biodiversity conservation in the BIOTA/FAPESP program. Biota Neotropica, 6, 1–23.

Mongkolvisut, W.; Sutthivaiyakit, S.; Leutbecher, H.; Mika, S.; Klaiber, I.; Möller, W.; Rösner, H.; Beifuss, U.; Conrad, J. 2006. Integerrimides A and B, cyclic heptapeptides from the latex of *Jatropha integerrima*. Journal of the Natural Products, 69, 1435–1441.

Morandim-Giannetti, A. A.; Cotinguiba, F.; Regasini, L. O.; Frigieri, M. C.; Eliana A.; Varanda, E. A.; Coqueiro, A.; Kato, M. J.; Bolzani, V. S.; Furlan, M. 2011. Study of *Salmonella typhimurium* mutagenicity assay of (*E*)-piplartine by the Ames test. African Journal of Biotechnology, 10, 5398–5401.

Morellato, L. P. C.; Haddad C. F. B. 2000. Introduction: the Brazilian Atlantic forest. Biotropica, 32, 786–792.

Mors, W. B. 1997. Natural Products research in Brazil, looking at the origins. Ciência e Cultura, 49, 310.

Morsingh, F.; Robinson, R. 1970. The syntheses of brazilin and haematoxylin. Tetrahedron, 26, 281–289.

Mosmann, T. 1983. Rapid colorimetric assay for cellular growth and survival: application to proliferation and cytotoxicity assays. Journal of Immunological Methods, 65, 55–63.

Mota, J. S.; Leite, A. C.; Batista Junior, J. M.; López, S. N.; Ambrósio, D. L.; Duó Passerini, G. D.; Kato, M. J.; Bolzani, V. S.; Cicarelli, R. M. B.; Furlan, M. 2009. *In vitro* trypanocidal activity of phenolic derivatives from *Peperomia obtusifolia*. Planta Medica, 75, 620–623.

Motidome, M.; Lecking, M. E.; Gottlieb, O. R. 1970. A química das Solanaceas brasileiras I. A presença de solamargina e de solasodina no juá e na lobeira. Anais da Academia Brasileira de Ciências, 42, 375–376.

Muelas-Serrano, S.; Nogal-Ruiz, J. J.; Gómez-Barrio, A. 2000. Setting of a colorimetric method to determine the viability of *Trypanosoma cruzi* epimastigote. Parasitology Research, 86, 999–1002.

Myers, N.; Mittermeier, R. A.; Mittermeier, C. G.; Fonseca, G. A. B.; Kent, J. 2000. Biodiversity hotspots for conservation priorities. Nature, 403, 853–858.

Nakano, T.; Djerassi, C. 1961. Terpenoids. XLVI. Copalic acid. Journal of Organic Chemistry, 26, 167–173.

National Cancer Institute – NCI. 2018. Dictionary of Cancer Terms. https://www.cancer.gov/publications/dictionaries/cancer-terms?expand=C.html.

National List of Medicinal Plants of Interest to the Unique System Health (SUS) – Brasil. 2009. Brasília: Ministry of Health of Brazil. http://bvsms.saude.gov.br/bvs/sus/pdf/marco/ms_relacao_plantas_medicinais_sus_0603.pdf.

Newman, D. J.; Cragg, G. M. 2016. Natural products as sources of new drugs from 1981 to 2014. Journal of Natural Products, 79, 629–661.

Newman, D. J.; Cragg, G. M.; Snader, K. M. 2000. The influence of natural products upon drug discovery. Natural Product Report, 17, 215–234.

Oberlies, N. H.; Burgess, J. P.; Navarro, H. A.; Pinos, R. E.; Fairchild, C. R.; Peterson, R. W.; Soejarto, D. D.; et al. 2002. Novel bioactive clerodane diterpenoids from the leaves and twigs of *Casearia Sylvestris*. Journal of Natural Products, 65, 95–99.

Odds, F. C. 2003. Synergy, antagonism, and what the chequerboard puts between them. Journal of Antimicrobial Chemotherapy, 52, 1.

Okumura, L. L.; Regasini, L. O.; Fernandes, D. C.; Silva, D. H. S.; Zanoni, M. V. B.; Bolzani, V. S. 2012. Fast screening for antioxidant properties of flavonoids from *Pterogyne nitens* using electrochemical methods. Journal of AOAC International, 95, 773–77.

Oliveira, A. M.; Santos, A. G.; Santos, R. A.; Csipak, A. R.; Olivato, C.; da Silva, I. C.; de Freitas, M. B.; et al. 2009. Ethanolic extract of *Casearia sylvestris* and its clerodane diterpen (caseargrewiin F) protect against DNA damage at low concentrations and cause DNA damage at high concentrations in mice's blood cells. Mutagenesis, 24, 501–506.

Ollis, W. D.; Ward, A. D.; Oliveira, H. M.; Zelnik, R. 1970. Andirobin. Tetrahedron, 26, 1637–1645.

Ott, J. 2002. Pharmahuasca, anahuasca e jurema preta: farmacologia humana da DMT via oral combinada com harmina. In B. C. Labate, W. S. Araújo eds. O Uso Ritual da ayahuasca. Campinas: Mercado das Letras – FAPESP.

Pachter, I. J. P.; Zacharias, D.; Ribeiro, O. 1959. Indole alkaloids of *Acer saccharinum* (the silver maple), *Dictyoloma incanescens*, *Piptadenia colubrina*, and *Mimosa hostilis*. The Journal of Organic Chemistry, 24, 1285–1287.

Pasquini-Netto, H.; Manente, F. A.; Moura, E. L.; Regasini, L. O.; Pinto, M. E. F.; Bolzani, V. S.; Oliveira, O. M. M. F.; Vellosa, J. C. R. 2012. Avaliação das atividades antioxidante, anti e pró-hemolítica do extrato etanólico das folhas de Pterogyne nitens Tul. (Fabaceae-Caesalpinioideae). Revista Brasileira de Plantas Medicinais, 14, 666–672.

Pera, F.; Mattias, P.; Detzer, K. 1977. Methods for determining the proliferation kinetics of cells by means of 5-bromodeoxyuridine. Cell Tissue Kinetics, 10, 255–264.

Pilon, A. C.; Valli, M.; Dametto, A. C.; Pinto, M. E. F.; Freire, R. T.; Castro-Gamboa, I.; Andricopulo, A. D.; Bolzani, V. S. 2017. NuBBEDB: an updated database to uncover chemical and biological information from Brazilian biodiversity. Scientific Reports, 7, 7215.

Pinto, A. C. 1995. O Brasil dos viajantes e dos exploradores e a química de produtos naturais brasileira. Química Nova, 18, 608–615.

Pinto, A. C.; Silva, D. H. S.; Bolzani, V. S.; Lopes, N. P.; Epifanio, R. A. 2002. Produtos Naturais: atualidade, desafios e perspectivas. Química Nova, 25, 45–61.

Pinto, M. E. F.; Batista, J. M.; Koehbach, J.; Gaur, P.; Sharma, A.; Nakabashi, M.; Cilli, E. M.; et al. 2015. Ribifolin, an orbitide from *Jatropha ribifolia*, and its potential antimalarial activity. Journal of Natural Products, 78, 374–380.

Polo, M. 1958/1982. The Travels of Marco Polo. Harmondsworth and New York: Penguin Books.

Pontual, E. V.; Pires-Neto, D. F.; Fraige, K.; Higino, T. M. M.; Carvalho, B. E. A.; Alves, N. M. P.; Lima, T. A.; et al. 2018. A trypsin inhibitor from *Moringa oleifera* flower extract is cytotoxic to *Trypanosoma cruzi* with high selectivity over mammalian cells. Natural Product Research, 32, 2940–2944.

Prieto, A. M.; Santos, A. G.; Csipak, A. R.; Caliri, C. M.; Silva, I. C.; Arbex, M. A.; Silva, F. S.; et al. 2012. Chemopreventive activity of compounds extracted from *Casearia sylvestris* (Salicaceae) Sw against DNA damage induced by particulate matter emitted by sugarcane burning near Araraquara, Brazil. Toxicology and Applied Pharmacology, 265, 368–372.

Prieto, A. M.; Santos, A. G.; Oliveira, A. P. S.; Cavalheiro, A. J.; Silva, D. H.; Bolzani, V. S.; Varanda, E. A.; Soares, C. P. 2013. Assessment of the chemopreventive effect of casearin B, a clerodane diterpene extracted from *Casearia sylvestris* (Salicaceae). Food and Chemical Toxicology, 53, 153–159.

Queiroz, M. M. F.; Marti, G.; Queiroz, E. F.; Marcourt, L.; Castro-Gamboa, I.; Bolzani, V. S.; Wolfender, J. L. 2015. LC-MS/MS quantitative determination of *Tetrapterys mucronata* alkaloids, a plant occasionally used in Ayahuasca preparation. Phytochemical Analysis, 26, 183–188.

Queiroz, M. M. F.; Queiroz, E. F.; Zeraik, M. L.; Ebrahimi, S. N.; Marcourt, L.; Cuendet, M.; Castro-Gamboa, I.; Hamburger, M.; Bolzani, V. S.; Wolfender, J. L. 2014a. Chemical composition of the bark of *Tetrapterys mucronata* and identification of acetylcholinesterase inhibitory constituents. Journal of Natural Products, 77, 650–656.

Queiroz, M. M. F.; Queiroz, E. F.; Zeraik, M. L.; Martia, G.; Favre-Godal, Q.; Simões-Pires, C.; Marcourt, L.; et al. 2014b. Antifungals and acetylcholinesterase inhibitors from the stem bark of *Croton heliotropiifolius*. Phytochemistry Letters, 10, lxxxviii–xciii.

Regasini, L. O.; Castro-Gamboa, I.; Silva, D. H. S.; Furlan, M.; Barreiro, E. J.; Ferreira, M. P.; Pessoa, C.; et al. 2009. Cytotoxic guanidine alkaloids from *Pterogyne nitens*. Journal of Natural Products, 72, 473–476.

Regasini, L. O.; Fernandes, D. C.; Castro-Gamboa, I.; Silva, D. H. S.; Furlan, M.; Bolzani, V. S.; Barreiro, E. J.; et al. 2008b. Constituintes químicos das flores de *Pterogyne nitens* (Caesalpinioideae). Química Nova, 31, 802–806.

Regasini, L. O.; Oliveira, C. M.; Vellosa, J. C. R.; Oliveira, M. M. F.; Silva, D. H. S.; Bolzani. V. S. 2008a. Free radical scavenging activity of *Pterogyne nitens* Tul. (Fabaceae). African Journal of Biotechnology, 7, 4609–4613.

Regasini, L. O.; Pivatto, M.; Scorzoni, L.; Benaducci, T.; Fusco-Almeida, A. M.; Giannini, M. J. S. M.; Barreiro, E. J.; Silva, D. H. S.; Bolzani, V. S. 2010. Antimicrobial activity of *Pterogyne nitens* Tul., Fabaceae, against opportunistic fungi. Brazilian Journal of Pharmacognosy, 20, 706–711.

Regasini, L. O.; Vellosa, J. C. R.; Silva, D. H. S.; Furlan, M.; Oliveira, O. M.; Khalil, N. M.; Brunetti, I. L.; Young, M. C.; Barreiro, E. J.; Bolzani, V. S. 2008c. Flavonols from *Pterogyne nitens* and their evaluation as myeloperoxidase inhibitors. Phytochemistry, 69, 1739–1744.

Repke, D. R.; Torres, C. M. 2006. Anadenanthera: Visionary Plant of Ancient South America. New York: Haworth Herbal Press.

Ribeiro, M. C.; Metzger, J. P.; Martensen, A. C.; Ponzoni, F. J.; Hirota, M. M. 2009. The Brazilian atlantic forest: how much is left, and how is the remaining forest distributed? Implications for conservation. Biological Conservation, 142, 1141–1153.

Russo, M.; Dugo, P.; Fanali, C.; Dugo, L.; Zoccali, M.; Mondello, L.; Gara, L. 2018. Use of an online extraction technique coupled to liquid chromatography for determination of caffeine in coffee, tea, and cocoa. Food Analytical Methods, 11, 1–8.

Sabandar, C. W.; Ahmat, N.; Jaafar, F. M.; Sahidin, I. 2013. Medicinal property, phytochemistry and pharmacology of several *Jatropha* species (Euphorbiaceae): a review. Phytochemistry, 85, 7–29.

Saini, R. K.; Sivanesan, I.; Keum, Y-S. 2016. Phytochemicals of *Moringa oleifera*: a review of their nutritional, therapeutic and industrial significance. 3 Biotech, 6, 203.

Sangalli-Leite, F.; Scorzoni, L.; Silva, A. C. A. P.; Silva, J. F.; Oliveira, H. C.; Singulani J. L.; Gullo, F. P.; et al. 2016. Synergistic effect of pedalitin and amphotericin B against *Cryptococcus neoformans* by *in vitro* and *in vivo* evaluation. International Journal of Antimicrobial Agents, 48, 504–511.

Santos, A. G.; Ferreira, P. M. P.; Vieira Júnior, G. M.; Perez, C. C.; Tininis, A. G.; Silva, G. H.; Bolzani, V. S.; Costa-Lotufo, L. V.; do Ó Pessoa, C.; Cavalheiro, A. J. 2010a. Casearin X, its degradation product and other clerodane diterpenes from leaves of *Casearia sylvestris*: evaluation of cytotoxicity against normal and tumor human cells. Chemistry & Biodiversity, 7, 205–215.

Santos, L. A.; Cavalheiro, A. J.; Tempone, A. G.; Correa, D. S.; Alexandre, T. R.; Quintiliano, N. F.; Rodrigues-Oliveira, A. F.; Oliveira-Silva, D.; Martins, R. C.; Lago, J. H. 2015. Antitrypanosomal acetylene fatty acid derivatives from the seeds of *Porcelia macrocarpa* (Annonaceae). Molecules, 20, 8168–8180.

Santos, N. P.; Pinto, A. C.; Alencastro, R. B. 1998. Theodoro Peckolt: naturalista e farmacêutico do Brasil imperial. Química Nova, 21, 666–670.

Santos, V. A. F. F. M.; Leite, K. M.; Siqueira, M. C.; Martinez, I.; Regasini, L. O.; Nogueira, C. T.; Galuppo, M. K.; et al. 2013. Antiprotozoal activity of quinonemethide triterpenes from *Maytenus ilicifolia* (Celastraceae). Molecules, 18, 1053–1062.

Santos, V. A. F. F. M.; Regasini, L. O.; Nogueira, C. R.; Passerini, G. D.; Martinez, I.; Bolzani, V. S.; Graminha, M. A.; Cicarelli, R. M.; Furlan, M. 2012. Antiprotozoal sesquiterpene pyridine alkaloids from *Maytenus ilicifolia*. Journal of Natural Products, 75, 991–995.

Santos, V. A. F. F. M.; Santos, D. P.; Castro-Gamboa, I.; Zanoni, M. V. B.; Furlan, M. 2010b. Evaluation of antioxidant capacity and synergistic associations of quinonemethide triterpenes and phenolic substances from *Maytenus ilicifolia* (Celastraceae). Molecules, 15, 6956–6973.

Schultes, R. E. 1969. The plant kingdom and hallucinogens (part II). https://www.unodc.org/unodc/en/data-and-analysis/bulletin/bulletin_1969-01-01_4_page004.html#s0005.

Schwab, W. 2003. Metabolome diversity: too few genes too many metabolites? Phytochemistry, 62,837–849.

Scorzoni, L.; Benaducci, T.; Almeida, A. M. F.; Silva, D. H. S.; Bolzani, V. S.; Gianinni, M. J. S. M. 2007. The use of standard methodology for determination of antifungal activity of natural products against medical yeasts *Candida* sp. and *Cryptococcus* sp. Brazilian Journal of Microbiology, 38, 391–397.

Seidler, R. 2001. Cocaine. Current Therapeutics, 42, 82–83.

Selvamuthukumaran, M.; Shi, J. 2017. Recent advances in extraction of antioxidants from plant by-products processing industries. Food Quality and Safety, 1, 61–81.

Shoham, S.; Levitz, S. M. 2005. The immune response to fungal infections. British Journal of Haematology, 129, 569–582.

Silva, D. H. S.; Castro-Gamboa, I.; Bolzani, V. S. 2010. Plant diversity from Brazilian Cerrado and atlantic forest as a tool for prospecting potential therapeutic drugs. In L. Mander; H. W. Lui, eds., Comprehensive Natural Products II Chemistry and Biology. Oxford: Elseier.

Silva, L. L. D.; Nascimento, M. S.; Cavalheiro, A. J.; Silva, D. H.; Castro-Gamboa, I.; Furlan, M.; Bolzani, V. S. 2008. Antibacterial activity of labdane diterpenoids from *Stemodia foliosa*. Journal of Natural Products, 71, 1291–1293.

Souza, A.; Vendramini, R. C.; Brunetti, I. L.; Regasini, L. O.; Bolzani, V. S.; Silva, D. H. S.; Pepato, M. T. 2009. Tratamento crônico com extrato alcoólico de *Pterogyne nitens* não melhora parâmetros clássicos do diabetes experimental. Revista Brasileira de Farmacognosia, 19, 412–417.

Souza, A.; Vendramini, R. C.; Brunetti, I. L.; Regasini, L. O.; Silva, D. H. S.; Bolzani, V. S.; Pepato, M. T. 2010. Alcohol extract of *Pterogyne nitens* leaves fails to reduce severity of streptozotocin-induced diabetes in rats. Journal of Medicinal Plants Research, 4, 802–808.

Stehmann, J. R.; Forzza, R. C.; Salino, A.; Sobral, M.; Costa, D. P.; Kamino, L. H. Y. 2009. Plantas da Floresta Atlântica (Plants in the Atlantic Forest), pp. 1–505. Rio de Janeiro: Jardim Botânico do Rio de Janeiro.

Tanaka, T.; Asai, F.; Iinuma, M. 1998. Phenolic compounds from *Peperomia obtusifolia*. Phytochemistry, 49, 229–232.

Tong, C.; Peng, M.; Tong, R.; Ma, R.; Guo, K.; Shi. S. 2018a. Use of an online extraction liquid chromatography quadrupoletime-of-flight tandem mass spectrometry method for thecharacterization of polyphenols in Citrus paradisi cv. Changshanhuyupeel. Journal of Chromatography A, 1533, 87–93.

Tong, R.; Peng, M.; Tong, C.; Guo, K.; Shi, S.; 2018b. Online extraction–high performance liquid chromatography–diode array detector–quadrupole time-of-flight tandem mass spectrometry for rapid flavonoid profiling of Fructus aurantii immaturus. Journal of Chromatography B, 1077–1078, 1–6.

Valli, M.; Altei, W.; Santos, R. N.; Lucca Jr., E. C.; Dessoy, M. A.; Pioli, R. M.; Cotinguiba, F.; et al. 2017. Synthetic analogue of the natural product piperlongumine as a potent inhibitor of breast cancer cell line migration. Journal of the Brazilian Chemical Society, 28, 475–484.

Valli, M.; Santos, R. N.; Figueira, L. D.; Nakajima, C. H.; Andricopulo, A. D.; Bolzani, V. S. 2013. Development of a natural products database from the biodiversity of Brazil. Journal of Natural Products, 76, 439–444.

Valli, M.; Young, M. C. M.; Bolzani, V. S. 2016. A beleza invisível da biodiversidade: o táxon Rubiaceae. Revista Virtual de Química, 8, 296–310.

Varma, V. R.; Hausdorff, J. M.; Studenski, S. A.; Rosano, C.; Camicioli, R.; Alexander, N. B.; Chen, W. G.; Lipsitz, L. A.; Carlson, M. C. 2016. Aging, the central nervous system, and mobility in older adults: interventions. The Journals of Gerontology Series A: Biological Sciences and Medical Sciences, 71, 1451–1458.

Veiga Jr., V. F.; Patitucci, M. L.; Pinto, A. C. 1997. Controle de autenticidade de óleos de copaíba comerciais por cromatografia gasosa de alta resolução. Química Nova, 20, 612–615.

Vellosa, J. C. R.; Regasini, L. O.; Belló, C.; Schemberger, J. A.; Khalil, N. M.; de Araújo, A. M. -G.; Bolzani, V. S.; Brunetti, I. L.; Oliveira, O. M. F. 2015. Preliminary *in vitro* and *ex vivo* evaluation of afzelin, kaempferitrin and pterogynoside action over free radicals and reactive oxygen species. Archives of Pharmacal Research, 38, 1168–1177.

Vellosa, J. C. R.; Regasini, L. O.; Khalil, N. M.; Bolzani, V. S.; Khalil, O. A. K.; Manente, F. A.; Pasquini Netto, H.; Oliveira, O. M. M. F. 2011. Antioxidant and cytotoxic studies for kaempferol, quercetin and isoquercitrin. Eclética Química, 36, 7–20.

Wang, M.; Carver, J. J.; Phelan, V. V.; et al. 2016. Sharing and community curation of mass spectrometry data with global natural products social molecular networking. Nature Biotechnology, 34, 828–837.

White, R. L.; Burgess, D. S.; Manduru, M.; Bosso, J. A. 1996. Comparison of three different in vitro methods of detecting synergy: time-kill, checkerboard, and E test. Antimicrobial Agents and Chemotherapy, 40, 1914–1918.

Wolfender, J. L.; Marti, G.; Thomas, A.; Bertrand, S. 2015. Current approaches and challenges for the metabolite profiling of complex natural extracts. Journal of Chromatography A, 1382, 136–164.

Wood, C. L.; Lafferty, K. D.; Deleo, G.; Young, H. S.; Hudson, P. J.; Kuris. A. M. 2014. Does biodiversity protect humans against infectious disease? Ecology, 95, 817–832.

World Health Organization – WHO. 2016. 19th WHO Model List of Essential Medicines http://www.who.int/medicines/publications/essentialmedicines/EML2015_8-May-15.pdf.

World Health Organization – WHO. 2018. http://www.who.int/news-room/fact-sheets/detail/cancer.html.

Yuliana, N. D.; Khatib, A.; Choi, Y. H.; Verpoorte. R. 2011. Metabolomics for bioactivity assessment of natural products. Phytotherapy Research, 25, 157–169.

Zani, C. L.; Carroll, A. R. 2017. Database for rapid dereplication of known natural products using data from MS and fast NMR experiments. Journal of Natural Products, 80, 1758–1766.

Zeraik, M. L.; Queiroz, E. F.; Castro-Gamboa, I.; Silva, D. H. S.; Cuendet, M.; Bolzani, V. S.; Wolfender, J-L.; Marcourt, L.; Ciclet, O. 2016b. Processo de extração e isolamento de substâncias ativas presentes na polpa do umbu, substâncias ativas, alimentos nutracêuticos e/ou funcionais compreendendo as referidas substâncias ativas e seu uso. Patent number: BR 102014025601-6 A2, Brazil.

Zeraik, M. L.; Queiroz, E. F.; Marcourt, L.; Ciclet, O.; Castro-Gamboa, I.; Silva, D. H. S.; Cuendet, M.; Bolzani, V. S.; Wolfender, J. L. 2016a. Antioxidants, quinone reductase inducers and acetylcholinesterase inhibitors from *Spondias tuberosa* fruits. Journal of Functional Foods, 21, 396–405.

10 Plants from the Caatinga Biome with Medicinal Properties[†]

Maria da Conceição Ferreira de Oliveira[a],
Mary Anne Sousa Lima[a], Francisco Geraldo Barbosa[a],
Jair Mafezoli[a], Mary Anne Medeiros Bandeira[b],
and Wellyda Rocha Aguiar[b]
[a]Department of Organic and Inorganic Chemistry, Federal
University of Ceará, Campus do Pici, Fortaleza, Brazil
[b]Medicinal Herb Garden "Francisco José de Abreu Matos", Federal
University of Ceará, Campus do Pici, Fortaleza, Brazil

CONTENTS

[†] Dedicated to Prof. Francisco José de Abreu Matos (*In Memoriam*), "father" of Farmácias Vivas (Living Pharmacies), a scientific and social ("local plants for local people") program with plants from Caatinga.

10.1   INTRODUCTION

Caatinga ("white forest" in indigenous Tupi language) is a semi-arid ecosystem found exclusively in Brazil, which is referred to as "a mosaic of scrubs and patches of seasonally dry forest" (Santos et al., 2011). This unique biome occupies a large geographic area (ca. 800,000 km²) of the country that spread from the state of Ceará to the north of the state of Minas Gerais, covering about 60% of the Northeast region (Figure 10.1). Despite being one of the largest Brazilian biomes, the scientific knowledge of this biome's biodiversity is still very poor (Coe and Souza, 2014; Maia, 2004; Prado 2003; Santos et al., 2011).

   Two very distinct seasons are present in Caatinga: the dry and rainy seasons. During the dry season, which occurs for most of the year (8-9 months), the vegetation has a light gray aspect after the leaves fall (Figure 10.2b), justifying the indigenous name "Caatinga" (Coe and Souza, 2014; Maia, 2004; Prado, 2003). Probably, because of this aspect, Martius, the famous German botanist, depicted Caatinga as "*silva aestu aphylla*" (forest without leaves in summer) (Prado, 2003). The rainy season is limited to few months and the total rainfall (irregular and poorly distributed) reaches 500-1,100 mm/year. In this period, the plant leaves reappear and the landscape is completely changed (Figure 10.2a). Notwithstanding the extreme stressing conditions found in Caatinga (high temperature, high solar radiation, low irregular rainfall and low humidity), this biome hides a great, but highly threatened, biodiversity of animals and plants (Coe and Souza, 2014).

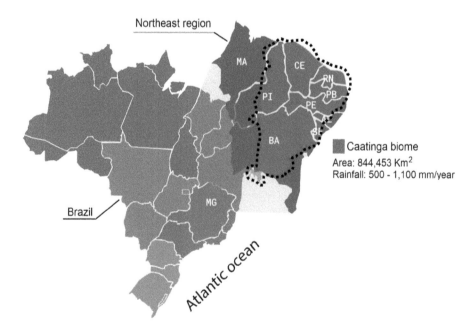

FIGURE 10.1   Map displaying the location of the Caatinga biome in the Northeast region of Brazil.

FIGURE 10.2   Representative pictures from the Caatinga biome during the dry (a) and rainy (b) seasons.

Because of the high-water deficit, plants from Caatinga have developed strategies to survive, and the xeromorphic vegetation is characterized by the presence of cacti and shrubs with spines or prickles (Figure 10.2). The diversity of plants is estimated in *ca.* 5,000 species that are distributed into eight ecoregions and 12 distinct types of vegetation. A noteworthy fact is that about 800 of the species are endemic to the biome (Coe and Souza, 2014).

Many of the plant species from Caatinga are used in folk medicine and their potential therapeutic benefits have been reported (Albuquerque et al., 2007; Viana et al., 2013). Albuquerque et al. reported a quantitative approach on medicinal plants from Caatinga. The authors calculated the ethnopharmacological relative importance index (RI) of 389 medicinal species used by rural and indigenous communities from the Northeast region of Brazil (Albuquerque et al., 2007). A similar study investigated 119 medicinal plants used by local population from Aiuaba, in Ceará state, and revealed some species as having great versatility of use (Cartaxo et al., 2010).

Thirteen endemic medicinal plants from Caatinga with RI ≥1 (Albuquerque et al., 2007) are listed in Table 10.1. Among them, *Maytenus rigida* (RI = 1.9), *Bauhinia cheilantha* (RI = 1.7) and *Cereus jamacaru* (RI = 1.7) are the species with the highest ethnopharmacological relative importance indexes. The broad spectrum of therapeutic properties reported for these plants, including anti-inflammatory, antiophidic, expectorant and calmative, and for treating problems such as diges-tive, headache, toothache, influenza and more is worth highlighting.

Table 10.2 displays 37 non-endemic medicinal plants from Caatinga with RI ≥1 (Albuquerque et al., 2007). The highest RI values are reported for *Amburana cearensis* (RI = 2.0), *Myracrodruon urundeuva* (RI = 2.0) and *Argemone mexicana* (RI = 1.8). Several uses and properties are also reported for these non-endemic medicinal plants, which include those described for the endemic plants.

Some of the species listed in Tables 10.1 and 10.2 are commercialized for manufacturing of herbal medicines and had their properties scientifically proven, such as *A. cearensis*, *Erythrina velutina*, *M. urundeuva* and *Sideroxylon obtusifolium* (Albuquerque et al., 2007).

The biological and therapeutic potential of seven medicinal plants from Caatinga (*A. cearensis*, *Anadenanthera colubrina*, *Anacardium occidentalis*, *Bauhinia forficata*, *Cissus sicyoides*,

TABLE 10.1

Endemic Medicinal Plants from Caatinga with RI ≥1 (Albuquerque et al., 2007)

RI	Species	Family	Common Names	Uses and Properties
1.9	*Maytenus rigida* Mart.	Celastraceae	Bom-nome	Renal problems, hepatic problems, pains in general, rheumatism, sexual impotence, menstrual disturbances, inflammations in general, asthma, cough, bronchitis, blow, injury, anemia, circulation problems, cardiac problems, ovarian inflammation, ovarian infection, skin ulcers and vaginal ulcers.
1.7	*Bauhinia cheilantha* (Bong.) Steud.[a]	Fabaceae	Mororó, pata-de-vaca and unha-de-vaca	Diabetes, high cholesterol levels, inflammations in general, spinal problems, cough, influenza, dysphonia, asthma, blood thinner, depurative, rheumatism, migraine, nervous disturbances, inappetence, helminthiasis, expectorant, calmative and tonic.
1.7	*Cereus jamacaru* DC	Cactaceae	Mandacaru and babão	Many problems, such as renal, hepatic, respiratory, spinal, urethral, syphilis, injury, influenza, cough, bronchitis, ulcers, constipation, hypertension, rheumatism, enteritis, fever, etc.
1.5	*Capparis jacobinae* Moric. ex Eichler	Capparaceae	Icó-preto, icó-verdadeiro and incó	Cough, pertussis, colds, digestive problems, skin diseases, abdominal pain, intoxication, fever, diabetes, lung inflammation, cardiac problems and emmenagogic.
1.5	*Erythrina velutina* Willd.[a]	Fabaceae	Mulungu	Tooth inflammation, odontalgia, headache, fever, maternal milk production, diabetes, hypertension, cough, bronchitis, nervous disturbances, insomnia, hemorrhoids, helminthiasis and calmative.
1.4	*Caesalpinia pyramidalis* Tul.	Fabaceae	Catingueira and catingueira-rasteira	Cough, bronchitis, respiratory infection, influenza, asthma, gastritis, colic, fever, heartburn, flatulence, diarrhea, collision, injury, diabetes, aphrodisiac, stomachache and expectorant.
1.4	*Spondias tuberosa* Arruda	Anacardiaceae	Umbuzeiro, imbu, and umbu	Ophthalmia, venereal diseases and digestive problems, intestinal problems, diabetes, renal infection, menstrual disturbances, placental delivery, throat problems, antiemetic and tonic.
1.3	*Ziziphus joazeiro* Mart.	Rhamnaceae	Joá and joazeiro	Teeth cleaning, dandruff, asthma, cough, influenza, pneumonia, tuberculosis, bronchitis, constipation, throat inflammation, indigestion, scabies, seborrheic dermatitis, itching, skin problems, head injuries, healing and expectorant.
1.2	*Commiphora leptophloeos* (Mart.) J.B. Gillett	Burseraceae	Imburana, emburana, emburana-de-cambão and umburana	Renal problems, influenza, cough, bronchitis, dysphonia, inflammations in general, odontalgia, colic, diarrhea, antiemetic, tonic and healing.

(Continued)

TABLE 10.1 *(Continued)*
Endemic Medicinal Plants from Caatinga with RI ≥1 (Albuquerque et al., 2007)

RI	Species	Family	Common Names	Uses and Properties
1.1	*Croton argyrophylloides* Mull. Arg.	Euphorbiaceae	Angolinha, marmeleiro-branco, and sacatinga	Stomach ache, diabetes, inflammations in general, venereal diseases, blood thinner, back pain, intoxication and headache.
1.0	*Capparis flexuosa* (L.) L.	Capparaceae	Feijão-bravo and feijão-de-boi	Cough, pneumonia, influenza, colds, digestive problems, skin diseases, abdominal pain, rheumatism and antiophidic.
1.0	*Cordia leucocephala* Moric.	Boraginaceae	Piçarra, moleque-duro and nego-duro	Bleeding, throat inflammation, rheumatism, indigestion, arthritis, rickets, calmative and tonic.

[a] Although not classified as endemic by Albuquerque et al. (2007), this species is also reported as endemic (Maia, 2004; Viana et al., 2013).

TABLE 10.2
Non-Endemic Medicinal Plants from Caatinga with RI ≥1 (Albuquerque et al., 2007)

RI	Species	Family	Common Names	Uses and Properties
2.0	*Amburana cearensis* (Freire Allemão) A. C. Smith	Fabaceae	Cumaru, emburana-de-cheiro and amburana.	Tooth inflammation, genital inflammation, colic, diarrhea, intestinal problems, placental delivery, pains in general, headache, influenza, cough, bronchitis, pertussis, sinusitis, asthma, heartburn, skin ulcers, urinary infection, antiophidic, antispasmodic, expectorant and tonic.
2.0	*Myracrodruon urundeuva* Allemão	Anacardiaceae	Aroeira and Aroeira-do-sertão	Inflammations, pains, infections, blow, injury, vaginal discharge, asthma, influenza, cough, bronchitis, tuberculosis, heartburn, gastritis, odontalgia, placental delivery, anemia, diphtheria, skin ulcers and uterine.
1.8	*Argemone mexicana* L.	Papaveraceae	Cardo santo and cadinho	Herpes labialis, fever, influenza, asthma, tonsillitis, bronchitis, pneumonia, uterine inflammation, ophthalmia, thrombosis, scrophula, physical weakness, inflammations in general, constipation, skin ulcers, conjunctivitis, cholagogue, laxative and digestive.
1.6	*Anadenanthera colubrina* (Vell.) Brenan	Mimosaceae	Angico, angico-de-caroço and angico-branco	Anemia, cough, asthma, bronchitis, pertussis, lung inflammation, influenza, constipation, inflammations in general, cancer, blood thinner, blow, injury, scrofula, diphtheria, fissures in foot, gastritis and expectorant.
1.6	*Operculina hamiltonii* (G. Don) D.F. Austin & Staples	Convolvulaceae	Batata-de-purga	Influenza, pneumonia, asthma, bronchitis, cough, cardiovascular problems, helminthiasis, rheumatism, constipation, inappetence, digestive problems, hydropsy, syphilis, amenorrhea, teething, inflammations in general, flatulence and laxative.

(Continued)

TABLE 10.2 *(Continued)*

Non-Endemic Medicinal Plants from Caatinga with RI ≥1 (Albuquerque et al., 2007)

RI	Species	Family	Common Names	Uses and Properties
1.6	*Rosmarinus officinalis* L.	Lamiaceae	Alecrim and alecrim-de-jardim	Menstrual disturbances, cough, influenza, asthma, colic, fever, pains in general, flatulence, stomachic, intestinal problems, hepatic problems, renal problems, injury, rheumatism, sedative, antispasmodic and cardiotonic.
1.5	*Aloe vera* (L.) Burm. f.	Liliaceae	Babosa	Hemorrhoid, heartburn, stomach problems, cancer, dandruff, hair loss, contusion, rheumatism, injury, helminthiasis, inflammations in general, emollient and emmenagogic.
1.4	*Chenopodium ambrosioides* L.	Chenopodiaceae	Mastruz	Influenza, cough, bronchitis, tuberculosis, helminthiasis, cancer, blow, digestive problems, headache, fever, antimicrobial, expectorant, hematoma, gastritis and stomachic.
1.4	*Cymbopogon citratus* (DC.) Stapf	Poaceae	Capim-santo and capim-caboclo	Stomach ache, gastritis, ulcer, diarrhea, indigestion, inappetence, colic, pains in general, hypertension, insomnia, fever, cough, calmative and diuretic.
1.4	*Leonotis nepetifolia* (L.) R. Br.	Lamiaceae	Cordão-de-são-francisco and cravinho	Menstrual colic, urinary retention, rheumatism, inflammations in general, fever, cystitis, dysentery, calculus of kidney, childbirth pains, paralysis, aphrodisiac, sedative, stomachic and healing.
1.3	*Cnidoscolus phyllacanthus* (Mull. Arg.) Pax & L. Hoffm.	Euphorbiaceae	Urtiga branca and favela	Hemorrhoid, renal problems, ophthalmic diseases, blow, injury, fractures, warts, skin problems, eye cleansing, urinary infection and inflammations in general.
1.3	*Helianthus annuus* L.	Asteraceae	Girassol	Weakness, thrombosis, fever, ulcer, neuralgiainjury, contusion, epistaxis, inappetence, hypercholesterolemia and emollient.
1.3	*Lippia alba* (Mill.) N.E. Br.	Verbenaceae	Cidreira, erva-cidreira, and melissa	Diabetes, hypertension, colic, diarrhea, heartburn, inappetence, fever, headache, anemia, cold, anticonceptive, calmative, cardiac problems and gastritis.
1.3	*Ruta graveolens* L.	Rutaceae	Arruda and arruda macho	Menstrual colic, headache, otalgia, conjunctivitis, arthritis, nevralgy, helminthiasis, amenorrhea, digestive problems and antispasmodic.
1.2	*Achillea millefolium* L.	Asteraceae	Novalgina and mil-folhas	Pains in general, fever, abscess, skin problems, headache, physical weakness, intoxication, injury, stomachic, antispasmodic and depurative.
1.2	*Anacardium occidentale* L.	Anacardiaceae	Cajú, cajueiro and caju-roxo	Diarrhea, inflammations in general, renal infection, heartburn, tuberculosis, diabetes, blow and antiseptic.

(Continued)

TABLE 10.2 *(Continued)*

Non-Endemic Medicinal Plants from Caatinga with RI ≥1 (Albuquerque et al., 2007)

RI	Species	Family	Common Names	Uses and Properties
1.2	*Caesalpinia ferrea* Mart.	Fabaceae	Jucá and pau-ferro	Inflammations in general, blow, throat afflictions, bronchitis, anemia, swelling, back pain, injury, labyrinthitis, renal problems, stress and fatigue.
1.2	*Hyptis suaveolens* (L.) Poit.	Lamiaceae	Alfazema-brava, alfavaca-caboclo and alfazema-de-caboclo	Digestive problems, menstrual colic, amenorrhea, odontalgia, headache, fever, influenza, respiratory problems in general, gout and eye cleansing.
1.2	*Momordica charantia* L.	Cucurbitaceae	Melão-de-são-caetano and melão-de-sabiá	Injury, STDs, lice, scabies, hemorrhoid, allergy, diabetes, helminthiasis, rheumatism, diarrhea and dandruff.
1.2	*Ocimum basilicum* L.	Lamiaceae	Manjericão and manjericão-roxo	Otalgia, bronchitis, cough, influenza, fever, tonsillitis, gingivitis, diarrhea, headache, stomachic, antispasmodic, diuretic and antiemetic.
1.2	*Plectranthus amboinicus* (Lour.) Spreng.	Lamiaceae	Hortelã da folha grande, hortelã-graúdo and hortelã da folha grossa	Otalgia, inflammations in general, cough, bronchitis, tonsillitis, pneumonia, influenza, constipation, hepatic problems, menstrual disturbances, dysphonia, stomachic and helminthiasis.
1.2	*Parkinsonia aculeata* L.	Fabaceae	Turco	Migraine, thrombosis, influenza, asthma, diabetes, hypertension, fever, epilepsy, malaria and antiophidic
1.2	*Sideroxylon obtusifolium* (Humb. ex Roem. & Schult.) T.D. Penn.	Sapotaceae	Quixaba, quixabeira and rompe-gibão	Blow, pains in general, duodenal ulcer, gastritis, heartburn, chronic inflammation, genital injury, ovarian inflammation, adnexitis, colic, renal problems, cardiac problems, diabetes and healing.
1.1	*Allium sativum* L.	Liliaceae	Alho	Insect bites, dysphonia, constipation, digestive problems, odontalgia, conjunctivitis, hypercholesterolemia, antibiotic, influenza and throat afflictions.
1.1	*Combretum leprosum* Mart.	Combretaceae	Mofumbo	Bronchitis, influenza, cough, pertussis, sweating, diphtheria, heartburn, calmative, hemostatic and expectorant.
1.1	*Croton rhamnifolius* Willd.	Euphorbiaceae	Pau-de-leite and velame	Influenza, cough, stomachache, menstrual disturbances, back pain, anemia, blood thinner, hypercholesterolemia and depurative.
1.1	*Hyptis pectinata* (L.) Poit.	Lamiaceae	Alfazema-brava, alfazema-de-caboclo, and sambacaitá	Headache, odontalgia, amenorrhea, hepatalgia, hepatic problems, flatulence, rheumatism, gastritis, ulcer, asthma, cough and bronchitis.
1.1	*Lippia* sp.	Verbenaceae	Alecrim de caboclo	Influenza, sinusitis, cough, nasal congestion, pains in general, headache, odontalgia, eliminate material after child delivery, calmative and expectorant.
1.1	*Mimosa tenuiflora* (Willd.) Poir.	Mimosaceae	Jurema-preta and jurema	Injury, odontalgia, inflammations in general, fever, menstrual colic, headache, hypertension, bronchitis and cough.

(Continued)

TABLE 10.2 *(Continued)*

Non-Endemic Medicinal Plants from Caatinga with RI ≥1 (Albuquerque et al., 2007)

RI	Species	Family	Common Names	Uses and Properties
1.1	*Ocimum gratissimum* L.	Lamiaceae	Alfavaca, alfavaca-branco and quioiô	Digestive, flatulence, digestive problems, influenza, cough, itching, expectorant, calming, stress, headache and fatigue.
1.1	*Scoparia dulcis* L.	Scrophulariaceae	Vassourinha and vassourinha-de-nossa-senhora	Tortion, bronchitis, influenza, cough, fever, inflammations in general, uterine problems, helminthiasis, diabetes, amenorrhea and emmenagogic.
1.0	*Acmella uliginosa* (Sw.) Cass.	Asteraceae	Agrião	Caries, asthma, inflammations in general, injury, hair loss, stomachic, depurative, tonic and expectorant.
1.0	*Heliotropium indicum* L.	Boraginaceae	Crista-de-galo and fedegoso	Foot swelling, influenza, cough, inflammations in general, hepatic problems, conjunctivitis, renal problems and diuretic.
1.0	*Kalanchoe brasiliensis* Cambess.	Crassulaceae	Pratudo and coirama	Pains in general, ulcer, gastritis, inflammations in general, injury, lung afflictions, asthma, kidney stones and emollient.
1.0	*Leucas martinicensis* (Jacq.) R. Br.	Lamiaceae	Cordão-de-são francisco and cordão-de-frade	Renal problems, rheumatism, inflammations in general, nevralgy, sweating, flatulence, tonic, antispasmodic and calmative.
1.0	*Marsypianthes chamaedrys* (Vahl) Kuntze	Lamiaceae	Bentônica-brava and hortelã-do-mato	Cough, bronchitis, flatulence, fever, articular rheumatism, antiophidic, stimulant and digestive.
1.0	*Tabebuia impetiginosa* (Mart. ex DC.) Standl.	Bignoniaceae	Pau-d'arco-roxo, ipê-roxo	Ulcer, surgical infections, inflammations in general, leucorrhea, cancer, gingivitis, cardiac problems, antimicrobial and antiseptic.

M. urundeuva and *Zingiber officinalis*) were reviewed by Silva et al. (2012). Additionally, Viana et al. (2013) reviewed 16 medicinal plants from Caatinga, including *A. cearensis*, *A. occidentalis*, *M. urundeuva* and *Z. officinalis*.

This chapter focuses on five medicinal plants from Caatinga (listed in alphabetical order), *B. cheilantha* (RI = 1.7), *Combretum leprosum* (RI = 1.1), *Egletes viscosa* (RI = 0.9), *E. velutina* (RI = 1.5) and *M. urundeuva* (RI = 2.0), Table 10.3, which were selected based not only on their ethnopharmacological

TABLE 10.3

Summary of the Names of All Plant Species, Their Respective Family and Common Names Discussed in This Chapter

Species	Family	Common Names
Bauhinia cheilantha (sin. *Pauletia cheilantha*)	Fabaceae	mororó, pata-de-vaca, unha-de-vaca
Combretum leprosum (sin. *Combretum hasslerianum*)	Combretaceae	mofumbo, mufumbo, pente de macaco
Egletes viscosa (sin. *Platystephium graveolens*)	Compositae	macela, macela-da-terra
Erythrina velutina (sin. *Erythrina splendida*)	Fabaceae	mulungu, suinão, canivete, corticeira, pau-de-coral, sanaduí, sananduva
Myracrodruon urundeuva	Anacardiaceae	aroeira, aroeira-do-sertão

RIs but also on the scientific validation of their popular uses. For each species, the botanic aspects and occurrence, ethnopharmacology, chemical studies and pharmacological studies are presented.

10.2   *B. CHEILANTHA* (BONG.) STEUD. (FABACEAE)

10.2.1   BOTANICAL ASPECTS AND OCCURRENCE

Bauhinia is a pantropical genus that comprises about 300-350 species (Sinou et al., 2009). In Brazil, this genus is represented by 91 species, which are commonly known as "mororó" (indigenous Tupi language) and "pata-de-vaca" (literally translated as cow nail); the latter because of the bifoliated leaves (Gutiérrez et al., 2011). A noteworthy detail is that the most common *Bauhinia* species in the state of Ceará (Northeast region of Brazil) are *Bauhinia ungulata* L. (found in the litoral) and *B. cheilantha* (Bong.) Steud. (found in Caatinga) (Viana et al., 2013).

B. cheilantha is a medicinal plant (Shrub above 1 m in height) with high ethnopharmacological relative importance index (RI = 1.7). This species is an endemic plant from Caatinga, also found in Cerrado bioma although with less frequency (Gutiérrez, 2010).

10.2.2   ETHNOPHARMACOLOGY

Several uses and properties are reported for *B. cheilantha*, such as hypoglycemia (treatment of diabetes), calmative, blood thinner, depurative, rheumatism, migraine, nervous disturbances, inappetence, helminthiasis, etc. (Table 10.1). Infusion or decoction from stem-barks of the plant is used as tonic, depurative and for treating diabetes (Agra et al., 2007). Additionally, roasted seeds from *B. cheilantha* are used in a hot beverage as a substitute for coffee (Lucena et al., 2007).

10.2.3   CHEMICAL STUDIES

Despite the vast popular use of *B. cheilantha* and the plant's high ethnopharmacological relative importance index, the chemical study of this species is still incipient. Phytochemical screening for identification of secondary metabolites in organic extracts from leaves, barks and roots revealed the presence of alkaloids, anthraquinones, steroids, flavonoids and xanthones (Luna et al., 2005).

The chemical study of the aqueous extract from dried leaves of *B. cheilantha* yielded the flavonoids afzelin (**1**, kaempferol-3-*O*-rhamnoside) and quercetrin (**2**, quercetin 3-*O*-rhaminoside) (Figure 10.3).

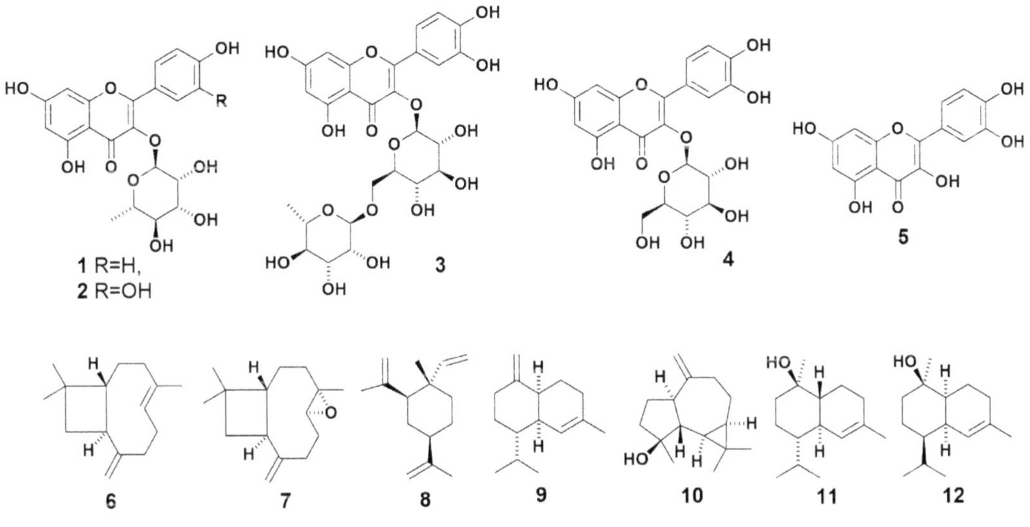

FIGURE 10.3   Chemical structures of the secondary metabolites from *B. cheilantha*.

Moreover, HPLC quantitative analyses of the aqueous extract from leaves of seven *B. cheilantha* specimens revealed the presence of the flavonoids rutin (**3**), isoquercitrin (**4**) and quercetin (**5**), shown in Figure 10.3, besides compound **2** (Brígido, 2001). The phytochemical investigation of the ethanol extract from the aerial parts of *B. cheilantha* also yielded compounds **1** and **2** (Oliveira, 2008).

The essential oils' (extracted by hydrodistillation) composition of the leaves from three specimens of *B. cheilantha* was investigated. Among the 39 compounds present in the oils, *trans*-caryophyllene (**6**), caryophyllene oxide (**7**), β-elemene (**8**), γ-muurolene (**9**), spathulenol (**10**), α-cadinol (**11**) and α-muurolol (**12**) were the major constituents (Brígido, 2001).

The fatty acid composition of the seeds oil of *B. cheilantha* revealed linoleic acid (C18:2 $\Delta^{9,12}$; 42.3%), palmitic acid (C16:0; 25.7%), octadec-7-enoic acid (C18:1 Δ^7; 15.3%) and stearic acid (C18:0; 10.4%) as the major constituents (Brígido, 2001). In addition, high protein content (58.9%) was found in the seeds of *B. cheilantha*, being even higher than those found in soybean (39.5-44.5%) and cowpea bean (19.5-26.1%) (Teixeira et al., 2013).

10.2.4 PHARMACOLOGICAL STUDIES

The hypoglycemic property of a methanol extract (600 mg kg^{-1}) from leaves of *B. cheilantha* was confirmed in alloxan-induced diabetic rats (Silva and Cechinel-Filho, 2002). The same activity was also observed when the ethanol extract (900 mg kg^{-1}) from the aerial parts was tested. The lack of acute toxicity of this extract when tested in concentrations of 1,000 and 2,000 mg kg^{-1} (ip), and 5,000 mg kg^{-1} (oral administration) in mice is worth highlighting. In this case, no mortality was observed after 48 hours of extract administration (Oliveira, 2008).

Ethanol extracts from the aerial parts of *B. cheilantha*, collected in three different periods, were assayed for their antioxidant activities (DPPH method). EC$_{50}$ values of 11.67 ± 1.43 (1st collection), 26.68 ± 0.23 (2nd collection) and 21.77 ± 0.13 μg mL^{-1} (3rd collection) were observed (Oliveira, 2008).

The antinociceptive activity of the aqueous extract from the bark of *B. cheilantha* was evaluated using three different tests (writhing induced by acetic acid −0.6% formalin 1% and hot plate). The extract (administered orally: 400 mg kg^{-1}) reduced nociception by 54.4% in the writhing test. The analgesic effect of the extract (100, 200 and 400 mg kg^{-1}) in the 1% formalin test was observed (57.4%, 46.1% and 46.2% inhibition of the pain reaction, respectively, in the 2nd phase). Finally, the analgesic properties of the extract were corroborated by the hot plate test. The extract (100, 200 and 400 mg kg^{-1}) increased the latency time by 39.8%, 30.7% and 32.8%, respectively. Additionally, no acute toxicity was found for the extract tested at doses up to 3 g kg^{-1} (Silva et al., 2005).

In summary, *B. cheilantha* is an endemic medicinal plant from Caatinga, which presents one of the highest RIs. Several uses are reported for the plant including the treatment of diabetes. Pharmacological studies support the species hypoglycemic and analgesic properties; the latter may be associated with the calming activity reported for the species. The lack of more comprehensive chemical and pharmacological studies that can validate the ethnopharmacological use of this species draw attention.

10.3   *C. LEPROSUM* MART. (COMBRETACEAE)

10.3.1 BOTANICAL ASPECTS AND OCCURRENCE

C. leprosum is a member of the family Combretaceae, constituted by circa 600 species in 18 genera, of which *Terminalia* and *Combretum* are the most important. This species is a popular medicinal plant, which grows wild as a shrub in the Northeast region of Brazil, mainly in the semi-arid regions of Ceará state known as "mofumbo", "mufumbo" or "pente de macaco" (Facundo et al., 1993, 2005, 2008).

10.3.2 Ethnopharmacology

Combretum species are used extensively in traditional medicine against inflammation, infections, malaria, bleeding, diarrhea, digestive disorders and diabetes, and as diuretic (Lima et al., 2012). Infusions or decoctions from bark, bast, leaves, flowers and roots of *C. leprosum* are used in the folk medicine against bronchitis, influenza, cough, pertussis, sweating, diphtheria and heartburn, and as calmative, hemostatic and expectorant. *C. leprosum* showed an RI of 1.1, suggesting for this species a very high ethnopharmacological versatility (Albuquerque et al., 2007).

10.3.3 Chemical Studies

The phytochemical investigation of the organic extract from leaves (hexane and ethanol) and roots (ethanol) of *C. leprosum* collected in Ceará state yielded the triterpenes $3\beta,6\beta,16\beta$-trihydroxylup-20(29)-ene (**13**), arjunolic acid (**14**) and mollic acid (**15**), besides the flavonoids 5,7,3′,4′-tetrahydroxy-3-methoxyflavone (3-*O*-methylquercetin, **16**) and 3-*O*-α-L-rhamnopyranosylquercetin (quercetrin, **2**), Figure 10.4 (Facundo et al., 1993). Independently, the chemical investigation of the ethanol extract from flowers of the same plant led to the isolation of **13-16**, together with the flavonoids 5,3′,4′-trihydroxy-3,7-dimethoxy-flavone (**17**) and 5,3′-dihydroxy-3,7,4′-trimethoxyflavone (**18**), the cycloartane triterpenes **19** and **20**, and the α-D-glucopiranoside-(3β)-stigmast-5-en-3-yl (**21**), Figure 10.4 (Facundo et al., 2008).

10.3.4 Pharmacological Studies

Pharmacological investigations of *C. leprosum* point to similar bioactivities already observed for other species of *Combretum*. Oral administration of the ethanol extract (200 mg kg⁻¹; 37.6% inhibition) from roots of *C. leprosum* as well as the secondary metabolite arjunolic acid (**14**, Figure 10.4; 100 mg kg⁻¹; 80.8% inhibition) revealed anti-inflammatory activity for both tested samples (extract and isolated compound) by reducing the paw edema induced by carrageenan. No activity, is worth mentioning, was observed for the ethanol extract from leaves of *C. leprosum* (Facundo et al., 2005).

The gastroprotective and anti-ulcerogenic effects of the ethanol extract from the stem bark of *C. leprosum* was reported (Nunes et al., 2009). The extract inhibited the gastric acid secretion and increased the mucosal defensive factors, such as mucus and prostaglandin.

The antiproliferative and anti-inflammatory properties of the ethanol extract from the flower of *C. leprosum* was demonstrated in models of skin inflammatory and hyperproliferative process

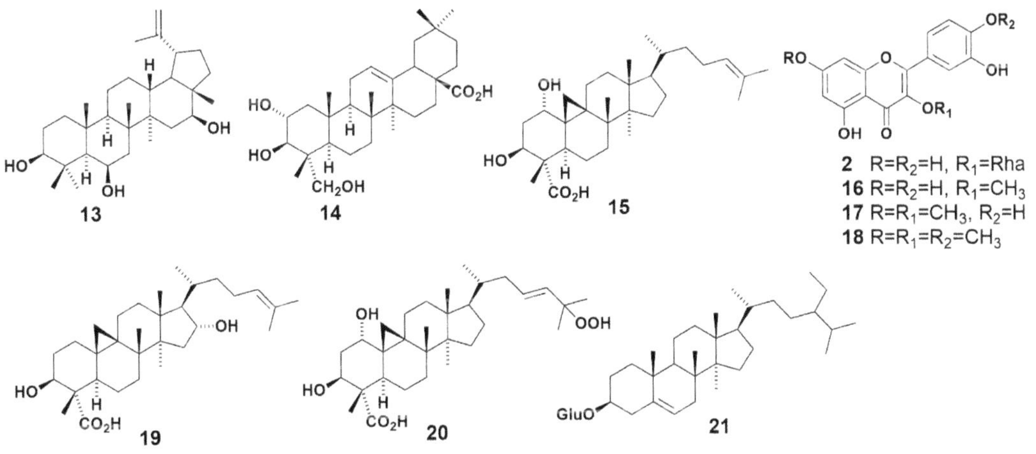

FIGURE 10.4   Chemical structures of the secondary metabolites from *C. leprosum*.

(Horinouchi et al., 2013). Additionally, the anti-inflammatory effect of the triterpene 3β,6β,16β-trihydroxylup-20(29)-ene (13; Figure 10.4), isolated from leaves of *C. leprosum*, was also demonstrated in mouse cutaneous wound healing model (Nascimento-Neto et al., 2015).

The ethanol extract obtained from roots of *C. leprosum* and its isolated compound arjunolic acid (14; Figure 10.4) showed the capacity of neutralizing critical points in the tissue damage process induced by *Bothrops jararacussu* and *B. jararaca* venoms. The results represented the first scientific validation for the popular use of the plant against snakebites in the Northeast region of Brazil (Fernandes et al., 2014).

The neuroprotective potential of the ethanol extract from *C. leprosum* to Parkinson's disease, a brain disorder associated with inflammatory processes, was investigated using a murine model induced by neurotoxin MPTB. The plant extract was able to improve motor deficits, by attenuating of the hyperlocomotion and similar to the control groups (no MPTB added). According to this study, the extract of the plant could be a new therapeutic approach for preventing Parkinson's disease, mainly by the preservation of the dopaminergic tonus (Moraes et al., 2016).

The antioxidant, cytotoxic and genotoxic properties of the ethanol extract from flowers of *C. leprosum* and is worth mentioning 5,3',4'-trihydroxy-3,7-dimethoxyflavone (17) and 5,3'-dihydroxy-3,7,4'-trimethoxyflavone (18) secondary metabolites, Figure 10.4, were also evaluated. In this case, the ethanol extract was mutagenic at high concentrations (500 μg mL^{-1}), while the flavonoid 5,3',4'-trihydroxy-3,7-dimethoxyflavone (17) induced an increase in DNA damage. Nevertheless, flavonoid 5,3'-dihydroxy-3,7,4'-trimethoxyflavone (18) showed a better antioxidant action with lower toxicity and absence of genotoxicity (Viau et al., 2016).

In summary, *C. leprosum* is a medicinal plant that occurs in the Northeast region of Brazil, especially in the state of Ceará, and which displays high ethnopharmacological RI of 1.1. This plant is used in folk medicine for treating inflammations, in which the crude extracts and isolated compounds were validated by pharmacological studies. Furthermore, the antiophidic potential of the ethanol extract also corroborated the popular use for treating snakebites.

10.4 *E. VISCOSA* LESS. (ASTERACEAE)

10.4.1 BOTANICAL ASPECTS AND OCCURRENCE

E. viscosa, one of the ten reported species of *Egletes*, is a medicinal flowering plant (wild and annual herb) native to the intertropical America. The species grows throughout the Northeast region of Brazil at margins of river banks and lakes as soon as the raining season ends. *E. viscosa* is popularly known as "macela" and sometimes as "macela-da-terra", the latter contrasting from *Achyrocline satureoides* (Lam.) DC. and *Chrysanthemum parthenium* L. (Berh.), which are native Asteraceae plants that occur in the South and Southeast regions of Brazil (Lorenzi, 2000).

10.4.2 ETHNOPHARMACOLOGY

Due to stomachic, antidiarrheal, emmenagogue and diaphoretic properties, the dried flower buds of *E. viscosa* are widely sold in herbal stores and supermarkets from the Northeast region of Brazil to be used in folk medicine as a tea or decoction (Braga, 1976; Corrêa, 1984).

10.4.3 CHEMICAL STUDIES

The major constituents of the flower buds of *E. viscosa* were reported (Cunha, 2003; Lima et al., 1996). Phytochemical investigation of the hexane extract of the plant yielded centipedic acid (22) and 12-acetoxy-hawtriwaic acid lactone (23), while the ethanol extract yielded ternatin (24), Figure 10.5 (Lima et al., 1996). Besides compounds 22-24, barbatol (25), tarapacol (26),

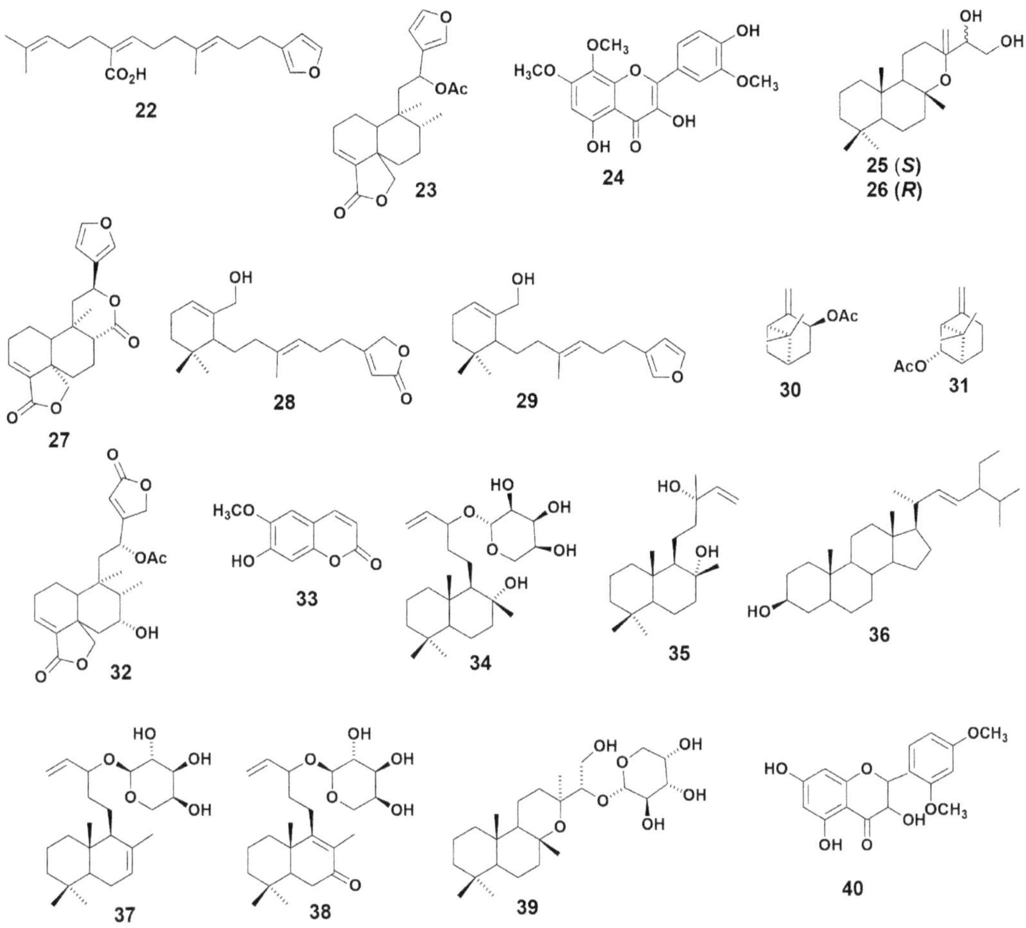

FIGURE 10.5   Chemical structures of the secondary metabolites from *E. viscosa*.

12-*epi*-bacchotricuneatin (**27**), 15,20-epoxy-9,10-seco-4,9(10),13(20),14-labda-tetraen-18-ol (**28**) and 9,10-*seco*-4,9(10),13(14)-labdatrien-15,20-olid-18-ol (**29**), Figure 10.5, were isolated from the chloroform extract of the plant (Cunha, 2003).

The essential oils from flower buds of *E. viscosa* specimens from different sites in the state of Ceará (Northeast of Brazil) were also investigated. The study revealed the existence of two chemotypes of the plant: *trans*-pynocarveil acetate (**30**) and *cis*-isopynocarveil acetate (**31**), Figure 10.5, based on the high content of these major volatile constituents in each essential oil (Cunha, 2003). From these findings, the chemical composition of lyophilized teas from the flower buds (cultivated specimens) of both chemotypes was investigated. From the *trans*-pynocarveil acetate chemotype compounds **22-24** were isolated, besides the diterpene 12-acetoxy-7-hydroxy-3,13(14)-clerodandien-18,19:15,16-diolide (**32**). The *cis*-isopinocarveyl chemotype yielded compounds **23-25**, **32** and scopoletin (**33**), Figure 10.5 (Vieira et al., 2006).

The chemical investigation of the aerial parts of *E. viscosa* led to the isolation of **24-26**, in addition to compounds 8α-hydroxylabd-14(15)-ene-13(*S*)-*O*-β-D-ribopyranoside (**34**), 13-*epi*-sclareol (**35**) and spinasterol (**36**), Figure 10.5 (Silva-Filho et al., 2007). The phytochemical study of the entire plant native from Peru revealed the presence of the following compounds in the plant: tarapacol (**26**), glycosides 13(*R*)-*O*-α-L-arabinopiranoside-13-hydroxy-labda-7,14-dien (**37**), 13*R*-*O*-α-L-arabinopyranoside-13-hydroxy-7-oxalabda-8,14-dien (**38**), tarapacol 14-*O*-α-L-arabinose (**39**) and 5,7-dihydroxy-2′,4′-dimethoxyflavonol (**40**), Figure 10.5 (Lee et al., 2005).

10.4.4 Pharmacological Studies

The essential oil from the flower heads of *E. viscosa* showed a significant dose-dependent analgesic activity induced by acetic acid and formalin tests, besides anticonvulsant activity induced by pentylenetetrazol in mice. The antibacterial activity of the oil was also reported against resistant strains of *Staphylococus aureus*, with the MIC ranging from 0.625 to 2.5 µL mL^{-1} (Souza, Rao et al., 1998).

Centipedic acid (**22**) and 12-acetoxy-hawtriwaic lactone (**23**) (Figure 10.5) were tested on acute and chronic models of mouse ear dermatitis induced by 12-*O*-tetradecanoylphorbol-13-acetate (TPA) and oxazolone. The assays indicated that these diterpenes are effective topical anti-inflammatory agents (Calou et al., 2008). Additionally, compound **23** also attenuated capsaicin-induced ear edema and hind paw nociception in mice, possibly involving capsaicin-receptive TRPV1 receptors, endogenous adenosine and ATP-sensitive potassium channels (Melo et al., 2006).

Centipedic acid (**22**; Figure 10.5) showed significant inhibition against the gastric ulceration induced by indomethacin, with the gastroprotective mechanism later elucidated using the mouse model of gastric mucosal damage induced by ethanol, indicating a cytoprotective role for the diterpene (Guedes et al., 2002, 2008).

The flavonoid ternatin (**24**; Figure 10.5) showed anti-inflammatory and anti-anaphylactic properties in modulating mouse passive cutaneous anaphylaxis (PCA) and rat carrageenan-induced pleurisy (Souza et al., 1992). The anti-inflammatory activity of **24** was reinforced when examined for the compound's possible influence on thioglycolate-elicited neutrophil influx into the rat peritoneal cavity *in vivo* and nitric oxide production in lipopolysaccharide (LPS)-activated mouse peritoneal macrophages *ex vivo* (Rao et al., 2003).

The hepatoprotective activity of ternatin (**24**) was investigated using different models (Rao et al., 1994, Rao et al., 1997; Souza, et al., 1997; Souza, et al., 1998). This flavonoid showed marked inhibition of CCl$_4$ serum enzymes and morbid histological changes when the model of liver injury induced by carbon tetrachloride in rats was used (Rao et al., 1994). Hepatoprotective activity was also confirmed through the antiperoxidative effect against lipid peroxidation. In this model, ternatin (**24**) showed significant inhibition in hind paw edema induced by adriamycin in a manner similar to vitamin E (Souza et al., 1997). Ternatin was also able to prevent significantly acetaminophen-induced acute increase in serum enzymes, inhibition of hepatocellular necrosis and bile duct proliferation induced by aflatoxin (Souza et al., 1998). The compound's gastroprotective and antidiarrheal properties were observed through significant inhibition on gastric mucosal damage induced by hypothermic restraint stress, ethanol and indomethacin, and in both intestinal transit and accumulation of intestinal fluids induced by castor oil, besides antagonized the contractile response evoked by acetylcholine, histamine, serotonin and barium chloride (Rao et al., 1997).

The antithrombotic activity of **24** was also evaluated on *in vitro* platelet aggregation induced by ADP and an *in vivo* mouse model of tail thrombosis induced by kappa-carrageenan (KC). Besides showing the ADP-induced platelet aggregation inhibition (concentration-dependent; IC$_{50}$ of 390 µM), the flavonoid showed marked protection of mice from thrombotic challenge with KC (Souza et al., 1994).

In summary, *E. viscosa* is a medicinal plant that occurs in the Northeast of Brazil (Caatinga bioma) known for stomachic, antidiarrheal, emmenagogue and diaphoretic properties. Dried flower buds are sold in herbal store and supermarkets, and local people use the plant as tea or decoction. The major chemical constituents from the flower buds extract of *E. viscosa*, centipedic acid (**22**), 12-acetoxy-hawtriwaic lactone acid (**23**) and ternatin (**24**), demonstrated several pharmacological activities related to gastrointestinal disorders, and these findings support the validation of the species ethnobotanical use.

10.5 *ERYTHRINA VELUTINA* WILLD. (FABACEAE)

10.5.1 BOTANIC ASPECTS AND OCCURRENCE

E. veluntina is one of the 110 species described for the *Erythrina* genus, which is distributed in tropical and subtropical regions worldwide. That about 70 of the *Erythrina* species are native to the Americas is noteworthy (Virtuoso et al., 2005). *E. veluntina* is a medium size tree plant endemic to the plain and riverbanks of the semi-arid regions of Northeast of Brazil, popularly known as "mulungu", "suinão", "canivete", "corticeira", "pau-de-coral", "sanaduí", "sananduva", and commonly used in gardens and parks as an ornamental plant (Lorenzi, 1992).

10.5.2 ETHNOPHARMACOLOGY

The aqueous extract or tincture from harvested stem bark of *E. velutina* have popular use against a number of central nervous system disorders, such as insomnia, convulsion, anxiety, nervous cough and rheumatism (Corrêa, 1984; Lorenzi and Matos, 2002). The dried fruit of the plant has local anesthetic action used in the form of cigarettes as a toothache remedy (Lorenzi and Matos, 2002).

10.5.3 CHEMICAL STUDIES

Chemical investigation of the ethanol extract from the stem bark of *E. velutina* yielded the flavonoids 4'-*O*-methylsigmoidin (**41**), eryvellutinone (**42**), homohesperetin (**43**), phaseolin (**44**), 5,7,3'-trihydroxy-5'-prenyl-6-methoxyisoflavone (**45**), phaseollidin (**46**), besides erythrodiol (**47**) and lupeol (**48**), Figure 10.6 (Cunha et al., 1996; Rabelo et al., 2001; Rodrigues, 2004).

Chemical studies of the methanol extract from seeds of *E. velutina* were also reported (Ozawa et al., 2008; Ozawa et al., 2009; Ozawa et al., 2011). The alkaloids hypaphorine (**49**), erysodine *N*-oxide (**50**), erythraline (**51**), 8-oxo-erythraline (**52**), erysotrine (**53**), erysodine (**54**), erysovine (**55**), glycoerysodine (**56**), sodium erysovine 15-*O*-sulfate (**57**), erysopine 15-*O*-sulfate (**58**), 16-*O*-β-glucopyranosyl coccoline (**59**), erymelanthine (**60**) and sodium erysovine *N*-oxy-15-*O*-sulfate (**61**), Figure 10.6, were isolated from this part of the plant.

10.5.4 PHARMACOLOGICAL STUDIES

The tranquilizing effects of the hydro-alcohol extracts from the stem bark of *E. velutina* were extensively investigated. The anxiolytic potential on the behavior of female mice submitted to the open-field tests and to elevated plus-maze after oral or intraperitoneal administration of the extract showed decreased locomotor activity and sedative effect (Vasconcelos et al., 2004). The effects of acute and chronic oral administration of the extract in rats submitted to the elevated T-maze model of anxiety suggested that the extract exerts anxiolytic-like effects on a specific subset of defensive behaviors, which have been associated with generalized anxiety disorder (Ribeiro et al., 2006). The extract also showed anticonvulsant effects in a strychnine-induced seizure model, with possible action in a glycine system and a potentiation of pentobarbital sleeping time, suggesting depressant action in the nervous central system (Vasconcelos et al., 2007). Studies with a mice model submitted to elevated plus-maze, forced swim, spontaneous locomotor activity and habituation of active chamber effects of acute and chronic treatments showed anxiolytic-like effect of the extract that could serve as a new approach for treatment anxiety, although at low doses the extract may have an amnesic effect (Raupp et al., 2008). Moreover, the neuroprotective property of the plant extract was suggested, reducing the excitatory amino acid concentrations and increased the inhibitory amino acid levels after ischemia (Rodrigues et al., 2017).

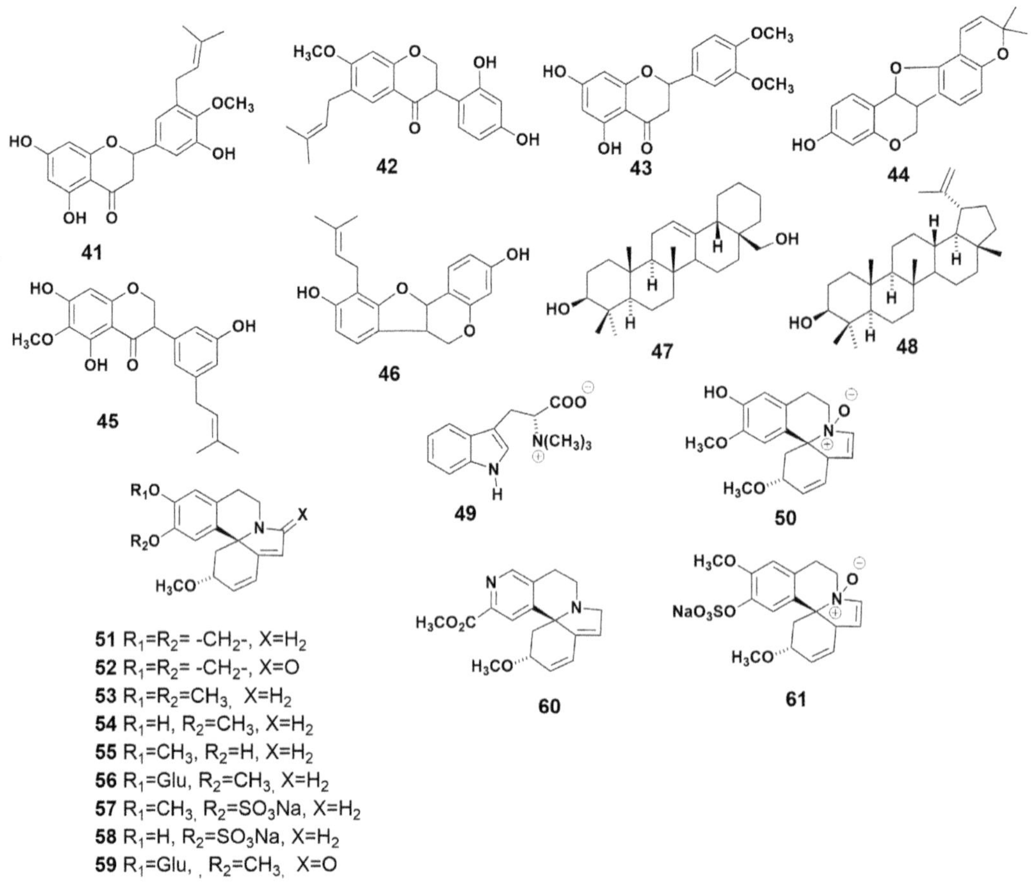

FIGURE 10.6 Chemical structures of the secondary metabolites from *E. velutina*.

The anti-inflammatory activity of the hydro-alcohol extract from the stem bark of *E. velutina* was also demonstrated in a dextran model. The extract decreased the paw edema at 1, 2, 3, 4 and 24 hours, probably by interfering in the inflammatory processes in which mast cells have an important role (Vasconcelos et al., 2011). Moreover, the antibacterial activity (disk diffusion method) of both the crude ethanol extract and hexane fraction from the plant against *S. aureus* and *Streptococcus pyogenes* was demonstrated in preliminary experiments (Virtuoso et al., 2005). A reasonable lack of toxicity of the aqueous extract of *E. velutina* leaves was observed on assays of acute toxicity in experimental animals (Craveiro et al., 2008).

An opiate-like analgesic effect of the aqueous extract from leaves in mice has been suggested, when the antinociceptive effect was reversed by pretreatment with the opiate antagonist naloxone (Dantas et al., 2004; Marchioro et al., 2005). However, the mechanism of action was elucidated definitively in experiments using terminal segments of the guinea pig ileum smooth muscle, the following was observed when contractile response involving $GABA_A$ receptor activation, acetylcholine release, muscarinic receptor activation, augmentation of Ca^+ entry through L-type calcium channels and calcium release from the intracellular stores (Carvalho et al., 2009). The extract also crossed the blood-brain barrier and inhibited cholinesterase in the central nervous system of mice, suggesting anticholinesterase activity in mouse brains (Santos et al., 2012).

The pharmacological activities of the alkaloids isolated from seeds of *E. velutina* were investigated (Ozawa et al., 2008; Ozawa et al., 2009). Hypaphorine (**49**; Figure 10.6) showed a significant increase on non-rapid eye movement (NREM) sleep time of mice during the first hour after treatment confirming the sleeping promoting property. The enhancing effects of the alkaloids erythraline (**51**), erysodine (**54**) and erysovine (**55**), Figure 10.6, on the cytotoxicity of TRAIL (tumor necrosis factor [TNF] related apoptosis-inducing ligand) against Jurkat cells were also reported. However, these compounds showed some cytotoxicity even when their activities were moderated in the absence of TRAIL. Compounds 8-oxo-erythraline (**52**), erysotrine (**53**) and glycoerysodine (**56**), Figure 10.6, exhibited no cytotoxicity by themselves, but they acted synergistically when combined by TRAIL. Erysotrine (**53**) showed the superior results among the alkaloids tested when combined with TRAIL, and no cytotoxicity by itself.

In summary, the pharmacological effects of the hydro-alcohol extract from stems bark of *E. velutina* in the central nervous system support the plant's popular use as tranquilizer in the Brazilian folk medicine. The low toxicity of the leaf extract and the antinociceptive, analgesic and anticholinesterase activities reported suggest new therapeutic uses for this plant.

10.6 *MYRACRODRUON URUNDEUVA* ALLEMÃO (ANACARDIACEAE)

10.6.1 BOTANICAL ASPECTS AND OCCURRENCE

Popularly known as "Aroeira-do-Sertão", *M. urundeuva* is a tree found in Brazil, mainly in the semi-arid vegetation of the Northeast region of Brazil (Cruz, 1979). The flowering period of the plant goes from June to August, when the tree is completely leafless, and the fruiting period occurs from August to November (Lorenzi and Matos, 2002).

10.6.2 ETHNOBOTANY AND ETHNOPHARMACOLOGY

Ethnobotanical records mentioned the innumerable uses of this plant. The bark is used for tanning of animal skins due to the high content of tannins. Additionally, the bark is considered one of the most resistant Brazilian woods, being used in civil construction girders and railway sleepers, with noteworthy emphasize to their resistance to crushing and to physical and biological agents, even in long-time contact with soil and water (Braga, 1976).

Ethnopharmacological studies refer to the use of the bark (deprived of inner bark), as one of the oldest plant remedies for gynecological problems used in folk medicine of Northeastern Brazil. The use in the oral treatment of diseases of the respiratory system, urinary tract, hemoptysis, metrorrhagia and diarrhea, in the form of infusions or decoctions is also reported. The plant has an excellent reputation in the home treatment of postpartum sequelae, and skin and mouth injury (oral and topical concomitant uses) (Braga, 1976).

10.6.3 CHEMICAL STUDIES

The chromatographic fractionation of the ethyl acetate extract from the inner bark of the plant yielded several fractions with pronounced anti-inflammatory activity. One of the bioactive fractions was composed by a mixture of the dimeric chalcones, urundeuvins A (**62**), B (**63**) and C (**64**), Figure 10.7. Additionally, tannins were identified as the main constituents in another active fraction (Bandeira, 2002). This preliminary work on *M. urundeuva* demonstrated that the therapeutic activity of this plant may be associated with a phytotherapeutic complex rather than simply a single active compound.

Liquid chromatography coupled to mass spectrometry (UPLC-ESI-QTOF MS/MS) analyses of the inner bark extract confirmed the presence of the chalcones **62-64** (Figure 10.7), besides other polyphenol compounds such as gallic acid derivatives tentatively identified (Galvão et al., 2018).

10.6.4 PHARMACOLOGICAL STUDIES

Studies on the medicinal properties of the inner bark of *M. urundeuva* were based on the vast folk-use of this plant (Bandeira, 2002). Both hydro-alcohol and aqueous extracts were subjected

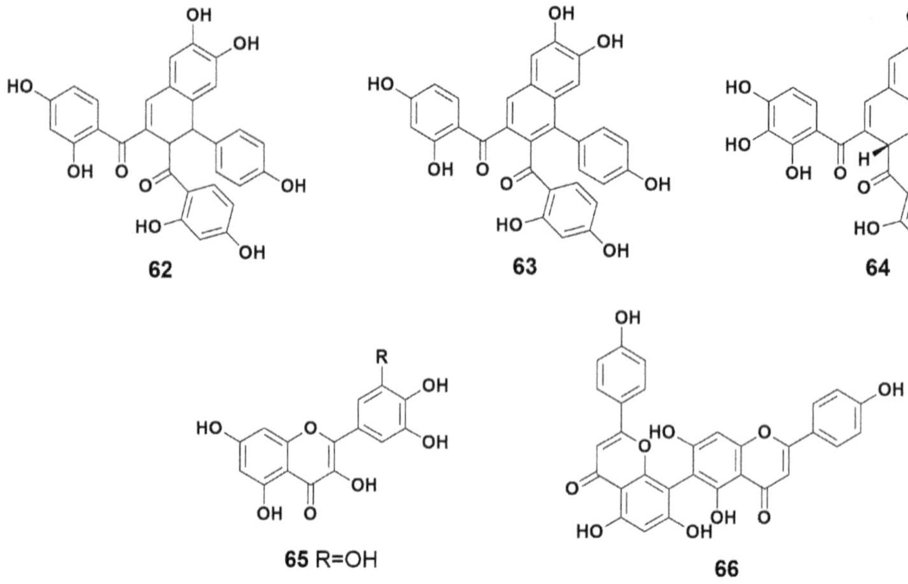

FIGURE 10.7   Chemical structures of the secondary metabolites from *M. urundeuva*.

to non-clinical pharmacological tests, and showed evident anti-inflammatory, analgesic, healing and antiulcer effects, together with antihistamine and anti-bradykinin actions. Toxicological studies have shown that these extracts are deprived practically of toxic effects orally, and point to the lack of teratogenic effects of the plant (Viana et al., 1995). However, when tested in pregnant rats, the extract induced bone malformations in the fetus. In this case, no anatomopathological alterations, decreased mating capacity or infertility were observed (Carlini et al., 2013).

Preliminary clinical studies in patients with a peptic ulcer and in patients with cervicitis and ectopia, using the experimental pharmaceutical preparations elixir and vaginal ointment, respectively, supported the clinical use of the plant in these pathologies (Bandeira, 2002; Viana et al., 1995).

The antiviral potential of the leaves from *M. urundeuva* was investigated (Cecílio et al., 2016). The flavonoid enriched fraction, obtained from the ethanol leaves' extract(adult tree), showed a pronounced action against rotavirus.

Studies revealed the pronounced anti-inflammatory, analgesic and antiulcerogenic activities of the fraction containing the dimeric chalcones urundeuvins A-C (**62-64**) (Albuquerque et al., 2004; Viana et al., 2003). This fraction was effective in reducing cell death induced by 6-hydroxydopamine (6-HODA), in addition to inhibiting lipid peroxidation and preventing 6-HODA-induced necrosis, suggesting that these compounds may be beneficial in neurodegenerative disorders, such as Parkinson's disease (Nobre-Júnior et al., 2009).

The anti-inflammatory and antiulcer properties of the enriched tannin fraction were evaluated through the formalin test in mice, in rats by the model of paw edema induced by carrageenan and in ulcer models (Souza et al., 2007). The results showed that this fraction significantly inhibited intraperitoneal indomethacin-induced gastric ulceration and significantly decreased gastric ulceration induced by indomethacin. The results showed that the tannins present in this species have anti-inflammatory and antiulcerogenic effects, characteristic and common to the tannins in general (Monteiro et al., 2005; Simões et al., 2016).

10.6.5   CHEMICAL AND PHARMACOLOGICAL STUDIES OF CULTIVATED *M. URUNDEUVA*

Due to the predatory extraction of the inner bark of *M. urundeuva* for medicinal purposes and the use of the wood in carpentry, this species is threatened with extinction (Galvão et al., 2018).

Since *M. urundeuva* exhibits secondary growth, the tissues of the inner bark and of the shoots exhibit intense metabolic activity; therefore, they must produce the same chemical constituents and show similar pharmacological actions. To prove this and to make a proposition to conserve this species, agronomic studies have been carried out integrating the pharmacological and chemical studies. These studies demonstrated that the cultivated species (shoots of 40 cm in height) maintains genetic characteristics regarding the pharmacological activity and produces qualitatively the same chemical markers of the inner bark, the dimeric chalcones, urundeuvines A (**62**), B (**63**) and C (**64**), Figure 10.7, and tannins (Bandeira, 2002; Souza et al., 2007). Therefore, representing a possible preservation and technological alternative to substitute the inner bark for shoots of the species studied.

Purification and isolation of the secondary metabolites from bioactive fractions resulted in the identification of the active dimeric chalcones **62** and **63** (Figure 10.7) from the 40 cm-tall shoot (Bandeira, 2002). Furthermore, the flavonoids quercetin (**5**), aromadendrinole (**65**) and agathisflavone (**66**), Figure 10.7, were identified in the leaves of the cultivated specimen (Bandeira, 2002). Flavonoids and chalcones have a known common biosynthetic precursor and both exhibit anti-inflammatory activity (Cecílio et al., 2016). Therefore, the presence of flavonoids enforces the medicinal use of the shoot leaves.

Similar to the inner bark of *M. urundeuva* (Bandeira, 2002), tannins were identified in all parts of the cultivated plant (stem and leaf). The quantification of total tannin content yielded 8.2% for the inner bark, 2.2% for shoot stems and 3.9% for shoot leaves (Aguiar, 2013). The data indicated that, for the substitution of the inner bark by the shoots' stem an amount of approximately four times greater of stem in relation to the inner bark is necessary. On the other hand, the high content of tannins in the shoots' leaves points to a possible medicinal use of this plant part (Aguiar, 2013). This result suggested that the young plant also has a phytotherapeutic complex, acting in an additive or synergistic form in the pharmacological effect (Carmona and Pereira, 2013).

Liquid chromatography coupled to mass spectrometry (UPLC-ESI-QTOF MS/MS) analyses of extracts from shoot stem and leaves of cultivated *M. urundeuva*, led to the tentative identification of 11 and 15 compounds, respectively (Aguiar Galvão et al., 2018). Noteworthy evidence demonstrated that the chemical profiles of these extracts were similar (in terms of chemical classes) to those from the inner bark of the non-cultivated plant. These findings reinforce the presence of a phytotherapeutic complex in *M. urundeuva* and point to the similarity between the chemical classes present in the different complex mixtures.

The neuroprotective activity of the fluid extracts prepared from stems and leaves of the cultivated *M. urundeuva* shoots in a Parkinson's disease model was evaluated (Calou et al., 2014). The oral extracts reversed the behavioral changes, as well as decreased dopamine and dihydrophenylacetic acid levels in rats, suggesting their potential in the prevention and treatment of neurodegenerative conditions. These effects are related, possibly, to the anti-inflammatory and antioxidant properties of the biologically active compounds presents in the plant shoots.

Studies on the thermal stability of the cultivated plant were performed aiming at the production of a vegetable drug from shoots of *M. urundeuva*. The plant's drying parameters were established to be performed in an oven with air circulation at 40°C in order to maintain higher levels of the chemical markers (dimeric chalcones and tannins). Residual moisture, total ash content and granulometry of the plantdrug's general quality specifications were possible to determine with the material dried at 40°C (Aguiar, 2013).

Gastroprotective and anti-inflammatory activities of the extracts (700 or 1,000 mg kg^{-1}) were assessed on ethanol-induced gastric lesions and croton oil-induced ear edema in rats, respectively. The pharmacological assays revealed that the fluid extracts obtained from shoot stems and leaves have pharmacological activities similar to that of the inner bark of the adult plant, as there was no significant difference between the treated groups. The insertion of shoot leaves, especially in the mixture with shoot stem (1:1) will be important not only from the pharmacochemical context, because flavonoids and tannins present in the extracts have relevant pharmacological actions (Hoensch and Oertel, 2015), but also from the point of view of the yield of the raw material. When deciding on the use of the leaves, a significant amount of green is possible to be generated (Aguiar Galvão et al., 2018).

In summary, the ethnopharmacological data of *M. urundeuva* are useful not only to direct the pharmacological studies on the inner bark but also as an approach to reverse the potential extinction of the plant. Chemical and pharmacological studies demonstrated that the therapeutic activity of this plant may be associated with the presence of a phytotherapeutic complex (dimeric chalcones and tannins) rather than simply an active compound. Recent studies suggest that the substitution of the inner bark of the adult tree by the stems and leaves of the shoots (cultivated plant), which preserve the same chemical and biological behavior of the adult specimen, might be an encouraging approach and worthy of further research

ACKNOWLEDGMENTS

The authors thank Fundação Cearense de Apoio a Pesquisa (FUNCAP; grant # PNP-0058-00137.01.00/11), Conselho Nacional do Desenvolvimento Científico e Tecnológico (CNPq; grant # 405001/2013-4) and Coordenação de Aperfeiçoamento de Pessoal de Nível Superior (CAPES) – Finance Code 001 for the financial support. M. C. F. Oliveira and M. A. S. Lima also thank CNPq for the research grants # 307667/2017-0 and 302804/2015-3, respectively. M. A. M. Bandeira thanks CNPq for the financial sponsor of the project "Production of shoots of aroeira do sertão (*Myracrodruon urudeuva* Allemão): reduction of anthropic pressure by predatory collection of bark, contribution to the species conservation and social inclusion of adolescents and their relatives" (grant # 557618/2009-6 – Call MCT/CNPq nº 29/2009 – Technological Extension for Social Inclusion/Announcement 29/2009 – Theme 2: Social Technologies for Agroecology).

REFERENCES

Agra, M. F.; Baracho, G. S.; Nurit, K.; Basílio, I. J. L. D.; Coelho, V. P. M. 2007. Medicinal and poisonous diversity of the flora of 'Cariri Paraibano', Brazil. Journal of Ethnopharmacology, 111, 383–395.

Aguiar, W. R. 2013. Desenvolvimento de técnicas farmacêuticas para obtenção de droga vegetal a partir de *Myracrodruon Urundeuva* Allemão (aroeira-do-sertão). Universidade Federal do Ceará.

Aguiar Galvão, W. R.; Braz Filho, R.; Canuto, K. M.; Ribeiro, P. R. V.; Campos, A. R.; Moreira, A. C. O. M.; Silva, S. O.; et al. 2018. Gastroprotective and anti-inflammatory activities integrated to chemical composition of *Myracrodruon urundeuva* Allemão – a conservationist proposal for the species. Journal of Ethnopharmacology, 222, 177–189.

Albuquerque, R. J. M.; Rodrigues, L. V.; Viana, G. S. B. 2004. Análise clínica e morfológica da conjuntivite alérgica induzida por ovalbumina e tratada com chalcona em cobaias. Acta Cirúrgica Brasileira, 19, 43–48.

Albuquerque, U. P.; Medeiros, P. M.; Almeida, A. L. S.; Monteiro, J. M.; Lins-Neto, E. M. F.; Melo, J. G.; Santos, J. P. 2007. Medicinal plants of the caatinga (semi-arid) vegetation of ne brazil: a quantitative approach. Journal of Ethnopharmacology, 114, 325–354.

Bandeira, M. A. M. 2002. Aroeira-do-sertão (*Myracrodruon Urundeuva* Fr. Allemão): constituintes químicos e ativos da planta em desenvolvimento e adulta. Universidade Federal do Ceará.

Braga, R. 1976. Plantas do nordeste, especialmente do ceará. 3a ed. Natal: Escola Superior de Agricultura de Mossoró.

Brígido, C. L. 2001. Estudo dos constituintes químicos de *Bauhinia cheilantha* (Bongard) Steudel. Universidade Federal do Ceará.

Calou, I.; Bandeira, M. A.; Aguiar-Galvao, W.; Cerqueira, G.; Siqueira, R.; Neves, K. R.; Brito, G. A.; Viana, G. S. B. 2014. Neuroprotective properties of a standardized extract from *Myracrodruon urundeuva* Fr. all. (aroeira-do-sertao), as evaluated by a Parkinson's disease model in rats. Parkinson's Disease, 2014, 11.

Calou, I. B. F.; Sousa, D. I. M.; Cunha, G. M. A.; Brito, G. A. C.; Silveira, E. R.; Rao, V. S.; Santos, F. A. 2008. Topically applied diterpenoids from egletes viscosa (asteraceae) attenuate the dermal inflammation in mouse ear induced by tetradecanoylphorbol 13-acetate- and oxazolone. Biological & Pharmaceutical Bulletin, 31, 1511–1516.

Carlini, E. A.; Duarte-Almeida, J. M.; Tabach, R. 2013. Assessment of the toxicity of the brazilian pepper trees *Schinus terebinthifolius* Raddi (aroeira-da-praia) and *Myracrodruon urundeuva* allemão (aroeira-do-sertão). Phytotherapy Research, 27, 692–698.

Carmona, F.; Pereira, A. M. S. 2013. Herbal medicines: old and new concepts, truths and misunderstandings. Brazilian Journal of Pharmacognosy, 23, 379–385.

Cartaxo, S. L.; Souza, M. M. A.; Albuquerque, U. P. 2010. Medicinal plants with bioprospecting potential used in semi-arid Northeastern Brazil. Journal of Ethnopharmacology, 131, 326–42.

Carvalho, A. C. C. S.; Almeida, D. S.; Melo, M. G. D.; Cavalcanti, S. C. H.; Marçal, R. M. 2009. Evidence of the mechanism of action of *Erythrina Velutina* Willd (Fabaceae) leaves aqueous extract. Journal of Ethnopharmacology, 122, 374–378.

Cecílio, A. B.; Oliveira, P. C.; Caldas, S.; Campana, P. R. V.; Francisco, F. L.; Duarte, M. G. R.; Mendonça, L. A. M.; de Almeida, V. L. 2016. Antiviral activity of *Myracrodruon urundeuva* against rotavirus. Brazilian Journal of Pharmacognosy, 26, 197–202.

Coe, H. H. G.; Souza, L. O. F. 2014. The Brazilian 'Caatinga': ecology and vegetal biodiversity of a semiarid region. In F. E. Greer, ed., Dry Forest: Ecology, Species Diversity and Sustainable Management, 1st ed. New York: Nova Science Publishers.

Corrêa, M. P. 1984. Dicionário das Plantas Úteis do Brasil e das Exóticas Cultivadas, vol. 5. Rio de Janeiro/Brasília: Imprensa Nacional Brasília.

Craveiro, A. C. S.; Carvalho, D. M. M.; Nunes, R. S.; Fakhouri, R.; Rodrigues, S. A.; Silva, F. T. 2008. Toxicidade aguda do extrato aquoso de folhas de erythrina velutina em animais experimentais. Brazilian Journal of Pharmacognosy, 18, 739–743.

Cruz, G. L. 1979. Dicionário das Plantas Úteis do Brasil, 1.ed. Rio de Janeiro: Editora Civilização Brasileira.

Cunha, A. N. 2003. Aspectos químicos do estudo estudo multidisciplinar (químico, farmacológico, botânico e agronômico) de *Egletes viscosa* Less. Universidade Federal do Ceará.

Cunha, E. V. L.; Dias, C.; Barbosa-Filho, J. M.; Gray, A. I. 1996. Eryvellutinone, an isoflavanone from the stem bark of *Erythrina vellutina*. Phytochemistry, 43, 1371–1373.

Dantas, M. C.; Oliveira, F. S.; Bandeira, S. M.; Batista, J. S.; Silva, C. D.; Alves, P. B.; Antoniolli, A. R.; Marchioro, M. 2004. Central nervous system effects of the crude extract of *Erythrina velutina* on rodents. Journal of Ethnopharmacology, 94, 129–133.

Facundo, V. A.; Andrade, C. H. S.; Silveira, E. R.; Braz-Filho, R.; Hufford, C. D. 1993. Triterpenes and flavonoids from *Combretum leprosum*. Phytochemistry, 32, 411–415.

Facundo, V. A.; Rios, K. A.; Medeiros, C. M.; Militão, J. S. L. T.; Miranda, A. L.P.; Epifanio, R. A.; Carvalho, M. P.; Andrade, A. T.; Pinto, A. C. 2005. Arjunolic acid in the ethanolic extract of *Combretum leprosum* root and its use as a potential multi-functional phytomedicine and drug for neurodegenerative disorders : anti-inflammatory and anticholinesterasic activities. Journal of the Brazilian Chemical Society, 16, 1309–1312.

Facundo, V. A.; Rios, K. A.; Moreira, L. S.; Militão, J. S. L. T.; Stebeli, R. G.; Braz-Filho, R.; Silveira, E. R. 2008. Two new cycloartanes from *Combretum leprosum* mart.(combretaceae). Revista Latinoamericana de Química, 36, 76–82.

Fernandes, F. F. A.; Tomaz, M. A.; El-Kik, C. Z.; Monteiro-Machado, M.; Strauch, M. A.; Cons, B. L.; Tavares-Henriques, M. S.; Cintra, A. C. O.; Facundo, V. A.; Melo, P. A. 2014. Counteraction of bothrops snake venoms by *Combretum leprosum* root extract and arjunolic acid. Journal of Ethnopharmacology, 155, 552–562.

Guedes, M. M.; Carvalho, A. C. S.; Lima, A. F.; Lira, S. R.; de Queiroz, S. S.; Silveira, E. R.; Santos, F. A.; Rao, V. S. 2008. Gastroprotective mechanisms of centipedic acid, a natural diterpene from *Egletes viscosa* LESS. Biological & Pharmaceutical Bulletin, 31, 1351–55.

Guedes, M. M.; Cunha, A. N.; Silveira, E. R.; Rao, V. S. 2002. Antinociceptive and gastroprotective effects of diterpenes from the flower buds of *Egletes viscosa*. Planta Medica, 68, 1044–1046.

Gutiérrez, I. E. M. 2010. Micropropagação de *Bauhinia cheilantha* (Bong.) Steud. (Fabaceae). Universidade Estadual de Feira de Santana.

Gutiérrez, I. E. M.; Nepomuceno, C. F.; Ledo, C. A. S.; Santana, J. R. F. 2011. Regeneração in vitro via organogênese direta de *Bauhinia cheilantha*. Ciência Rural, 41, 260–265.

Hoensch, H. P.; Oertel, R. 2015. The value of flavonoids for the human nutrition: short review and perspectives. Clinical Nutrition Experimental, 3, 8–14.

Horinouchi, C. D. S.; Mendes, D. A. G. B.; Soley, B. S.; Pietrovski, E. F.; Facundo, V. A.; Santos, A. R. S.; Cabrini, D. A.; Otuki, M. F. 2013. *Combretum leprosum* mart. (combretaceae): potential as an antiproliferative and anti-inflammatory agent. Journal of Ethnopharmacology, 145, 311–319.

Lee, D.; Li, C.; Graf, T. N.; Vigo, J. S.; Graham, J. C.; Cabieses, F.; Farnsworth, N. R.; et al. 2005. Diterpene glycosides from *Egletes viscosa*. Planta Medica, 71, 792–794.

Lima, G. R. M.; Sales, I. R. P.; Filho, M. R. D. C.; Jesus, N. Z. T.; Falcão, H. S.; Barbosa-Filho, J. M.; Cabral, A. G. S.; Souto, A. L.; Tavares, J. F.; Batista, L. M. 2012. Bioactivities of the genus *Combretum* (combretaceae): a review. Molecules, 17, 9142–9206.

Lima, M. A. S.; Silveira, E. R.; Marques, M. S. L.; Santos, R. H. A.; Gambardela, M. T. P. 1996. Biologically active flavonoids and terpenoids from *Egletes viscosa*. Phytochemistry, 41, 217–223.

Lorenzi, H. 1992. Arvores Brasileiras: Manual de Identificação e Cultivo de Plantas Arbóreas Nativas do Brasil, 1st ed. São Paulo: Editora Plantarum.

Lorenzi, H. 2000. Plantas Daninhas do Brasil: Terrestres, Aquáticas, Parasitas e Tóxicas, 3rd ed. Nova Odessa: Instituto Plantarum.

Lorenzi, H.; Matos, F. J. A. 2002. Plantas Medicinais No Brasil: Nativas e Exóticas Cultivadas, 2nd ed. São Paulo: Editora Plantarum.

Lucena, R. F. P.; Albuquerque, U. P.; Monteiro, J. M.; Almeida, C. F. C. B. R.; Florentino, A. T. N.; Ferraz, J. S. F. 2007. Useful plants of the semi-arid Northeastern region of Brazil – a look at their conservation and sustainable use. Environmental Monitoring and Assessment, 125, 281–90.

Luna, J. S.; Santos, A. F.; Lima, M. R. F.; Omena, M. C.; Mendonça, F. A. C.; Bieber, L. W.; Sant'Ana, A. E. G. 2005. A study of the larvicidal and molluscicidal activities of some medicinal plants from Northeast Brazil. Journal of Ethnopharmacology, 97, 199–206.

Maia, G. N. 2004. Caatinga: Árvores e Arbustos e Suas Utilidades, 1st ed. São Paulo: D&Z Computação gráfica e editora.

Marchioro, M.; Blank, M. F. A.; Mourão, R. H. V.; Antoniolli, A. R. 2005. Anti-nociceptive activity of the aqueous extract of *Erythrina velutina* leaves. Fitoterapia, 76, 637–642.

Melo, C. M.; Maia, J. L.; Cavalcante, I. J. M.; Lima, M. A. S.; Vieira, G. A. B.; Silveira, E. R.; Rao, V. S.; Santos, F. A. 2006. 12-Acetoxyhawtriwaic acid lactone, a diterpene from *Egletes viscosa*, attenuates capsaicin-induced ear edema and hindpaw nociception in mice: possible mechanisms. Planta Medica, 72, 584–589.

Monteiro, J. M.; Lins Neto, E. M. F.; Amorim, E. L. C.; Strattmann, R. R.; Araújo, E. L.; Albuquerque, U. P. 2005. Tannin concentration in three simpatric medicinal plants. Sociedade de Investigações Florestais, 29, 999–1005.

Moraes, L. S.; Rohor, B. Z.; Areal, L. B.; Pereira, E. V.; Santos, A. M. C.; Facundo, V. A.; Santos, A. R. S.; Pires, R. G. W.; Martins-Silva, C. 2016. Medicinal plant *Combretum leprosum* Mart ameliorates motor, biochemical and molecular alterations in a Parkinson's disease model induced by MPTP. Journal of Ethnopharmacology, 185, 68–76.

Nascimento-Neto, L. G.; Evaristo, F. F. V.; Alves, M. F. A.; Albuquerque, M. R. J. R.; Santos, H. S.; Bandeira, P. N.; Arruda, F. V. S.; Teixeira, E. H. 2015. Effect of the triterpene 3β, 6β, 16β-trihydroxylup-20(29)-ene isolated from the leaves of *Combretum leprosum* mart. on cutaneous wounds in mice. Journal of Ethnopharmacology, 171, 116–120.

Nobre-Júnior, H. V.; Oliveira, R. A.; Maia, F. D.; Nogueira, M. A. S.; Moraes, M. O.; Bandeira, M. A. M.; Andrade, G. M.; Viana, G. S. B. 2009. Neuroprotective effects of chalcones from *Myracrodruon urundeuva* on 6-hydroxydopamine-induced cytotoxicity in rat mesencephalic cells. Neurochemical Research, 34, 1066–1075.

Nunes, P. H. M.; Cavalcanti, P. M. S.; Galvão, S. M. P.; Martins, M. C. C. 2009. Antiulcerogenic activity of *Combretum leprosum*. Pharmazie, 64, 58–62.

Oliveira, A. M. F. 2008. Estudo Químico e avaliação da atividade hipoglicemiante e antioxidante de *Bauhinia cheilantha* (Bong.) Steudel. Universidade Federal da Paraíba.

Ozawa, M.; Etoh, T.; Hayashi, M.; Komiyama, K.; Kishida, A.; Ohsaki, A. 2009. TRAIL-enhancing activity of erythrinan alkaloids from *Erythrina velutina*. Bioorganic and Medicinal Chemistry Letters, 19, 234–236.

Ozawa, M.; Honda, K.; Nakai, I.; Kishida, A.; Ohsaki, A. 2008. Hypaphorine, an indole alkaloid from *Erythrina velutina*, induced sleep on normal mice. Bioorganic & Medicinal Chemistry Letters, 18, 3992–3994.

Ozawa, M.; Kishida, A.; Ohsaki, A. 2011. *Erythrinan* alkaloids from seeds of *Erythrina velutina*. Chemical and Pharmaceutical Bulletin, 59, 564–567.

Prado, D. E. 2003. As caatingas da América do sul. In I. R. Leal; M. Tabarelli; J. M. C. da Silva, eds., Ecologia e Conservação da Caatinga, 1st ed. Recife: Editora Universitária da UFPE.

Rabelo, L. A.; Agra, M. F.; Cunha, E. V. L.; Silva, M. S.; Barbosa-filho, J. M. 2001. Homohesperetin and phaseollidin from *Erythrina velutina*. Biochemical Systematics and Ecology, 29, 543–544.

Rao, V. S.; Figueiredo, E. G.; Melo, C. L.; Viana, G. S. B.; Menezes, D. B.; Matos, M. S. F.; Silveira, E. R. 1994. Protective effect of ternatin, a flavonoid isolated from *Egletes viscosa* less. in experimental liver injury. Pharmacology, 48, 392–397.

Rao, V. S.; Paiva, L. A. F.; Souza, M. F.; Campos, A. R.; Ribeiro, R. A.; Brito, G. A. C.; Teixeira, M. J.; Silveira, E. R. 2003. Ternatin, an anti-inflammatory flavonoid, inhibits thioglycolate-elicited rat peritoneal neutrophil accumulation and LPS-activated nitric oxide production in murine macrophages. Planta Medica, 69, 851–853.

Rao, V. S.; Santos, F. A.; Sobreira, T. T.; Souza, M. F.; Melo, C. L.; Silveira, E. R. 1997. Investigations on the gastroprotective and antidiarrhoeal properties of ternatin, a tetramethoxyflavone from *Egletes viscosa*. Planta Medica, 63, 146–149.

Raupp, I. M.; Sereniki, A.; Virtuoso, S.; Ghislandi, C.; Cavalcanti e Silva, E. L.; Trebien, H. A.; Miguel, O. G.; Andreatini, R. 2008. Anxiolytic-like effect of chronic treatment with *Erythrina velutina* extract in the elevated plus-maze test. Journal of Ethnopharmacology, 118, 295–299.

Ribeiro, M. D.; Onusic, G. M.; Poltronieri, S. C.; Viana, M. B. 2006. Effect of *Erythrina velutina* and *Erythrina mulungu* in rats submitted to animal models of anxiety and depression. Brazilian Journal of Medical and Biological Research, 39, 263–270.

Rodrigues, A. C. P. 2004. Contribuição ao conhecimento químico de plantas do nordeste: estudo quimico e farmacoloógico de *Anona squamosa* e *Erythrina velutina* Willd. Universidade Federal do Ceará.

Rodrigues, F. T. S.; Sousa, C. N. S.; Ximenes, N. C.; Almeida, A. B.; Cabral, L. M.; Patrocínio, C. F. V.; Silva, A. H.; et al. 2017. Effects of standard ethanolic extract from *Erythrina velutina* in acute cerebral ischemia in mice. Biomedicine and Pharmacotherapy, 96, 1230–1239.

Santos, J. C.; Leal, I. R.; Almeida-Cortez, J. S.; Fernandes, G. W.; Tabarelli, M. 2011. Caatinga: the scientific negligence experienced by a dry tropical forest. Tropical Conservation Science, 4, 276–286.

Santos, W. P.; Carvalho, A. C. S.; Estevam, C. S.; Santana, A. E. G.; Marçal, R. M. 2012. *In vitro* and *ex vivo* anticholinesterase activities of *Erythrina velutina* leaf extracts. Pharmaceutical Biology, 50, 919–924.

Silva, A. M. O.; Teixeira-Silva, F.; Nunes, R. S.; Marçal, R. M.; Cavalcanti, S. C. H.; Antoniolli, A. R. 2005. Antinociceptive activity of the aqueous extract of *Bauhinia cheilantha* (bong.) Steud. (leguminosae: caesalpinioideae). Biologia Geral e Experimental, 5, 10–15.

Silva, K. L.; Cechinel-Filho, V. 2002. Plantas do gênero bauhinia: composição química e potencial farmacológico. Química Nova, 25, 449–454.

Silva, M. I. G.; Melo, C. T. V.; Vasconcelos, L. F.; Carvalho, A. M. R.; Sousa, F. C. F. 2012. Bioactivity and potential therapeutic benefits of some medicinal plants from the caatinga (semi-arid) vegetation of Northeast Brazil: a review of the literature. Brazilian Journal of Pharmacognosy, 22, 193–207.

Silva-Filho, F. A.; Lima, M. A. S.; Bezerra, A. M. E.; Braz Filho, R.; Silveira, E. R. 2007. A labdane diterpene from the aerial parts of *Egletes viscosa* less. Journal of the Brazilian Chemical Society, 18, 1374–1378.

Simões, C. M. O.; Schenkel, E. P.; Mello, J. C. P.; Mentz, L. A.; Petrovick, P. R. 2016. Farmacognosia: do Produto Natural ao Medicamento, 1st ed. Porto Alegre: Artmed Editora.

Sinou, C.; Forest, F.; Lewis, G. P.; Bruneau, A. 2009. The genus *Bauhinia* s.l. (leguminosae): a phylogeny based on the plastid *trn*L–*trn*F region. Botany, 87, 947–960.

Souza, M. F.; Cunha, G. M. A.; Fontenele, J. B.; Viana, G. S. B.; Rao, V. S.; Silveira, E. R. 1994. Antithrombotic activity of ternatin, a tetramethoxy flavone from *Egletes viscosa* less. Phytotherapy Research, 8, 478–481.

Souza, M. F.; Rao, V. S.; Silveira, E. R. 1992. Anti-anaphylactic and anti-inflammatory effects of ternatin, a flavonoid isolated from *Egletes viscosa* less. Brazilian Journal of Medical and Biological Research, 25, 1029–1032.

Souza, M. F.; Rao, V. S.; Silveira, E. R. 1997. Inhibition of lipid peroxidation by ternatin, a tetramethoxyflavone from *Egletes viscosa* L. Phytomedicine, 4, 27–31.

Souza, M. F.; Rao, V. S.; Silveira, E. R. 1998. Prevention of acetaminophen-induced hepatotoxicity by ternatin, a bioflavonoid from *Egletes viscosa* less. Phytotherapy Research, 12, 557–561.

Souza, M. F.; Santos, F. A.; Rao, V. S.; Sidrim, J. J. C.; Matos, F. J. A.; Machedo, M. I. L.; Silveira, E. R. 1998. Antinociceptive, anticonvulsant and antibacterial effects of the essential oil from the flower heads of *Egletes viscosa* L. Phytotherapy Research, 12, 28–31.

Souza, S. M. C.; Aquino, L. C. M.; Milach, A. C.; Bandeira, M. A. M.; Nobre, M. E. P.; Viana, G. S. B. 2007. Antiinflammatory and antiulcer properties of tannins from *Myracrodruon urundeuva* allemão (anacardiaceae) in rodents. Phytotherapy Research, 21, 220–225.

Teixeira, D. C.; Farias, D. F.; Carvalho, A. F. U.; Arantes, M. R.; Oliveira, J. T. A.; Sousa, D. O. B.; Pereira, M. L.; Oliveira, H. D.; Andrade-Neto, M.; Vasconcelos, I. M. 2013. Chemical composition, nutritive value, and toxicological evaluation of *Bauhinia cheilantha* seeds: a legume from semiarid regions widely used in folk medicine. BioMed Research International, 2013, 1–7.

Vasconcelos, S. M. M.; Lima, N. M.; Sales, G. T. M.; Cunha, G. M. A.; Aguiar, L. M. V.; Silveira, E. R.; Rodrigues, A. C.; et al. 2007. Anticonvulsant activity of hydroalcoholic extracts from *Erythrina velutina* and *Erythrina mulungu*. Journal of Ethnopharmacology, 110, 271–274.

Vasconcelos, S. M. M.; Macedo, D. S.; Melo, C. T. V.; Monteiro, A. P.; Cunha, G. M. A.; Sousa, F. C. F.; Viana, G. S. B.; Rodrigues, A. C. P.; Silveira, E. R. 2004. Central activity of hydroalcoholic extracts from *Erythrina velutina* and *Erythrina mulungu* in mice. Journal of Pharmacy and Pharmacology, 56, 389–393.

Vasconcelos, S. M. M.; Sales, G. T. M.; Lima, N.; Lobato, R. F. G.; Macêdo, D. S.; Barbosa-filho, J. M.; Leal, L. K. A. M.; et al. 2011. Anti-inflammatory activities of the hydroalcoholic extracts from *Erythrina velutina* and *Erythrina mulungu* in mice. Brazilian Journal of Pharmacognosy, 21, 1155–1158.

Viana, G. S. B.; Bandeira, M. A. M.; Matos, F. J. A. 2003. Analgesic and antiinflammatory effects of chalcones isolated from *Myracrodruon urundeuva* Allemão. Phytomedicine, 10, 189–195.

Viana, G. S. B.; Leal, L. K. A. M.; Vasconcelos, S. M. M. 2013. Plantas Medicinais Da Caatinga: Atividades Biológicas e Potencial Terapêutico, 1st ed. Fortaleza: Expressão Gráfica e Editora.

Viana, G. S. B.; Matos, F. J. A.; Bandeira, M. A. M.; Rao, V. S. 1995. Aroeira-Do-Sertão: Estudo BotâNico, Farmacognóstico, Químico e Farmacológico, 2nd ed. Fortaleza: Edições UFC.

Viau, C. M.; Moura, D. J.; Pflüger, P.; Facundo, V. A.; Saffi, J. 2016. Structural aspects of antioxidant and genotoxic activities of two flavonoids obtained from ethanolic extract of *Combretum leprosum*. Evidence-Based Complementary and Alternative Medicine, 2016, 10.

Vieira, G. A. B.; Lima, M. A. S.; Bezerra, A. M. E.; Silveira, E. R. 2006. Chemical composition of teas from two cultivated chemotypes of *Egletes viscosa* ('macela-da-terra'). Journal of the Brazilian Chemical Society, 17, 43–47.

Virtuoso, S.; Davet, A.; Dias, J. F. G.; Cunico, M. M.; Miguel, M. D.; Oliveira, A. B.; Miguel, O. G. 2005. Estudo preliminar da atividade antibacteriana das cascas de *Erythrina velutina* willd.; fabaceae (leguminosae). Brazilian Journal of Pharmacognosy, 15, 137–142.

11 Natural Products Structures and Analysis of the Cerrado Flora in Goiás

Lucilia Kato[a], Vanessa Gisele Pasqualotto Severino[a],
Aristônio Magalhães Teles[b], Aline Pereira Moraes[a],
Vinicius Galvão Wakui[a],
Núbia Alves Mariano Teixeira Pires Gomides[c],
Rita de Cássia Lemos Lima[d], and Cecilia Maria Alves de Oliveira[a]
[a]Instituto de Química, [b] Instituto de Ciências Biológicas,
Universidade Federal de Goias, Goiânia, Brazil
[c]Unidade Acadêmica Especial de Biotecnologia,
Universidade Federal de Goias, Catalão, Brazil
[d]Department of Drug Design and Pharmacology,
University of Copenhagen, Copenhagen, Denmark

CONTENTS

11.1 THE BRAZILIAN CERRADO

The Cerrado is the second largest biome of South America, occupying more than 20% of the Brazilian territory (MMA, 2018). In terms of area, only the vegetation formation in Brazil of the Amazon forest exceeds the Cerrado territory (Ratter et al., 1997). This biome naturally predominates almost exclusively in the Central Plateau of Brazil, although other biomes also are present, but in a smaller proportion and covers about 2 million km^2 (Ratter et al., 1997).

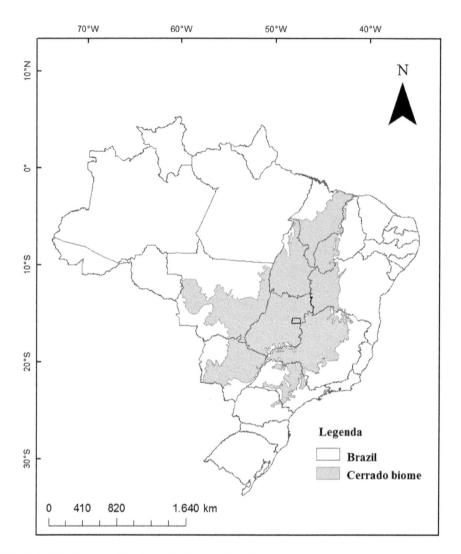

FIGURE 11.1   Distribution of the Cerrado biome in Brazil.

The Cerrado covers a continuous area in the states of Goiás, Tocantins, Mato Grosso, Mato Grosso do Sul, Minas Gerais, Bahia, Maranhão, Piauí, Rondônia, Paraná, São Paulo and Federal District (Figure 11.1) (Ribeiro and Walter, 2008). Small outlying distinct areas occur in Amapá, Roraima, and Amazonas States (Eiten, 1972). The Cerrado is considered one of the 36 most important terrestrial hotspots of the world due to its high diversity and endemic species, coupled with the strong threat, especially anthropic, (Carmignotto et al., 2012; CEPF, 2018; Myers et al., 2000). This biome is still listed as a World Natural Heritage Site by the United Nations Educational, Scientific and Cultural Organization (UNESCO, 2018).

Although the Cerrado is considered a "vegetation complex" with a phytophysiognomic and eco-logic relation with other American tropical savanna, and from other continents such as Africa and Australia, the species have their own unique characteristics (Walter et al., 2008). The vegetation presents numerous distinct phytophysiognomies (Figure 11.2), ranging from dense grassland, with a sparse covering of shrubs and small trees, to nearly closed woodland, which may or may not be associated with watercourses (Ratter et al., 1997).

The flora is surprisingly rich and presents characteristics, which are specific from the adja-cent biomes, although some physiognomies share species with other biomes (Oliveira-Filho and

FIGURE 11.2 Main phytophysiognomies of Cerrado biome found in Goiás state, Brazil. (**A**) Campo Limpo (Clean Field). (**B**) Campo Sujo (Dirty Field). (**C**) Vereda (Brazilian Palm Swamps). (**D**) Mata Ciliar (Riparian Forest). (**E**) Campo Rupestre (Rupestrian Field). (**F**) Mata Seca (Dry Forest).

Ratter, 1995). In terms of biologic diversity, the Brazilian Cerrado is considered the richest savanna in the world with 11,627 species of vascular plants (Mendonça et al., 2008). According to Mendonça et al. (2008), the families of seed plants richest in species number in Cerrado are Leguminosae (1,174 spp.), Asteraceae (1,074 spp.) and Orchidaceae (666 spp.).

The Cerrado biome is predominant in Goiás state (Figure 11.1). The Flora do Brasil (2020) (in construction) reporting on flora of Cerrado of Goiás an amount of 5,766 species of seed plants (angiosperms and gymnosperms) grouped in 1,210 genera and 165 families. These numbers correspond to 51% of the species of seed plants reported in the Cerrado biome. This review presents an outlook into future perspectives of research in natural products from the Cerrado in Goiás. Considering how the development of state-of-the-art strategies has emerged during the past decade, this allows us to assess and better understand the chemical biodiversity from Cerrado flora. This research includes *the identification of active metabolites targeting Rubiaceae species*, which are recognized as a source of potential active alkaloids. The research studies include DESI-MSI, a rapid and efficient approach to obtain mass spectrometric imaging directly from samples, without extraction being employed to accelerate the chemical investigation of plant species directly from leaves.

11.2 ETHNOBOTANICAL STUDIES

The use of herbal medicine is increasing globally, since many species from Cerrado flora are used in popular medicine for treatment of a broad set of diseases. The flora is an important part of health care, mainly for rural areas in the country. Therefore, traditional medicine has been among the several targeted strategies to choose a plant for study with one of the approaches being Ethnobotany.

11.2.1   The Significance of Ethnobotany to the Research in Natural Products Chemistry

Ethnobotany can be defined as the study of interrelations between human beings and nature (Alcorn, 1995; Alexiades and Sheldon, 1996). Furthermore, this considers both how some social groups classify plants and employs them (Di Stasi, 1996). Another inclusive characteristic from this scientific program is the concern in relation to the planning of the strategies of biological conservation, which may also merge human development with cultural survival.

The interaction between human beings and natural resources leads to the ethnobotanical knowledge, which can be noted in everyday life by means of actions that support knowhow, in several human activities, for instance, fishing, cattle farming, agriculture, religious parties, bathing in rivers, going to the field, calm talk and meal preparation (Guarim Neto, 2006).

The comprehension of use of specific plants is achieved when ecological, chemical and cultural aspects are studied. They can be represented through language, human cognition, cultural history, beliefs, religions, social networks and access to information (Maffi, 2005).

Moreover, religious and cultural manifestations help in the selection and confirmation of plant species, which present some potential for various uses. For this reason alone, we believe many researchers ought to respect and validate the traditional knowledge. Undoubtedly, this is a great source for researchers and future generation (Radomski, 2003). Finally, they should focus on the devolution of information that was delivered by members for the community itself.

However, one of the commitments of Ethnobotany should be a two way road of sharing traditional knowledge that generates scientific findings and partaking with those communities that contribute to improve the quality of life of those populations (Lima, 1996).

Medicinal plants used by some communities can provide important information with environmental and sustainability studies and biotechnological developments, and among them, there are pharmacology, chemistry, agronomy considerations. Therefore, working together with the scientific community much traditional knowledge will be recovered (Brasileiro et al., 2008).

Moreover, the more information recorded on vplant species which becomes available, the more the research can deliver some findings, which are identified as a means of assessing efficacy and safety in relation to their use. Considering the important impact of the pharmaceutical and cosmetic industry, for example, traditional knowledge on the use of medicinal plants may support the development of new products through rational research, which seeks to find new lead substances. The cost-benefit may be improved by approximately 50 times (Di Stasi, 1996). Thereupon, the research starts by means of a planning based on empirical knowledge that has been established through a consistent practice; however, that should be evaluated on a scientific basis (Brasileiro et al., 2008).

An outstanding amount of members of the community do not know the benefits and dangers of all medicinal plants. Among species that are employed and indicated by society, there are some that may frequently be used for more than one illness. Several herbs can also be used in an isolated manner or in combination in order to handle a specific problem. The selection of a medicinal plant is made by means of combination between experience and belief (Pasa, 2011).

Therefore, by considering folk therapeutic knowledge into account, many species can be identified through classes related to systems in accordance with the International Statistical Classification of Diseases and Related Health Problems (ICD) (2018), which is recognized by the World Health Organization (WHO, 2018).

Together with an ethnobotanical survey, quantitative data can be achieved, for example, the relative importance (RI) of each species, the informant consensus factor (ICF) and the relative frequency citation (RFC), which are presented as chief indicators in relation to the selection of promising species to be studied in the light of chemistry and biology.

The RI shows its preponderance, based on the number of medicinal properties, which are attributed to similar aspects by inquiry. The maximum value of RI is 2. The latter expresses which vegetal species demonstrate great versatility relative to their uses by the indigenous population. According to Bennett and Prance (2000), that number can be acquired through the formula:

$$IR = NSC + NP$$

where,

NSC: the number of body systems treated by a given species (NCSS) divided by the total number of body systems treated by the most versatile species (NSCSV);

NP: relationship between the number of properties attributed to a species (NPS) divided by the total number of properties attributed to the most versatile species (NPSV).

Through the ICF it is possible to identify the body systems, which present the highest local RI by means of popular appointment in key categories (Trotter and Logan, 1986). Accordingly, the formula for the factor is:

$$ICF = nur - nt/nur - 1$$

In this regard, nur: number of use citatîns made by inquirers to a category of ailments; nt: number of species used.

The maximum value of ICF is 1, when there is consensus among inquirers about medicinal plants within the ailment category.

The RFC indicates how a given species can be highlighted in relation to the others (Begossi, 1996), indicating, however, its importance to the community under study. RFC may be derived from the formula RFC = FC/N (0 < RFC < 1).

In this case, FC: the Frequency of Citation (FC) is the number of inquirers that speaks about the use of species. Beyond this aspect, N: the total number of informants who participate in ethnobotanical survey, not considering some categories of use.

In this context, the study of listed plant species can be investigated through recovering and recording of popular knowledge, since those species have not only a great pharmacological, food, agronomical potential, but also the presence of substances with structural variety and diversity.

11.2.2 THE CONTEXT OF ETHNOBOTANY IN BRAZIL AND GOIÁS STATE

The ethnobotanical outlook in Latin America is characterized by a far-reaching quantity of research, which are led by foreign researchers. Nevertheless, the substantial quantity of research made by national researchers appears as a positive feature for Brazil and consolidates Ethnobotany into the Brazilian scientific community (Oliveira et al., 2009).

Through a survey made by Oliveira et al. (2009), in a database from Lattes Platform, from the Ministry of Science and Technology and the National Council for Scientific and Technological Development, using the word "Etnobotânica", studies by 469 researchers with a PhD degree and 964 researchers with masters, graduates, students and technical students were evaluated.

The authors who wrote this chapter completed a search, using the same keyword, in March 2018, in Lattes Platform. Thereupon, a big increase over ten years of PhDs (1,692) and other researchers (2,284) were uncovered. Therefore, throughout the past nine years, the number of Brazilian scientists who are involved with this subject has increased considerably, and hence a new ongoing tendency of ethnobotanical studies was revealed.

In Brazil, biological diversity represents an aspect that explains the predominance of medicinal plants, which can be estimated at 46,000 vegetable species that are known, including algae, bryophytes, ferns, lycophytes, fungi, angiosperms and gymnosperms. The latter correspond to 72% of the total species (Flora do Brasil, 2020 [unpublished]).

In particular, the Goiás state population could be expressed by multiple origin and culture as people from rural communities, the "quilombola" families, Afro-Brazilians, "caiçaras", among others whose identities are linked to the flora of bioma Cerrado, with a wide use of medicinal plants, which have a great level of endemism, because 4,400 vegetal species, among 10,000 known, are endemic, representing 1,5% of plants around the world (Novaes et al., 2013). This context captures interest from Brazilian researchers in relation to knowledge and ways of conducting research.

Thus, some examples of organizations in Goiás will be listed, such as The Moinho Community in Alto Paraíso, Kalunga in Cavalcante, Monte Alegre and Teresina, Flores Velha in Flores de Goiás and Rufino Francisco in Niquelândia city.

Moreover, a known fact is that the Southeastern region of Goiás state contains about 21 rural communities, which present a form of organization based on family farming and the use of natural available resources (Mendes, 2005). In particular, one of these rural communities is the so-called Coqueiros, located in Catalão, where knowledge about medicinal plants is very important to this research. This community has been a relevant contributor for the study of medicinal plants performed by a specific group from the Laboratory of Natural Products and Organic Synthesis, from the Institute of Chemistry, which is part of the Federal University of Goiás.

11.2.3 THE COQUEIROS COMMUNITY

The existence of small agglomerations that are concentrated can be identified as a rural community, in which tradition, moral, ethical and religious values safeguard them, constituting the local residents' life and permeating a history that combines land, work, family and daily routine (Mendes, 2005).

Within the municipality of Catalão, in Goiás state, families who went to work in the construction of the railroad around the end of the nineteenth century established a community region, seeking a better quality of life.

There was the appearance of communities in the Catalão region, where the first local residents of the Coqueiros community were migrants (Mendes, 2005). The members of Arcanjo and Tomés families were the first local residents, aside from the Duarte family, which came from Portugal and Minas Gerais. The members of these families married among themselves and then a new family was constituted. As a result, almost all members have a family relationship and contribute to the transfer of intrinsic knowledge throughout the community.

First and foremost, families from the Coqueiros community were countless, and they had an average of nine children per couple. Considering the way of work on the farm, fathers looked after cattle and crops, while mothers were responsible for undertaking domestic activities, educating children and other works, such as producing flour, tapioca flour and growing vegetables.

In these communities, sons began to work by the age of eight. They worked with their parents. Girls worked with their mothers, helping them with everyday tasks. Therefore, the community's formation has deep roots, the value of work, as well as the relation of love and respect relative to the land, a tradition handed down from father to son.

The children's dedication could be identified as a tool to be employed in the structuring of customs linked to many aspects of work in the countryside. This idea presents an argument for the cultural formation of honest people. This social organization guaranteed that both knowledge and the way of life were safeguarded and passed on to future generations.

In the current context, where this rural community integrates, there is now a drop in the number of children, averaging two-three per couple. Younger children do not expect to stay in the field, because of low income. Consequently, they move to the city, seeking some job opportunity and student improvement (Mendes, 2005). Notwithstanding this migration, the community's people have aged (Mendes, 2005).

In this respect, in the Coqueiros community, former knowledge tends to get lost, because younger generations have other interests and way of life, which are different from their parents. In face of this, Ethnobotany is a crucial tool to recover such knowledge, highlighting the use of medicinal plants, which can be recorded though interviews and participation in the population's daily activities.

FIGURE 11.3 A photo of a household from the Coqueiros community (**3a**). The second photo presents some medicinal plants grown in farmyards from the Coqueiros community (**3b**). (Photo taken from authors who wrote this chapter.)

The Coqueiros community is formed of 38 homes, where midwives, people who bless, and woods-men are living. These social actors have a great medicinal knowledge and they are very important to the Coqueiros community's history. Recording their botanical knowledge retrieved from them to be investigated by the scientific community and returned to the community is essential.

The Coqueiros community (Figure 11.3) is located in the southeastern of state of Goiás, between 47°17′ and 48°12′ West longitude and between 17°28′ and 18°30′ South latitude, in the Centre-north of Catalão, far away from its municipal area 15 km (Figure 11.4).

Through an ethnobotanical survey that was undertaken in the Coqueiros community, it was found that all interviewees know and employ medicinal herbs. A total of 109 plant species were

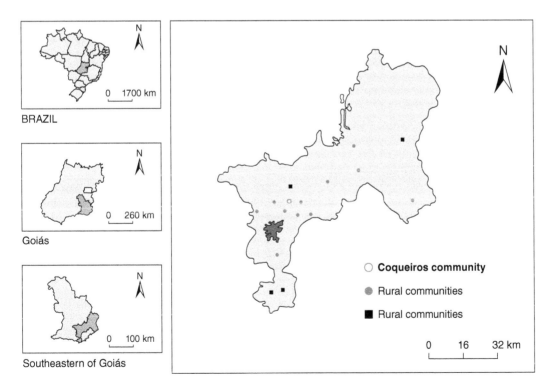

FIGURE 11.4 A map of the Coqueiros community, located in the southeastern of Goiás state, in the Centre-north of Catalão city. (Adapted from Silva and Hespanhol, 2016.)

identified, which were distributed among 12 categories related to the body systems, according to ICD, such as respiratory diseases, the nervous, the genitourinary, the digestive, the circulatory and the musculoskeletal systems. Furthermore, there are other diseases, for instance, injury, poisoning and certain other consequences of external causes; certain infectious and parasitic diseases; diseases of connective tissue, the skin and subcutaneous tissue; endocrine, nutritional and metabolic diseases; diseases of the blood and blood-forming organs and certain disorders involving the immune mechanism; and unknown affections or pains.

From this survey, 109 plant species were placed in 57 botanical families. The most represented families are Fabaceae, Lamiaceae, Myrtaceae, Annonaceae, Anacardiaceae, Rutaceae, Asteraceae, Apocynaceae, Solanaceae, Rubiaceae, Euphorbiaceae and Apiaceae.

Considering specialized literature, among all vegetal species identified, only nine plant species' biological activity has not been registered. Nevertheless, by means of surveys, only few species showed some biological potential, what underscores the importance of continuing this research.

Finally, the targeting of bioprospecting studies has been possible, to some extent, due to the ethnobotanical knowledge that was concluded through this research, which identified the potential plant species linked to possible biotechnological development. Therefore, the group from the Laboratory of Natural Products and Organic Synthesis has developed its research in light of popular knowledge, as well as some plant species studied, such as, *Hymenaea stigonocarpa* Hayne (Fabaceae), *Kielmeyera coriacea* Mart. & Zucc (Calophyllaceae) and *Annona coriacea* Mart. (Annonaceae). Besides these, many other plant species have already been studied before. The main chemical and biological data of these plant species are presented herein. All plant species focused on in this chapter are listed in Table 11.1.

11.2.4 DOURADINHA AND DOURADÃO EXAMPLES OF MEDICINAL PLANTS USED BY THE GOIÁS POPULATION AND THEIR PHYTOCHEMICAL STUDIES

Douradinha or *congonha do campo*, a plant of the Rubiaceae family identified as *Palicourea coriacea* (Cham.) K. Shum., is the object of study of several research groups in Goiás, since *Psychotria coriacea* is known as a cure for several diseases including treatment of kidney stones. Freitas et al. (2011) was the first to show this plant as a diuretic agent. The diuretic effects presented by the *P. coriacea* could be explained by the presence of several compounds, but mainly ursolic acid (**1**), well known for the diuretic effects. In parallel, the toxicity was studied by Passos et al. (2010) who evaluated the possible cytotoxic, genotoxic and antigenotoxic effects of the aqueous extract of *P. coriaceae* in somatic cells of *Drosophila melanogaster* (Meigen, 1830). The results indicated no cytotoxicity suggesting that the leaves are safe for tea consumption.

Douradinha is a classic example of a medicinal plant of Cerrado studied by different approaches. Backed by ethnobotanical information, chemical and biological evaluation corroborate the popular use by traditional communities in country Goiás, but further studies are still necessary to evaluate the mechanisms involved in its biological activity and safety.

There is information also available about *P. coriacea*, a medicinal plant, from the Rubiaceae family. The initial approach chosen by some researchers was based on traditional work using phytochemical analyses as the main goal to identify the main compounds present in the plant. Nascimento et al. (2006) described a tetrahydro β-carboline trisaccharide (**2**) isolated from the roots of *P. coriacea*, and its structure was elucidated using spectral 2 D (Nuclear Magnetic Resonance (NMR) methods: Correlation Spectroscopy (COSY), heteronuclear multiple-quantum correlation (HMQC), heteronuclear multiple-bond correlation (HMBC) and Nuclear Overhauser Effect Spectroscopy (NOESY). The aglycone was deduced by analysis of COSY, HMQC and HMBC connectivities, and analysis of the cross-peak in the COSY and NOESY NMR assignments confirmed the first example of strictosidinic acid incorporating a sucrose unit. The known alkaloids, strictosidinic acid (**3**), epi-strictosidinic acid (**4**), ketone strictosidinic (**5**)

TABLE 11.1

Summary of the Names of All Plant Species, Their Respective Family and Common Names Discussed in This Chapter

Scientific Name	Family	Common Names
Amaioua guianensis (sin. *Duhamelia glabra*)	Rubiaceae	
Annona coriacea (sin. *Annona geraensis*)	Annonaceae	Araticum
Banisteriopsis caapi (sin. *Banisteria inebrians*)	Malpighiaceae	Ayahuasca, caapi, yagé
Coffea arabica (sin. *Coffea bourbonica*)	Rubiaceae	Coffee, mountain coffee, café arábica
Dysoxylum gotadhora (sin. *Dysoxylum binectariferum*)	Meliaceae	
Galianthe ramosa	Rubiaceae	
Hymenaea stigonocarpa	Fabaceae	Jatobá-do-Cerrado
Ixora brevifolia (sin. *Ixora glaziovii*)	Rubiaceae	
Kielmeyera coriacea (sin. *Bonnetia coriacea*)	Calophyllaceae	Pau santo
Palicourea coriacea (sin. *Uragoga xanthophylla*)	Rubiaceae	Douradinha, congonha do campo
Palicourea rigida (sin. *Uragoga rigida*)	Rubiaceae	Bate-caixa, gritadeira, chapéu-de-couro, douradão
Psychotria gracilenta (sin. *Psychotria brachybotrya*)	Rubiaceae	
Psychotria capitata (sin. *Palicourea capitata*)	Rubiaceae	
Psychotria colorata (sin. *Psychotria calviflora*)	Rubiaceae	
Psychotria goyazensis (sin. *Psychotria argoviensis*)	Rubiaceae	
Psychotria henryi	Rubiaceae	
Psychotria hoffmannseggiana (sin. *Palicourea hoffmannseggiana*)	Rubiaceae	
Psychotria ipecacuanha (sin. *Carapichea ipecacuanha*)	Rubiaceae	Ipecac
Psychotria myriantha (sin. *Psychotria myriantha*)	Rubiaceae	
Psychotria pilifera	Rubiaceae	
Psychotria prunifolia (sin. *Psychotria xanthocephala*)	Rubiaceae	
Psychotria umbellata (sin. *Uragoga calva*)	Rubiaceae	
Psychotria ulviformis (sin. *Palicourea alba*)	Rubiaceae	
Psychotria viridis (sin. *Psychotria glomerata*)	Rubiaceae	
Uncaria tomentosa (sin. *Nauclea tomentosa*)	Rubiaceae	Cat's claw, unha de gato

and calycanthine (**6**) were isolated by Nascimento et al. (2006) using acid-base fractionation of leaves and roots extracts (Figure 11.5). Calycanthine (**6**) was obtained as a monocrystal and was elucidated by X-ray diffraction (Vencato, 2004). Silva et al. (2008) described the isolation of 11 compounds belonging to several classes of chemicals, showing that *P. coriacea* is not an exclusive source of alkaloids.

Another medicinal plant used in folk medicine is *Palicourea rigida*, called *douradão*, *bate-caixa* that has been traditionally used in the treatment of urinary tract disorders. In some studies, cytotoxic activity was observed (Rosa et al., 2012) as well as using various phytochemical screens to show the presence of common phytosterols (stigmasterol, campesterol and sitosterol), flavonoids, coumarin, iridoids and alkaloids like strictosidinic acid (**3**) and vallesiachotamine (**7**) (Alves et al., 2017; Bolzani et al., 1992; Lopes et al., 2004; Rosa et al., 2010; Vencato et al., 2006).

P. rigida is not endemic to Goiás Cerrado but occurs in the entire Brazilian Cerrado region. A peculiar fact of this species is that the isolation of the alkaloid from *P. rigida* was just for the first time isolated from the species collected in Goiás Cerrado (Vencato et al., 2006).

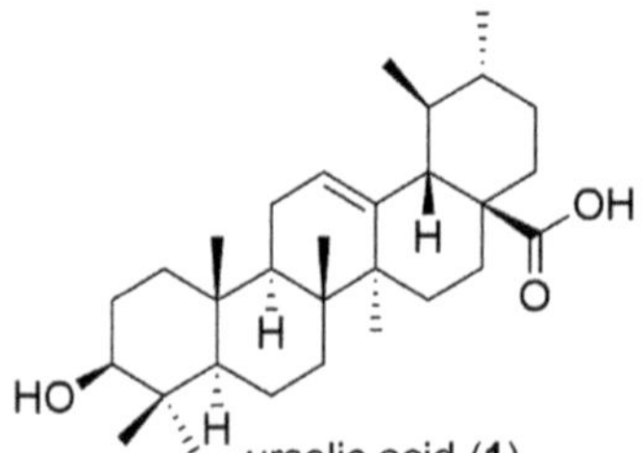

FIGURE 11.5   Alkaloids and ursolic acid isolated from *Palicourea coriacea* extracts.

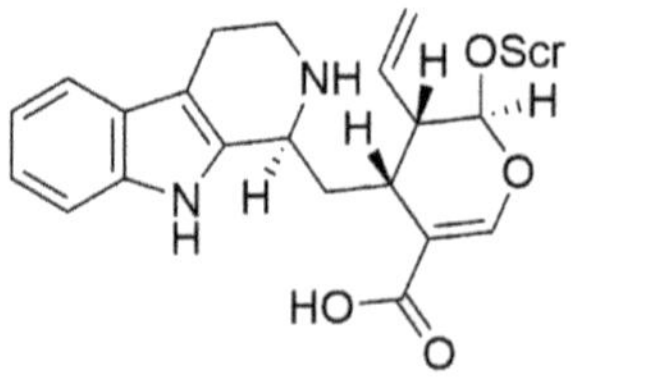

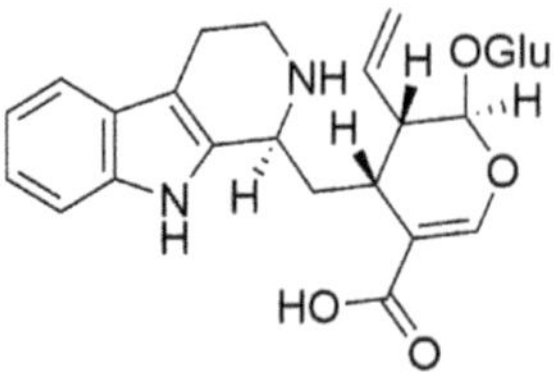

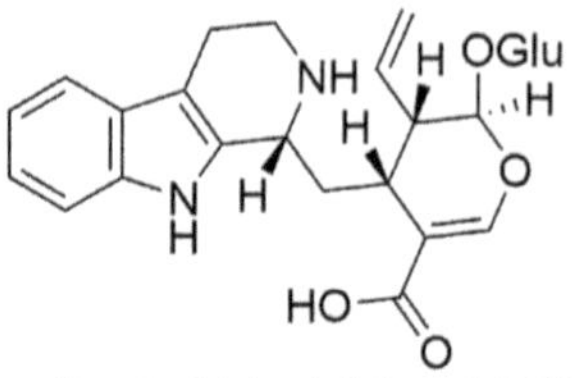

Vallesiachotamine (**7**) is an indolic monoterpenic alkaloid that exhibits inhibitory activity against human melanoma cells SKMEL37 with promotion of G0/G1 cell cycle arrest, apoptosis and necrosis (Soares et al., 2012).

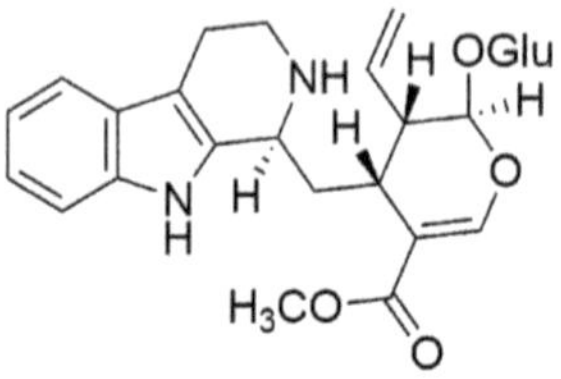

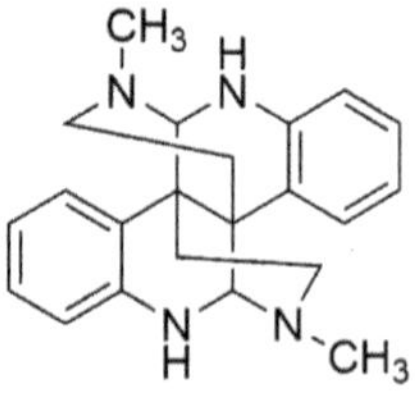

Morel et al. (2011) showed the quantification of loganin (iridoid) in individuals plants collected in random areas of the Brazilian Cerrado, suggesting the great variation of loganin, which was most abundant in the plants collected in Luziania city (Goiás) whereas the lowest yields were in the plants collected in Jaguara city (Minas Gerais). Based on these finding the authors observed the variation on the loganin production occurred inside and among populations and they suggest that the concentration of iridoids in *P. rigida* plants are influenced by both genetic and environmental variability, supporting the great diversity from Cerrado plants due to environmental stress.

11.3   ALKALOIDS AS A METABOLIC TARGET

The strategy used by some researchers is based on chemotaxonomy, searching for specific metabolites that characterize some family plants. The Rubiaceae family is the fourth largest flowering plant family and is estimated to contain around 600 genera and between 6,000 and 13,143 species and is a predominantly tropical family, with biomass and diversity concentrated in the tropics and subtropics. The genus *Psychotria* L. and *Palicourea* Aubl. are among the ten largest (by species number) genera in Rubiaceae, and *Psychotria* is still the largest genus with 1,834 species (Davis, 2009). Several phytochemical studies corroborate that the Rubiaceae family is a well-known and a prolific source of alkaloids with great structural diversity and pharmacological properties. As a known example the genus *Coffea* L. is one of the most economically important, mainly the species *Coffea arabica* L., popularly known as coffee, which has caffeine as one of the principal chemical components.

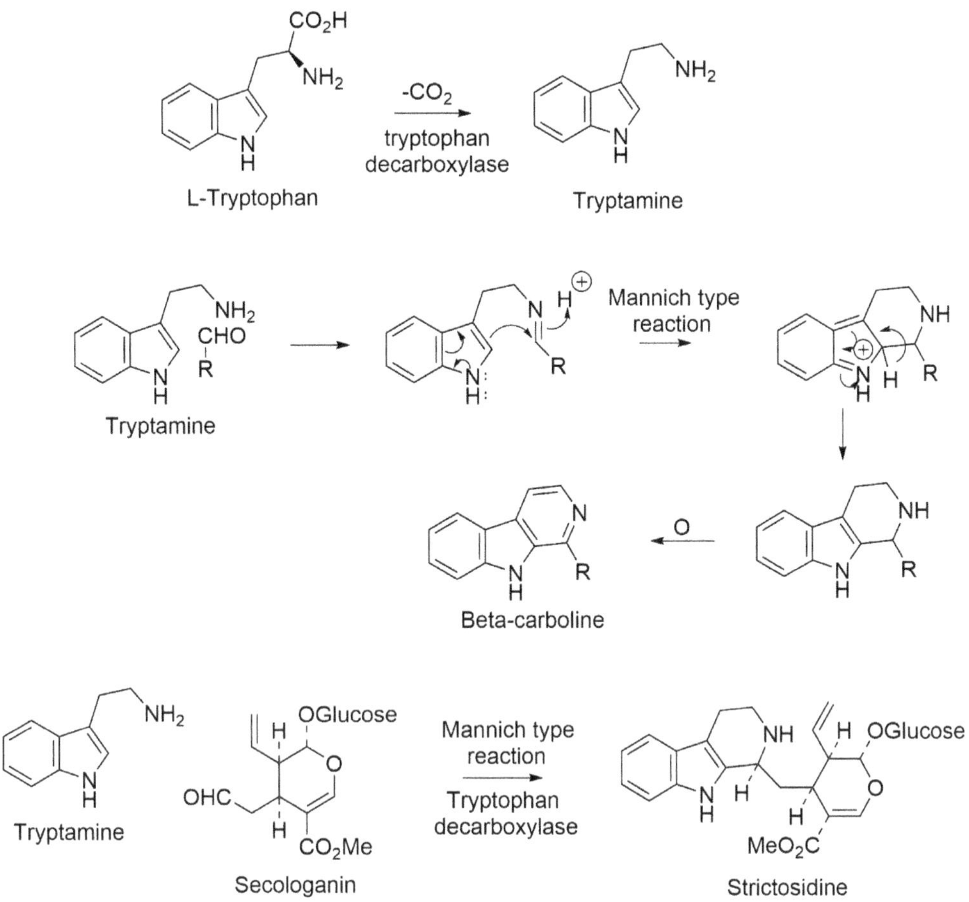

FIGURE 11.6 Biosynthetic approach proposed for strictosidine (β-carboline monoterpenic alkaloid, Dewick, 2009).

The well-known phytotherapic *Uncaria tomentosa* (Willd.) DC. (Rubiaceae), popularly called "unha de gato", is used in folk Brazilian medicine and studies have shown that alkaloids isolated from this plant have immunostimulant and antitumor activity (Nunez and Martins, 2016).

In the case of alkaloids found in *Psychotria*, most of them belong to the indole, monoterpene indole (MIA-type) or pyrrolidinoindoline subclass (Moraes, 2013). These alkaloids are derived from the amino acid tryptophan, which is decarboxylated by the enzyme tryptophan-decarboxylase to form tryptamine. Once the indole ring of the tryptamine is a nucleophilic system, β-carboline alkaloids are formed through a Mannich/Pictet-Spengler type reaction. This reaction also allows the condensation of secologanin with tryptamine, producing strictosidine which contains a β-carboline system and it is a precursor of approximately 3,000 MIA-type alkaloids (Figure 11.6, Dewick, 2009).

11.3.1 ALKALOIDS FROM *PSYCHOTRIA* GENUS

Psychotria L. is the largest genus in the Rubiaceae family, comprising approximately 1,600 species. This genus is well known through the species *P. viridis* Ruiz and Pav. or *P. carthagenensis* Jacq., together with *Banisteriopsis caapi* (Spruce ex Griseb.) C.V. Morton (Malpighiaceae), in the preparation of the psychoactive plant tea "ayahuasca", which has been used since pre-Colombian

TABLE 11.2

New Alkaloids Isolated from *Psychotria* Species Reported Since 2013

Species	Compound	Subclass
P *Psychotria henryi*	Alkaloid (**8**); Alkaloid (**9**)	Indole
	Psychohenin (**10**)	
P *Psychotria umbellata*	3,4-Dehydro-18,19-beta-epoxy-psychollatine (**11**)	Monoterpene indole
	N_4-[1-((R)-2-Hydroxypropyl)]-psychollatine (**12**)	Monoterpene indole
	N_4-[1-(S)-2-Hydroxypropyl)]-psychollatine (**13**)	Monoterpene indole
P *Psychotria brachybotrya*	Brachybotryne (**14**)	Simple indole
	Brachybotryne *bis-N*-oxide (**15**)	Simple indole
P *Psychotria pilifera*	16,17,19,20-Tetrahydro-2,16-dehydro-18-deoxyisostrychnine (**16**)	Indole

times for medical and religious purposes. More recently, in the last century this tea was used by syncretic religious groups in Brazil, particularly "Santo Daime", "União do Vegetal" and "Barquinha" (Riba et al., 2004). Furthermore, some other species are used by different traditional communities with a variety of pharmacological purposes: in Amazonia the "caboclos" use the flowers of *P. colorata* (Willd. ex Roem. and Schult.) Müll. Arg. as an analgesic and to treat earache and stomach ache; the Wayapi Indians use *P. ulviformis* Steyerm. in an antipyretic bath and as an analgesic too; *P. ipecacuanha* (Brot.) Stokes is used as a stimulant and to treat intoxication (Porto et al., 2009; Santos et al., 2017).

The traditional use of *Psychotria* species has encouraged the phytochemical study of a great number of species of this genus. These studies have shown that alkaloids are the main metabolite identified/isolated in *Psychotria*. In fact, nine new alkaloids have been identified in the genus (Table 11.2; Figure 11.7) since the publication of the most recent reviews on *Psychotria* metabolites (Calixto et al., 2016; Klein-Júnior et al., 2014). Liu et al. (2013) studied *P. henryi* H. Lev., which is used in the traditional Chinese medicine to relieve pain and eliminate dampness, and isolated two novel dimeric indole alkaloids (**8** and **9**). Interestingly, compound (**9**) contains an unusual decacyclic ring, which has not been described for any other dimeric alkaloid.

Important biological activities have been attributed to the secondary metabolites, mainly alkaloids, found in *Psychotria* species, such as, analgesic, anti-inflammatory, anxiolytic, antidepressant, antioxidant, antimutagenic and cytotoxic activity (Calixto et al, 2016; Magedans et al., 2017; Moller and Wink, 2007). *Psychotria* alkaloids are also known for their effects on the central nervous system as demonstrated by various reports of *Psychotria* species of Brazil.

These studies regard the inhibition of the enzymes, monoamine oxidases A and B (MAO-A and MAO-B) and acetyl cholinesterase A and B (AchE and BchE), which are important targets in the treatment of neurodegenerative disorders such as Parkinson's disease and Alzheimer disease (Repsold et al., 2018). Phytochemical study of the leaves of *P. prunifolia* (Kunth) Steyerm. resulted in the isolation of the MIA-type alkaloids prunifoleine (**17**) and 14-oxoprunifoleine (**18**) (Figure 11.8), which inhibited both cholinesterases (AChE and BChE) and MAO-A. These compounds exhibit noncompetitive inhibition with IC_{50} values of 10 μM and 3.39 μM for AChE, and 100 μM and 11 μM for BChE, respectively. Furthermore, these compounds exhibited MAO-A selectivity with IC_{50} values of 7.41 μM and 6.92 μM. However, both compounds showed a time-dependent MAO inhibition, indicating that they can act as irreversible inhibitors, which can be dangerous since they may cause cardiovascular toxic effects (Passos et al., 2013).

These results agree with the literature (Hamid et al., 2017; Rüben et al., 2015), which shows that the quaternary β-carboline scaffold displays an important role in the selectivity for some enzymes such as AChE and MAO-A. In fact, several quaternary β-carboline derivatives have been

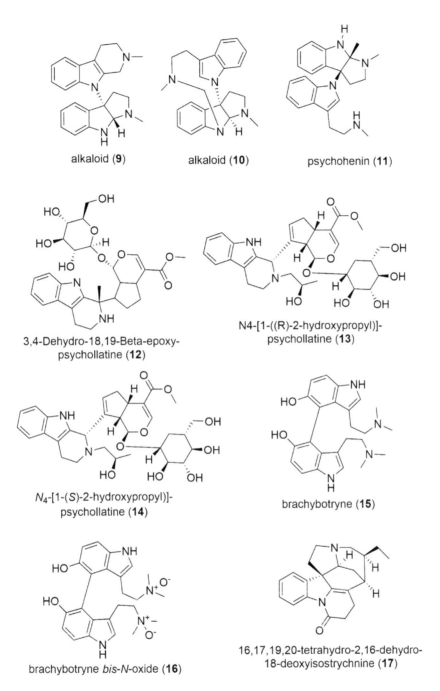

alkaloid (**9**) alkaloid (**10**) psychohenin (**11**)

3,4-Dehydro-18,19-Beta-epoxy-psychollatine (**12**)

N4-[1-((R)-2-hydroxypropyl)]-psychollatine (**13**)

N_4-[1-(S)-2-hydroxypropyl)]-psychollatine (**14**)

brachybotryne (**15**)

brachybotryne *bis-N*-oxide (**16**)

16,17,19,20-tetrahydro-2,16-dehydro-18-deoxyisostrychnine (**17**)

FIGURE 11.7 Chemical structure of the new alkaloids isolated from *Psychotria* species published since 2013.

synthetized to act as multitarget compounds aiming at monoamine and cholinesterase enzymes (Revenga et al., 2015; Santillo et al., 2014).

Furthermore, numerous studies have shown that indole alkaloids display important antiprotozoal activities (Bharate et al., 2013; Pereira et al., 2017). Prunifoleine (**17**) and 14-oxoprunifoleine (**18**) were tested against *Leishmania amazonensis* Laison & Shaw, 1972, promastigotes with IC_{50} values of 16.0 and 40.7 µg mL^{-1}. Also, other MIA-type alkaloids like 10-hydroxyisodeppeaninol (**19**) and

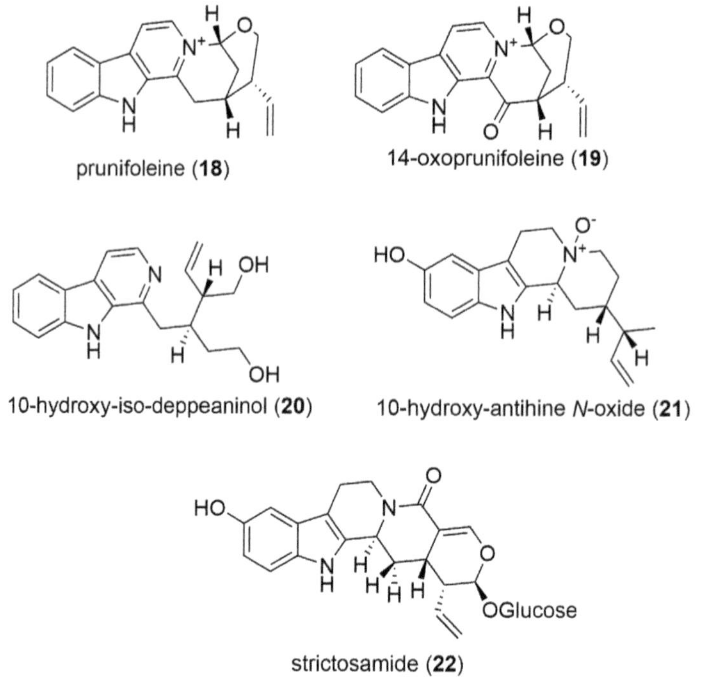

prunifoleine (**18**)

14-oxoprunifoleine (**19**)

10-hydroxy-iso-deppeaninol (**20**)

10-hydroxy-antihine *N*-oxide (**21**)

strictosamide (**22**)

FIGURE 11.8 Chemical structures of the alkaloids isolated from *Psychotria prunifolia* (Kunth) Steyerm.

10-hydroxyantirhine *N*-oxide (**20a**), together with the known strictosamide (**21**) were isolated from the roots and branches of *P. prunifolia* (Kunth) Steyerm. (Kato et al., 2012).

From the leaves of *P. hoffmannseggiana* (Schult.) Müll. Arg. several alkaloids were isolated, such as *N*-methyltryptamine (**22**), harmane (**23**), *N*-methyl-1,2,3,4-tetrahydro-β-carboline (**24**), (+) chimonantine (**25**) and the major alkaloid strictosidinic acid (**3**) (Naves, 2014). Strictosidinic acid (**3**), first isolated from *P. myriantha* Müll. Arg., was assayed on rat hippocampus showing a decrease in the serotonin (5-HT) levels, possibly indicating an inhibition of the precursor enzymes of the 5-HT biosynthesis. Tryptamine and β-carboline type alkaloids were also isolated from the leaves of *P. capitata* Ruiz & Pav., with bufotenine (**26**) and its *N*-oxide derivative (**27**) as the major alkaloids (Wakui, 2015). Although early studies (Moraes et al., 2011) have shown the presence of β-carboline alkaloids in *P. capitata* ethanol extract, only 6-hydroxy-2-methyl-1,2,3,4-tetrahydro-β-carboline (**28**) was identified. From *P. goyazensis* Müll. Arg., the quinolone alkaloid calycanthine (**6**) was isolated together with strictosidinic acid (**3**) and harmane (**23**), so far, the first report of the occurrence of a monoterpene indole and a quinolinic type alkaloid in the same species of *Psychotria* (Januário, 2015) (Figure 11.9).

The β-carboline alkaloids from *P. prunifolia* and from Rubiaceae and Apocynaceae species have been assayed against inhibitors of malate synthase, an important enzyme from *Paracoccidioides* spp. (*Pb*MSL) (Costa et al., 2015). The β-carboline alkaloids are crucial for stability in the binding pocket of *Pb*MSL and were chosen as candidate molecules after virtual screening and molecular docking studies were obtained through studies of receptor-ligand interactions. In addition, the alkaloids **29**, **30**, **31** and **21** (Figure 11.10) were assayed and the alkaloids **29** and **30** showed no cytotoxicity in A549 and MRC5 cells. This result is concomitant with bioassays described by Costa et al. (2015) where alkaloid **29**, isolated from *Galianthe ramosa* E.L. Cabral (Freitas et al., 2014), another Rubiaceae species, is a good candidate for antifungal development.

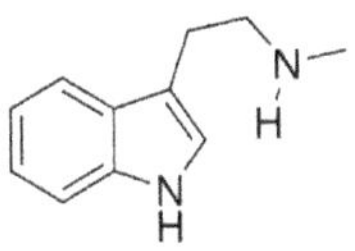

N-methyltryptamine (**23**) 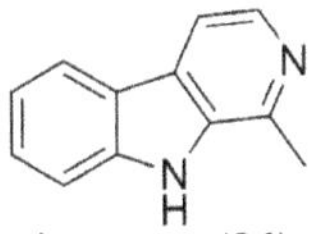 harmane (**24**) 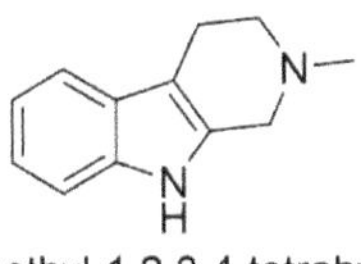 N-methyl-1,2,3,4-tetrahydro-beta-carboline (**25**)

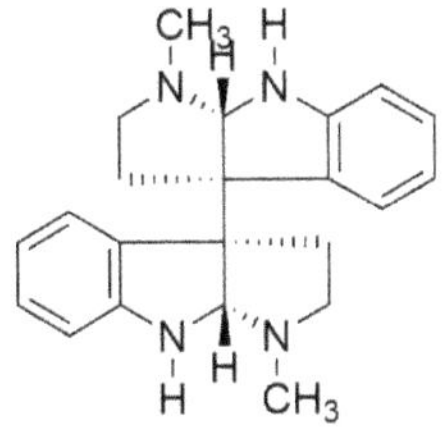

(+) chimonantine (**26**)

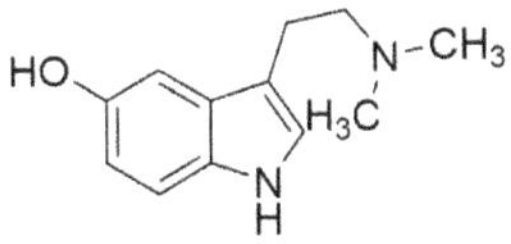

bufotenine (**27**)

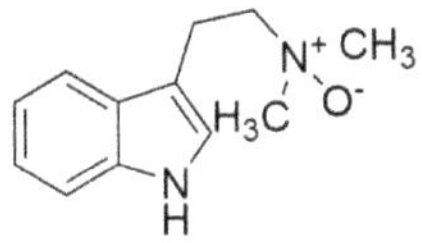

bufotenine N-oxide (**28**) 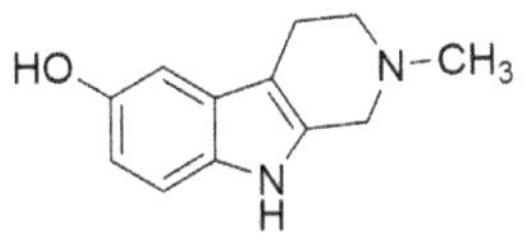 6-hydroxide-2-methyl-1,2,3,4-tetrahydro-beta-carboline (**29**)

FIGURE 11.9 Chemical structures of alkaloids isolated from *P. hoffmannseggiana*, *P. capitata* and *P. goyazensis*.

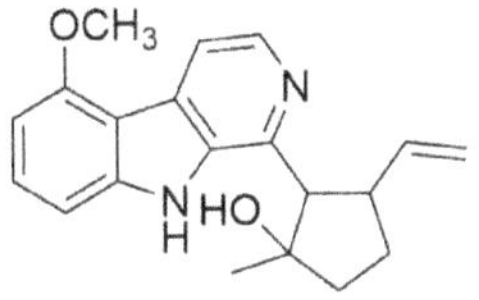

cyclopentanol, 3-ethenyl-1,-methyl-2-(5-methoxy-9H-pyrido[4,3-b]indol-4-yl) (**30**)

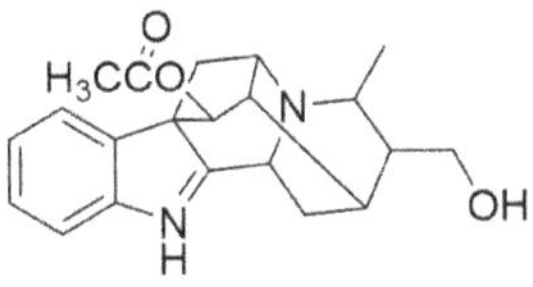

18-norajmalan-17,19-diol,1,2-didehydro-1-demethyl-21-methyl-,17-acetate (**31**)

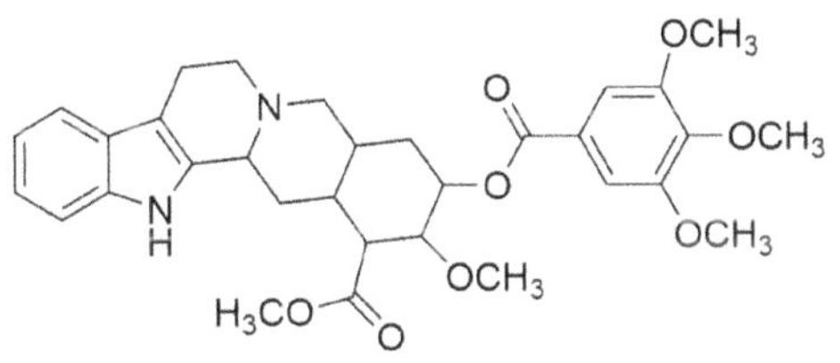

reserpine (**32**)

FIGURE 11.10 Structure of alkaloid assayed against enzyme isolated from *Paracoccidioides* spp (PbMSL).

11.3.2 Cyclopeptide Alkaloids from *Amaioua guianensis* Aubl. and *Ixora brevifolia* Benth

The cyclopeptide alkaloids are polyamide based composed of 13-, 14- or 15-membered macrocyclic rings in which a 10- or 12-membered peptide-type bridge spans the 1,3 or 1,4 positions of a benzene ring. Typically, the molecule contains two amino acids and one styrylamine unit. The 14-membered

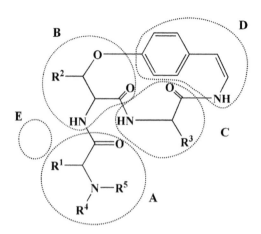

FIGURE 11.11   Basic structure of 14-membered cyclopeptide: **(A)** basic terminal (end) amino acid; **(B)** β-hydroxy amino acid; **(C)** ring-bound amino acid; **(D)** hydroxystyrylamine unit. Sometimes between the **(A)** and **(B)** unit an additional (intermediary) amino acid is present and is designated as **(E)**.

rings are more common kinds of cyclopeptides and contain four building blocks A (basic terminal end amino acid), B (β-hydroxyamino acid); C (a ring bonded amino acid taking part in the macro-cycle ring) and D (hydroxylstyrylamine) as illustrated in Figure 11.11.

The cyclopeptides showed broad biological activity as well as antibacterial effects (Giacomelli et al., 2004; Morel et al., 2002; 2005), antifungal activity (Gournelis et al., 1997) and cytotoxicity against leukemia cells L12100 and KB murine human cells (Liu et al., 1997) and they exert activity on the central nervous system (Tuenter et al., 2017).

One example of Rubiaceae species as a source of cyclopeptide alkaloid is *Amaioua guianensis* Aubl. (Rubiaceae). Initially, Oliveira et al. (2009) began this phytochemical study by examination of promising antioxidant active extracts of *Amaioua guianensis* Aubl., which exhibited a moder-ate antioxidant activity (IC_{50} 70 µg mL^{-1}). This extract was subjected to solvent partitioning and chromatographic separation to provide a new cyclopeptide alkaloid (**32**), a mixture of two known proanthocyanidins and iridoids.

The complete elucidation of the cyclopeptide alkaloid amaiouine (**32**) (Figure 11.12) was achieved after crystallization in EtOAc of this compound as colorless needles. The NMR analy-sis enabled elucidation of the amino-acid portions and some linkages among them were secured after HMBC and NOESY experiments. Although the cyclopeptide skeleton was discovered after X-ray diffraction analysis, the Oak Ridge Thermal Ellipsoid Plot (ORTEP) diagram of the crystal structure confirmed that the skeleton was made up of proline bearing a cinnamoyl and styrylami-nine moiety together with two units of phenylalanine. The X-ray analysis also confirmed the trans

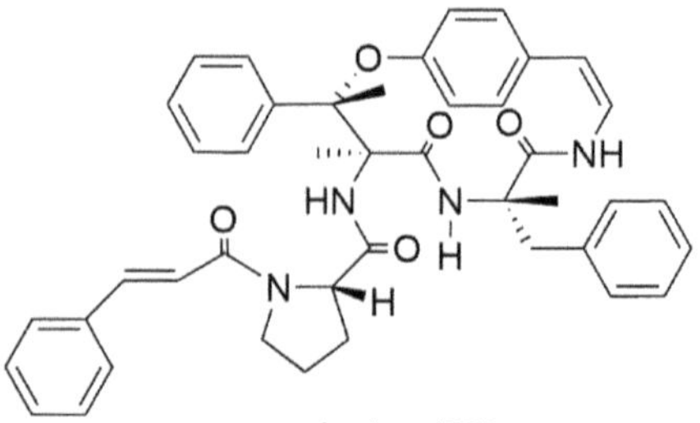

amaiouine (**33**)

FIGURE 11.12   Structure of cyclopeptide amaiouine isolated from *Amaioua guianensis.*

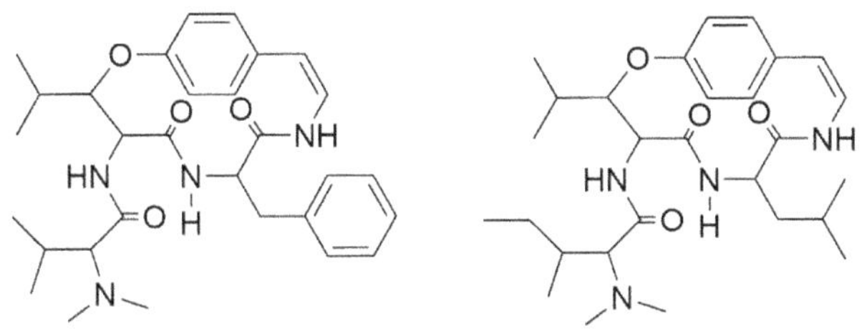

ixorine (**34**)   frangulanine (**35**)

FIGURE 11.13   Structures of cyclopeptides ixorine e frangulanine.

configuration (3*S**, 4*S**) between H-3 and H-4 and was used to decisively assign the orientation (7*S**, 25*S**) to the hydrogens at H-7 and H-25. Since this was the first occurrence of this cyclopeptide it was named amaiouine (**32**).

Other cyclopeptides reported for the first time in the literature from Rubiaceae species are ixorine (**33**), isolated together with the known frangulanine (**34**) from *Ixora brevifolia* branches (Figure 11.13). Medina et al. (2016) described the complete structural elucidation by NMR analysis including stereochemical finding by analysis of NOESY experiments attributing the *L-erythro* configuration to the β-substituted leucine moiety.

Since the cyclopeptides exhibited antibacterial activity, the alkaloidal mixture of ixorine (**33**) and frangulanine (**34**) was assayed against the bacteria *Escherichia coli* Meigen, 1830, *Pseudomonas aeruginosa* Schroeter, 1872, *Staphylococcus aureus* Rosenbach, 1884 and fungi *Candida albicans* (C.-P. Robin) Berkhout, 1923, but no significant biological activity was observed. However, the anti-protozoal activity *in vitro* against promastigotes of *Leishmania amazonensis* inhibited the parasite growth with IC_{50} value of 54.16 µg mL^{-1}. Leishmaniasis is regarded as a neglected disease, and Medina et al. (2016) described for the first time the positive results in assays with these or related cyclic peptides against this protozoal (Medina et al., 2016).

11.4 IMAGING MASS SPECTROMETRY OF ALKALOIDS FROM *PSYCHOTRIA* AND *PALICOUREA* LEAVES

All studies involving phytochemical analysis of *Palicourea* or *Psychotria* are a product of extraction procedures, which result in information loss on the spatial localization of these alkaloids in the plant. Visualizing the Also, this knowledge is desirable for understanding the physiological functions and their biosynthesis. Also, the presence of alkaloids on the leaves' surface have been related to the defense system since this surface offers the opportunity to display a chemical barrier to phytophagous insects and mites.

A review (Bjarnholt et al., 2014) showed mass spectrometry imaging (MSI) techniques as an efficient tool to analyze metabolite plant distribution. The report also indicates that DESI-MSI provides sensitivity and selectivity for this technique and shows this tool has become attractive due to the simplicity requiring few sample preparation steps compared with vacuum imaging techniques. An overview of research involving MSI of various types of plant material demonstrated a lack of study involving DESI-MSI in leaves, mainly due to the analysis of the spatial distribution of compounds in tissues using MS imaging is usually performed by tissue sections obtained by cryosectioning, and that is a barrier when looking to analyze leaves or petals. For these kinds of tissues, DESI-MSI has recently been performed using an indirect approach with porous Teflon imprints (Thunig et al., 2011) or TLC blotting (Cabral et al., 2013). Direct imaging of the plant has been published and requires very carefully optimization and depends on the nature of the leaf (Li et al., 2013). Kumara et al. (2015)

---

* relative configuration

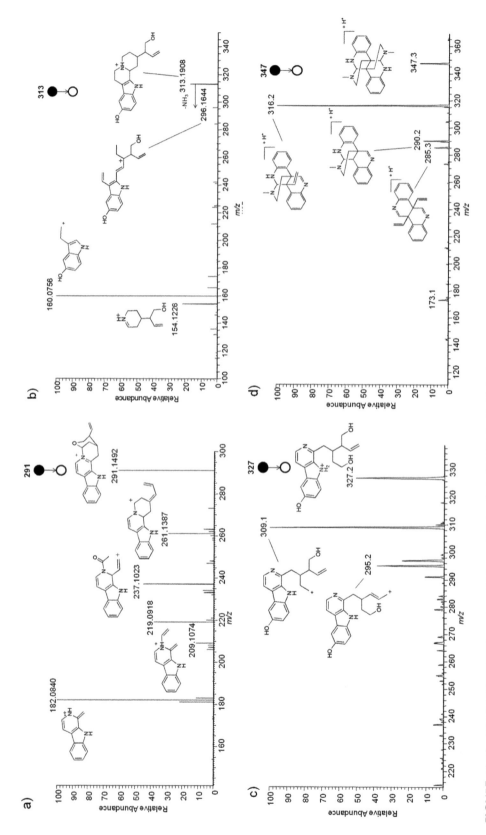

FIGURE 11.14 ESI-(+)-HRMS/MS spectra of (**a**) prunifoleine (m/z 291.1942), (**b**) 10-hydroxy-isodeppeaninol (m/z 327 and (**d**) calycanthine (m/z 347). (From Kato et al., 2018.) 10-hydroxyantirhine (m/z 313.1917); and DESI-(+)-MS/MS of (**c**) 10-hydroxy-

successfully described the spatial and temporal distribution of the alkaloid rohitukine in *Dysoxylum binectariferum* (Roxb.) Hook. f. ex Bedd. seeds from the Meliaceae family.

The increasing number of phytochemical studies of *Psychotria* species has supported ethnobotanical, pharmacological and chemotaxonomic studies. However, the spatial distribution of the alkaloids in this genus is not known. Kato et al. (2018) explored the utilization of DESI-MS imaging as a tool for visualization of this alkaloid distribution in leaves. As ambient DESI imaging requires a flat surface for a good representation of molecular distribution and since reports have described that an imprint on Teflon results on a stable, good and intense surface with porous Teflon (Li et al., 2013), the same methodology was followed of just washing the porous Teflon with drops of methanol before making the imprints.

For imaging, the resolution was limited by the size of leaf (ca: area 1,500 mm²) and was performed at 300 mm without implications (not to increase the rastering time series).

Teflon imprint DESI-MSI from leaf revealed the alkaloids have a heterogeneous distribution on the leaf surface.

Among the alkaloids isolated from *P. prunifolia*, DESI experiments showed the presence of the major alkaloids, prunifoleine (*m/z* 291, **17**), together with the alkaloids 10-hydroxyisodeppeaninol (*m/z* 327, **19**) and 10-hydroxyantirhine (*m/z* 313, **20b**). Since there was no report about DESI fragmentation for these alkaloids, the authors proposed a mass fragmentation pattern for these alkaloids under DESI/ESI conditions (Figure 11.14).

Imaging DESI experiments for imprint leaves of *P. prunifolia* resulted in well-defined images of *m/z* 291, 313 and 327 amu. The localization of prunifoleine (*m/z* 291amu) predominantly in the midrib; 10-hydroxyisodeppeaninol (*m/z* 313 amu), in the midrib but concentrated close to petiole and roughly a distribution of 10-hydroxyantirhine (*m/z* 327amu) in the whole leaf can be observed (Figure 11.15). The imprint image of *P. coriacea* shows the homogeneous distribution of calycanthine (*m/z* 347amu) in the whole leaf (Figure 11.16).

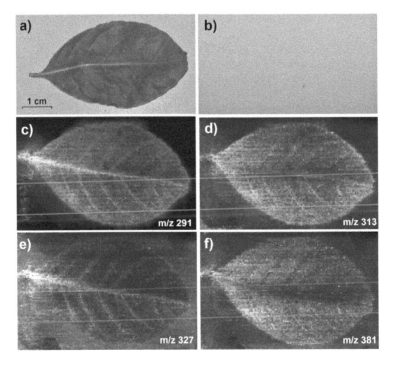

FIGURE 11.15 Imaging mass spectrometry approach applied for *Psychoria prunifolia* leaves. (a) Photography of *Psychotria prunifolia* leaf after imprint, (b) imprint in Teflon from *P. prunifolia* leaf, (c) DESI-MS images of the ions of *m/z* 291 = [prunifoleine]⁺, (d) *m/z* 327 = [10-hydroxyisodeppeaninol + H]⁺, (e) *m/z* 313 = [10-hydroxyantirhine + H]⁺ and (**F**) *m/z* 381 = [sucrose + K]⁺. (From Kato et al, 2018.)

  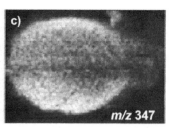

FIGURE 11.16 (a) Photography of *Palicourea coriacea* leaf after imprint; (b) imprint in Teflon from *P. coriacea* leaf, (c) DESI-MS image of the ion of m/z 347 = [calycanthine + H]$^+$.

This was the first study findings of different localization of various alkaloids in plant leaves using DESI-MSI, and the imaging findings are in complete accordance with the LC analyses and histochemical results (Kato et al., 2018). The DESI-MSI is undoubtedly a powerful tool to give us information about the spatial localization of the metabolites of plants, and these results consolidate this technology for use in plants.

ACKNOWLEDGMENTS

The author thanks R. N. Ribeiro for photos in Figures 11.1 and 11.2 and A. M. Uemura for adaptation of Figure 11.4. Some of the data presented were obtained with financial support from Conselho Nacional de Desenvolvimento Científico e Tecnológico (CNPq), Fundação de Amparo à Pesquisa do Estado de Goiás (FAPEG) and the Coordenação de Aperfeiçoamento de Pessoal de Nível Superior - Brasil (CAPES) - Finance Code 001.

REFERENCES

Alcorn, J. B. 1995. The scope and aims of ethnobotany in a developing world. In R. E. Schultes and S. V. Reis (Eds.) Ethnobotany: Evolution of a Discipline. Cambridge: Timber Press.

Alexiades, M. N.; Sheldon, J. W. 1996. Ethnobotanical Research: A Field Manual. New York: The New York Botanical Garden.

Alves, V. G.; Schuquel, I. T. A.; Ferreira, H. D.; Santin, S. M. O.; Silva, C. C. da. 2017. Coumarins from roots of *Palicourea rigida*. Chemistry of Natural Compounds, 53, 1157–1159.

Begossi, A. 1996. Use of ecological methods in ethnobotany: diversity indices. Economic Botany, 50, 280–289.

Bennett, B. C.; Prance, G. T. 2000. Introduced plants in the indigenous pharmacopoeia of Northern South America. Economic Botany, 54, 90–102.

Bharate, S. B.; Yadav, R. R.; Khan, S. I.; Tekwani, B. L.; Jacob, M. R.; Khan, I. A.; Vishwakarma, R. A. 2013. Meridianin G and its analogs as antimalarial agents. Medicinal Chemistry Communications, 4, 1042–1048.

Bjarnholt, N.; Li, B.; D'Alvise, J.; Janfelt, C. 2014. Mass spectrometry imaging of plant metabolites – principles and possibilities. Natural Products Report, 31, 818–837.

Bolzani, V.S.; Trevisan, L. M. V.; Young, M. C. 1992. Triterpenes of *Palicourea rigida* H.B.K. Revista Latinoamericana de Quimica, 23, 20–21.

Brasileiro, B. G.; Pizziolo, V. R.; Matos, D. S.; Germano, A. M.; Jamal, C. M. 2008. Plantas medicinais utilizadas pela população atendida no "programa de saúde da família", governador valadares, MG, Brasil. Brazilian Journal of Pharmaceutical Sciences, 44, 629–636.

Cabral, E. C.; Mirabelli, M. F.; Perez, C. J.; Ifa, D. R. 2013. Blotting assisted by heating and solvent extraction for DESI-MS imaging. Journal of the American Society for Mass Spectrometry, 24, 956–965.

Calixto, N. O.; Pinto, M. E. F.; Ramalho, S. D.; Burger, M. C. M.; Bobey, A. F.; Young, M. C. M.; Bolzani, V. S.; Pinto, A. C. 2016. The genus *Psychotria*: phytochemistry, chemotaxonomy, ethnopharmacology and biological properties. Journal of Brazilian Chemical Society, 27, 1355–1378.

Carmignotto, A. P.; Vivo, M.; Langguth, A. 2012. Mammals of the cerrado and caatinga: distribution patters of the tropical open biomes of the Central America. In B. D. Patterson and E. Costa (Eds.) Bones Clones and Biomes. Chicago: University of Chicago Press.

Costa, F. G.; Neto, B. R. da; Gonçalves, R. L.; da Silva, R. A.; de Oliveira, C. M.; Kato, L.; Freitas, C. dos S.; et al. 2015. Alkaloids as inhibitors of malate synthase from *Paracoccidioides* spp: receptor-ligand interaction-based virtual screening and molecular docking studies, antifungal activity, and the adhesion process. Antimicrobial Agents and Chemotherapy, 59, 5581–5594.

Critical Partnership Fund, The – CEPF. 2018. Hotspots defined. https://www.cepf.net/our-work/biodiversity-hotspots/hotspots-defined.

Davis, A. P.; Govaerts, R.; Bridson, D. M.; Ruhsam, M.; Moat, J.; Brummitt, N. A. 2009. A global assessment of distribution, diversity, endemism, and taxonomic effort in the rubiaceae. Annals of the Missouri Botanical Garden, 96, 68–78.

Dewick, M. P. 2009. Medicinal Natural Products: A Biosynthetic Approach, 3rd ed. Chichester: John Wiley & Sons Ltd.

Di Stasi, L.C. 1996. Plantas Medicinais: Arte e Ciência. Um Guia de Estudo Interdisciplinar. São Paulo: Editora da Universidade Estadual Paulista.

Eiten, G. 1972. The cerrado vegetation of Brazil. The Botanical Review, 38, 201–341.

Flora do Brasil. 2020. Jardim Botânico do Rio de Janeiro. http://floradobrasil.jbrj.gov.br/.

Freitas, C. S. de; Kato, L.; Oliveira, C. M. A.; de Queiroz, Jr, L. H. K.; Santana, M. J.; Schuquel, I. T.; Delprete, P. G.; et al. 2014. β-carboline alkaloids from *Galianthe ramosa* inhibit malate synthase from Paracoccidioides spp. Planta Medica, 80, 1746–1752.

Freitas, P. C. M.; Pucci, L. L.; Vieira, M. S.; Lino, R. S. Jr.; Oliveira, C. M.; Cunha, L. C.; Paula, J. R.; Valadares, M. C. 2011. Diuretic activity and acute oral toxicity of *Palicourea coriacea* (Cham.) K Schum. Journal of Ethnopharmacology, 134, 501–503.

Giacomelli, S. R.; Maldaner, G.; Gonzaga, W. A.; Garcia, C. M.; da Silva, U. F.; Dalcol, I. I.; Morel, A. F. 2004. Cyclic peptide alkaloids from the bark of *Discaria Americana*. Phytochemistry, 65, 933–937.

Gournelis, D. C.; Laskaris, G. G.; Verpoorte, R. 1997. Cyclopeptide alkaloids. Natural Product Reports, 14, 75–82.

Guarim Neto, G. 2006. O saber tradicional pantaneiro: as plantas medicinais e a educação ambiental. Revista Eletrônica do Mestrado em Educação Ambiental, 17, 71–89.

Hamid, H. A.; Ramli, A. N. M.; Yusoff, M. M. 2017. Indole alkaloids from plants as potential leads for anti-depressant drugs: a mini review. Frontiers in Pharmacology, 8, 1–7.

International Statistical Classification of Diseases and Related Health Problems – ICD. 2018. Classification of Diseases. http://www.who.int/classifications/icd/.

Januário, M. A. P. 2015. Contribuição do estudo fitoquímico de espécies de Psychotria L. (Rubiaceae): *Psychotria goyazensis* Müll. Arg. Master's thesis, Federal University of Goiás.

Kato, L.; Moraes, A. P.; de Oliveira, C. M. A.; de Almeida, G. L.; Silva, E. C. E.; Janfelt, C. 2018. The spatial distribution of alkaloids in *Psychotria prunifolia* (Kunth) steyerm and *Palicourea coriacea* (Cham.) K. Schum leaves analysed by desorption electrospray ionisation mass spectrometry imaging. Phytochemistry Analysis, 29, 69–76.

Kato, L.; Oliveira, C. M. A. de; Faria, E. O.; et al. 2012. Antiprotozoal alkaloids from *Psychotria prunifolia* (Kunth) steyerm. Journal of Brazilian Chemical Society, 23, 355–360.

Klein-Júnior, L. C.; Passos, C. dos S.; Moraes, A. P.; Wakui, V. G.; Konrath, E. L.; Nurisso, A.; Carrupt, P.-A.; Oliveira, C. M. A. de; Kato, L.; Henriques, A. T. 2014. Indole alkaloids and semisynthetic indole derivatives as multifunctional scaffolds aiming the inhibition of enzymes related to neurodegenerative diseases – a focus on *Psychotria* L. Genus. Current Topics in Medicinal Chemistry, 14, 1056–1075.

Kumara, P. M.; Srimany, A.; Ravikanth, G.; Shaanker, R. U. R.; Pradeep, T. 2015. Ambient ionization mass spectrometry imaging of rohitukine, a chromone anti-cancer alkaloid, during seed development in *Dysoxylum binectariferum* Hook (Meliaceae). Phytochemistry, 116, 104–110.

Li, B.; Hansen, S. H.; Janfelt, C. 2013. Direct imaging of plant metabolites in leaves and petal by desorption electrospray ionization mass spectrometry. International Journal Mass Spectrometry, 348, 15–22.

Lima, R. X. 1996. Estudos Etnobotânicos em Comunidades Continentais da Área de Proteção Ambiental de Guaraqueçaba – Paraná – Brasil. Master's thesis, Federal University of Paraná.

Liu, R.; Zhang, P.; Gan, T.; Cook, J. M. 1997. Regiospecific bromination of 3-methylindoles with NBS and its application to the concise synthesis of optically active unusual tryptophans present in marine cyclic peptides. Journal Organic Chemistry, 62, 7447–7456.

Liu, Y.; Wong, J.-S.; Wang, X.-B.; Kong, L.-Y. 2013. Two novel dimeric alkaloids from the leaves and twigs of *Psychotria henryi*. Fitoterapia, 86, 178–182.

Lopes, S.; Poser, G. L.; Kerber, V. A.; Farias, F. M.; Konrath, E. L.; Moreno, P.; Sobral, E. S.; Zuanazzi, J. A. S.; Henriques, A. T. 2004. Taxonomic significance of alkaloids and iridoid glucosides in the tribe Psychotrieae (Rubiaceae). Biochemical Systematics and Ecology, 32, 1187–1195.

Maffi, L. 2005. Linguistic, cultural and biological diversity. Annual Review of Anthropology, 29, 599–617.

Magedans, Y. V. da S.; Matsura, H. N.; Tasca, R. A. J. C.; Wairich, A.; Junkes, C. F. O.; Costa, F. de; Fett-Neto, A. G. 2017. Accumulation of the antioxidant alkaloid brachycerine from *Psychotria brachyceras* Müll. Arg. is increased by heat and contributes to oxidative stress mitigation. Environmental and Experimental Botany, 143, 185–193.

Medina, R. P.; Schuquel, I. T. A.; Pomini, A. M.; Silva, C. C.; Oliveira, C. M. A. de; Kato, L.; Nakamura, C. V.; Santin, S. M. O. 2016. Ixorine, a new cyclopeptide alkaloid from the branches of *Ixora brevifolia*. Journal Brazilian Chemical Society, 27, 753–758.

Mendes, E. P. P. 2005. A produção rural familiar em Goiás: as comunidades rurais no município de Catalão. Phd thesis., Paulista State University.

Mendonça, R. C.; Felfili, J. M.; Walter, B. M. T.; Silva Júnior, M. C.; Resende, A. V.; Filgueiras, T. S.; Nogueira, P. E.; Fagg, C. W. 2008. Flora vascular do bioma cerrado: checklist com 12.356 espécies. In S. M. Sano; S. P. Almeida and J. F. Ribeiro (Eds.) Cerrado: Ecologia e Flora. Vol. 2. Brasilia: Embrapa Cerrados, 421–1279.

Ministério do Meio Ambiente – MMA. 2018. O Bioma Cerrado. http://www.mma.gov.br/biomas/cerrado.

Moller, M.; Wink, M. 2007. Characteristics of apoptosis induction by the alkaloid emetine in human tumor cell lines. Planta Medica, 73, 1389–1396.

Moraes, A. P. 2013. Alcaloides indólicos das partes aéreas de *Psychotria* sp. (Rubiaceae) e síntese de tiohidantoínas e tioureias derivadas de aminoácidos e do *R*-(+)-limoneno. Master's thesis, Federal University of Goiás.

Moraes, T. M. da S.; Araújo, M. H. de; Bernades, N. R.; Oliveira, D. B de; Lasunskaia, E. B.; Muzitano, M. F.; Cunha, M. da. 2011. Antimycobacterial activity and alkaloid prospection of Psychotria species (Rubiaceae) from the Brazilian atlantic rainforest. Planta Medica, 77, 964–970.

Morel, A. F.; Araujo, C. A.; Silva, U. F.; Hoelzel, S. C. S. M.; Záchia, R.; Bastos, N. R. 2002. Antibacterial cyclopeptide alkaloids from the bark of *Condalia buxifolia*. Phytochemistry, 61, 561–566.

Morel, A. F.; Maldaner, G.; Ilha, V.; Missau, F.; Silva, U. F.; Dalcol, I. I. 2005. Cyclopeptide alkaloids from scutia bruxifolia reiss and their antimicrobial activity. Phytochemistry, 66, 2571–2576.

Morel, L. J. F.; Baratto, D. M.; Pereira, P. S. P.; Contini, S. H. T.; Momm, H. G.; Bertoni, B. W. B.; França, S. C.; Pereira, M. S. 2011. Loganin production in Palicourea rigida H.B.K. (Rubiaceae) from populations native to Brazilian cerrado. Journal of Medicinal Plants Research, 5, 2559–2565.

Myers, N.; Mittermeier, R. A.; Mittermeier, C. G.; Fonseca, G. A. B.; Kent, J. 2000. Biodiversity hotspots for conservation priorities. Nature, 403, 853–858.

Nascimento, C. A.; Gomes, M. S.; Lião, L. M.; Oliveira, C. M. A. et al. 2006. Alkaloids from *Palicourea coriacea* (Cham.) K. Schum. Zeitschrift für Naturforschung B, 61, 1443–1446.

Naves, R. F. 2014. Estudo Fitoquímico das Folhas de *Psychotria hoffmannseggiana* Roem. & Schult (Rubiaceae). Master's thesis, Federal University of Goiás.

Novaes, P.; Molinillo, J. M. G.; Varela, R. M.; Macías, F. A. 2013. Ecological phytochemistry of cerrado (Brazilian savanna) plants. Phytochemistry Reviews, 12, 1–17.

Nunez, C. V.; Martins, D. 2016. Secondary metabolites from rubiaceae species. Molecules, 20, 13422–13495.

Oliveira, F. C.; Albuquerque, U. P.; Fonseca-Kruel, V. S.; Hanazaki, N. 2009. Avanços nas pesquisas etnobotânicas no Brasil. Acta Botanica Brasilica, 23, 1–16.

Oliveira, P. L.; Tanaka, C. M. A.; Kato, L.; Silva, C. C. da; Medina, R. P.; Moraes, A. P.; Sabino, J. R.; Oliveira, C. M. A. de. 2009. Amaiouine, a Cyclopeptide Alkaloid from the Leaves of *Amaioua guianensis*. Journal of Natural Products, 72, 1195–1197.

Oliveira-Filho, A. T.; Ratter, J. A. 1995. A study of the origin of central Brazilian forests by the analysis of plant species distribution patterns. Edinburgh Journal of Botany, 52, 141–194.

Pasa, M. C. 2011. Saber local e medicina popular: a etnobotânica em cuiabá, mato grosso, Brasil. Boletim do Museu Paraense Emílio Goeldi. Ciências Humanas, 6, 179–196.

Passos, C. S.; Simões-Pires, C. A.; Nurisso, A.; Soldi, T. C.; Kato, L.; Oliveira, C. M.; Faria, E. O.; et al. 2013. Indole alkaloids of *Psychotria* as multifunctional cholinesterases and monoamine oxidases inhibitors. Phytochemistry, 86, 8–20.

Passos, D. C.; Ferreira, H. D.; Vieira, I. L. F. B.; Nunes, W. B.; Felício, L. P.; Silva, E. M.; Vale, C. R.; Duarte, S. R.; Silva, E. S.; Carvalho, S. 2010. Modulatory effect of *Palicourea coriacea* (Rubiaceae) against damage induced by doxorubicin in somatic cells of *Drosophila melanogaster*. Genetics and Molecular Research, 9, 1153–1162.

Pereira, M. D. P.; Silva, T. da; Aguiar, A. C. C.; Olivia, G.; Guido, R. V. C.; Yokoyama-Yasunaka, J. K. U.; Uliana, S. R. B.; Lopes, L. M. X. 2017. Chemical composition, antiprotozoal and cytotoxic activities of indole alkaloids and benzofuran neolignan of *Aristolochia cordigera*. Planta Medica, 83, 912–920.

Porto, D. D.; Henriques, A. T.; Fett-Neto, A. G. 2009. Bioactive alkaloids from South American *Psychotria* and related species. The Open Bioactive Compounds Journal, 2, 29–36.

Radomski, M. I. 2003. Plantas Medicinais – Tradição e Ciência. https://ainfo.cnptia.embrapa.br/digital/bitstream/item/50923/1/Radomski.pdf.

Ratter, J.A.; Ribeiro, J. F.; Bridgewater, S. 1997. The Brazilian cerrado vegetation and threats to its biodiversity. Annals of Botany, 80, 223–230.

Repsold, B. P.; Malan, S. F.; Joubert, J.; Oliver, D. W. 2018. Multi-targeted directed ligands for Alzheimer's disease: design of novel lead coumarin conjugates. SAR and QSAR in Environmental Research, 29, 231–255.

Revenga, M. de la F.; Pérez, C.; Morales-García, J. A.; Alonso-Gil, S.; Pérez-Castillo, A.; Caignard, D.-H.; Yáñez, M.; Gamo, A. M.; Rodríguez-Franco, M. I. 2015. Neurogenic potential assessment and pharmacological characterization of 6-methoxy-1,2,3,4-tetrahydro-β-*carboline (Pinoline) and Melatonin-Pinoline Hybrids*. ACS Chemical Neuroscience, 6, 800–810.

Riba, J.; Anderer, P.; Jané, F.; Saletu, B.; Barbanoj, M. J. 2004. Effects of the South American psychoactive beverage ayahuasca on regional brain electrical activity in humans: a functional neuroimaging study using low-resolution electromagnetic tomography. Neuropsychobiology, 50, 89–101.

Ribeiro, J. F.; Walter, B. M. T. 2008. As principais fitofisionomias do bioma Cerrado. In S. M. Sano; S. P. Almeida and J. F. Ribeiro (Eds.) Cerrado: Ecologia e Flora. Vol. 1, 151–212. Embrapa Cerrados, Brasília.

Rosa, E. A. da; Silva, B. C.; Silva, F. M.; Tanaka, C. M. A.; Peralta, R. M.; Oliveira, C. M. A. de; Kato, L.; Ferreira, H. D.; Silva, C. C. da. 2010. Flavonoides e atividade antioxidante em *Palicourea rigida* kunth, rubiaceae. Brazilian Journal of Pharmacognosy, 20, 484–488.

Rüben, K.; Wurzlbauer, A.; Walte, A.; Sippl, W.; Bracher, F.; Becker, W. 2015. Selectivity profiling and biological activity of novel β-carbolines as potent and selective DYRK1 kinase inhibitors. PLoS ONE, 10, 1–18.

Santillo, M. F.; Liu, Y.; Ferguson, M.; Vohra, S. N.; Wiesenfeld, P. L. 2014. Inhibition of monoamine oxidase (MAO) by b-carbolines and their interactions in live neuronal (PC12) and liver (HuH-7 and MH1C1) cells. Toxicology in Vitro, 28, 403–410.

Santos, A. de F. A.; Vieira, A. L. S.; Pic-Taylor, A.; Caldas, E. D. 2017. Reproductive effects of the psychoactive beverage ayahuasca in male wistar rats after chronic exposure. Revista Brasileira de Farmacognosia, 27, 353–360.

Silva, J. M.; Hespanhol, R. A. M. 2016. Discussão sobre comunidade e características das comunidades rurais no município de Catalão – GO. Sociedade e Natureza, 28, 361–374.

Silva, V. C.; Carvalho, M. G.; Alves, A. N. 2008. Chemical constituents from leaves of *Palicourea coriacea* (Rubiaceae). Journal of Natural Medicines, 62, 356–357.

Shen, B. A. 2015. New golden age of natural products drug discovery. Cell, 163, 1297–1300.

Soares, P. R. O.; Oliveira, P. L.; Oliveira, C. M. A. de; Kato, L. 2012. In vitro antiproliferative effects of the indole alkaloid vallesiachotamine on human melanoma cells. Archives of Pharmacal Research, 35, 565–571.

Thunig, J.; Hansen, S. H.; Janfelt, J. 2011. Analysis of secondary plant metabolites by indirect desorption electrospray ionization imaging mass spectrometry. Analytical Chemistry, 83, 3256–3259.

Trotter, R.; Logan, M. 1986. Informant consensus: a new approach for identifying potentially effective medicinal plants. In Plants in Indigenous Medicine and Diet: Biobehavioural Approaches. Bedford Hills, New York: Redgrave Publishers.

Tuenter, E.; Exarchou, V.; Apers, S.; Pieter, L. 2017. Cyclopeptide alkaloids. Phytochemistry Reviews, 16, 623–637.

UNESCO. 2018. World Heritage List. http://whc.unesco.org/en/list/.

Vencato, I.; Lariucci, C.; Oliveira, C. M. A. de; Kato, L.; Nascimento, C. A. do. 2004. Acta Crystallographica Section E, 60, 1023–1025.

Vencato, I.; Silva, F. M da; Oliveira, C. M. A. de; Kato, L.; Tanaka, C. M. A.; Silva, C. C. da; Sabino, J. R. 2006. Vallesiachotamine. Acta Crystallographica Section E, 62, 429–431.

Wakui, V. G. 2015. Alcaloides de *Psychotria capitata* Ruiz & Pav. (Rubiaceae): Determinação Estrutural e Atividade Biológica. Master's thesis, Federal University of Goiás.

Walter, B. M. T.; Carvalho, A. M.; Ribeiro, J. F. 2008. O conceito de Savana e de seu componente Cerrado. In S. M. Sano; S. P. Almeida and J. F. Ribeiro (Eds.) Cerrado: Ecologia e Flora. Vol. 1. Brasília: Embrapa Cerrados, 19–42.

World Health Organization – WHO. 2018. Classification of diseases. http://www.who.int/classifications/icd/.

12 Total Synthesis of Some Important Natural Products from Brazilian Flora

Leonardo da Silva Neto[a,b], Breno Germano de Freitas Oliveira[a], Wellington Alves de Barros[a], Rosemeire Brondi Alves[a], Adão Aparecido Sabino[a], and Ângelo de Fátima[a]
[a]Grupo de Estudos em Química Orgânica e Biológica (GEQOB), Departamento de Química, Instituto de Ciências Exatas, Universidade Federal de Minas Gerais *UFMG), Belo Horizonte, Brazil
[b]Instituto Federal Farroupilha, Alegrete, Brazil

CONTENTS

12.1   INTRODUCTION

Since ancient times, mankind has taken advantage of natural products to treat and/or prevent many diseases and dysfunctions, either as original compounds or after modifications (Lachance et al., 2012; Newman and Cragg, 2012). Indeed, nature contains a vast source of natural products that exhibit a plethora of biological activities. The diversity of chemical structure makes natural products very valuable to pharmaceutical industries and agricultural segments as well (Modolo et al., 2015a). Natural products from plants have been a great source of inspiration for improving the quality of human and animal life as disease therapeutics and for increasing food resources (Cragg and Newman, 2013; Dayan et al., 2009; de Fátima et al., 2008; de Fátima et al., 2014; Rates, 2001; Rice et al., 1998; Silva et al., 2014).

Brazil is one of the largest countries (8.5 million km^2) in the world and the largest in all Latin America, in addition to a marine area of more than 4.5 million km^2. In terms of natural resources, Brazil has five important continental biomes and the largest river system in the world, standing out at the global level because it has the richest continental biota on the planet (Prates and Irving, 2015). Brazil contains between 15% and 20% of all world biodiversity (Barreiro and Bolzani, 2009). In addition, it has the largest number of endemic species, the largest tropical forest (the Amazon), and two of the 19 hotspots worldwide (the Atlantic Forest and the *Cerrado*) places Brazil to the first place in the list of megadiverse countries. Brazil is home to 13.2% of the world biota, which means approximately 207,000 known species and 1.8 million projected species, including those yet to be discovered (Prates and Irving, 2015). These characteristics make the Brazilian territory an important resource of natural products that are valuable for the development of new drugs for improving the quality of human and animal life.

However, various natural products that have therapeutic properties are not available in sufficient amounts for sustainable use (de Fátima et al., 2014). Moreover, obtaining a renewable supply of active compounds from biological sources may be problematic, especially with respect to perennial plant species. The complexity of many natural products can also limit the scope of chemical modifications necessary to optimize therapeutic use (de Fátima et al., 2006). Despite these barriers, the total synthesis of various bioactive natural products and analogs has proven that organic synthesis is a powerful tool for increasing the availability of valuable natural products of limited supply or very complex structures (Burns et al., 2009; Mayer et al., 2010; Mendoza et al., 2012; Mickel et al., 2004; Nicolaou and Snyder, 2003; Nicolaou and Sorensen, 1996; Shi et al., 2011; Su et al., 2011).

In this context, we present herein some total syntheses of selected Brazilian plant-derived natural products of pharmacological and agricultural interest, including examples of coumarins, flavonoids, alkaloids, terpenes, and lignoids, among others.

12.2 TOTAL SYNTHESIS OF BRAZILIAN PLANT-DERIVED NATURAL PRODUCTS

12.2.1 COUMARINS: A PRIVILEGED PYRONE-PHENYL IN NATURAL PRODUCTS

In the plant kingdom, coumarins are secondary metabolites responsible for a range of functions, including controlling pathogen dissemination, free radical scavenging, and protection from abiotic stress. The backbone structure of coumarins is constituted by a pyrone-phenyl system, and there is evidence that their biosynthesis in plants occurs from the shikimic acid pathway or from mixed routes (shikimic acid and acetate pathways) (Modolo et al., 2015b). The coumarin pyrone-phenyl system can be prepared by methods such as the von Pechmann reaction (Chenera et al., 1993), Claisen-Cope rearrangement (Cairns et al., 1994), Perkin reaction (Federsel, 2000), Knoevenagel condensation (Shaabani et al., 2009), and Wittig reaction (Demyttenaere et al., 2004).

In 2004, Demyttenaere et al. described a total synthesis of 7-(2,3-epoxy-3-methylbutoxy)-6-methoxycoumarin (**1**; Figure 12.1) from 2,4,5-trimethoxybenzaldehyde in four steps and a 30% overall yield (Figure 12.1). Coumarin **1** is a substance that was first isolated from *Conyza obscura* DC (Asteraceae) (Bohlmann and Jakupovic, 1979) and from petroleum Et_2O extracts of the aerial parts of *Pterocaulon balansae* Chodat (Asteraceae), a plant species from the southern and southeastern Brazilian states (Magalhães et al., 1981). The key feature of Demyttenaere et al. synthetic approach to **1** is the access to the pyrone-phenyl system scopoletin, employing the Wittig reaction followed by an intramolecular transesterification.

Robustic acid (**3**; Figure 12.2) is a pyranocoumarin that was first isolated from *Derris robusta* (Fabaceae), an Indian tree, by Harper (1942), and only after approximately 20 years was its chemical structure elucidated by Johnson and Pelter (1964). In 2001, researchers also isolated this coumarin from petrol and dichloromethane extracts of *Deguelia hatschbachii* (Fabaceae), roots, a Brazilian native species (Magalhães et al., 2001). Donnelly et al. (1995) reported a total synthesis of robustic acid from methyl 2,4,6-trihydroxyphenyl ketone in nine steps and an 18% overall yield (Figure 12.2). In this study, the authors employed the Claisen reaction followed by an intramolecular transesterification to construct the pyrone-phenyl system of compound **5**. A regioselective

Natural product

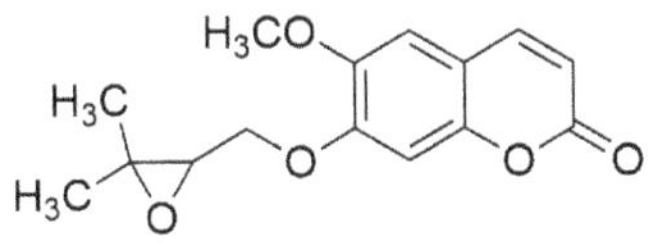

7-(2,3-epoxy-3-methylbutoxy)-6-methoxycoumarin (1)

Total synthesis: 4 steps and 30% overall yield

Brazilian natural source: Pterocaulon balansae Chodat.
(family: Asteraceae, genus: Pterocaulon)

(1)

Key reaction:

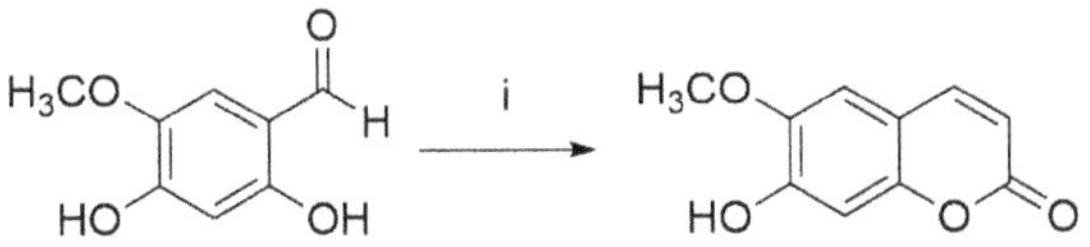

(2)   (scopoletin)

Reagents and conditions: i) (C$_6$H$_5$)$_3$P=CHCOOEt, Et$_2$NC$_6$H$_5$; 190°C to rt, 4 days, 62%

FIGURE 12.1   Synthesis of the 7-(2,3-epoxy-3-methylbutoxy)-6-methoxycoumarin (1) (Demyttenaere et al., 2004).

Natural product

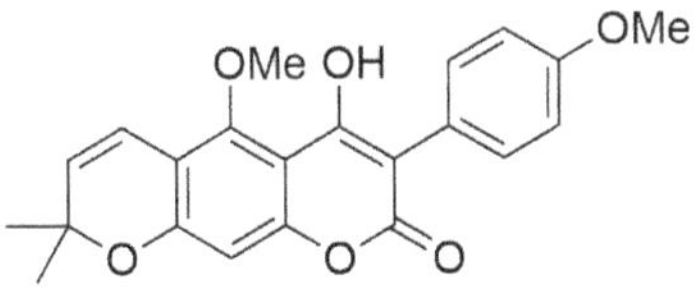

Robustic acid (3)

Total synthesis: 9 steps and 18% overall yield

Brazilian natural source: Deguelia Hastchbachii A. M. G.Azevedo
(family: Leguminosae, genus: Deguelia)

(3)

Key reaction:

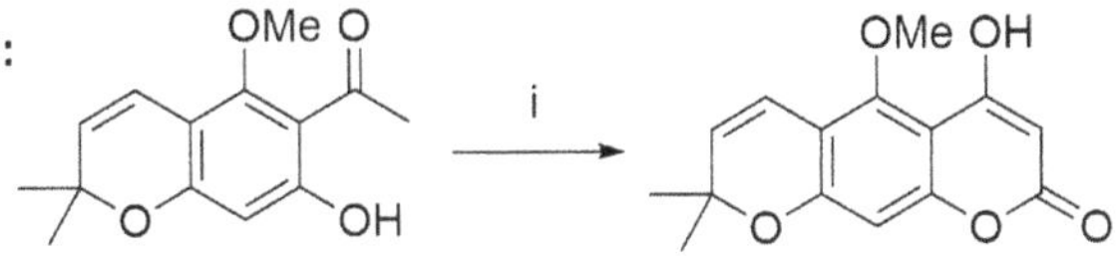

(4)   (5)

Reagents and conditions: i) NaH, Et$_2$CO$_3$, 45°C, 30 min, 80%

FIGURE 12.2   Synthesis of robustic acid (3) (Donnelly et al., 1995).

arylation reaction was employed using an appropriate aryl lead triacetate, in part inspired from earlier work (Barton et al., 1989; Donnelly et al., 1993). Using this methodology in the final step, Donnelly et al. were able to prepare robustic acid, a 3-aryl-pyranocoumarin derivative.

(+)-Calanolide A (**6**; Figure 12.3) is a pyranocoumarin derivative found in two main *Calophyllum* species: *Calophyllum brasiliense* (Calophyllaceae; *guanandi, jacareúba*, or *landim*) (Flora do Brasil, 2020a; Huerta-Reyes et al., 2004) and *Calophyllum lanigerum* (Calophyllaceae) (Kashman et al., 1992). (+)-Calanolide A and other similar naturally occurring pyranocoumarins, called cala-nolides, are well-known to exhibit anti-tuberculotic activity and possess potent anti-HIV activity (Brahmachari, 2015). Therefore, calanolides occupy an important position in the coumarin class of compounds, and many researchers are interested in new methodologies of their total synthesis (Brahmachari, 2015). (+)-Calanolide A was first isolated from *C. lanigerum*, a tropical rainfor-est tree (Kashman et al., 1992), and, in the early 2000s, from hexane extracts of *C. brasiliense* (Calophyllaceae; *guanandi, jacareúba*, and *landim*) leaves, a tree widely distributed in Brazilian

Natural product

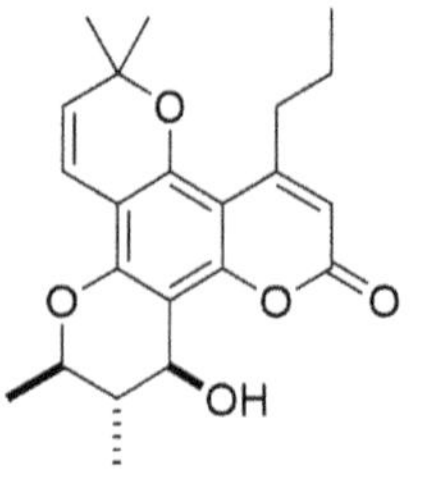

(+)-Calanolide A (6)

Total synthesis: 8 steps and 14% overall yield

Brazilian natural source: Calophyllum brasiliense Cambess.
(family: Calophyllaceae, genus: Calophyllum)

(6)

Key reaction:

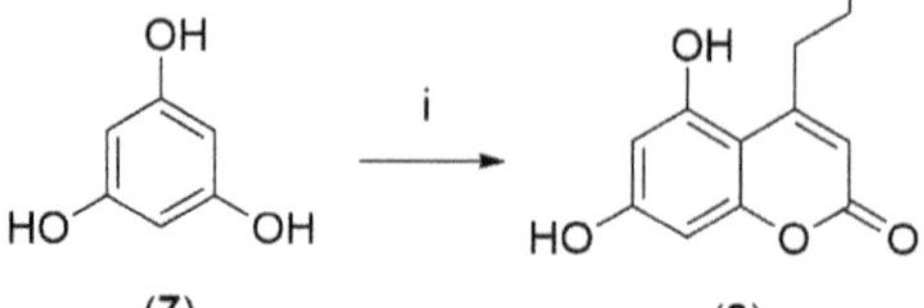

(7)   (8)

Reagents and conditions: i) ethyl 3-propyl-3-oxopropionic acid, TFA, rt, 12 h, 86%

FIGURE 12.3   Synthesis of (+)-calanolide A (6) (Sekino et al., 2004).

territory (Huerta-Reyes et al., 2004). The first enantioselective total synthesis of (+)-calanolide A was reported by Deshpande et al. (1995). Later, in 2004, Sekino et al. published a total synthesis of (+)-calanolide A using phloroglucinol **7** as the starting material (Figure 12.3). To obtain the pyrone-phenyl system of intermediate **8**, the authors employed the von Pechmann reaction using the starting material **7**, ethyl 3-propyl-3-oxopropionic acid, and acidic conditions. This reaction is one of the main synthetic strategies to obtain coumarins, first reported in 1883 by Hans von Pechmann and Duisberg. It consists of a transesterification, an electrophilic attack and a dehydration (not necessarily in this order) in the presence of protic or Lewis acid (Daru and Stirling, 2011). Sekino et al. (2004) performed the total synthesis of (+)-calanolide A from phloroglucinol **7** in eight steps and in a 14% overall yield (Figure 12.3).

12.2.2   FLAVONOIDS: A DIETARY NATURAL PRODUCT WITH HEALTH BENEFITS

Flavonoids are polyphenol compounds and they are more commonly found in vegetal species. These compounds possess the three-ring structures, and are often color responsive in many plants, and participate in insect defense mechanisms (Jaisankar et al., 2014). Pharmacologically, flavonoids are widely known for their antioxidant properties; they are biochemically synthesized from cinnamoyl-CoA and three malonyl-CoA molecules to form a polyketide that is enzymatically converted to the basic flavonoid building block (Dewick, 2002).

In 2006, Lee and Kim described an elegant synthesis for the pyranoflavone isolonchocarpin (**9**; Figure 12.4) in five steps with an overall yield of 48%. Isolonchocarpin **9** is a pyranoflavone isolated from several plant species (Canzi et al., 2014; Garcez et al., 1988; Huo et al., 2015; Wang et al., 2015) and was isolated by Magalhães et al. (1996) from the petroleum extract of the roots of *Dahlstedtia floribunda* (Fabaceae, basionym: *Lonchocarpus subglaucescens*; *embira-de-sapo*), a species of plant native to Brazil, where the name of the substance is derived from the genus name of the plant (Flora do Brasil, 2020b). One of the key steps of the strategy presented by Lee and Kim is a one-pot Knoevenagel electrocyclic reaction of diketone **10** with 3-methyl-2-butenal, which provides a key starting material **11** for the construction of natural pyranoflavones. Various solvents and catalysts were evaluated, obtaining the best yield in benzene in the presence of 10 mol% ethylamine diacetate (Figure 12.4). Although the route presented by Lee and Kim shows a good yield at all

Natural Product

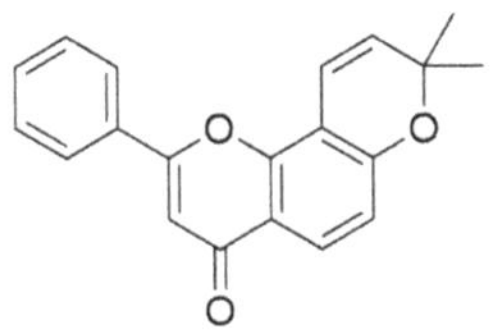

Isolonchocarpin (9)

Total synthesis: 5 steps and 48% of overall yield

Brazilian natural source: Dahlstedtia floribunda
basionym: Lonchocarpus subglaucescens
(family: Fabaceae, genus: Dahlstedtia)

(9)

Key Reaction:

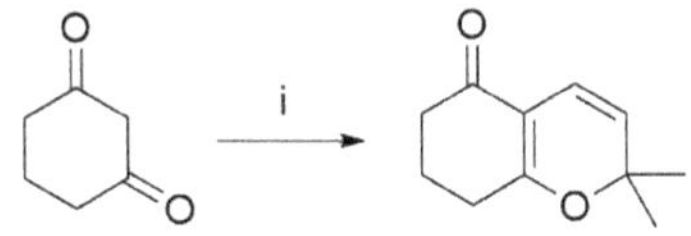

(10)   (11)

Reagents and conditions: i) 3-methyl-2-butenal, ethylenediamine diacetate 10 mol%, benzene, reflux, 3 h, 92%

FIGURE 12.4   Synthesis of isolonchocarpin (9) (Lee and Kim, 2006).

stages (>70%), in the key stage, the best yield was in benzene, a highly toxic solvent. As presented by the authors, this problem can be overcome by using fewer toxic solvents with still high yields, such as CH_2Cl_2 and 88% yield (Lee and Kim, 2006).

In 2007, Urgaonkar and Shaw described a synthesis of the glycosylated flavonoid kaempferitrin (12; Figure 12.5) in nine steps with an overall yield of 5% (Figure 12.5). Kaempferitrin 12 is a glycosylated flavonoid derived from kaempferol. This compound was isolated from various plants (Euler and Alam, 1982; Yang et al., 2010), including the ethanolic extract from *Pterogyne nitens* (Fabaceae; *amendoinzeiro, amendoim-bravo, cocal, tipá, viraró, madeira-nova,* or *vilão*), a common plant species in Brazil (Flora do Brasil, 2020c; Regasini et al., 2008). The synthetic strategy of Urgaonkar and Shaw to prepare 12 starts with a conversion of 7 into a functionalized ketone by

Natural Product

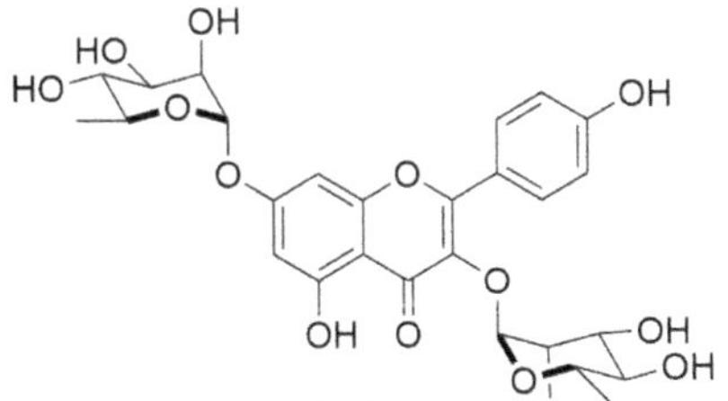

Kaempferitrin (12)

Total synthesis: 9 steps and 5% of overall yield

Brazilian natural source: Pterogyne nitens
(family: Fabaceae, genus: Pterogyne)

(12)

Key Reactions:

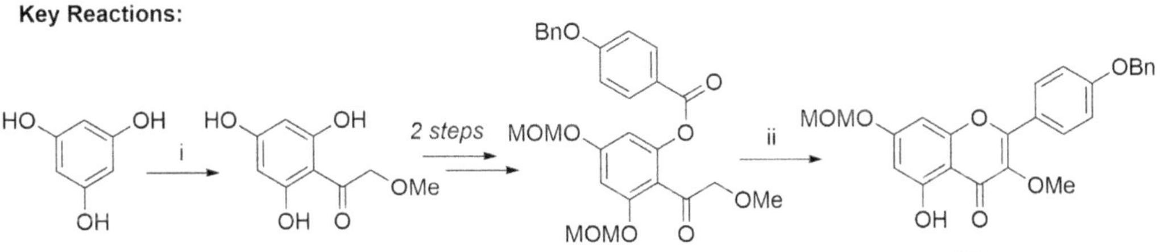

(7)   (13)   (14)   (15)

Reagents and conditions: i) methoxyacetonitrile, $ZnCl_2$, HCl (g), 0°C, 56 h, 76%; ii) K_2CO_3, pyridine, reflux, 1 h, 42%

FIGURE 12.5   Synthesis of kaempferitrin (12) (Urgaonkar and Shaw, 2007).

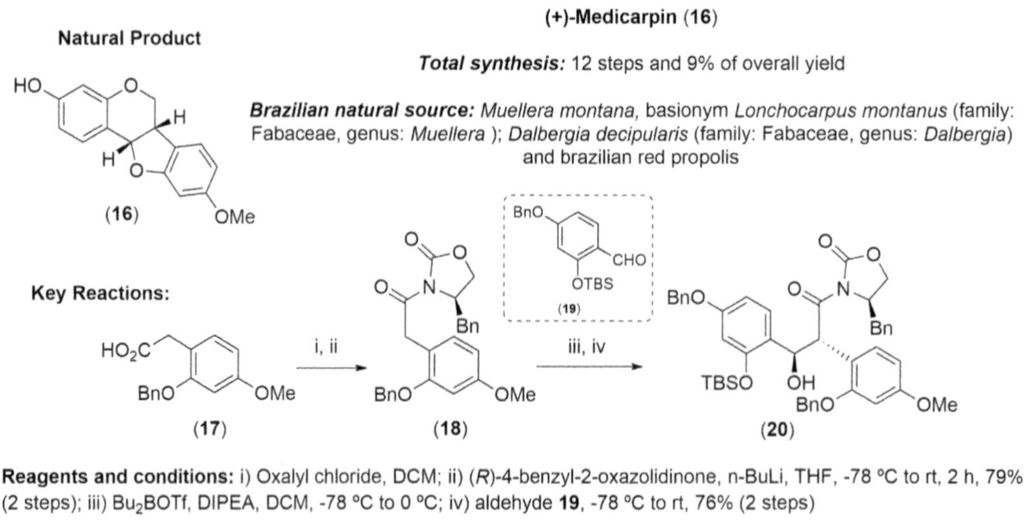

(+)-Medicarpin (16)

Natural Product

Total synthesis: 12 steps and 9% of overall yield

Brazilian natural source: Muellera montana, basionym *Lonchocarpus montanus* (family: Fabaceae, genus: *Muellera*); *Dalbergia decipularis* (family: Fabaceae, genus: *Dalbergia*) and brazilian red propolis

Key Reactions:

Reagents and conditions: i) Oxalyl chloride, DCM; ii) (*R*)-4-benzyl-2-oxazolidinone, n-BuLi, THF, -78 °C to rt, 2 h, 79% (2 steps); iii) Bu$_2$BOTf, DIPEA, DCM, -78 °C to 0 °C; iv) aldehyde **19**, -78 °C to rt, 76% (2 steps)

FIGURE 12.6   Synthesis of (+)-medicarpin (**16**) (Yang et al., 2017).

a Houben-Hoesch reaction. Intermediate **14** was obtained by protection and benzoylation of **13**. Another important step, this strategy is the cyclization reaction of **14** to obtain the key intermediate **15**, the core of the flavonoid skeleton. With the flavonoid core **15**, product **12** can be obtained from bisglycosylation. The authors tried two approaches to glycosylation, using both a rhamnose source and an activator, noting that when acidic conditions were used, decomposition of intermediate **15** occurred. Urgaonkar and Shaw described a total synthesis from simple reagents; however, the strategy requires many steps of protection and deprotection, reducing the overall yield of the route.

In 2017, Yang et al. presented an interesting example of an asymmetric synthesis for the pterocarpan (+)-medicarpin (**16**; Figure 12.6), a type of isoflavonoid; the synthesis was performed in 12 steps with a 9% overall yield. (+)-Medicarpin **16** has already been isolated from several species of plants, among them some of Brazilian occurrence, and can be found in the petroleum extract of the roots of *Muellera montana* (Fabaceae; basionym: *Lonchocarpus montanus*; *cabelouro*, or *carrancudo*) (Magalhães et al., 2007; Santos, Braga, et al., 2009), in the benzene extract of the core of *Dalbergia decipularis* (Fabaceae; *Sebastião-de-arruda*) (de Alencar et al., 1972), and found in the Brazilian red propolis (Li et al., 2008). The strategy presented by Yang et al. for the asymmetric synthesis of **16** made use of a chiral oxazolidone auxiliary group to construct the two chiral centers in one step; reacting **17** with (*R*)-4-benzyl-2-oxazolidinone using n-butyllithium as a base gave **18** in good yield, which was then converted to **20** *via* an Evans asymmetric aldol addition with the aldehyde **19**.

12.2.3   Alkaloids: Ubiquitous Bioactive Natural Products

Alkaloids are an important class of substances that contain at least one basic heterocyclic nitrogen and are obtained from α-amino acids (AAs) or their immediate derivatives through related biosynthetic pathways (true alkaloids). Many alkaloids have elevated pharmacologic activity acting primarily on the central nervous system, with relatively high toxicity for humans. Alkaloids are distributed in a large number of subtypes, and the majority are derived from the following AAs: L-ornithine, L-lysine, L-phenylalanine, L-tyrosine, anthranilic acid, L-tryptophan, L-histidine, and nicotinic acid. Despite being found in animals, insects, and microorganisms, the main source of alkaloids is in the plant kingdom, especially among angiosperms (Cordell, 1981; Hegnauer, 1963; Roberts and Wink, 1998).

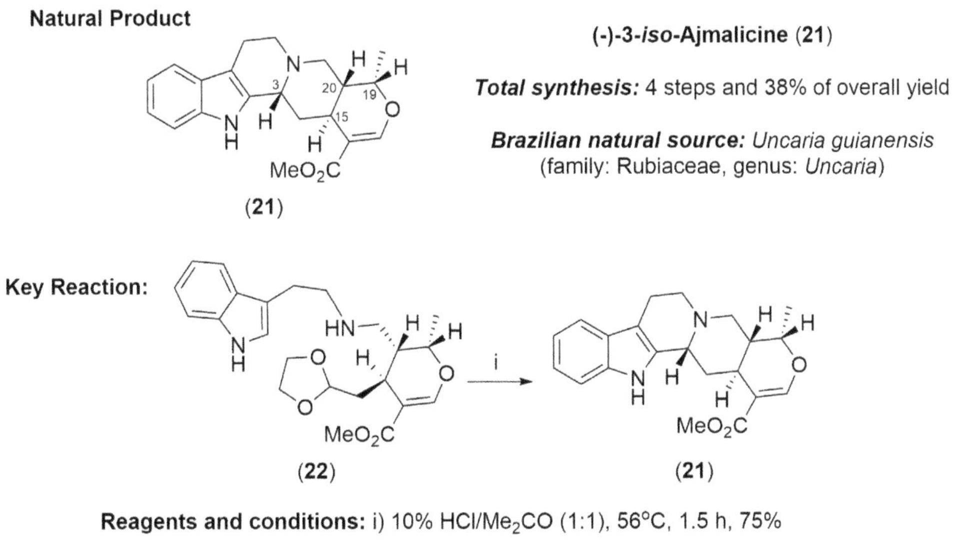

Natural Product

(21)

(-)-3-*iso*-Ajmalicine (21)

Total synthesis: 4 steps and 38% of overall yield

Brazilian natural source: *Uncaria guianensis*
(family: Rubiaceae, genus: *Uncaria*)

Key Reaction:

(22)   **(21)**

Reagents and conditions: i) 10% HCl/Me₂CO (1:1), 56°C, 1.5 h, 75%

FIGURE 12.7   Synthesis of (−)-3-iso-ajmalicine (**21**) (Brown et al., 2002).

In 2002, Brown et al. described an approach to access three heteroyohimbine alkaloids from secologanin, a natural biogenetic source of indole alkaloids (Contin et al., 1998). There are many pharmacological properties reported for heteroyohimbine alkaloids, a group repeatedly found in the Rubiaceae family, mainly in the *Uncaria* genus. In South America, *Uncaria tomentosa* (Rubiaceae; *unha-de-gato*) and *Uncaria guianensis* (Rubiaceae; *unha-de-gato*), known as cat's claw, are commonly used in local popular medicine (Sandoval et al., 2002). Working with the extracts of powdered leaves of *U. guianensis* (Rubiaceae), collected from the Viro forest reserve in Pará state, Brazil, Bolzani and coworkers (2004) isolated some oxindole alkaloids, including the less common 3-*iso*-ajmalicine **21**. The synthesis of this alkaloid was reported by Brown et al. in 2002 in four steps and a 38% overall yield (Figure 12.7). The key step of this proposal was the final step, where the two central rings were constructed, and the five-ring system was all connected. This process was accomplished by employing a Pictet-Spengler condensation, which occurs when compound **22** is treated in acidic media to afford the desired alkaloid **21**. This reaction sequence can be inverted, with the Pictet-Spengler step being performed in the initial steps, as previously reported by Brown and Leonard (1979); however, the route described herein allowed better control over the C-3 stereochemistry, although a minor amount (~10%) of the C-3 epimer was observed on some occasions.

Another oxindole alkaloid is diaboline **23**, which belongs to the same family as strychnine. In 1984, Nicoletti et al. identified this alkaloid from Brazilian biodiversity during their studies with the *Strychnos* genus. Indeed, *Strychnos pseudoquina* (Longaniacea; *quina-do-campo* or *falsa-quina*), a plant that grows in the Brazilian Cerrado, locally known as "quina do campo" or "falsa quina" and used in regional medicine for the treatment against malaria and other diseases, afforded the (+)-diaboline **23** alkaloid from ethanolic extracts of its ground and dried leaves. The structural complexity of this molecule was evidenced by the synthetic approach undertaken by Ohshima et al. (2004) that accessed this alkaloid in 28 steps and an approximately 1% overall yield (Figure 12.8). The key step comprised a catalytic and enantioselective Michael reaction between malonate **24** and cyclic enone **25**, mediated by the AlLibis(binaphthoxide) complex (ALB) (Shimizu et al., 1998; Takayoshi et al., 1996); the improved methodology in this work allowed the Michael adduct **26** to be obtained on a multi-gram scale and in high enantiomeric excess (*e.e.*). Another very well-delineated strategy was the construction of the BCDE ring system through the domino cyclization process starting from compound **27** in three steps and 66% yield, affording compound **28,** which already has the basic core of the desired product. Finally, the (+)-diaboline **23** was achieved in an elegant synthetic

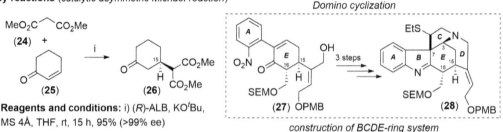

Natural Product

(+)-Diaboline (23)

Total synthesis: 28 steps and 1% of overall yield

Brazilian natural source: *Strychnos pseudoquina*
(family: Loganiaceae, genus: *Strychnos*)

(23)

Key reactions (*catalytic asymmetric Michael reaction*)

Domino cyclization

Reagents and conditions: i) (*R*)-ALB, KO*t*Bu,
MS 4Å, THF, rt, 15 h, 95% (>99% ee)

construction of BCDE-ring system

FIGURE 12.8   Synthesis of (+)-diaboline (**23**) (Ohshima et al., 2004).

route, likely the most efficient from the literature, and can be used in the synthesis of other advanced *Strychnos* alkaloids. However, some drawbacks can be noted in this approach, such as the formation of a mixture of isomers in some steps, which decreases the overall yield, and an unexpected epimerization of the C-16 stereocenter during the E-ring formation.

Another indole alkaloid found in South American plants was (+)-affinisine **29**, which exhibits relevant pharmacological activities and has been isolated from ethanolic extracts of ground whole plant of *Peschiera affinis* (Apocynaceae; *grão-de-galo*) collected in northeastern Brazil (Santos, Magalhães, et al., 2009; Weisbach et al., 1963). This compound is also found in *Tabernaemontana hystrix* (Apocynaceae; *esperta*), a species of plant native to Southeastern Brazil, known as "esperta" (Monnerat et al., 2005). Although the alkaloid affinisine **29** possesses important biological activities, there are no syntheses described to date. However, Liu and coworkers have shown an elegant synthetic approach to obtain the (–)-enantiomer of affinisine in a total of nine steps and a 22% overall yield (Figure 12.9) (Liu et al., 2000). Starting from L-tryptophan derivative **30**, asymmetric

Natural Product

(-)-Affinisine

Total synthesis: 9 steps and 22% of overall yield

Brazilian natural source (+)-Affinisine: *Peschiera affinis*
(family: Apocynaceae, genus: *Peschiera*)

(+)-Affinisine (29)

Key Reaction:

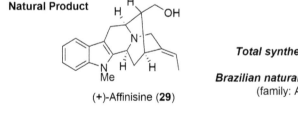

(30)   (31)

Reagents and conditions: i) (MeO)₂CHCH₂CH₂CO₂Me, TFA, DCM, rt, 7 days, 90%

FIGURE 12.9   Synthesis of the enantiomer of (+)-affinisine (**29**) (Liu et al., 2000).

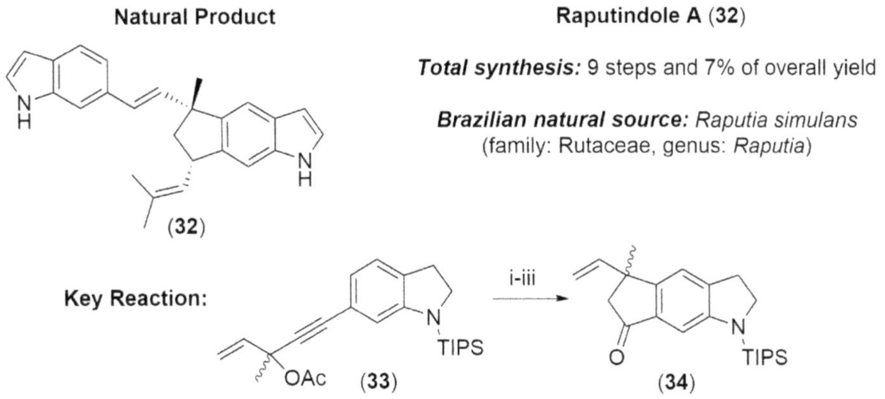

Natural Product							**Raputindole A (32)**

(32)

Total synthesis: 9 steps and 7% of overall yield

Brazilian natural source: Raputia simulans
(family: Rutaceae, genus: *Raputia*)

Key Reaction:

Reagents and conditions: i) AuCl(PPh$_3$) 3 mol%, AgClO$_4$ 3 mol%, DCM, rt, sonication, 10 min, then celite filtration; ii) 13 in DCM, rt, 24 h; iii) NaOMe, MeOH, rt, 10 min, 85% (3 steps)

FIGURE 12.10   Synthesis of raputindole A (**32**) (Kock et al., 2017).

Pictet-Spengler cyclization was employed to afford adduct **31** with 100% diastereoselectivity. This process is the key transformation and one of more used approaches to access other indole alkaloids (Li et al., 1999; Wang et al., 1998). As stated before, although there is no synthesis to natural alkaloid **29**, an enantiomer from an intermediate posterior to compound **31** was already obtained in previous works (Li et al., 1999; Wang et al., 1998; Yu and Cook, 1998), suggesting the possibility of accessing the natural (+)-affinisine (**29**).

A new class of uncommon alkaloids, bisindoles, was established by Vougogiannopoulou et al. (2010), who isolated four new structures from an Amazonian plant. From these structures, raputindole A **32** was isolated from dichloromethane (DCM) extracts of dried and powdered roots of *Raputia simulans* Kallunki (Rutaceae), a plant found in the extreme west of the Brazilian Amazon.

In 2017, Kock et al. reported the first total synthesis of raputindole A in nine steps and a 7% overall yield (Figure 12.10). In fact, starting from the advanced propargyl ester adduct **33**, a cyclopentannulation reaction was employed, catalyzed by Au(I), which resulted in the cyclopenta[*f*]indoline **34** with regioselectivity and good yield. Indeed, this key reaction had been previously reported by Marsch et al. (2016), whose cyclization process was first described by Marion et al. in 2006. With tricyclic core **34** constructed, the bisindole alkaloid raputindole A was accessed in an additional five steps and 56% overall yield. However, the alkaloid was obtained in the racemic form together with one epimer in a nearly 1:1 ratio, requiring a subsequent purification step with semipreparative chiral HPLC to afford the natural (+)-raputindole A **32**; obtaining the pure compound allowed the determination of the absolute configuration through calculation of the ECD spectrum (Kock et al., 2017).

In 2014, L'Homme et al. reported a concise approach for the total synthesis of the tetracyclic alkaloid erysotrine **35** (Figure 12.11) in nine steps and 0.1% overall yield. Erysotrine **35** is an alkaloid commonly isolated from plants of the *Erythrina* genus, a species that grows in tropical countries such as Brazil. Indeed, Sarragiotto et al. (1981) have isolated several alkaloids representative of the *Erythrina* species, including erysotrine **35**, from methanolic extracts of finely ground flowers from *Erythrina mulungu* (Leguminosae; *mulungu* or *mulungu-coral*), a plant collected in the Santa Elisa farm from Campinas, Brazil (de Lima et al., 2006). For the synthetic odyssey proposed by Canesi and coworkers, a key step is the dearomatization of the phenolic A-ring of compound **36**, promoted by hypervalent iodine with the aim of improving the reactivity of the enone **37**. In sequence, the B-ring was constructed by an aza-Michael cyclization, followed by a Pictet-Spengler cyclization to afford the tetracyclic core (ABCD ring) of the desired alkaloid (Figure 12.11) (L'Homme et al., 2014). Despite the simple approach and short route, this strategy provides a very low overall yield.

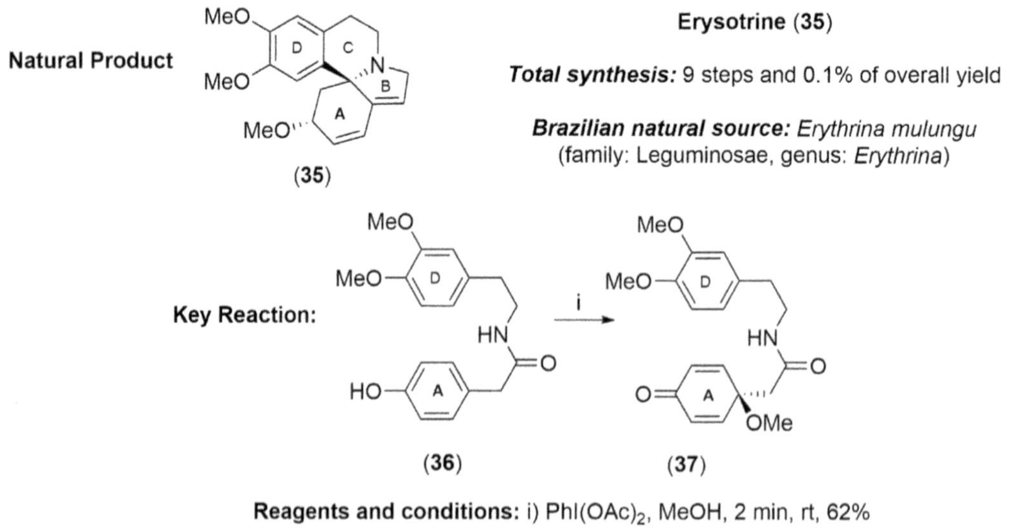
Natural Product
MeO
MeO
MeO
D
C
N
B
A
(35)
Erysotrine (35)
Total synthesis: 9 steps and 0.1% of overall yield
Brazilian natural source: Erythrina mulungu
(family: Leguminosae, genus: Erythrina)
Key Reaction:
MeO
MeO
D
HN
O
HO
A
i
OMe
(36)
(37)
Reagents and conditions: i) PhI(OAc)2, MeOH, 2 min, rt, 62%

FIGURE 12.11 Synthesis of erysotrine (**35**) (L'Homme et al, 2014).

12.2.4 Terpenes: A Diverse Class of Natural Products with Valuable Bioactivities

The terpenoids are an important class of secondary metabolites produced by plants. These compounds are formed by the union of isoprene units, and the terpenes are classified according to the number of isoprene units. In this way, the hemiterpenes are formed by an isoprene unit (C_5), monoterpenes by two (C_{10}), sesquiterpenes by three (C_{15}), diterpenes by four (C_{20}), sesterterpenes by five (C_{25}), triterpenes by six (C_{30}), and tetraterpenes by eight (C_{40}). The terpenes have structural diversity since after the union of the isoprene units and cyclization of these units, several structural modifications can occur during the biosynthetic process (Dewick, 2002).

In 1993, Wang et al. described the stereoselective total synthesis of 7β-acetoxyvouacapane (**38**; Figure 12.12) in nine steps and an overall yield of approximately 18%. 7β-Acetoxyvouacapane

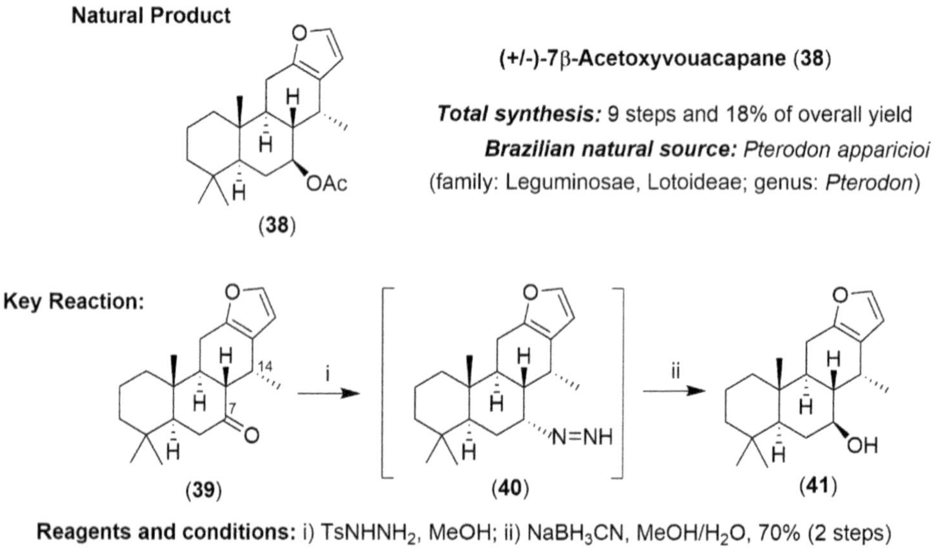
Natural Product
O
H
H
H
OAc
(38)
(+/-)-7β-Acetoxyvouacapane (38)
Total synthesis: 9 steps and 18% of overall yield
Brazilian natural source: Pterodon apparicioi
(family: Leguminosae, Lotoideae; genus: Pterodon)
Key Reaction:
14
7
i
ii
N=NH
OH
(39)
(40)
(41)
Reagents and conditions: i) TsNHNH2, MeOH; ii) NaBH3CN, MeOH/H2O, 70% (2 steps)

FIGURE 12.12 Synthesis of 7β-acetoxyvouacapane (**38**) (Wang et al., 1993).

Natural Product

(42)

(-)-Seychellene (42)

Total synthesis: 18 steps and 13% of overall yield

Brazilian natural source: Cedrela odorata and *Toona ciliate*
(family: Meliaceae, genus: *Cedrela* and *Toona*)

Key Reactions:

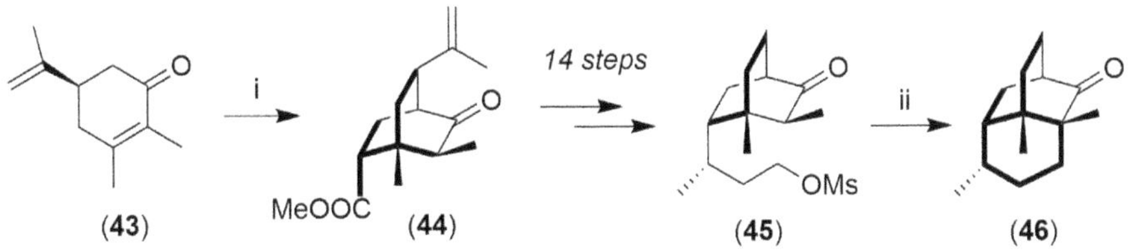

(43)   MeOOC (44)   14 steps   (45) OMs   (46)

Reagents and conditions: i) LiHMDS/Hexane, $H_2C=CHCO_2Me$, rt, 66%; ii) NaH/THF, reflux, 76%

FIGURE 12.13   Synthesis of (–)-seychellene (**42**) (Srikrishna and Ravi, 2008).

is a tetracyclic furano diterpene isolated from plant species of the genus *Pterodon*, such as *Pterodon apparicioi* (Leguminosae; *sucupira*), a plant species from the banks of the Rio Cipó, state of Minas Gerais, Brazil (Fascio et al., 1976; Hansen et al., 2010). The key feature of Wang and cowork-ers' synthetic approach to **38** is access to the intermediary **39** using the stereoselective reductive hydrolysis of a tosylhydrazone **40**. This reduction step does not work when traditional methods are used (LiAlH$_4$/THF; Li(*t*-BuO)$_3$AlH/THF; NaBH$_4$/MeOH; Raney nickel/MeOH) because of steric hindrance of the 7-carbonyl group due to the 14α-methyl group (Figure 12.12) (Wang et al., 1993). The main disadvantage of the synthesis route for 7β-acetoxyvouacapane proposed by Wang and coworkers is that the furano diterpene is obtained in racemic form.

In 2008, Srikrishna and Ravi described the stereoselective total synthesis of (–)-seychellene **42** from (*R*)-carvone in 18 steps and a 13% overall yield (Figure 12.13). (–)-Seychellene is a tricyclic sesquiterpene hydrocarbon isolated from *Cedrela odorata* (Meliaceae; *cedro, cedro-branco, cedro-rosa,* or *cedro-vermelho*) and *Toona ciliata* (Meliaceae; *cedro australiano*), two plant species from Viçosa, state of Minas Gerais, Brazil (Flora do Brasil, 2020d; Maia et al., 2000). The key features of Srikrishna and Ravi's synthetic approach to **42** are the generation of intermediate **44** via tandem intermolecular Michael addition-intramolecular Michael addition and the access of two vicinal qua-ternary carbon atoms in the tricyclic structure **46** (Figure 12.13). In addition, the authors obtained optically pure (–)-seychellene **42,** and they used the readily available monoterpene (*R*)-carvone **43** as the starting material (Figure 12.13) (Srikrishna and Ravi, 2008). The main disadvantage of the synthetic route to (–)-seychellene **42** proposed by Srikrishna and Ravi is the large number of steps required.

In 2009, Surendra and Corey described a short enantioselective total synthesis of the pentacyclic triterpene lupeol (**47**; Figure 12.14) in eight steps and a 10% overall yield. Lupeol is a pentacyclic triterpene isolated from several plants, including the plant species *Cordiera macrophylla* (Rubiaceae; basionym: *Alibertia macrophylla; marmelada-de-cachorro*) herbs, shrubs, or trees from the Atlantic Forest and Cerrado of São Paulo, state of São Paulo, Brazil (Bolzani et al., 1991; Flora do Brasil, 2020e). The key feature of Surendra and Corey's synthetic approach to **47** is the access to epoxide **50** with correct stereochemistry, which is essential for stereocontrolled cation olefin polycyclization (**50** for **51**; Figure 12.14). This synthesis was possible due to the careful choice of the starting material.

Natural Product

Lupeol (47)

Total synthesis: 8 steps and 10% of overall yield

Brazilian natural source: *Cordiera macrophylla*, basionym: *Alibertia macrophylla* (family: Rubiaceae, genus: *Cordiera*)

Key Reactions:

Reagents and conditions: i) Li$_2$CuCl$_4$, THF, 0°C, 65%; ii) MeAlCl$_2$, Me$_2$AlCl, DCM, -78°C; iii) TBAF, THF, 0°C, 43% (2 steps)

FIGURE 12.14   Synthesis of lupeol (**47**) (Surendra and Corey, 2009).

In 2014, Tran and Cramer described the biomimetic synthesis of (+)-viridiflorol **52** from (+)-2-carene in seven steps and an overall yield of approximately 20% (Figure 12.15). (+)-Viridiflorol is a sesquiterpene alcohol isolated from *Cedrela odorata* (Meliaceae; *cedro, cedro branco, cedro rosa,* or *cedro vermelho*) and *Toona ciliata* (Meliaceae; *cedro australiano*), two plant species from Viçosa, state of Minas Gerais, Brazil (Flora do Brasil, 2020d; Maia et al., 2000). One of the key features of Tran and Cramer's synthetic approach to **52** is access to intermediate **55** using stereoselective olefination from ketoaldehyde **53** (Figure 12.15). This step does not work well when a

Natural Product

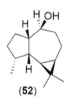

(+)-Viridiflorol (52)

Total synthesis: 7 steps and 20% of overall yield

Brazilian natural sources: *Cedrela odorata* and *Toona ciliata* (family: Meliaceae; genus: *Cedrela* and *Toona*)

Key Reactions:

Reagents and conditions: i) phosphonium salt **54**, BuLi, LiBr, -100°C, 69%; ii) pTsOH, acetone, 94%; iii) SmI$_2$, THF (10 mM), 70°C, 61%

FIGURE 12.15   Synthesis of (+)-viridiflorol (**52**) (Tran and Cramer, 2014).

high reaction temperature or an alteration of the base or solvent is employed, resulting in a loss of diastereoselectivity (Figure 12.15) (Tran and Cramer, 2014). Another critical step in this synthesis to obtain (+)-viridiflorol is the cyclization of intermediate **56**. The first attempt of direct cyclization of the olefin **56** using McMurry conditions or related couplings failed. Thus, the aldehyde was unprotected, and cyclization was performed under McMurry conditions with complete diastereoselectivity (Figure 12.15) (Tran and Cramer, 2014).

12.2.5 Lignans: Phenylpropane Derivatives Widely Distributed in Higher Plants

Formed from two units of phenylpropanoid derived from cinnamic acid, in which biosynthesis is catalyzed by different enzymes, lignans are dimeric structures with high diversity in plants; different structural derivatives of cinnamic acid are found in this class of compounds (Dewick, 2002). Some authors hold that the term lignans should be employed only when the dimers are coupled from the C3 side chain, as the nomenclature used for another type of coupling is neolignans (Dewick, 2002). There are currently more than 7,000 natural varieties of lignans, and one of the pioneers and major contributors in Brazil in the isolation and identification of this class of compounds was Dr. Otto Gottlieb (1978) (Fazary et al., 2016). Within medicinal chemistry, lignans exhibit cardiovascular (Ghisalberti, 1997), antioxidant (Lee et al., 2004), anti-inflammatory (Kim et al., 2009), anticancer (Lee et al., 2014), antibacterial (Bai et al., 2018), and antiviral (Charlton, 1998) activities. Due to the diverse pharmacological properties associated with this class, many lignan natural products isolated from plants have become desired synthetic targets in the search for new structures with different applications (Xu et al., 2018).

Plants of the Lauraceae family, common in Latin America and found in the Amazon region, are sources of bioactive neolignoids, such as eusiderin A **58** and B **59** (Figure 12.16), isolated from the species *Licaria aurea* (Lauraceae; *folha-de-ouro* or *folha-dourada*) and *Licaria rigida* (Lauraceae; *louro-fígado-de-galinha*), respectively, which have antiviral activity (Braz-Filho et al., 1981; Flora do Brasil, 2020f; Marques et al., 1992; Pilkington et al., 2018). In 2012, Pilkington and Barker reported an enantioselective divergent synthesis of these structures in 11 steps with a yield of 0.7% for eusiderin A and 0.5% for eusiderin B, in addition to the absolute determination of the stereochemistry of eusiderins.

To synthesize different eusiderins, the authors traced a strategy involving a general route, varying only the reagent used in the last step responsible for the side chain and the substituents of the attached aromatic ring. The synthesis has two key steps: the first key step is a Mitsunobu reaction between the phenolic derived from the previously synthesized *o*-vanillin and (*S*)-ethyl lactate, thus fixing the first chiral center in the structure, followed by a reduction with DIBAL at −78°C that results in the formation of aldehyde **62** (Figure 12.16), a common intermediate for the other eusiderins, which undergoes an addition of an aromatic organometallic reagent that differentiates eusiderin A **58** from eusiderin B **59**; after a reduction and cyclization in Amberlyst 15, formation of the intermediate occurs (**63, 64**), which undergoes the second key step of eusiderin synthesis (Suzuki reaction). Then, the eusiderins are synthesized from the intermediates **63** and **64**, and the boronate ester side chain derived from organoboron is used, since the Suzuki reaction is a carbon-carbon cross-coupling reaction. A disadvantage faced in the synthetic route proposed by Pilkington and Barker (2012) is the step involving the formation of the second stereogenic center of the 1,4-benzodioxane ring, wherein the cyclization generates the *cis* and *trans* isomers in a ratio of 1:5, resulting in a reduction in the reaction yield due to the need to separate the products for the last step of the synthesis.

Another bioactive lignan is fargesin (**65**; Figure 12.17) that has anti-inflammatory (Yue et al., 2018) and antimycobacterial (Jiménez-Arellanes et al., 2012) activities, which can be found in Brazilian plants of the *Aristolochia labiata* (Aristolochiaceae; basionym *Aristolochia galeata; angelicó, buta, crista-de-galo, milhomens, papo-de-peru,* or *peru-bosta*) and *Virola flexuosa* (Myristicaceae; *ucuuba* or *ucuuba-folha-grande*) species, which are from Brazilian Cerrado and

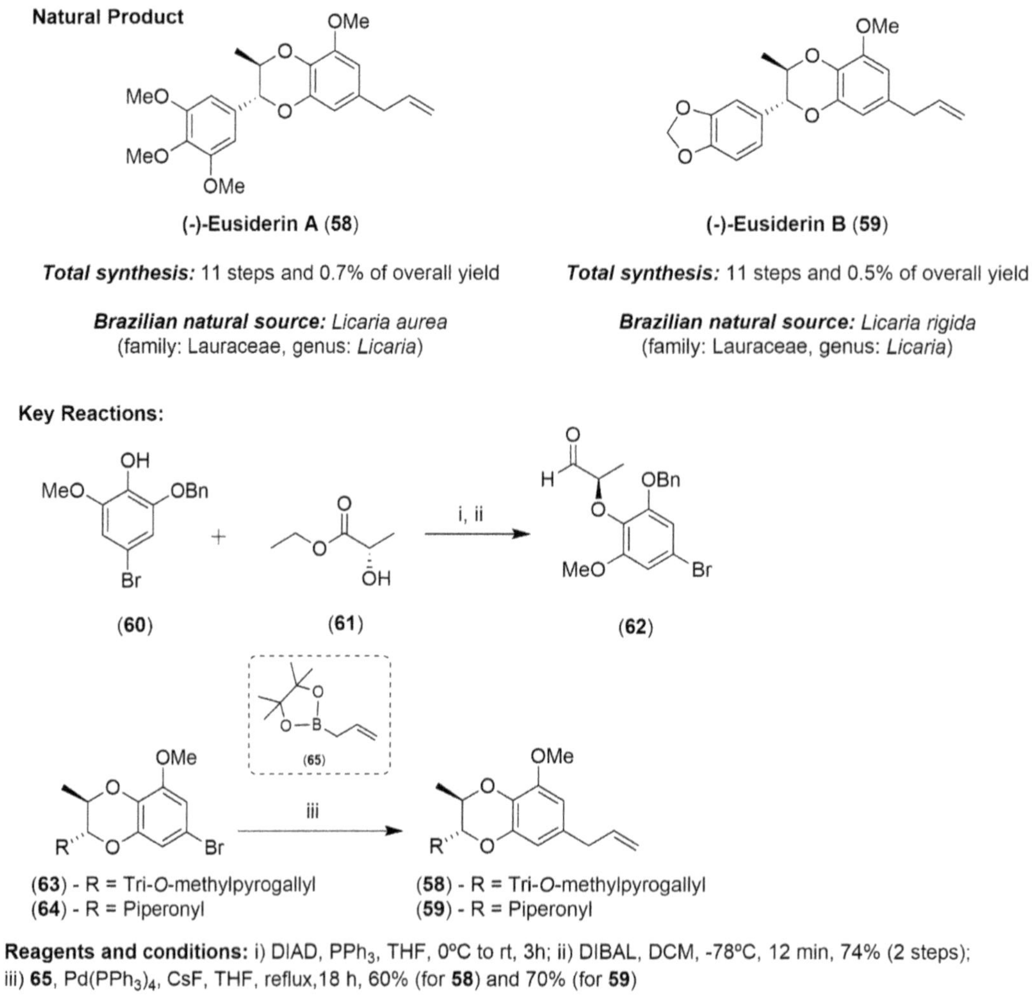
Natural Product
OMe
MeO
MeO
OMe
(-)-Eusiderin A (58)
Total synthesis: 11 steps and 0.7% of overall yield
Brazilian natural source: Licaria aurea
(family: Lauraceae, genus: Licaria)
OMe
(-)-Eusiderin B (59)
Total synthesis: 11 steps and 0.5% of overall yield
Brazilian natural source: Licaria rigida
(family: Lauraceae, genus: Licaria)
Key Reactions:
OH
MeO
OBn
Br
(60)
OH
(61)
i, ii
H
OBn
MeO
Br
(62)
(65)
OMe
Br
iii
OMe
(63) - R = Tri-O-methylpyrogallyl
(64) - R = Piperonyl
(58) - R = Tri-O-methylpyrogallyl
(59) - R = Piperonyl
Reagents and conditions: i) DIAD, PPh3, THF, 0°C to rt, 3h; ii) DIBAL, DCM, -78°C, 12 min, 74% (2 steps); iii) 65, Pd(PPh3)4, CsF, THF, reflux,18 h, 60% (for 58) and 70% (for 59)

FIGURE 12.16 Synthesis of (–)-eusiderin A (**58**) and (–)-eusiderin B (**59**) (Pilkington and Barker, 2012).

Amazonian regions, respectively (Cavalcante et al., 1985; da Silva et al., 2011; Flora do Brasil, 2020g, 2019h; Lopes and Bolzani, 1988). In order to find methods for the synthesis of bioactive lignans, Yoda et al. (2005) propose a synthetic route of furofuran lignans starting from the monoterpene lactone with eight steps and 6.4% overall yield. The synthetic strategy is based on the opening of the lactam ring forming the first stereogenic center, which after oxidation undergoes a stereoselective Grignard reaction. The main characteristic of the synthetic approach of Yoda et al. is the production of tetrol **70** from the coupling reaction between **69** and the corresponding aldehyde followed by deprotection and reduction reactions; in the hold of tetrol, the authors show the method of forming the furofuran ring from mesylation. However, a negative point of the fargesin synthesis is that the cyclization final step forms the methyl piperitol diastereoisomer as the major product, soon abruptly reducing the overall yield for these syntheses.

12.2.6 Some Miscellaneous Synthetic Examples

Many other interesting examples of the total synthesis of Brazilian plant-derived natural products of pharmacological and agricultural interest are reported elsewhere. For instance, rotenone

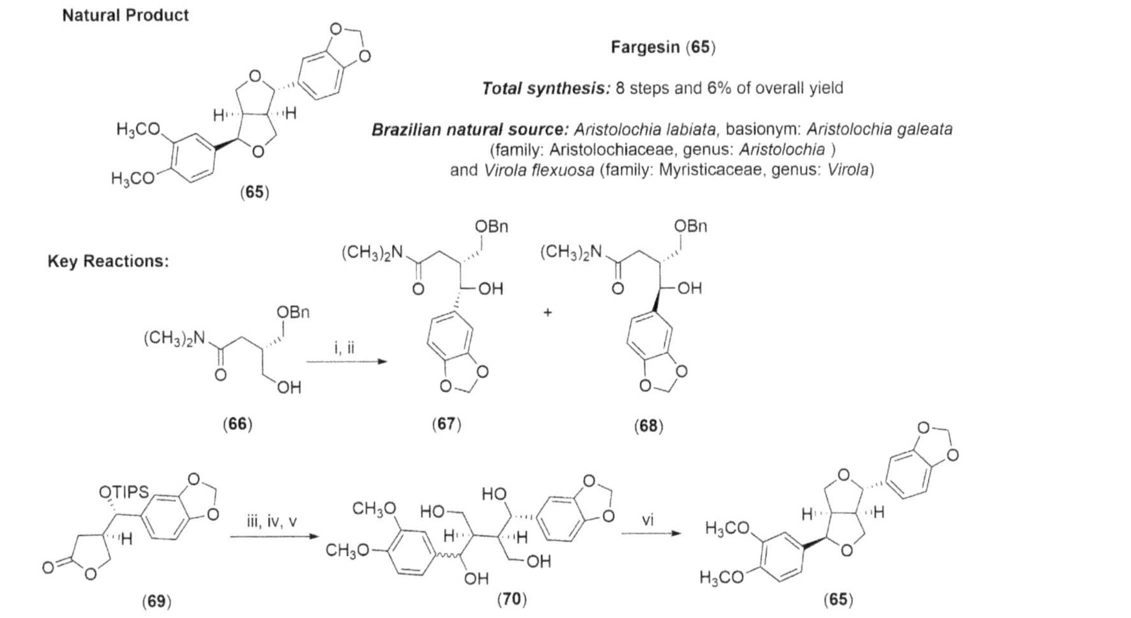
Natural Product
Fargesin (65)
Total synthesis: 8 steps and 6% of overall yield
Brazilian natural source: Aristolochia labiata, basionym: Aristolochia galeata
(family: Aristolochiaceae, genus: Aristolochia)
and Virola flexuosa (family: Myristicaceae, genus: Virola)
(65)
Key Reactions:
(66)
(67)
(68)
i, ii
(69)
iii, iv, v
(70)
vi
(65)

Reagents and conditions: i) $(COCl)_2$, $(C_2H_5)_3N$, -78 to -45 °C, DMSO/THF; ii) $C_7H_5BrMgO_2$, 0 °C, THF, 75% for **67**, 3% for **68** (2 steps); iii) LiHMDS, $C_9H_{10}O_3$, -78 °C; iv) $(C_4H_9)_4NF$, THF; v) $LiAlH_4$, 64% (3 steps); vi) MsCl, pyridine, DCM, 24%.

FIGURE 12.17 Synthesis of fargesin (**65**) (Yoda et al., 2005).

(**71**; Figure 12.18) and tephrosin (**76**; Figure 12.19), members of a class of secondary metabolites of plant origin known as rutenoids, were isolated from plants of the Fabaceae family, both exhibiting pesticide and insecticide activities (Garcia et al., 2010). Rotenone (**71**; Figure 12.18) is a metabolite isolated from numerous plants that is marketed as an insecticide and shows activities such as cytotoxicity, genotoxicity, and larvicidal activity against the dengue vector *Aedes aegypti* (Estrella-Parra et al., 2014; Huang et al., 2009; Vasconcelos et al., 2012). The fact that rotenone is marketed as an insecticide generates health warnings since studies indicate that this metabolite can lead to the manifestation of Parkinson's disease and is even used as a Parkinson's inducer for scientific models

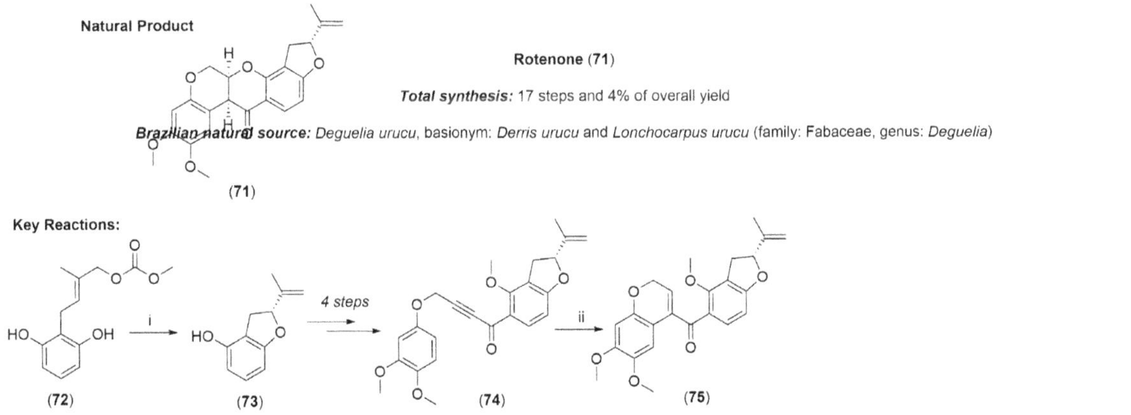
Natural Product
Rotenone (71)
Total synthesis: 17 steps and 4% of overall yield
Brazilian natural source: Deguelia urucu, basionym: Derris urucu and Lonchocarpus urucu (family: Fabaceae, genus: Deguelia)
(71)
Key Reactions:
(72)
i
(73)
4 steps
(74)
ii
(75)

Reagents and conditions: i) 2 mol% $Pd(dba)_2$, 6 mol% R,R'-Trost ligand, 1 eq AcOH, DCM, rt, 18 h, 98% (92% *ee*); ii) cat. $PtCl_2$ toluene, 70°C, 2 h, 77%

FIGURE 12.18 Synthesis of rotenone (**71**) (Georgiou et al., 2017).

Natural Product

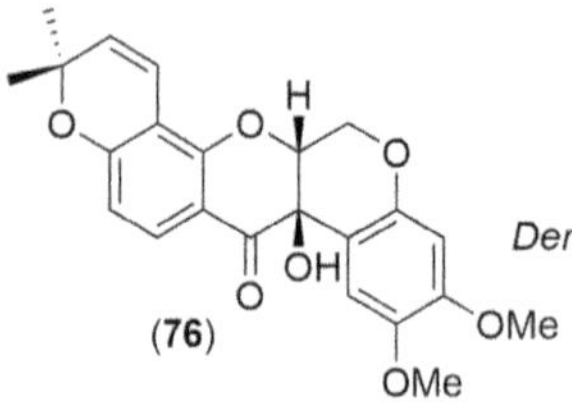

Tephrosin (76)

Total synthesis: 7 steps and 8% of overall yield

Brazilian natural source: Deguelia urucu, basionym:
Derris urucu and *Lonchocarpus urucu* (family: Fabaceae, genus: *Deguelia*) and
Tephrosia candida (family: Fabaceae, genus: *Tephrosia*)

(76)

Key Reaction:

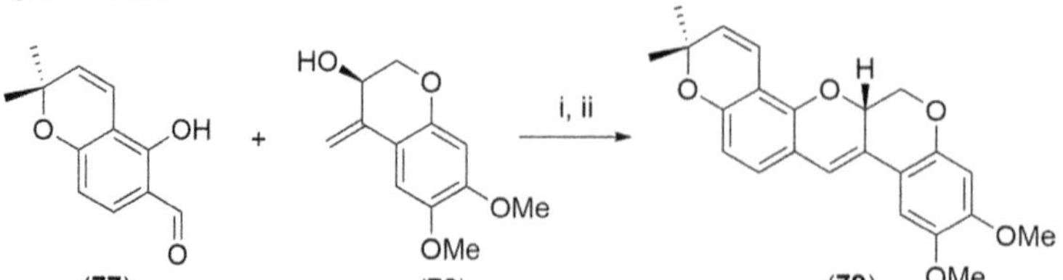

(77)   +   **(78)**   i, ii   **(79)**

Reagents and conditions: i) PS-DEAD (2.0 eq), PPh$_3$ (2.0 eq.), Et$_3$N (2.0 eq), DCM, 23°C, 24 h, 63%; ii) cat. Grubbs II (0.10 eq), DCM, 23°C, 4 h, 78%

FIGURE 12.19   Synthesis of Tephrosin (**76**) (Garcia et al., 2010).

(Lin et al., 2018; Maturana et al., 2014). Rotenone (**71**) can be found in the plant *Deguelia urucu* (Fabaceae; basionym: *Derris urucu* and *Lonchocarpus urucu*; *timbó-urucu*) of occurrence in Brazil (Fang and Casida, 1999; Flora do Brasil, 2020i). In 2017, Georgiou et al. published the first total stereoselective total synthesis for rotenone in 17 steps with a 4% overall yield. The presented strategy involved two key transformations: The first transformation was previously reported by Pelly et al. (2007), consisting of the Pd π-allyl-mediated cyclization of **72** to obtain the dihydrobenzofuran skeleton **73**. And, the second key transformation was a 6-end hydroarylation of intermediate **74** to construct the chromene **75** precursor for rotenone (Georgiou et al., 2017).

Tephrosin (**76**; Figure 12.19) was isolated by Braz-Filho et al. (1975) from the ethanolic extract of the aerial wood of *D. urucu* (Fabaceae; basionym: *Derris urucu* and *Lonchocarpus urucu*; *timbó-urucu*) and by Parmar et al. (1988) from the seeds of *Tephrosia candida* (Fabaceae; *tefrósia* or *anil branco*) (ANVISA, 2010; Flora do Brasil, 2020i). In 2010, Garcia et al. reported a convergent total synthesis with seven steps starting from 3,4-dimethoxyphenol with an 8% overall yield. The main step for the synthesis is the convergence step between the routes, a coupling between the key intermediates **77** and **78** under Mitsunobu conditions followed by cyclization with Grubb's second-generation catalyst to afford intermediate **79** (Garcia et al., 2010).

One example of α-pyrone total synthesis was reported in 2018 by Vaithegi and Prasad when they described, for the first time, the total synthesis of cryptopyranmoscatone B2 (**80**; Figure 12.20) in 16 steps and a 0.9% overall yield. The cryptopyranmoscatone B2 **80** is one of the six cryptopyranmoscatones isolated from the branch of stem bark of *Cryptocarya moschata* (Lauraceae; *canela-batalha* or *canela*), a tree found in the Atlantic woods in the southeastern region of Brazil (Cavalheiro and Yoshida, 2000). Obtaining cryptopyranmoscatone B2 **80** is not a simple synthesis, and the low overall yield is justified because the molecule possesses five stereogenic centers. Among the 16 steps involved in the product preparation, the olefin metathesis and Brown's allylation stand out, being that the latter forms homoallylic alcohol as the unique diastereoisomer. However, Vaithegi and Prasad highlight the formation of the pyran ring as a key synthetic step since the optimized conditions by the authors favored the formation of *cis* tetrahydropyran as the major product to fix the stereochemistry of the pyran ring. Some disadvantages of the total synthesis proposed by Vaithegi and Prasad are steps that have yields of less than 50%, such as the selective protection of

Natural Product

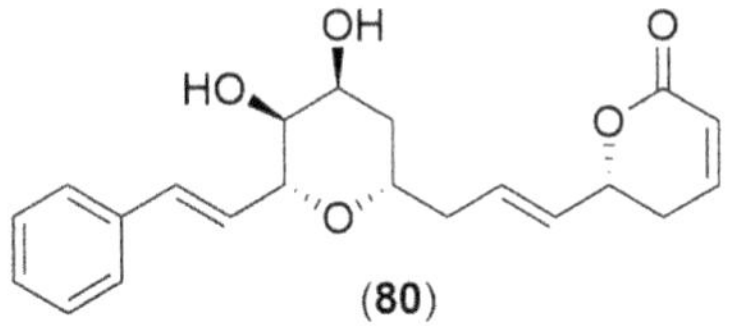

Cryptopyranmoscatone B2 (80)

Total synthesis: 16 steps and 0.9% of overall yield

Brazilian natural source: *Cryptocarya moschata*
(family: Lauraceae, genus: *Cryptocarya*)

Key Reactions:

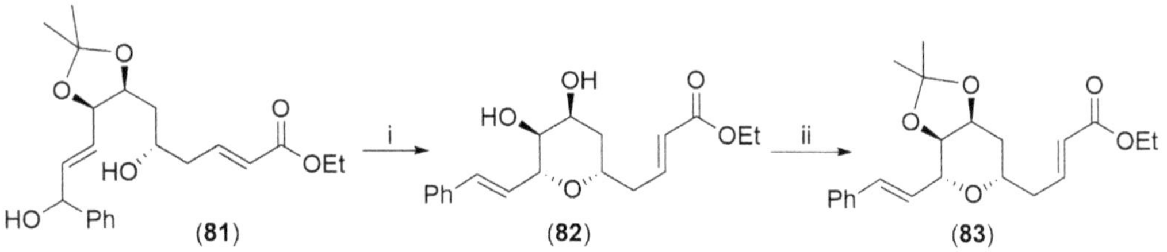

Reagents and conditions: i) FeCl$_3$ .6 H$_2$O (30 mol%), DCM , rt, 5 h; ii) 2-methoxypropene, CSA, DCM , 0°C, 10 min, 64% (2 steps)

FIGURE 12.20   Synthesis of cryptopyranmoscatone B2 (**80**) (Vaithegi and Prasad, 2018).

primary alcohol with TES and acryloylation, a key step for lactam ring formation, in addition to performing another protection step due to the experimental conditions for **82** formation.

In 2012, Peng et al. described a short total synthesis of the pyranochalcone lonchocarpin (**84**; Figure 12.21) in three steps and a 24% overall yield. Lonchocarpin is a flavonoid isolated from *Derris floribunda* (Fabaceae; *timbó-venenoso*) collected in the vicinity of Manaus, state of Amazonas, Brazil (Braz-Filho et al., 1975). The synthetic approach of Peng et al. to access **84** is quite short and simple. First, the key intermediate **87** was obtained by the condensation of 2,4-dihydroxylacetophenone **86** with 3-methylcrotonaldehyde **85**, as described in Figure 12.21; then, compound **86** was protected by MOM-Cl, and Claisen-Schmidt condensation between the protected compound and benzaldehyde led to **84** in a good yield.

The name of the Brazilian plant species referred to as in this chapter and additional information are summarized in Table 12.1.

Natural Product

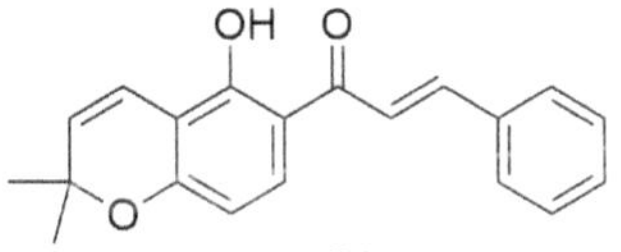

Lonchocarpin (84)

Total synthesis: 3 steps and 24% of overall yield

Brazilian natural source: *Derris floribunda*
(family: Fabaceae, genus: *Derris)*

Key Reaction:

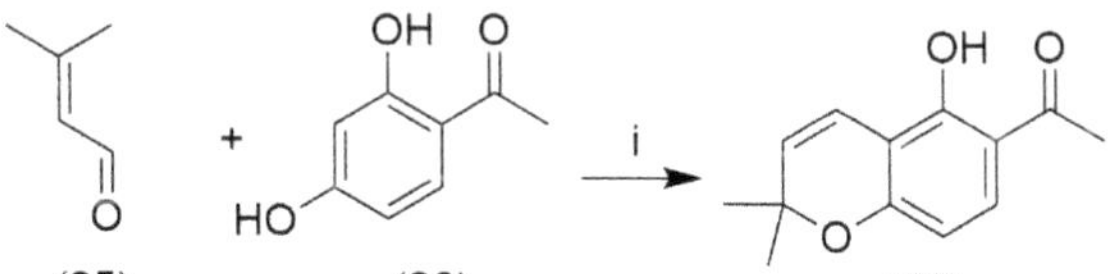

Reagents and conditions: i) pyridine, reflux, 120°C, 12 h, 50%

FIGURE 12.21   Synthesis of lonchocarpin (**84**) (Peng et al., 2012).

TABLE 12.1

The Names of All Plant Species Including Family and Common Names from Which Some Natural Products Originated and Are Presented in This Chapter

Scientific Name	Family	Common Names
Aristolochia labiata (basionym *Aristolochia galeata*)	Aristolochiaceae	*Angelicó, buta, crista-de-galo, milhomens, papo-de-peru* or *peru-bosta*
Calophyllum brasiliense (sin. *Calophyllum lucidum*)	Calophyllaceae	*Guanandi, jacareúba* or *landim*
Calophyllum lanigerum (sin. *Calophyllum frutescens*)	Calophyllaceae	
Cedrela odorata (*sin. Cedrela angustifolia*)	Meliaceae	*Cedro, cedro-branco, cedro-rosa* or *cedro-vermelho*
Conyza obscura (sin. *Webbia kraussii*)	Asteraceae	
Cordiera macrophylla (sin. *Alibertia macrophylla*)	Rubiaceae	*Marmelada-de-cachorro*
Cryptocarya moschata (sin. *Cryptocarya moschata f. angustifolia*)	Lauraceae	*Canela-batalha* or *canela*
Dahlstedtia floribunda (sin. *Lonchocarpus subglaucescens*)	Fabaceae	*Embira-de-sapo*
Dalbergia decipularis	Fabaceae	*Sebastião-de-arruda*
Deguelia hatschbachii	Fabaceae	
Deguelia urucu (basionyms: *Derris urucu; Lonchocarpus urucu*)	Fabaceae	*Timbó-urucu*
Derris floribunda	Fabaceae	*Timbó-venenoso*
Derris robusta (sin. *Brachypterum robustum*)	Fabaceae	
Erythrina verna (*sin. Erythrina mulungu*)	Leguminosae	*Mulungu* or *mulungu-coral*
Licaria aurea (sin. *Acrodiclidium aureum*)	Lauraceae	*Folha-de-ouro* or *folha-dourada*
Licaria chrysophylla (sin. *Licaria rigida*)	Lauraceae	*Louro-fígado-de-galinha*
Muellera montana, (basionym: *Lonchocarpus montanus*)	Fabaceae	*Cabelouro* or *carrancudo*
Pterocaulon balansae Chodat (sin. *Pterocaulon paniculatum*)	Asteraceae	
Pterodon emarginatus (*sin. Pterodon apparicioi*)	Leguminosae	*Sucupira*
Pterogyne nitens (sin. *Pterogyne nitens* f. *parvifolia*)	Fabaceae	*Amendoinzeiro, amendoim-bravo, cocal, tipá, viraró, madeira-nova* or *vilão*
Raputia simulans Kallunki	Rutaceae	
Strychnos pseudoquina	Longaniacea	*Quina-do-campo* or *falsa-quina*
Tabernaemontana catharinensis (sin. *Peschiera affinis*)	Apocynaceae	*Grão-de-galo*
Tabernaemontana hystrix (sin. *Tabernaemontana gracillima*)	Apocynaceae	*Esperta*
Tephrosia candida (sin. *Robinia candida*)	Fabaceae	*Tefrósia* or *anil branco*
Toona ciliata (*sin. Toona microcarpa*)	Meliaceae	*Cedro australiano*
Uncaria guianensis (sin. *Ourouparia guianensis*)	Rubiaceae	*Unha-de-gato*
Uncaria tomentosa (sin. *Nauclea polycephala*)	Rubiaceae	*Unha-de-gato*
Virola flexuosa	Myristicaceae	*Ucuuba* or *ucuuba-folha-grande*

12.3   CONCLUDING REMARKS

Brazilian flora is undoubtedly one of the most plentiful sources of inspiration for the development of new drugs. Nevertheless, few products have emerged from this rich source of chemical diversity, and few have their total synthesis described so far, especially by Brazilian organic synthetic groups. The total synthesis of various bioactive natural products and analogs from Brazilian flora has proven that organic synthesis is a powerful tool for increasing the availability of valuable natural products of limited supply for confirmation and/or correction of the structure of such natural products. Moreover, the unique carbon-carbon scaffolds of such natural products represent a great opportunity and a challenge for the synthetic organic chemist, with a push to develop new synthetic methodologies to synthesize such interesting natural products. In general, Brazil has the potential to be one of the most important countries in the world in the field of fine chemicals, pharmaceuticals, and agrochemicals, and it is critical to respond to contemporary challenges by having effective and continuous supportive policies.

REFERENCES

Agência Nacional de Vigilância Sanitária (ANVISA) – Brasil. 2010. Resolução RE n° 4.479 de 30/09/10. http://portal.anvisa.gov.br/documents/111215/117782/T63.pdf/2c979355-3eca-4200-93ed-907fe615a861.

Bai, M.; Wu, S. -Y.; Zhang, W. -F.; Song, X.-P.; Han, C.-R.; Zheng, C.-J.; Chen, G.-Y. 2018. One new lignan derivative from the fruiting bodies of *Ganoderma lipsiense*. Natural Product Research, 16, 1–5.

Barreiro, E. J.; Bolzani, V. S. 2009. Biodiversity: potential source for drug discovery. Quimica Nova, 32, 679–688.

Barton, D. H. R.; Donnelly, D. M. X.; Finet, J. -P.; Guiry, P. J. 1989. A facile synthesis of 3-aryl-4-hydroxycoumarins. Tetrahedron Letters, 30, 1539–1542.

Bohlmann, F.; Jakupovic, J. 1979. 8-Oxo-α-selinen und neue scopoletin-derivate aus *Conyza*-arten. Phytochemistry, 18, 1367–1370.

Bolzani, V. S.; Trevisan, L. M. V.; Young, M. C. M. 1991. Caffeic acid esters and triterpenes of *Alibertia Macrophylla*. Phytochemistry, 30, 2089–2091.

Brahmachari, G. 2015. Bioactive natural products: chemistry and biology. In G. Brahmachari, ed., Naturally Occurring Calanolides: Chemistry and Biology, 1st ed., pp. 349–374. Weinheim: Wiley-VCH Verlag GmbH & Co. KGaA.

Braz-Filho, R. B.; de Carvalho, M. G.; Gottlieb, O. R.; Maia, J. G. S.; Da Silva, M. L. 1981. Neolignans from *Licaria rigida*. Phytochemistry, 20, 2049–2050.

Braz-Filho, R.; Gottlieb, O. R.; Mourão, A. P.; da Rocha, A. I.; Oliveira, F. S. 1975. Flavonoids from *Derris* species. Phytochemistry, 14, 1454–1456.

Brown, R. T.; Dauda, B. E. N.; Pratt, S. B.; Richards, P. 2002. Short stereoselective synthesis of (–)-ajmalicine, (–)-3-iso-ajmalicine and their 5-methoxycarbonyl derivatives from secologanin. Heterocycles, 56, 51–58.

Brown, R. T.; Leonard, J. 1979. Biomimetic synthesis of cathenamine and 19-epicathenamhe, key intermediates to heteroyohimbine alkaloids. Journal of the Chemical Society, Chemical Communications, 20, 877–879.

Burns, N. Z.; Krylova, I. N.; Hannoush, R. N.; Baran, P. S. 2009. Scalable total synthesis and biological evaluation of haouamine A and its atropoisomer. Journal of American Chemical Society, 131, 9172–9173.

Cairns, N.; Harwood, L. M.; Astles, D. P. 1994. Tandem thermal Claisen-Cope rearrangements of coumarate derivatives. Total syntheses of the naturally occurring coumarins: suberosin, demethylsuberosin, ostruthin, balsamiferone and gravelliferone. Journal of Chemical Society, Perkin Transactions, 1(21), 3101–3107.

Canzi, E. F.; Marques, F. A.; Teixeira, S. D.; Tozzi, A. M. G. A.; Silva, M. J.; Duarte, R. M. T.; Duarte, M. C. T.; et al. 2014. Prenylated flavonoids from roots of *Dahlstedia glaziovii* (Fabaceae). Journal of the Brazilian Chemical Society, 25, 995–1001.

Carbonezi, C. A.; Hamerski, L.; Otavio, A. F.; Furlan, M.; Bolzani, V. S. 2004. Determinação por RMN das configurações relativas e conformações de alcalóides oxindólicos isolados de *Uncaria guianensis*. Química Nova, 27, 878–881.

Cavalcante, S. de H.; Fernandes, D.; Paulino Fo, H. F.; Yoshida, M.; Gottlieb, O. R. 1985. Lignoids from the fruit of three *Virola* species. Phytochemistry, 24, 1865–1866.

Cavalheiro, A. J.; Yoshida, M. 2000. 6-[ω-arylalkenyl]-5,6-dihydro-α-pyrones from *Cryptocarya moschata* (Lauraceae). Phytochemistry, 53, 811–819.

Charlton, J. L. 1998. Antiviral activity of lignans. Journal of Natural Products, 61, 1447–1451.

Chenera, B.; West, M. L.; Finkelstein, J. A.; Dreyer, G. B. 1993. Total synthesis of (±)-calanolide A, a non-nucleoside inhibitor of HIV-1 reverse transcriptase. The Journal of Organic Chemistry, 58, 5605–5606.

Contin, A.; van der Heijden, R.; Lefeber, A. W. M.; Verpoorte, R. 1998. The iridoid glucoside secologanin is derived from the novel triose phosphate/pyruvate pathway in a *Catharanthus roseus* cell culture. FEBS Letters, 434, 413–416.

Cordell, G. A. 1981. Introduction to Alkaloids: A Biogenetic Approach, 1st ed. New York: Wiley.

Cragg, G. M.; Newman, D. J. 2013. Natural products: a continuing source of novel drugs leads. Biochimica et Biophisica Acta – General Subjects, 1830, 3670–3695.

da Silva, A. D.; Borghetti, F.; Thompson, K.; Pritchard, H.; Grime, J. P. 2011. Underdeveloped embryos and germination in *Aristolochia galeata* seeds. Plant Biology, 13, 104–108.

Daru, J.; Stirling, A. 2011. Mechanism of the Pechmann reaction: a theoretical study. The Journal of Organic Chemistry, 76, 8749–8755.

Dayan, F. E.; Cantrell, C. L.; Duke, S. O. 2009. Natural products in crop protection. Bioorganic & Medicinal Chemistry, 17, 4022–4034.

de Alencar, R.; Filho, R. B.; Gottlieb, O. R. 1972. Pterocarpanoids from *Dalbergia decipularis*. Phytochemistry, 11, 1517.

de Fátima, A.; Modolo, L. V.; Conegero. L. S.; Pilli, R. A.; Ferreira, C. V.; Kohn, L. K.; de Carvalho, J. E. 2006. Styryl lactones and their derivatives: biological activities, mechanisms of action and potential leads for drug design. Current Medicinal Chemistry, 13, 3371–3384.

de Fátima, A.; Modolo, L. V.; Sanches, A. C.; Porto, R. R. 2008. Wound healing agents: the role of natural and non-natural products in drug development. Mini-Reviews in Medicinal Chemistry, 8, 879–888.

de Fátima, A.; Terra, B. S.; da Silva, C. M.; da Silva, D. L.; Araujo, D. P.; Silva-Neto, L.; de Aquino, R. A. N. 2014. From nature to market: examples of natural products that became drugs. Recent Patents on Biotechnology, 8, 76–88.

de Lima, M. R. F.; Luna, J. S.; dos Santos, A. F.; Andrade, M. C. C.; Sant'Ana, A. E. G.; Genet, J. P.; Marquez, B.; Neuville, L.; Moreau, N. 2006. Anti-bacterial activity of some Brazilian medicinal plants. Journal of Ethnopharmacology, 105, 137–147.

Demyttenaere, J.; Vervisch, S.; Debenedetti, S.; Coussio, J.; Maes, D.; Kimpe, N. 2004. Synthesis of virgatol and virgatenol, two naturally occurring coumarins from *Pterocaulon virgatum* (L.) DC, and 7-(2,3-epoxy-3-methylbutoxy)-6-methoxycoumarin, isolated from *Conyza obscura* DC. Synthesis, 11, 1844–1848.

Deshpande, P. P.; Tagliaferri, F.; Victory, S. F.; Victory, S. F.; Yan, S.; Baker, D. C. 1995. Synthesis of optically active calanolides A and B. The Journal of Organic Chemistry, 60, 2964–2965.

Dewick, P. M. 2002. Medicinal Natural Products: A Biosynthetic Approach, 2nd ed. New York: John Wiley & Sons.

Donnelly, D. M. X.; Finet, J. -P.; Rattigan, B. A. 1993. Organolead-mediated arylation of Allyl b-ketoesters: a selective synthesis of isoflavanones and isoflavones. Journal of Chemical Society, Perkin Transactions 1, 15, 1729–1735.

Donnelly, D. M. X.; Molloy, D. J.; Reilly, J. P.; Finet, J. 1995. Aryllead-mediated synthesis of linear 3-arylpyranocoumarins: synthesis of robustin and robustic acid. Journal of Chemical Society, Perkin Transactions 1, 20, 2531–2534.

Estrella-Parra, E. A.; Gomes-Verjan, J. C.; Gonzáles-Sánchez, I.; Vázquez-Martínez, E. R.; Vergara-Casstañeda, E.; Cerbón, M. A.; Alavez-Solano, D.; Ryes-Chilpa, R. 2014. Rotenone isolated from *Pachyrhyzus erosus* displays cytotoxicity and genotoxicity in K562 cells. Natural Product Research, 28, 1780–1785.

Euler, K. L.; Alam, M. 1982. Isolation of kaempferitrin from *Justicia specigera*. Journal of Natural Products, 45, 211–212.

Fang, N.; Casida, J. E. 1999. Cube' resin insecticide: identification and biological activity of 29 rotenoid constituents. Journal of Agricultural Food Chemistry, 47, 2130–2136.

Fascio, M.; Mors, W. B.; Gilbert, B.; Mahajan, J. R.; Monteiro, M. B.; Dos Santos Filho, D.; Vichnewski, W. 1976. Diterpenoid furans from *Pterodon* species. Phytochemistry, 15, 201–203.

Fazary, A. E.; Alfaifi, M. Y.; Saleh, K. A.; Alshehri, M. A.; Elbehairi, S. E. I. 2016. Bioactive lignans: a survey report on their chemical structures? Natural Products Chemistry & Research, 4, 1–15.

Federsel, H.-J. 2000. Development of a process for a chiral aminochroman antidepressant: a case story. Organic Process Research & Development, 4, 362–369.

Flora do Brasil. 2020a. em construção. Jardim Botânico do Rio de Janeiro. *Calophyllum brasiliense* Cambess. http://reflora.jbrj.gov.br/reflora/floradobrasil/FB6827, accessed on September 3, 2018.

Flora do Brasil. 2020b. em construção - Jardim Botânico do Rio de Janeiro. *Dahlstedtia floribunda* (Vogel) M.J. Silva & A.M.G. Azevedo. http://reflora.jbrj.gov.br/reflora/floradobrasil/FB135557, accessed on September 3, 2018.

Flora do Brasil. 2020c. em construção - Jardim Botânico do Rio de Janeiro. *Pterogyne nitens* Tul. http://reflora.jbrj.gov.br/reflora/floradobrasil/FB28161, accessed on September 3, 2018.

Flora do Brasil. 2020d. em construção - Jardim Botânico do Rio de Janeiro. *Cedrela odorata* L. http://floradobrasil.jbrj.gov.br/jabot/floradobrasil/FB9992, accessed on September 3, 2018.

Flora do Brasil. 2020e. em construção - Jardim Botânico do Rio de Janeiro. *Cordiera macrophylla* (K.Schum.) Kuntze. http://floradobrasil.jbrj.gov.br/jabot/floradobrasil/FB38700, accessed on September 3, 2018.

Flora do Brasil. 2020f. em construção - Jardim Botânico do Rio de Janeiro. *Licaria aurea* (Huber) Kosterm. http://floradobrasil.jbrj.gov.br/jabot/floradobrasil/FB23371, accessed on September 3, 2018.

Flora do Brasil. 2020g. em construção - Jardim Botânico do Rio de Janeiro. *Aristolochia labiata* Willd. http://floradobrasil.jbrj.gov.br/jabot/floradobrasil/FB15758, accessed on September 3, 2018.

Flora do Brasil. 2020h. em construção - Jardim Botânico do Rio de Janeiro. *Virola flexuosa* A.C.Sm. http://www.floradobrasil.jbrj.gov.br/reflora/floradobrasil/FB79429, accessed on September 3, 2018.

Flora do Brasil. 2020i. em construção - Jardim Botânico do Rio de Janeiro. *Deguelia urucu* (Killip & A.C.Sm.) A.M.G. Azevedo & R.A. Camargo. http://www.floradobrasil.jbrj.gov.br/reflora/floradobrasil/FB129251, accessed on September 3, 2018.

Garcez, F. R.; Scramin, S.; do Nascimento, M. C.; Mors, W. B. 1988. Phrenylated flavonoids as evolutionary indicators in the genus *Dahlstedtia*. Phytochemistry, 27, 1079–1083.

Garcia, J.; Barluega, S.; Beebe, K.; Neckers, L.; Winssinger, N. 2010. Concise modular asymmetric synthesis of deguelin, tephrosin and investigation into their mode of action. Chemistry: A European Journal, 16, 9767–9771.

Georgiou, K. H.; Pelly, S. C.; Koning, C. B. 2017. The first stereoselective synthesis of the natural product, rotenone. Tetrahedron, 73, 853–858.

Ghisalberti, E. L. 1997. Cardiovascular activity of naturally occurring lignans. Phytomedicine, 4, 151–166.

Gottlieb, O. R. 1978. Neolignans. In W. Herz; H. Grisebach; G. W. Kirby, eds., Fortschritte der Chemie Organischer Naturstoffe/Progress in the Chemistry of Organic Natural Products, 1st ed., pp. 1–72. New York: Springer.

Hansen, D.; Haraguchi, M.; Alonso, A. 2010. Pharmaceutical properties of 'sucupira' (*Pterodon* spp.). Brazilian Journal of Pharmaceutical Sciences, 46, 607–616.

Harper, S. H. 1942. The active principles of leguminous fish-poison plants. Part VI. Robustic Acid. Journal of the Chemical Society, 181–182.

Hegnauer, R. 1963. The taxonomic significance of alkaloids. In T. Swain, ed., Chemical Plant Taxonomy, 1st ed., pp. 389–427. New York: Academic Press.

Huang, J. G.; Zhou L. J.; Xu, H. H.; Li, W. O. 2009. Insecticidal and cytotoxic activities of extracts of *Cacalia tangutica* and its to active ingredients against *Musca domestica* and *Aedes albopictus*. Biological and Microbial Control, 102, 1444–1447.

Huerta-Reyes, M.; Basualdo, M. C.; Abe, F.; Jimenez-Estrada, M.; Soler, C.; Reyes-Chilpa, R. 2004. HIV-inhibitory compounds from *Calophyllum brasiliense* leaves. Biological and Pharmaceutical Bulletin, 27, 1471–1475.

Huo, X. H.; Zhang, L. Z.; Gao, L.; Zhang, L. Z.; Li, L.; Si, J.; Cao, L. 2015. Antiinflammatory and analgesic activities of ethanol extract and isolated compounds from *Millettia pulchra*. Biological and Pharmaceutical Bulletin, 38, 1328–1336.

Jaisankar, P.; Gajbhiye, R. L.; Mahato, S. K.; Nandi, D. 2014. Flavonoid natural products: chemistry and biological benefits on human health: a review. Asian Journal of Advanced Basic Sciences, 3, 164–178.

Jiménez-Arellanes, A.; León-Díaz, R.; Meckes, M.; Tapia, A.; Molina-Salinas, G. M.; Luna-Herrera, J.; Yépez-Mulia, L. 2012. Antiprotozoal and antimycobacterial activities of pure compounds from *Aristolochia elegans* rhizomes. Evidence-Based Complementary and Alternative Medicine, 2012, 1–7.

Johnson, A. P.; Pelter, A. 1964. The structure of robustic acid. Tetrahedron Letters, 20, 1267–1274.

Kashman, Y.; Gustafson, K. R.; Fuller, R. W.; Cardellina, J. H.; McMahon, J. B.; Currens, M. J.; Buckheit, R. W. Jr.; Hughes, S. H.; Cragg, G. M.; Boyd, M. R. 1992. The calanolides, a novel HIV-inhibitor class of coumarin derivatives from the tropical rainforest tree, *Calophyllum lanigerum*. Journal of Medicinal Chemistry, 35, 2735–2743.

Kim, J. Y.; Lim, H. J.; Lee, D. Y.; Kim, J. S.; Kim, D. H.; Lee, H. J.; Jeon, R.; Ryu, J. H. 2009. In vitro anti-inflammatory activity of lignans isolated from *Magnolia fargesii*. Bioorganic & Medicinal Chemistry Letters, 19, 937–940.

Kock, M.; Jones, P. G.; Lindel, T. 2017. Total synthesis and absolute configuration of raputindole A. Organic Letters, 19, 6296–6299.

L'Homme, C.; Ménard, M.; Canesi, S. 2014. Synthesis of the erythrina alkaloid erysotramidine. The Journal of Organic Chemistry, 79, 8481–8485.

Lachance, H.; Wetzel, S.; Kumar, K.; Waldmann, H. 2012. Charting, navigating, and populating natural products chemical space for drug discovery. Journal of Natural Products, 55, 5989–6001.

Lee, J.; Lee, Y.; Oh, S. M.; Yi, J. M.; Kim, N.; Bang, O. S. 2014. Bioactive compounds from the roots of *Asiasarum heterotropoides*. Molecules, 19, 122–138.

Lee, S.; Son, D.; Ryu, J.; Lee, Y. S.; Jung, S. H.; Kang, J.; Lee, S. Y.; Kim, H. S.; Shin, K. H. 2004. Anti-oxidant activities of *Acanthopanax senticosus* stems and their lignan components. Archives of Pharmacal Research, 27, 106–110.

Lee, Y. R.; Kim, D. H. 2006. A new route to the synthesis of pyranoflavone and pyranochalcone natural products and their derivatives. Synthesis, 4, 603–608.

Li, F.; Awale, S.; Tezuka, Y.; Kadota, S. 2008. Cytotoxic constituents from Brazilian red propolis and their structure-activity releationship. Bioorganic & Medicinal Chemistry, 16, 5434–5440.

Li, J.; Wang, T.; Yu, P.; Peterson, A.; Weber, R.; Soerens, D.; Grubisha, D.; Bennett, D.; Cook J. M. 1999. General approach for the synthesis of ajmaline/sarpagine indole alkaloids: enantiospecific total synthesis of (+)-ajmaline, alkaloid g, and norsuaveoline via the asymmetric Pictet-Spengler reaction. Journal of the American Chemical Society, 121, 6998–7010.

Lin, D.; Liang, Y.; Zheng, D.; Chen, Y.; Jing, X.; Lei, M.; Zeng, Z. 2018. Novel biomolecular information in rotenone-induced cellular model of Parkinson's disease. Gene, 647, 244–260.

Liu, X.; Wang, T.; Xu, Q.; Ma, C.; Cook, J. M. 2000. Enantiospecific total synthesis of the enantiomer of the indole alkaloid affinisine. Tetrahedron Letters, 41, 6299–6303.

Lopes, L. M. X.; Bolzani, V. D. S. 1988. Lignans and diterpenes of three *Aristolochia* species. Phytochemistry, 27, 2265–2268.

Magalhães, A. F.; Magalhães, E. G.; Leitão Filho, H. F.; Frighetto, R. 1981. Coumarins from *Pterocaulon balansae* and *P. lanatum*. Phytochemistry, 20, 1369–1371.

Magalhães, A. F.; Tozzi, A. M. G. A.; Magalhães, E. G.; Moraes, V. R. S. 2001. Prenylated flavonoids from *Deguelia hatschbachii* and their systematic significance in *Deguelia*. Phytochemistry, 57, 77–89.

Magalhães, A. F.; Tozzi, A. M. G. A.; Magalhães, E. G.; Sanmomiya, M.; Sriano, M. D. P. C.; Perez, M. A. F. 2007. Flavonoids of *Lonchocarpus montanus* AMG Azevedo and biological activity. Annals of the Brazilian Academy of Sciences, 79, 351–367.

Magalhães, A. F.; Tozzi, A. M. G. A.; Sales, B. H. L. N.; Magalhães, E. G. 1996. Twety-three flavonoids from *Lonchocarpus subglaucescens*. Phytochemistry, 42, 1459–1471.

Maia, B. H. L. N. S.; de Paula, J. R.; Sant'Ana, J.; da Silva, M. F. G. F.; Fernandes, J. B.; Vieira, P. C.; Costa, M. S. S.; Ohashi, O. S.; Silva, J. N. M. 2000. Essential oils of *Toona* and *Cedrela Species* (Meliaceae): taxonomic and ecological implications. Journal of Brazilian Chemical Society, 11, 629–639.

Marion, N.; Díez-Gonzáles, S.; de-Fremónt, P.; Noble, A. R.; Nolan, S. P. 2006. Au(I)-catalyzed tandem [3,3] rearrangement-intramolecular hydroarylation: mild and efficient formation of substituted indenes. Angewandte Chemie International Edition, 45, 3647–3650.

Marques, M. O. M.; Yoshida, M.; Gottlieb, O. R.; Maia, J. G. S. 1992. Neolignans from *Licaria aurea*. Phytochemistry, 31, 360–361.

Marsch, N.; Kock, M.; Lindel, T. 2016. Study on the synthesis of the cyclopenta[*f*]indole core of raputindole A. Beilstein Journal of Organic Chemistry, 12, 334–342.

Maturana, M. V.; Pinheiro, A. S.; Souza, T. L. F.; Follmer, C. 2014. Unveiling the role of the pesticides paraquat and rotenone on α-synuclein *in vitro*. Neurotoxicology, 46, 35–43.

Mayer, A. M. S.; Glaser, K. B.; Cuevas, C.; Jacobs, R. S.; Kem, W.; Little, R. D.; McIntosh, J. M.; Newman, D. J.; Potts, B. C.; Shuster, D. E. 2010. The odyssey of marine pharmaceuticals: a current pipeline perspective. Trends in Pharmacological Sciences, 31, 255–265.

Mendoza, A.; Ishihara, Y.; Baran, P. S. 2012. Scalable enantioselective total synthesis of taxanes. Nature Chemistry, 4, 21–25.

Mickel, S. J.; Niederer, D.; Daefller, R.; Osmani, A.; Kuester, E.; Schmid, E.; Schaer, K.; et al. 2004. Large-scale synthesis of the anti-cancer marine natural product (+)-discodermolide. part 5: linkage of fragments C1-6 and C7-24 and finale. Organic Process Research & Development, 8, 122–130.

Modolo, L. V.; da Silva, C. J.; da Silva, F. G.; da Silva-Neto, L.; de Fátima, A. 2015b. Bioactive natural products: chemistry and biology. In G. Brahmachari, ed., Introduction to the Biosynthesis and Biological Activities of Phenylpropanoids, pp. 387–408. Weinheim: Wiley-VCH Verlag GmbH & Co. KGaA.

Modolo, L. V.; de Souza, A. X.; Horta, L. P.; Araujo, D. P.; de Fátima, A. 2015a. An overview on the potential of natural products as urease inhibitors: a review. Journal of Advanced Research, 6, 35–44.

Monnerat C. S.; de Souza, J. J.; Mathias, L.; Braz-Filho, R.; Vieira, J. C. 2005. A new indole alkaloid isolated from *Tabernaemontana hystrix* steud (apocynaceae). Journal of the Brazilian Chemical Society, 16, 1331–1335.

Newman, D. J.; Cragg, G. M. 2012. Natural products as sources of new drugs over the 30 years from 1981 to 2010. Journal of Natural Products, 75, 311–335.

Nicolaou, K. C.; Snyder, S. A. 2003. Classics in Total Synthesis II: More Targets, Strategies, Methods, 1st ed. New York: Wiley & Sons.

Nicolaou, K. C.; Sorensen, E. J. 1996. Classics in Total Synthesis: Targets, Strategies, Methods, 1st ed. New York: VCH.

Nicoletti, M.; Goulart, M. O. F.; de Lima, R. A.; Goulart, A. E.; Monache, F. D.; Bettolo, G. B. M. 1984. Flavonoids and alkaloids from *Strychnos pseudoquina*. Journal of Natural Products, 47, 953–957.

Ohshima, T.; Xu, Y.; Takita, R.; Shibasaki, M. 2004. Enantioselective total synthesis of (-)-strychnine: development of a; highly practical catalytic asymmetric carbon-carbon bond formation and domino cyclization. Tetrahedron, 60, 9569–9588.

Parmar, V. S.; Jain, R.; Gupta, S. R. 1988. Phytochemical investigation of *Tephrosia candida*: hplc separation of tephrosin and 12a-hydroxyrotenone. Brief Reports, 51, 185.

Pelly, S. C.; Govender, S.; Fernandes, M. A.; Schmalz, H.; Koning, C. B. 2007. Stereoselective syntheses of the 2-Isopropenyl-2,3-dihydrobenzofuran nucleus: potential chiral building blocks for the syntheses of tremetone, hydroxytremetone, and rotenone. Journal of Organic Chemistry, 72, 2857–2864.

Peng, F.; Wang, G.; Li, X.; Cao, D.; Yang, Z.; Ma, L.; Ye, H.; et al. 2012. Rational design, synthesis, and pharmacological properties of pyranochalcone derivatives as potent anti-inflammatory agents. European Journal of Medicinal Chemistry, 54, 272–280.

Pilkington, L. I.; Barker, D. 2012. Asymmetric synthesis and CD investigation of the 1,4-benzodioxane lignans eusiderins A, B, C, G, L, and M. The Journal of Organic Chemistry, 77, 8156–8166.

Pilkington, L. I.; Wagoner, J.; Kline, T.; Polyak, S. J.; Barker, D. 2018. 1,4-benzodioxane lignans: an efficient, asymmetric synthesis of flavonolignans and study of neolignan cytotoxicity and antiviral profiles. Journal of Natural Products, 81, 2630–2637.

Prates, A. P. L.; Irving, M. A. 2015. Biodiversity conservation and public policies for protected areas in Brazil: challenges and trends from the origin of the CBD until the Aichi targets. Revista Brasileira de Política Internacional, 5, 28–57.

Rates, S. M. K. 2001. Plants as source of drugs. Toxicon, 39, 603–613.

Regasini, L. O.; Vellosa, J. C. R.; Silva, D. H. S.; Furlan, M.; de Oliveira, O. M. M.; Khalil, N. M.; Brunetti, I. L.; Young, M. C. M.; Barreiro, E. J.; Bolzani, V. S. 2008. Flavonols from *Pterogyne nitens* and their evaluation as myeloperoxidase inhibitor. Phytochemistry, 69, 1739–1777.

Rice, M. J.; Legg, M.; Powell, K. A. 1998. Natural products in agriculture – a view from the industry. Pesticide Science, 52, 184–188.

Roberts, M. F.; Wink, M. 1998. Alkaloids – Biochemistry, Ecological and Medical Applications, 1st ed. New York: Plenum Press.

Sandoval, M.; Okuhama, N. N.; Zhang, X. J.; Condezo, L. A.; Lao, J.; Angeles, F. M.; Musah, R. A.; Bobrowski, P.; Miller, M. J. S. 2002. Anti-inflammatory and antioxidant activities of cat's claw *(Uncaria tomentosa* and *Uncaria guianensis)* are independent of their alkaloid contente. Phytomedicine, 9, 325–337.

Santos, A. K. L.; Magalhães, T. S.; Monte, F. J. Q.; de Mattos, M. C.; de Oliveira, M. C. F.; Almeida, M. M. B.; Lemos, T. L. G.; Braz-Filho, R. 2009. Alcaloides iboga de *Peschiera affinis* (apocynaceae) – atribuição inequívoca dos deslocamentos químicos dos átomos de hidrogênio e carbono. atividade antioxidante. Química Nova, 32, 1834–1838.

Santos, D. A. P.; Braga, P. A. C.; da Silva, M. F. G. F.; Fernandes, J. B.; Vieira, P. C.; Magalhães, A. F.; Magalhães, E. G. 2009. Anti-African trypanocidal and atimalarial activity of natural flavonoids, dibenzoylmethanes and synthetic analogues. Journal of Pharmacy and Pharmacology, 61, 257–266.

Sarragiotto, M. H.; Filho, H. L.; Marsaioli, A. J. 1981. Erysotrine-*N*-oxide and erythrartine-*N*-oxide, two novel alkaloids from *Erythrina mulungu*. Canadian Journal of Chemistry, 59, 2771–2775.

Sekino, E.; Kumamoto, T.; Tanaka, T.; Ikeda, T.; Ishikawa, T. 2004. Concise synthesis of anti-HIV-1 active (+)-inophyllum B and (+)-calanolide A by application of (-)-quinine-catalyzed intramolecular oxo-Michael addition. The Journal of Organic Chemistry, 69, 2760–2767.

Shaabani, A.; Ghadari, R.; Rahmati, A.; Rezayan, A. H. 2009. Coumarin synthesis *via* knoevenagel condensation reaction in 1,1,3,3-*N,N,N',N'*-tetramethylguanidinium trifluoroacetate ionic liquid. Journal of the Iranian Chemical Society, 6, 710–714.

Shi, J.; Manolikakes, G.; Yeh, C. H.; Guerrero, C. A.; Shenvi, R. A.; Shigehisa, H.; Baran, P. S. 2011. Scalable synthesis of cortistatin a and related structures. Journal of American Chemical Society, 133, 8014–8027.

Shimizu, S.; Ohori, K.; Arai, T.; Sasai, H.; Shibasaki, M.; Shibasaki, M. 1998. A catalytic asymmetric synthesis of tubifolidine. Journal of Organic Chemistry, 63, 7547–7551.

Silva, F. G.; Horta, L. P.; Faria, R. O.; Stehmann, J. R.; Modolo, L. V. 2014. Stressing conditions as tools to boost the biosynthesis of valuable plant natural products. Recent Patents on Biotechnology, 8, 89–101.

Srikrishna, A.; Ravi, G. A. 2008. Stereoselective total synthesis of (-)-seychellene. Tetrahedron, 64, 2565–2571.

Su, S.; Rodrigues, R. A.; Baran, P. S. 2011. Scalable, stereocontrolled total synthesis of (±)-axinellamines A and B. Journal of American Chemical Society, 133, 13922–13925.

Surendra, K.; Corey, E. J. 2009. A short enantioselective total synthesis of the fundamental pentacyclic triterpene lupeol. Journal of American Chemical Society, 131, 13928–13929.

Takayoshi, A.; Sasai, H.; Aoe, K.; Okamura, K.; Date, T.; Shibasaki, M. 1996. A new multifunctional hetero-bimetallic asymmetric catalyst for Michael additions and tandem Michael–aldol reactions. Angewandte Chemie International Edition in English, 35, 104–106.

Tran, D. N.; Cramer, N. 2014. Biomimetic synthesis of (+)-ledene, (+)-viridiflorol, (-)-palustrol, (+)-spathulenol, and psiguadial A, C, and D via the platform terpene (+)-bicyclogermacrene. Chemistry: A European Journal, 20, 1–7.

Urgaonkar, S.; Shaw, J. T. 2007. Synthesis of kaempferitrin. Journal of Organic Chemistry, 72, 4582–4585.

Vaithegi, K.; Prasad, K. R. 2018. Enantiospecific total synthesis of the putative structure of cryptopyran-moscatone B2. Tetrahedron, 74, 2627–2633.

Vasconcelos, J. N.; Santiago, G. M. P.; Lima, Q. J.; Mafezoli, J.; Lemos, T. L. G.; Silva, F. R. L.; Lima, M. A. S. 2012. Rotenoids from *Tephrosia toxicaria* with larvicidal activity against *Aedes aegypti*, the main vector of dengue fever Química Nova, 35, 1097–1100.

von Pechmann, H.; Duisberg, C. 1883. Ueber eine neue bildungsweise des benzoylessigäthers. Berichte Der Deutschen Chemischen Gesellschaft, 16, 2119–2128.

Vougogiannopoulou, K.; Fokialakis, N.; Aligiannis, N.; Cantrell, C.; Skaltsounis, A. L. 2010. The raputin-doles: novel cyclopentyl bisindole alkaloids from *Raputia simulans*. Organic Letters, 12, 1908–1911.

Wang, F.; Chiba, K.; Tada, M. 1993. Stereoselective synthesis of (+/-)-7β-acetoxyvouacapane. Chemistry Letters, 22, 2117–2120.

Wang, T.; Yu, P.; Li, J.; Cook, J. M. 1998. The enantiospecific total synthesis of norsuaveoline. Tetrahedron Letters, 39, 8009–8012.

Wang, W.; Wang, J.; Li, N.; Zhang, X.; Zhao, W.; Li, J.; Si, Y. 2015. Chemopreventive flavonoids from *Millettia pulchra* Kurz van-laxior (Dunn) Z.Wei (Yulangsan) function as Michael reaction acceptor. Bioorganic and Medicinal Chemistry Letters, 25, 1078–1081.

Weisbach, J. A.; Raffauf, R. F.; Macko, E.; Douglas, B.; Ribeiro, O. 1963. Problems in chemotaxonomy I. Alkaloids of Peschiera affinis. Journal of Pharmaceutical Sciences, 52, 350–353.

Xu, W. H.; Zhao, P.; Wang, M.; Liang, Q. 2018. Naturally occurring furofuran lignans: structural diversity and biological activities. Natural Product Research, 16, 1–17.

Yang, F.; Su, Y.; Bi, Y.; Xu, J.; Zhu, Z.; Tu, G.; Gao, X. 2010. Three new kaempferol glycosides from *Cardamine leucantha*. Helvetica Chimica Acta, 94, 536–541.

Yang, X.; Zhao, Y.; Hsieh, M.; Xin, G.; Wu, R.; Hsu, P.; Horng, L.; Sung, H.; Cheng, C.; Lee, K. 2017. Total synthesis of (+)-medicarpin. Jounal of Natural Products, 80, 3284–3288.

Yoda, H.; Suzuki, Y.; Matsuura, D.; Takabe, K. 2005. A new synthetic entry to furofuranoid lignans, methyl piperitol and fargesin. Heterocycles, 65, 519–522.

Yu, P.; Cook, J. M. 1998. Enantiospecific total synthesis of the sarpagine related indole alkaloids talpinine and talcarpine: the oxyanion-cope approach. Journal of Organic Chemistry, 63, 9160–9161.

Yue, B.; Ren, Y. J.; Zhang, J. J.; Luo, X. P.; Yu, Z. L.; Ren, G. Y.; Sun, A. N.; Deng, C.; Dou, W. 2018. Anti-inflammatory effects of fargesin on chemically induced inflammatory bowel disease in mice. Molecules, 23, 1380–1393.

Index

Note: *Italicized* page numbers refer to figures, **bold** page numbers refer to tables

9 781032 085036